Global Issues in Water Policy

Volume 31

Global Issues in Water Policy is now indexed in SCOPUS.

Policy work in the water sector has grown tremendously over the past two decades, following the Rio Declaration of 1992. The existing volume of water-related literature is becoming dominant in professional outlets, including books and journals. Because the field of water resources is interdisciplinary in nature, covering physical, economic, institutional, legal, environmental, social and political aspects, this diversification leads in many cases to partial treatment of the water issues, or incomplete analysis of the various issues at stake. Therefore, treating a whole host of a country's water resources issues in one set of pages will be a significant contribution to scholars, students, and other interested public. This book series is expected to address both the current practice of fragmented treatment of water policy analyses, and the need to have water policy being communicated to all interested parties in an integrated manner but in a non-technical language. The purpose of this book series is to make existing knowledge and experience in water policy accessible to a wider audience that has a strong stake and interest in water resources. The series will consist of books that address issues in water policy in specific countries, covering both the generic and specific issues within a common and pre-designed framework.

Daniela Rivera • Guillermo Donoso
Editors

Droughts in Chile

Impacts, Monitoring, and Adaptation (Management) Policies

Springer

Editors
Daniela Rivera
Faculty of Law and the Center for Water Law and Management
Pontificia Universidad Católica de Chile
Santiago, Chile

Guillermo Donoso
Faculty of Agronomy and Natural Systems and the Center for Water Law and Management
Pontifica Universidad Católica de Chile
Santiago, Chile

ISSN 2211-0631 ISSN 2211-0658 (electronic)
Global Issues in Water Policy
ISBN 978-3-031-85039-4 ISBN 978-3-031-85040-0 (eBook)
https://doi.org/10.1007/978-3-031-85040-0

This Springer imprint is published by the registered company Springer Nature Switzerland AG
The registered company address is: Gewerbestrasse 11, 6330 Cham, Switzerland

To our families and to all those who work every day for water security.

Foreword

Droughts in Chile: Impacts, Monitoring, and Adaptation Policies, edited by Daniela Rivera and Guillermo Donoso, is a timely and welcome addition to the body of literature on drought management and policy. The book captures the status of drought management in Chile and presents new options for improved drought management.

Globally, there have been significant advances in integrated drought management in the past decade. These advances have augmented our monitoring capability for the early detection of drought and its severity. Improved methods and tools for communicating this information to decision makers and the public in a timely manner have also provided opportunities for impact reduction. In addition, risk or vulnerability assessment methodologies have been developed and applied in many countries. These assessments are aimed at determining who and what is at risk and why, a key step toward risk reduction. Applying the results of these assessments can be instructive for the implementation of proactive measures that target those sectors most at risk in advance of future drought events.

Much of the progress over the past decade can be traced to the High-level Meeting on National Drought Policy (HMNDP), held in March 2013. This meeting served as a catalyst to promote the development of national drought policies, and the concepts associated with integrated drought management. Convened by three organizations of the United Nations—the World Meteorological Organizational (WMO), the Food and Agriculture Organization (FAO), and the Convention to Combat Desertification (UNCCD)—this collaboration facilitated a global dialogue on the importance of formulating national drought policies aimed at reducing the risks and, therefore, the vulnerability of all nations to this insidious natural hazard. Our changing climate, the increased frequency, severity, and duration of extreme weather and climate events, and the spiraling impacts of drought are significant motivating factors for all drought-prone nations to lessen societal vulnerability to drought by adopting a paradigm shift emphasizing integrated drought management.

What emerged from HMNDP was a regional and national emphasis on the need to develop national drought management policies. The Integrated Drought Management Programme (IDMP), established during the HMNDP by WMO and the Global Water Partnership (GWP), adopted the widely accepted 10-step drought planning methodology, commonly referred to the three pillars of drought policy and preparedness, as a mechanism to improve drought preparedness by shifting the paradigm for drought management from a reactive approach to a proactive risk management approach, i.e., transitioning from managing disasters to managing risk.[1]

I applaud the efforts of Daniela Rivera and Guillermo Donoso to document the current, albeit reactive, approach to drought management and preparedness in Chile. However, this book also presents new options for drought management in Chile through the development of comprehensive monitoring and early warning systems and the adoption of integrated water management techniques and other policy options to reduce the economic, social, environmental impacts of drought to achieve greater drought resilience.

The publication of this volume could not be more timely given the international meeting on Drought Resilience, i.e., the High-level Meeting on National Drought Policy+10, that was held in the fall of 2024. This meeting brought together scientists, policy makers, government officials, and stakeholders to assess progress on drought management over the past 10 years and identify key pathways moving forward. While the lessons learned from the chapters in *Drought in Chile* provide critical information to water managers and other officials in Chile and beyond, these lessons and recommendations may also help to inform outcomes from HMNDP+10.

Professor Emeritus, Drought Management and Policy Donald A. Wilhite
University of Nebraska
Lincoln, Nebraska, USA

Founding Director, U.S. National Drought Mitigation Center
University of Nebraska
Lincoln, Nebraska,, USA

[1] World Meteorological Organization (WMO) and Global Water Partnership (GWP) (2014) National Drought Management Policy Guidelines: A Template for Action (D.A. Wilhite). Integrated Drought Management Programme (IDMP) Tools and Guidelines Series 1. WMO, Geneva, Switzerland and GWP, Stockholm, Sweden.

Acknowledgments

This text originates from the research project awarded and funded by the National Agency for Research and Development (Agencia Nacional de Investigación y Desarrollo, ANID) of Chile, ANID Sequía FSEQ210018, entitled "Integrating drought monitoring and public policy design towards the proactive management of drought."

This project was executed by an interdisciplinary team of 16 professors from the Pontificia Universidad Católica de Chile: as principal investigators, Jorge Gironás (director), Sebastián Vicuña, Daniela Rivera, María Molinos and Oscar Melo; and, as associate investigators, Aurora Gaxiola, Camilo del Río, Francisco Suárez, Marcelo Miranda, David Morales, Francisco Meza, Guillermo Donoso, Eduardo Leiva, Pablo Pastén, Alonso Pérez, and Álvaro Lorca. In addition, an important number of postdoctoral researchers (Felipe Lobos and Cristian Jordán); support professionals (Marialina Núñez, Carolina Córdova, Nicolás Millie, Valentina Bravo, Paula Toledo, Catalina Aránguiz, Diego González, Javier Vargas, Sebastián González, Mauricio Montecinos, Alejandra Vega, José Antonio Díaz, Carolina Rodríguez, Francisca Contardo, Daniela Madrazo, Sara Acevedo, Fernanda Cepeda, Bastián Silva, Felipe Suárez, Kevin Unda, and María Valenzuela); and students (Fernando González, Rosa Vargas, Klaus Keim, Vicente Espinoza, Eduardo Viollier, Juan Figueroa, Antonia Ávila, Pamela Pimentel, Victoria Armas, Antonia Durán, and Marcel Favereau) participated. Most of them are involved as authors or co-authors of this book.

We would especially like to thank the National Agency for Research and Development (Agencia Nacional de Investigación y Desarrollo, ANID) of Chile, the authors who are part of this collective publication, all those who participated in the development of the aforementioned project FSEQ210018, and Daniela Madrazo for her support in editing this work. We hope that this book will be a contribution to review, evaluate, and improve drought management policies in Chile and other countries, given that this is an extreme hydrological phenomenon whose rigorous and proactive treatment is one of the key factors to advance in water security.

Contents

Editors and Authors

About the Editors

Daniela Rivera has a bachelor's degree in Legal Sciences, a master's degree in Legal Science, and a PhD in Law. She is part of the faculty members of the Center for Water Law and Management of the Pontificia Universidad Católica de Chile, created in July 2015, integrated by five faculties of the same university, and of which she was director between July 2017 and July 2023. She is an Assistant Professor of the Faculty of Law of the Pontificia Universidad Católica de Chile. Her research focuses on water law and management, a field in which she is the author of several publications, has participated in various academic seminars, in Chile and abroad, and develops several projects funded by ANID, FONDECYT, CORFO, and other public and private entities.

Guillermo Donoso is an Agricultural Economist from the Pontificia Universidad Católica de Chile and PhD of the Department of Agricultural and Resource Economics, University of Maryland, College Park. He is Full Professor of Agricultural Economics Department and Director of the Water Law and Management Center of the Pontificia Universidad Católica de Chile. He has held the position of Dean of the Facultad de Agronomía e Ingeniería Forestal of the Pontificia Universidad Católica de Chile from 1998 to 2007. His publications include books, book chapters, and publications in Water Resources Management, Water Policy, Environmental Science and Policy, Environmental Science and Pollution Research, Water Science and Technology, Futures Markets, and Agricultural Water Resources Management, among others.

About the Authors

Sebastián Aedo is a Research Associate at the Latin America Center of the Stockholm Environmental Institute. His main area of research is in climate change and integrated water resources modeling, with experience in basins in Chile, Colombia, Argentina, and Bolivia, and transboundary basins in El Salvador, Honduras, and Guatemala. Sebastián has also collaborated in hydroinformatics, climate change, and water resources modeling courses.

Antonia Ávila has an undergraduate degree in Industrial Engineering with a mention in Environmental Engineering and a master's in Engineering Sciences, both from the Pontificia Universidad Católica de Chile. She is a teacher assistant of the water resources management course since 2022. She has presented at the symposium held by the CR2met and will soon present at the Chilean Congress on Hydraulic Engineering held by the Chilean Society of Hydraulic Engineering (SOCHID, for its acronym in Spanish).

Valentina Bravo is a Forest Engineer and Master in Natural Resources from the Pontifical Catholic University of Chile. Her research focuses on modeling the effects of droughts and heatwaves on the dynamics of primary productivity and phenology in Mediterranean and temperate forests of South America. To achieve these objectives, she has solid expertise in forest ecology, geomatics, mathematical modeling based on machine learning, and programming. Currently, she works as an independent research professional on various projects associated with universities and companies in Chile and abroad.

Camilo del Río is a Geographer and Master in Geography and Geomatics from the Pontificia Universidad Católica (PUC) of Chile, and PhD in Natural Sciences from the University of Heidelberg, Germany. He is an Associate Professor of the Institute of Geography at the PUC. He is currently Director of the UC Atacama Desert Center. He has published more than 25 articles in scientific journals related to fog, fog water, and water resources.

José A. Díaz is a Civil Industrial Engineer with a specialization in Environmental Engineering from the Pontificia Universidad Católica de Chile. He has a Master of Science in Engineering from the same institution. He is a Project Engineer at the Center for Sustainable Urban Development (CEDEUS).

Juan de Dios Guzmán is a Hydraulic Engineer from the Pontificia Universidad Católica de Chile. He has participated in projects associated with hydrological modeling and the processing, analysis, and generation of both point and spatially distributed climate data.

Antonia Durán is an Industrial Environmental Engineer and MSc in Electrical Engineering of the Pontificia Universidad Católica de Chile. Her experience is related to the study of multipurpose reservoirs for hydropower and irrigation, and her interests are in water management, renewable energy, and the environment.

Marcel Favereau is an Assistant Professor and Researcher at the Department of Engineering Sciences at Universidad Andrés Bello. He holds a PhD in Engineering Sciences from Pontificia Universidad Católica de Chile, a master's in Engineering Sciences with a specialization in Logistics and Operations Management, and a bachelor's in Industrial Engineering from Universidad Andrés Bello. He has extensive teaching experience in undergraduate and graduate courses in systems simulation, operations research, and decision-making under uncertainty. His publications span journals such as *Computers and Operations Research*, *Annals of Operations Research*, *Journal of Water Resources Planning and Management*, *Applied Energy*, and *Natural Hazards*, focusing on hydroelectric resources management and disaster evacuation modeling.

Francisco J. Fernández is an Associate Professor at the Faculty of Economics and Government of Universidad San Sebastián (USS). He holds a PhD in Agricultural and Environmental Economics from the Technical University of Madrid (UPM) (2017), a Master of Agricultural and Environmental Economics degree (2012) from UPM, and a Bachelor of Agronomy degree (2007) from the Pontificia Universidad Católica de Valparaíso, Chile. His research centers on the modeling of economic agents' decisions through mathematical programming models, focusing primarily on the agricultural sector, water resources economics, climate change economics, and environmental economics. He has authored several scientific articles in high-impact economics and multidisciplinary journals, including *Water Resource Management*, *Journal of Environmental Management*, and *Environmental Research Letters*.

Aurora Gaxiola is a biologist from the Universidad Nacional Autónoma de México and PhD in Ecology and Ecosystem Sciences from the University of Cambridge. She is currently an Associate Professor in the Faculty of Biological Sciences at the Pontificia Universidad Católica de Chile, Principal Researcher at the Instituto de Ecología y Biodiversidad, and Scientific Director of the Centro Interdisciplinario de Cambio Global UC. She also serves as Vice President of the Sociedad Chilena de Ecología. From 2015 to 2018, she was selected as a Young Research Fellow for the "Jóvenes Ciencia de Frontera" program of the Academia Chilena de Ciencias. Her research focuses on plant–soil interactions, ecosystem dynamics, and the impacts of climate change on biodiversity. She has published over 40 scientific articles and book chapters in these fields.

Jorge Gironás is an Associate Professor in the Departamento de Ingeniería Hidráulica y Ambiental at PUC, where he teaches undergraduate and graduate courses in hydrology and water systems. His main area of research is in hydrology and water resources, with specific interests in urban hydrology, stormwater modeling, basin morphology, extreme events, and climate change. Dr. Gironás has led several research projects and participated in multiple interdisciplinary initiatives. In addition to more than 65 peer-reviewed papers, he is the main author of the EPA-SWMM model application's manual, co-author of the Chilean Urban Drainage Manual, and the chapter on drought characterization in the 2nd edition of the *Handbook of Applied Hydrology* published by McGraw-Hill. In 2021, he was co-editor of the book *Water Resources of Chile*, published by Springer in 2021.

Jeanine Jones is the Interstate Resources Manager for the California Department of Water Resources. She is a member and past Chair of the Western States Water Council, a Designee on the Colorado River Board of California, and a registered civil engineer in California and Nevada. She has more than 40 years of experience in water resources management, with an extensive background in drought preparedness and response. She has written numerous publications on drought in California, and served on many committees and advisory committees, including the Border Governors' Water Worktable and NOAA Climate Working Group.

Cristian Jordán holds a forestry degree for the University of Talca, Chile, and a PhD in Biosciences Engineering (Economics) for Ghent University, Belgium. He was a postdoctoral researcher in the ANID Sequía FSEQ210018 project "Integrating drought monitoring and public policy design towards the proactive management of drought". He has a broad experience as a public servant for several agricultural and water related institutions, Currently, he works as a consultant at FAO-RLC within the National Adaptation Plan, for the water resources sector.

Rodrigo Labatut is an Aquacultural Engineer from Universidad Católica del Norte, and MS and PhD in Biological and Environmental Engineering from Cornell University. Currently, he is an Assistant Professor at Universidad Católica de Chile, where he teaches courses in water and wastewater treatment technologies. Earlier in his career, he focused on developing zero-discharge aquaculture systems combining biological and advanced oxidation processes, as well as characterizing aquaculture reactors using computational fluid dynamics. For the last 17 years, Dr. Labatut has been primarily working on anaerobic digestion and related fermentation-based processes to develop more sustainable sludge and wastewater treatment systems. Currently, he is working on the biological degradation of siloxanes and other emerging pollutants derived from industrial wastewaters using anaerobic/microaerobic microbially-based systems, and also he is exploring the integration of thermo- and bio-chemical processes to produce bio-oil and to recover nutrients and chemicals while recovering water from sludge and wastewater streams. Dr. Labatut is an associate researcher at the technological consortium of water (COTH2O) and has published several papers, book chapters, and holds a patent on the subject of wastewater treatment technologies.

Eduardo Leiva is a Biochemist with a master's and PhD from the Pontificia Universidad Católica de Chile, 2013. He is currently an Associate Professor in the Department of Hydraulic and Environmental Engineering and the Faculty of Chemistry and Pharmacy. Between 2018 and 2019, he was a Visiting Professor and Postdoctoral Researcher at the University of Notre Dame (USA), where he conducted research on water treatment and nitrogen removal. His research focuses on water resource management, environmental sustainability, nanotechnology, water treatment and reuse, and environmental pollution, with over 10 years of experience. He is the author of more than 35 scientific articles published in WOS journals. Dr. Leiva is also a technology consultant in areas related to engineering, biotechnology, and environmental chemistry. He has participated in projects focused on technological developments applied to industry, aimed at developing and transferring R&D-based technological products to the market for environmental challenges related to water resources. Currently, Dr. Leiva is leading research projects focused on developing technological solutions for the reuse of wastewater and projects related to the treatment and removal of nitrogen compounds from water. He is also an associate researcher at the Water Technological Consortium (CoTH2O) and an alternate director of the ADAPT-Action Technological Program: Actions for the Adaptive Development of Landscapes, Territories, and Tourist Attractions.

Jesús López-Santiago currently serves as a Professor in the School of Agronomy, Food, and Biosystems Engineering at the University Polytechnic of Madrid since 2007. He earned his bachelor's degree in Agriculture Engineering from the University Polytechnic of Madrid and holds a master's degree in Agri-environmental and Agri-food sciences from the Autonomous University of Madrid and the National University of Distance Education. Additionally, he obtained a master's degree in Business Administration from Comillas Pontifical University. Dr. Santiago pursued his doctoral studies in Rural Planning and his PhD in Agro-engineering from the University Polytechnic of Madrid. With a wealth of expertise, he has successfully managed his own enterprise specializing in environment and food safety consulting for the past seventeen years. His professional journey includes significant roles as a general manager in rural development public institutions, coupled with extensive consultancy experience in rural development in Peru and Bolivia. Dr. Santiago's global academic contributions extend to being a Visiting Professor at the University of Kuala Lumpur (Malaysia) and the University Technological Metropolitan (Chile). Actively involved in research, he is a valued member of the Buildings, Infrastructures, and Projects for Rural and Environmental Engineering Research Group at the University Polytechnic of Madrid. His research focus encompasses food safety systems and environmental management systems, with notable publications in high-impact journals dedicated to these fields.

Álvaro Lorca is an Associate Professor in the Department of Electrical Engineering and the Department of Industrial and Systems Engineering at the Pontificia Universidad Católica de Chile. He is also the Director of the Energy and Complex Systems Laboratory and the Technology Director at Vinken-Dictuc. He earned his

PhD in Operations Research from the Georgia Institute of Technology and holds a Master of Science in Engineering and an Industrial Civil Engineering degree with a diploma in Mathematical Engineering from the Pontificia Universidad Católica de Chile. His research and professional activities focus on the development and application of mathematical and computational methods for energy systems management, including energy planning, power system operations, renewable energy integration, smart grids, and resilience. He has published his research in various specialized scientific journals, including *Operations Research*, *European Journal of Operational Research*, *Production and Operations Management*, *IEEE Transactions on Power Systems*, *IEEE Transactions on Smart Grid*, *Energy*, *Energy Economics*, *Applied Energy*, *Energy Policy*, and *Reliability Engineering and System Safety*, among others. Professor Lorca has collaborated with ISO New England, the World Bank, the Environmental Defense Fund, the Chilean National Electric Coordinator, the National Energy Commission, Chile's Ministry of Energy, Transelec, and Fraunhofer Chile Research, among others. He was a Fulbright Scholar and has received several awards for his research, including the Best Paper in Energy Award from the Institute for Operations Research and Management Science and the Best Paper Award from the IEEE Power and Energy Society

Óscar Melo is an Agricultural Engineer, from the Pontificia Universidad Católica de Chile, MSc and PhD in Agricultural and Resources Economics from the University of Maryland in the USA. He is currently an Associate Professor in the Department of Agricultural Economics at the Pontificia Universidad Católica de Chile. He was previously Director of the master's program in agricultural economics, Director of the Department of Agricultural Economics between 2007 and 2014. He has been an international consultant for the World Bank, Inter-American Development Bank, FAO, and ECLAC, working in several countries in Latin America and other continents. He has also assessed various Ministries in Chile and other institutions and companies. His research has focused on the valuation and regulation of natural resources, the functioning of water markets, the competitiveness and development of agriculture, the effects of climate change, and the characterization of food demand. He is a member of the Center for Global Change and the Center for Water Law and Management, both from the Catholic University. He teaches agricultural and environmental economics, econometrics, and natural resource economics, for undergraduate and graduate students. His latest publications have focused on the use of water option contracts to face climate change, the effects of climate change on people, and the energy-water-food nexus.

Francisco Javier Meza is an Agronomist from the Pontificia Universidad Católica de Chile, Master of Science in Water Resources from the Pontificia Universidad Católica de Chile and PhD in Atmospheric Sciences from Cornell University. He is a Full Professor of Ecosystems and Environment Department of the Pontificia Universidad Católica de Chile. He was the founder and first Director of the Centro de Cambio Global (Center for Global Change) and Director of the Instituto para el Desarrollo Sustentable (Institute for Sustainable Development). He is a Co-chair of the scientific steering committee of the Agricultural Intercomparison and Improvement Project. He has been lead author of the IPCC reports (5th and 6th Assessment) and the chapter of desertification of the IPCC special report on land. His publications include books, book chapters, and publications in climate, water resources, and agricultural modeling. He has supervised more than 25 graduate students (5 PhDs) and developed the academic program in Natural Resources Engineering. His research interest cover from climate variability and change, adaptation to climate change, risk assessment, and agricultural modelling.

Marcelo D. Miranda is a Forest Engineering graduate from Universidad de Chile, with an MSc and PhD from Université Laval, Canada. He is an Associate Professor of Geomatics and Remote Sensing at Pontificia Universidad Católica de Chile in Santiago. His research focuses on vegetation productivity and landscape dynamics in forest and agricultural ecosystems. He applies mathematical models, spatial analysis methods, and remote sensing to understand environmental complexity across diverse spatial and temporal scales. His scientific publications span various fields, including plant and animal genetics, vegetation productivity dynamics in natural ecosystems, and landscape ecology.

María Molinos-Senante holds a bachelor's degree in Environmental Science and PhD on Local Development and Territory from the University of Valencia. She is a Senior Postdoc at the Institute of Sustainable Processes of the University of Valladolid. Previously, she was an Associate Professor at Engineering School of the Pontificia Universidad Católica de Chile. Her research topics include water economics, benchmarking of water utilities, water-energy-carbon nexus, and sustainability assessment in the urban water cycle. She has published more than 200 WoS articles and participated in more than 25 national and international research projects.

Mauricio Montecinos is a Civil Industrial Engineer with a specialization in Environmental Engineering from the Pontificia Universidad Católica de Chile and holds a PhD in Engineering Sciences from the same institution. He currently works as a Project Engineer at the Center for Sustainable Urban Development (CEDEUS)

David Morales is an Agricultural Engineer from the Pontificia Universidad Católica de Chile, MSc in Natural Resources, specializing in Global Change. He is with the Department of Ecosystems and Environment, Pontificia Universidad Católica de Chile. He has worked in project development as well as research related to data processing, analysis, and Geographic Information Systems (GIS) management. He has led projects on environmental information management in GIS and statistics, including the Update of the National Water Balance, the creation of geospatial analysis platforms for climate information, and geographic analysis for various sectors such as agriculture, water resources, energy, among others. In recent years, he has focused on the analysis and processing of statistical and spatial data of a climatic, hydrological, and environmental nature, as well as the design and creation of various spatial outputs (layers, maps, images) and the development of web visualization platforms for monitoring climate indicators, among other tasks.

Felipe Morandé holds a PhD in Economics from the University of Minnesota and a bachelor's degree also in Economics from the Universidad Católica de Chile (PUC). Currently he is Full Professor at the School of Government of PUC where he teaches Macroeconomics. He has held several academic and professional positions, among others, Dean of the School of Economics and Business of the University of Chile, Chief Economist of the Central Bank of Chile, Minister of Transport and Telecommunications, and Ambassador of Chile to the OECD. He has also been Visiting Scholar at Stanford University and Assistant Professor at the University of Houston. He has authored and edited several books and book chapters, and published papers in *Economics Letters*, *Journal of Economic Development*, *Journal of International Money and Finance*, *Journal of Real Estate Finance and Economics*, *Emerging Market Review*, Análisis Económico, and Economía Chilena.

Matías Negrete-Pincetic is currently the Director and Associate Professor of the Department of Electrical Engineering at UC and a Senior Researcher at the Institute for Complex Engineering Systems (ISCI). He is Co-director of the Energy and Complex Systems Laboratory (ECS-Lab) at UC. He is the founder of Vinken, where he provides consulting and technological development for the energy industry. He has led various projects on topics related to power and energy systems, electricity markets, smart grids, and electricity regulation. Matías holds a PhD in Electrical and Computer Engineering and an MSc in Physics from the University of Illinois at Urbana-Champaign. He was also a postdoctoral researcher at the University of California, Berkeley. Additionally, he is an Electrical Engineer and holds a master's in Physics from the Pontificia Universidad Católica de Chile (UC). He is a Fulbright Scholar.

Marialina Núñez holds a master's in Engineering Sciences from the Pontificia Universidad Católica de Chile and is Project Engineer Department of Hydraulic and Environmental Engineering. She conducted work related to drought forecating, calculation of drought indices, and hydrological modelling.

Jairo Paizano-Potoy is a Humanitarian and Development Interventions Expert with more than 15 years of professional experience in the humanitarian and development sectors. Jairo P. has been engaged in identifying, implementing, monitoring, and evaluating food security and livelihoods interventions in Africa (Niger, Mauritania, Senegal, Nigeria); America (Guatemala, Honduras, Nicaragua, Venezuela); Asia (Bangladesh, Philippines); Middle East (Palestine, Tunisia, Libya); and Spain. He contributed to business development and projects implementation in different areas such as local economic development, microenterprise and startups development, creative economy, protection and recovery of livelihoods, women economic empowerment, cash transfer programming, and capacity building. Currently, Jairo P. is working with NRC as Livelihoods and Food Security

(LFS) Specialist in Venezuela. Under this position, he oversees technical development of LFS, strategic direction, and quality assurance. He is responsible for mentoring, training, and capacity building staff on LFS technical knowledge and represents NRC in relevant forums/clusters, including with national authorities and donors. Jairo P. is proficient in Spanish, English and French, and holds a master's degree in Science Economic Development and Public Policies. Jairo P is cofounder of https://imperialdeometepe.com/

Pilar Paneque Salgado is a Professor of Human Geography at the Universidad Pablo de Olavide in Seville, where she leads the Global Change Research Lab. She joined this university in 2000 and has held positions such as Director of Postgraduate Studies, Vice-Rector of Planning and Quality, and Director of the Board at the Universidad Internacional de Andalucía (UNIA). Additionally, she directed the Campus of International Excellence in Environment, Biodiversity, and Global Change, coordinating seven Andalusian universities and the CSIC. Her research focuses on water and land policy, with an emphasis on hydrological risks such as droughts and social and institutional vulnerability to climate change. She has led 14 competitive research projects, including seven at the European level, and supervised over 20 research technicians. She has also participated in 10 other research projects. Her contributions have been recognized with five six-year research and transfer awards granted by ANECA. In the academic sphere, she has supervised 11 doctoral theses across multiple disciplines and currently supervises eight more. Since February 2023, she has served as the Director of the National Agency for Quality Assessment and Accreditation (ANECA).

Pablo Pastén is a Civil Engineer (Pontificia Universidad Católica de Chile) and PhD (Northwestern University), full time faculty at the Department of Hydraulic and Environmental Engineering at Pontificia Universidad Católica de Chile. He is the author and coauthor of numerous research articles and book chapters, and his work combines cutting-edge research in environmental geochemistry, environmental engineering, and sustainability, with impact on the praxis and policy-making toward societal challenges. He is the Associate Editor of *Applied Geochemistry.* Pastén was recognized as a Latin American Trailblazer chemist by the American Chemical Society's *C&EN* magazine in 2024.

David Pereira is a Full Professor in the Department of Agroengineering at the Universidad Politécnica de Madrid. He obtained his PhD in Environmental Assessment, and since then has taught subjects such as environmental impact assessment, strategic environmental assessment, spatial planning, rural planning, and land use planning. He has participated in six European projects on environmental and economic aspects of climate change, and on new transdisciplinary research methodologies. He is a member of the observatory for the right to food in Spain and of the Center for Innovation in Technologies for Development (UPMitd) and belongs to the research group on Building, Infrastructure, and Projects in Rural and Environmental Engineering and to the group of experts on spatial planning of the

Ministry of Ecological Transition and Population Challenge (MITECO). He has participated in the elaboration of development plans for the Common Agricultural Policy and the Rural Development Strategy at both national and regional level, as well as in the impact measurement and impact assessment of plans and strategies in the fields of rural development, agricultural, energy, and mobility policies. He is an expert in sustainability indicators, spatial data analysis, landscape planning, rural development, and environmental assessment.

Francisco Suárez is an Associate Professor in the Departamento de Ingeniería Hidráulica y Ambiental at Pontificia Universidad Católica de Chile. Dr. Suárez is an Associate Researcher at the centers of excellence CEDEUS (FONDAP/1522A0002) and CEGA (FONDAP/15110020) and is a member of the UC Desierto de Atacama Center. His research focuses on the interaction between surface and groundwater, with a focus on hydrology and hydrogeology of arid zones. Dr. Suárez has led various research and extension projects. He has published more than 70 peer-reviewed papers and participated in more than 120 conference presentations. He also advised the Chilean Ministry of Foreign Affairs for more than 6 years by providing scientific evidence to demonstrate the international nature of a transboundary river basin shared by Bolivia and Chile. This evidence was presented at the International Court of Justice in The Hague, which allowed legal certainties in the use of these waters.

Julia Urquijo-Reguera is an Assistant Professor at the School of Agronomy, Food, and Biosystems Engineering at the University Polytechnic of Madrid since 2018. She has developed her PhD studies in the area of projects and rural planning on drought management evaluation (UPM, 2015), and obtained the Extraordinary Prize of Doctoral Thesis of the UPM 2015-2016. She holds a degree in Agricultural Engineer by the Polytechnic University of Madrid (UPM, 2005), a master's degree in Evaluation of programs and public policies by the Complutense University of Madrid (UCM, 2012). She has a professional experience of more than 18 years in the field of international cooperation, environment, and sustainable development, both in Spain and at European level, and with field work experience in Latin and Central America. He has specialized in the development of participatory methodologies, sustainability indicators, and evaluation of public programs and policies through different approaches and the design of different evaluation and analysis frameworks. She has worked for several public and private institutions at international and national levels such as FAO, AECID, SEGIB, and REDS, among others, as well as several Spanish public administrations. She is an active member of different groups/organizations such as the CEIGRAM (UPM), the EELISA DISCOVERY Community (UPM), the Observatory for the Right to Food in Spain (ODA-E), and the Iberian Association of Professionals for Evaluation (APROEVAL).

Jesús Vargas Molina is a Graduate in Environmental Sciences from the National University of Distance Education (UNED) and PhD in Environmental Studies from Pablo de Olavide University in Seville. Since 2021, he has been a full-time Professor in the Department of Geography at the University of Malaga and a researcher at the

Global Change Research Lab of Pablo de Olavide University in Seville. He specializes in water resources management and governance, as well as hydroclimatic risks in the context of climate change. He has participated in and led several national and international research projects and is part of various multidisciplinary and transdisciplinary teams focused on the mitigation and adaptation of natural risks. He has published numerous book chapters and scientific articles in high-impact international journals and publishers. His recent work focuses on methodologies and tools for citizen science and geographic information technologies for a better understanding of hydroclimatic risk dynamics.

Javier Vargas is a Civil Engineer in Biotechnology with an Environmental Focus from the Pontifical Catholic University of Chile, with experience in industrial engineering project management, systems engineering, software development, and environmental management. He is skilled in programming, data analysis, and interactive data visualization, recognized for strong teamwork abilities, research skills, and innovative solution development. He is experienced in scientific research and collaboration on climate change-related projects and is currently serving as a Project Engineer at the UC Center for Global Change, contributing to the development and execution of research initiatives in this field.

Ignacio T. Vargas is a Civil Engineer (Pontificia Universidad Católica de Chile, PUC) and holds a master's in Environmental Engineering and a PhD in Engineering Sciences, both from the PUC. Dr. Vargas is an Associate Professor in the Department of Hydraulic and Environmental Engineering at PUC and leads the Laboratory of Biotechnology and Environmental Microbiology. Dr. Vargas is also in charge of the environmental engineering undergraduate program, the Major of Engineering and Environmental Sciences, in the School of Engineering at PUC. His research has focused on the study of microbial biofilms, microbial electrochemical technologies, biocorrosion, biogeochemistry, and water reuse. Currently, he is Associate Researcher of the Center of Research in Sustainable Urban Development (CEDEUS), the Marine Energy Research & Innovation Center (MERIC), and the Technological Consortium of Water (COTH2O).

Felipe Vásquez-Lavín is a tenured Professor at the Faculty of Business and Economics of the Universidad del Desarrollo (UDD). He holds a PhD in Agricultural and Natural Resources Economics from the University of California at Berkeley (2007), a master's in Environmental Economics and Natural Resources (1998), and a Bachelor of Economics degree (1995) from the University of Concepción, Chile. His research focuses on the statistical modeling of economic agents' decisions using discrete and discrete-continuous econometric models, particularly in Water Resources Economics, Climate Change Economics, and Environmental Economics. He has authored more than 80 scientific articles in top-tier economics and multidisciplinary journals such as *Energy Economics*, *Energy Policy*, *Journal of Environmental Economics and Management*, *World Development*, *Resource and Energy Economics*, *Ecological Economics*, *Marine Resource Economics*, *Water*

Resources and Economics, *Water Resources Research*, *Water Resource Management*, *Climatic Change*, *Scientific Reports*, *Conservation Letters*, *Journal of Environmental Management*, and *Environmental Research Letters*.

Alejandra Vega is a Civil Industrial Engineer with a specialization in Hydraulics from the Pontificia Universidad Católica de Chile. She holds a master's and PhD in Engineering Sciences from the same institution. She is a Researcher at the Centro de Desarrollo Urbano Sustentable (CEDEUS).

Sebastián Vicuña is an Associate Professor at School of Engineering and Director of Centro de Cambio Global at the Pontificia Universidad Católica de Chile. His research interests relate to water resources, hydrologic modeling, climate change adaptation, and integrated watershed management. Sebastian has participated as lead author and task group president in three IPCC Assessment Cycles. He is an Environmental Engineer from the Pontificia Universidad Católica de Chile. In 2004 he earned a master's in Public Policy and a master's in Environmental Engineering from the University of California at Berkeley. In 2007 he obtained a PhD in Environmental Engineering from the same university.

Sarah Ann Wheeler is a Professor of Water Economics in the School of Economics and Public Policy at the University of Adelaide and is also on the university's Environment Institute's executive management board. She is a past President and a Distinguished Fellow of the Australasian Agricultural and Resource Economics Society (AARES), and a Fellow of the Academy of the Social Sciences in Australia. She graduated with her PhD in 2007, was an ARC Future Fellow (2014–2018), and has over 140 peer-reviewed outputs in the research areas of irrigated farming, climate change, Murray-Darling Basin, organic farming, water markets, water scarcity, mental health, and food waste. She is an Associate Editor of *Agricultural Economics*, and has been past editor of the *Australasian Journal of Water Resources*, and past Associate Editor of *Water Resources and Economics*, the *Australian Journal of Agricultural and Resource Economics* and guest editor for a special issue of *Agricultural Water Management*. Sarah is on a number of editorial boards, including: *International Review of Environmental and Resource Economics*, *Australian Journal of Agricultural and Resource Economics* and *Water*. In 2023, Sarah was awarded the Jill Hudson award for environmental protection at the SA Environment Awards.

Donald Wilhite is a Professor and Director emeritus of Applied Climate Science, School of Natural Resources at the University of Nebraska-Lincoln, USA. Prior to August 2012, he served as director of the School of Natural Resources, a position he held from 2007 to 2012. Don founded the US National Drought Mitigation Center at the University of Nebraska-Lincoln in 1995 and served as its director until 2007. He was elected fellow of the American Meteorological Society in 2013. Don's research and outreach activities have focused on issues of drought monitoring, planning, mitigation, and policy; the use of climate information in decision-making; and climate change. He has collaborated with and led many efforts by US states, federal agencies, foreign governments, and United Nations agencies to improve drought management following a risk-based management approach he developed, the 10-step planning process. This process has become the international standard for drought planning. Don is the editor or co-editor of numerous books, including his most recent book, *Drought and Water Crises: Integrating Science, Management, and Policy*, 2nd edition (CRC Press, 2018). He is also coauthor of two reports on the implications of climate change for Nebraska, published in 2014 and 2016, and coeditor of the *Atlas of Nebraska* (University of Nebraska Press, 2017).

Alec Zuo is an Applied Economist specializing in agricultural, environmental, and resource economics. Alec's research has addressed issues and questions that are fundamental to UN World Development Goals and Australian National Research Priorities. His research has contributed significantly to understanding the impact of climate change and resources on irrigation farming, farmer adaptation strategies, and market-based instruments for water resource management. His research has been published in leading international journals (e.g., *Global Environment Change*, *American Journal of Agricultural Economics*, *European Review of Agricultural Economics*, *Energy Economics*, etc.) and he has collaboration with researchers from over 20 organizations worldwide on agricultural, resource, environment, and development issues across 11 countries in four continents. Alec has substantial experience in initiating and managing large research projects. His research has been funded by the Australian Research Council (an ARC Future Fellow from 2023 to 2027), the Australian National Commission for UNESCO, the National Climate Change Adaptation Research Facility, the federal Department of Agriculture, Fisheries and Forestry, the Australian Centre for International Agricultural Research, the Australian Consumer & Competition Commission, and the Murray Darling Basin Authority.

Abbreviations

AAO	Antarctic Oscillation
ACADES	Asociación Chilena de Desalinización—Chilean Desalination Association
ADI	Aggregate Drought Index
AEMET	Agencia Estatal de Meteorología
AFP	Pension Fund Administrators
AGRITEMPO	Agrometeorological Information System
Al	Aluminum
AMD	Acid mine drainage
AMP	Productivity amplitude
ANID	Agencia Nacional de Investigación
AOPs	Advanced oxidation processes
APEC	Asia-Pacific Economic Cooperation
ARB	Antibiotic resistance genes
ARG	Argentina
As	Arsenic
AUD	Australian dollars
AVHRR	Advanced Very High-Resolution Radiometer
AWS	Amount of water in storage
BOD5	Biochemical oxygen demand over 5 days
BOL	Bolivia
BRA	Brazil
BVP	Base productivity value
CAPEX	Capital expenditure
CASEN	Encuesta de Caracterización Socioeconómica Nacional—National Socioeconomic Characterization Survey
CAZALAC	Centro del Agua para Zonas Áridas y Semiáridas de América Latina y el Caribe—Water Center for Arid and Semi-arid Zones of Latin America and the Caribbean
CC	Climate change
Cd	Cadmium

CDI	Combined Drought Indicator
CDPMN	Caribbean Drought and Precipitation Monitoring Network
CDWR	California Department of Water Resources
CECs	Contaminants of emerging concern
CEDEX	Centro de Estudios y Experimentación de Obras Públicas
CF	Coliformes Fecales—Fecal coliforms
CFU	Colony-forming unit
CIAU	Management of the Integral Urban Water Cycle
CIIFEN	Centro Internacional para la Investigación del Fenómeno de El Niño—International Center for the Investigation of El Niño Phenomenon
CLP	Chilean pesos
CMF	Comisión para el Mercado Financiero—Financial Market Commission
CMI	Crop Moisture Index
CMIP5	Coupled Model Intercomparison Project version 5
CMIP6	Coupled Model Intercomparison Project version 6
CNE	National Energy Commission
CNR	National Irrigation Commission
Co	Cobalt
CODELCO	Corporación Nacional del Cobre de Chile—National Copper Corporation of Chile
COL	Colombia
Cr	Chromium
CR2	Centro de Ciencia del Clima y la Resiliencia—Climate and Resilience Science Center
CRS-OCA	Regional Climate Center for Western South America
CRC-OSA	Regional Climate Center for Western South America
CRC-SAS	Regional Climate Center for South America
CSDI	Crop Specific Drought Index
CSIRO	Commonwealth Scientific and Industrial Research Organisation
CTCN	UN Climate Technology Centre and Network
Cu	Copper
CVP	Central Valley Project
DFL	Decreto Fuerza Ley—Decree Force of Law
DGA	Dirección General de Aguas—General Water Directorate
DIPRES	Dirección de Presupuesto—Budget Directorate
DIRECTEMAR	Maritime Water Agency
DPI	Department of Primary Industries
DMC	Dirección Meteorológica de Chile—Meteorological Directorate of Chile
DOH	Dirección de Obras Hidráulicas—Department of Hydraulics
DRAMP	Drought Resilience, Adaptation and Management Policy
DRM	Disaster Risk Management
DRR	Disaster Risk Reduction

DSI	Drought Severity Index
DWTP	Drinking Water Treatment Plant
EC	European Commission
ECOLEX	Information Service on Environmental Law
ECs	Exceptional circumstances
ECU	Ecuador
EDO	European Drought Observatory
EDII	European Drought Impact Inventory
ELAW	Environmental Law Alliance Worldwide
EM-DAT	The International Disaster Database. Centre for Research on the Epidemiology of Disasters
ENAP	Empresa Nacional del Petróleo—National Petroleum Company
ENDESA	Chile's National Electricity Company
ENSO	El Niño-Southern Oscillation
EOS	End of season
ESVA	Water Supply and Sanitation Firm for Viña del Mar and Valparaíso
ETDI	Evapotranspiration Deficit Index
Euroclima+	European Union's flagship regional cooperation programme
EVI	Enhanced Vegetation Index
EWS	Early Warning Systems
FAL	Autumn productivity
FAO	Food and Agriculture Organization
FC	Fecal coliforms
Fe	Iron
FE	Fixed Effects Panel Data
FMD	Farm management deposit
FONASA	Fondo Nacional de Salud—National Health Fund
g/L	Gram per liter
GDP	Gross domestic product
GFCF	Gross fixed capital formation
GHG	Greenhouse gas emissions
GMC	Global climate model
GOES	Geostationary Operational Environmental Satellites
GUY	Guyana
GWP	Global Water Partnership
HAB	Harmful algal blooms
Hg	Mercury
HMNDP	High-level Meeting on National Drought Policy
HTSP	Hydrothermal scheduling problem
hm	Hectometer
hm^3	Cubic hectometer
HMNDP	High-level Meeting on National Drought Policy
HTSP	Hydrothermal scheduling problem
IC	Infrastructure capacity

IED	Income Equalisation Deposit
IMF	International Monetary Fund
INA	National Water Institute
INMHI	National Institute of Meteorology and Hydrology
INDAP	Agricultural Development Institute
INE	Instituto Nacional de Estadísticas—National Institute of Statistics
INIA	Agricultural Research Institute
IPCC	Intergovernmental Panel on Climate Change
IPO	Interdecadal Pacific Oscillation
IPR	Indirect potable reuse
IRI	Climate and Society Map Room
ISAPRE	Instituciones de Salud Previsional—Social Security Health Institutions
IWA	International Water Association
K^+	Potassium
$KgCO_2eq/m^3$	Kilogram of carbon dioxide equivalent per kilogram per cubic meter
km^2	Square kilometer
KWh/year	Kilowatt-hour per year
l/s	Liters per second
LAC	Latin America and the Caribbean Region
LCDP	Liters per capita
LIN	Seasonal integral
LOS	Length of season
M2	Money supply
MCM	Million cubic meters
MD	Megadrought
MED	Multiple effect distillation
MEI	Multivariate ENSO Index
MERCOSUR	Mercado Común del Sur
MILP	Mixed-integer linear programming
MINAGRI	Ministry of Agriculture
MINDEF	Ministry of National Defense
MINEN	Ministry of Energy
MINVU	Ministerio de Vivienda y Urbanismo—Ministry of Housing and Urban Planning
MITERD	Ministry for Ecological Transition and Demographic Challenge
MJO	Madden-Jullian Oscillation
Mn	Manganese
MODFLOW	US Geological Survey Modular Finite-Difference Flow Model
MSDF	Ministerio de Desarrollo Social y Familia—Ministry of Social Development and Family of Chile
MMA	Ministry of Environment
MODIS	Moderate Resolution Imaging Spectroradiometer

MOP	Ministerio de Obras Públicas—Ministry of Public Works
MSF	Multistage flash distillation
MVA	Maximum productivity value
MW	Megawatts
MWh	Megawatts hour
NADAM	North American Drought Monitor
NDA	National Drought Agreement
NDRA	Natural Disaster Relief Arrangements
NDVI	Normalized Difference Vegetation Index
NDWI	Normalized Difference Water Index
NIDIS	National Integrated Drought Information System
NH_4^+	Ammonium
NO_2^-	Nitrogen oxides
NO_3^-	Nitrates
NOAA	National Oceanic and Atmospheric Administration
NSCA	Normas de Calidad Secundaria de Agua—Secondary Water Quality Standards
NSW	New South Wales
OECD	Organisation for Economic Co-operation and Development
OLI	Operational Land Imager
OLS	Ordinary least squares
OMEGA	Office for Monitoring Agricultural Emergencies
OPEX	Operation expenditure
PAHO	Pan American Health Organization
Pb	Lead
PDI	Policía de Investigaciones—Investigation Police
PDO	Pacific Decadal Oscillation
PER	Peru
PET	Peak of time
PH	Potential of Hidrogen—Acidity measure
PDSI	Palmer Drought Severity Index
PEM	Plan Especial de Emergencia—Emergency Plan for Drought Situations
PERH	Planes Estratégicos de Recursos Hídricos—Strategic Plans for Water Resources
PES	Special warning and eventual drought plan
PGU	Pensión Garantizada Universal—Universal Guaranteed Pension
PLP	Long-term planning
PMCCA	Observation Network in Surveillance Programs of NSCA
PMP	Positive mathematical programming
PO_4^{3-}	Phosphate
PP	Primary productivity
PPP	Purchasing power parity
Pywr	Open-source Python library for water resources simulator
PRY	Paraguay

RAI	Rainfall Anomaly Index
RBD	River Basin Districts
RCL	Revenue contingent loan
RCP	Representative Concentration Pathway
RDI	Reclamation Drought Index
RO	Seawater reverse osmosis
RSDI	Regional Streamflow Deficiency Index
S-WEI	S-Water Exploitation Index
S2ID	Integrated Disaster Information System
SDDP	Stochastic dual dynamic programming
SDDiP	Stochastic Dual Dynamic Integer Program
SEIA	Environmental Impact Assessment System
SEN	National Electric System
SENAMIH	National Meteorology and Hydrology Service of Peru
SENAPRED	National Disaster Prevention and Response Service
SEREMI	Regional Ministerial Secretariats
SFI	Standardized Flow Index
SIAT	Information and Early Warning System of the Southwest of Buenos Aires
SIC	Central Interconnected System
SISS	Superintendencia de Servicios Sanitarios
SISSA	Information System on Drought for Southern South American Region
SGWI	Standardized Groundwater Storage Index
SMA	Superintendency of Environment
SMDI	Soil Moisture Deficit Index
SNAPP	Science for Nature and People Partnership
SOI	Southern Oscillation Index
SOS	Start of season
SP	Spring productivity
SPEI	Standardized Precipitation Evapotranspiration Index
SPI	Standardized Precipitation Index
SRI	Standardized Runoff Index
SSI	Standardized Streamflow Index
SSMI	Standardized Soil Moisture Index
SUBTEL	Subsecretaria de telecomunicaciones—Undersecretariat of Telecommunications
SWP	State Water Project
SWSI	Surface Water Supply Index
TAO/TRITON	Tropical Atmosphere Ocean Array
TCI	Temperature Condition Index
tCO2eq	Quantity of greenhouse gases expressed as the product of the weight of the greenhouse gases in metric tonnes multiplied by their global warming potential
TM	Thematic Mapper

ToE	Time-of-Emergence
TP	Total phosphorus
TSS	Total suspended solids
UAE	United Arab Emirates
UF	Unidad de Fomento—Inflation Index
UNCCD	United Nations Convention to Combat Desertification
UNDRR	United Nations Office for Disaster Risk Reduction
UNESCO	United Nations Economic and Social Council
UNFCCC	United Nations Framework Convention on Climate Change
UNISDR	United Nations International Strategy for Disaster Reduction
UQM	Unbiased Quantile Mapping Method
USACE	US Army Corps of Engineers
USBR	US Bureau of Reclamation
USD	United States dollar
USDM	United States Drought Monitor
USEPA	United States Environmental Protection Agency
UTE	Shortage Territorial Units
UTS	Prolonged Drought Territorial Units
UV	Ultraviolet
VAR	Vector autoregressive
VCI	Vegetation Condition Index
VegDRI	Vegetation Drought Response Index
VEN	Venezuela
VES	End of season productivity value
VHI	Vegetation Health Index
VIDECI	Vice Ministry of Civil Defense
VSS	Start of season productivity value
WB	World Bank
WEAP	Water Evaluation And Planning System
WEI+	Water Exploitation Index+
WFD	Water Framework Directive
WMO	World Meteorological Organization
WR	Water Rights
WRI	World Resources Institute
WRM	Water resources models
WUOs	Water User Organizations
WWTP	Wastewater treatment plant
zcNDVI	Standardized Deviation of the Accumulated Normalized Difference Vegetation Index
Zn	Zinc

Chapter 1
Introduction

Daniela Rivera and Guillermo Donoso

Abstract Drought is a major natural hazard affecting millions worldwide, causing significant economic losses. It is a phenomenon, with unclear spatial and temporal boundaries, and its impacts are often silent. These effects can accumulate over years and even persist long after the drought has ended, worsening due to climate change. Managing drought is challenging due to its multidimensional nature, yet essential as water becomes an increasingly scarce resource critical to society and the environment. Chile, a country highly vulnerable to climate change, has faced an unprecedented 14-year drought. The vision and treatment of drought in Chile is based mainly on a reactive policy, in which the central purpose is to reduce, as far as possible, the damage caused. A shift to a proactive, preventive strategy is needed, emphasizing decentralized management, local capacity building, and stakeholder involvement. This book explores approaches to characterizing drought, monitoring systems, national and sectoral policies, and the legal framework's performance. Its objective is to analyze the current situation and propose proactive public policies to improve drought management in Chile. Effective management requires an interdisciplinary approach, integrating hydrological, economic, legal, environmental, social, and political aspects. This text provides an integrated perspective on Chile's drought policy, addressing gaps in existing strategies to foster sustainable and comprehensive solutions.

Keywords Reactive drought management · Proactive drought management strategies · Drought risk · Crisis response · Drought governance

D. Rivera (✉)
Faculty of Law and the Center for Water Law and Management, Pontificia Universidad Católica de Chile, Santiago, Chile
e-mail: dirivera@uc.cl

G. Donoso
Faculty of Agronomy and Natural Systems and the Center for Water Law and Management, Pontificia Universidad Católica de Chile, Santiago, Chile

D. Rivera, G. Donoso (eds.), *Droughts in Chile*, Global Issues in Water Policy 31,
https://doi.org/10.1007/978-3-031-85040-0_1

1.1 Introduction

Chile's diverse geography spans over 4200 km, ranging from the arid Atacama Desert in the north to subpolar regions in the south. While southern Chile holds abundant water resources, the central and northern zones face critical water scarcity exacerbated by a prolonged megadrought and climate change. Influenced by the Pacific Ocean, the Andes Mountains, and the Humboldt Current, Chile's climate is shaped by complex interactions that determine rainfall and temperature. Seasonal phenomena like El Niño and La Niña significantly affect precipitation patterns, with La Niña contributing to drier conditions.

Central Chile has seen a clear decline in precipitation, challenging to separate from the impacts of the ongoing megadrought, which began in 2009. Studies link these trends to anthropogenic climate change, with future projections indicating a 3–30% decrease in rainfall and temperature increases of up to 3.5 °C by the century's end. Droughts, driven by reduced precipitation and large-scale ocean-atmosphere interactions, have increasingly severe effects on ecosystems, agriculture, and water systems, highlighting the growing urgency for sustainable management.

The current megadrought has had profound socioeconomic and ecological consequences, marking it as Chile's longest recorded drought. Agricultural productivity has declined, with reductions in irrigated land and shifts toward water-efficient crops. Urban water supply systems have faced rising costs, amounting to $1.44 billion during the 2010–2020 period. Adaptive investments in urban water infrastructure have proven cost-effective, mitigating some of the economic impacts. However, rural areas reliant on agriculture have suffered significantly.

Drought also affects water quality, with reduced streamflow, higher temperatures, and hydrochemical changes degrading resources essential for agriculture, aquaculture, and drinking water. Increased salinity harms crops and soil health, while harmful algal blooms disrupt aquatic ecosystems. These challenges underscore the need for proactive and integrated approaches to manage water resources under changing climatic conditions.

To address the multifaceted nature of drought, Chile requires interdisciplinary strategies that integrate hydrological, economic, institutional, environmental, and social perspectives. Shifting from reactive crisis management to preventive risk management is essential. Strengthened infrastructure, effective governance, and proactive public policies are critical for ensuring sustainable water use and mitigating the long-term impacts of drought and climate change.

The objective of this book is to develop an integrated and proactive drought management strategy in Chile that addresses the multifaceted nature of droughts by transitioning from reactive crisis management to preventive risk management, ensuring sustainable water resource use and minimizing socioeconomic and environmental impacts. To evaluate the performance of Chile's current legal and institutional frameworks in addressing drought, it identifies gaps and recommends reforms to strengthen governance and resilience. The book provides evidence-based, proactive public policy recommendations aimed at reducing the long-term effects of droughts and addressing climate-related water scarcity in Chile through an interdisciplinary approach.

The discussion begins with an analysis of drought as a significant climate phenomenon affecting ecosystems, societies, and economies on a global scale. Chap. 2 emphasizes the stochastic and slow-developing nature of drought, which makes it challenging to define and predict. It underscores the various forms of drought, including meteorological, agricultural, hydrological, socioeconomic, and ecological types. The identification and characterization of droughts requires tools such as drought indices and run theory, which provide quantitative frameworks for evaluating their severity, duration, and impacts. These methodologies, combined with continuous monitoring, enable better preparation for temporary droughts and recognition of permanent climatic shifts. Emphasis is placed on integrating drought indices into decision-making to facilitate proactive measures and mitigate effects on water resources, agriculture, and societal well-being. The chapter underscores the need for adaptive strategies, particularly in regions like Chile, where climatic shifts may signal a transition to more arid conditions.

The following sections include a description of the main content of the various chapters.

1.2 Description of Chapters in Part I: Setting and Country Context: Drought Impacts in Chile

Chapter 3 provides an overview of Chile's economy, detailing its historical trajectory, current state, and challenges. Chile transitioned from a state-controlled, isolated economy in the 1970s to an open market economy driven by privatization and liberalization under the military government from 1973–1989. Subsequent democratic governments, maintained market-oriented policies while prioritizing social equity, significantly reducing poverty and improving social indicators such as literacy and life expectancy. Despite its progress, Chile faces challenges such as productivity stagnation and unequal income distribution. The country remains well-managed macroeconomically, with stable inflation and robust financial systems, supported by autonomous institutions like the Central Bank. Looking forward, opportunities lie in leveraging its abundant natural resources for renewable energy and green technologies, addressing climate change while boosting economic growth. Yet, achieving developed nation status requires improved public policies in education, research and development, investment incentives, and the adoption of adaptive strategies to overcome the middle-income trap and reduce disparities.

Chile's climate is highly diverse, ranging from the Atacama Desert, the world's driest, to the wetter and cooler regions of Patagonia. This diversity is driven by its latitudinal span, topography, and oceanic influences such as the Humboldt Current and the Pacific Anticyclone. Key climatic patterns include Mediterranean conditions in central Chile, supporting agriculture and hydropower, and significant variability tied to phenomena like ENSO, which influences rainfall and droughts. Trends show rising temperatures inland, declining precipitation in central Chile, and more frequent and severe droughts, including the decade-long Megadrought linked to climate change. Chapter 4 forecasts ongoing warming and less precipitation, worsening water scarcity and affecting agriculture, hydropower, and ecosystems.

Chapter 5 examines the economic impacts of droughts in Chile, particularly the megadrought that began in 2010, focusing on agriculture and urban water supply. It highlights the severe reductions in rainfall, leading to significant economic losses and the adaptation strategies employed by various sectors. Agriculture has faced notable challenges, including reduced crop yields, shifts to higher-value but water-intensive crops like fruits, and increased reliance on groundwater. While these measures have mitigated some impacts, the financial burden on farmers has escalated due to groundwater overexploitation and rising operational costs. The urban water supply sector has also experienced rising costs, with water companies investing heavily in infrastructure improvements to maintain supply amidst declining water resources. Despite substantial resilience and adaptation initiatives, the chapter highlights critical tipping points in water scarcity that could significantly increase economic costs. The findings emphasize the importance of proactive planning and collaboration to safeguard water-dependent sectors and ensure long-term economic sustainability.

An in-depth analysis of the diverse effects of droughts on water quality is presented in Chap. 6. The multifaceted impacts of droughts on water quality are analyzed, emphasizing how reduced precipitation, elevated temperatures, and declining streamflows alter hydrochemical parameters in Chilean Andean watersheds. Key consequences include increased salinity, heightened concentrations of nutrients and metals, and reduced dissolved oxygen levels due to prolonged water residence times and limited dilution capacity. The interactions between drought and water quality are shaped by diverse drivers such as snowmelt, erosion, and nutrient cycling, alongside human activities like agriculture, mining, and urbanization. Monitoring and managing water quality during droughts in Chile is constrained by limited datasets and infrastructure. Strengthening water quality networks and expanding data collection is crucial for understanding site-specific impacts and guiding adaptive strategies. These changes create significant challenges for water use in agriculture, drinking supply, and ecosystem health, especially under prolonged drought conditions exacerbated by climate change.

The last chapter in this section critically examines Chile's institutional and governance framework for drought management, highlighting significant fragmentation and reactive approaches. Chapter 7 identifies 17 key entities, including the General Directorate of Water (Dirección General de Aguas, DGA) and the Ministry of Agriculture (Ministerio de Agricultura, MINAGRI), responsible for drought-related functions, but notes a lack of coordination and strategic integration. The primary tools, such as declarations of water scarcity zones and agricultural emergencies, remain reactive, treating droughts as extraordinary events rather than integrating them into regular water resource planning. Legal gaps, limited data integration, and the absence of preventive measures exacerbate the inefficiency of current policies.

The chapter emphasizes the importance of addressing social vulnerabilities and fostering resilience through integrated policies, moving beyond emergency responses to ensure water security and sustainable adaptation to climate variability.

1.3 Description of Chapters in Part II: Comparative Experiences in Drought Management

Countries such as Spain, Australia, some of South America and the state of California in the United States, are transitioning from reactive to proactive management approaches, emphasizing the importance of anticipating and mitigating impacts through adaptive strategies. The chapters in this section present an analysis of policy instruments, regulatory tools, and drought management plans of these countries.

Chapter 8 provides an overview of drought policies in South America, emphasizing the need for a proactive approach to address increasing drought frequency and severity driven by climate change. It highlights the region's shift from reactive crisis management to risk-based approaches, with initiatives such as monitoring systems, early warning systems (EWS), and vulnerability assessments. While countries like Argentina, Bolivia, and Ecuador have developed specific drought plans integrating proactive strategies, others, including Venezuela and Uruguay, lag behind, relying primarily on reactive measures. Drought policy frameworks in many countries remain fragmented, focusing heavily on agriculture while overlooking broader societal impacts. However, despite progress in drought monitoring and planning, significant gaps persist, particularly in vulnerability and impact assessments. Limited integration of academic research into policymaking hinders effective drought management. The chapter underscores the importance of regional collaboration and knowledge-sharing are recommended to harmonize methodologies, improve monitoring, and strengthen resilience across the region. These efforts align with global frameworks such as the Sendai Framework and the Paris Agreement, offering opportunities for South America to enhance its drought management capabilities.

Spain's drought policy and governance is analyzed in Chap. 9. The chapter emphasizes the transition from crisis management to risk management in the context of increasing drought frequency due to climate change. Spain has made progress through the development of Special Warning and Eventual Drought Plans and Drought Emergency Plans for urban water supply, aiming to anticipate, prevent, and mitigate drought impacts. However, the water management model, heavily reliant on infrastructure like dams and reservoirs, has placed significant pressure on resources, exacerbating vulnerability during droughts. Despite adopting the European Union's Water Framework Directive (WFD), which promotes integrated and sustainable water management, the country struggles with inconsistencies in implementing proactive drought strategies, particularly in regions like the Canary Islands and among smaller municipalities. The chapter also highlights the importance of public participation and scientific advancements in drought monitoring, such as Spain's Meteorological Drought Monitor, to improve understanding and communication of drought risks.

Chapter 10 analyzes California's drought management, emphasizing the state's reliance on extensive water infrastructure and adaptive strategies to address its highly variable climate. Historical droughts like the 1976–77 and 2012–16 events shaped institutional responses, highlighting lessons such as the need for diversified water sources and improved groundwater management. Recent droughts highlight the escalating effects of climate change, characterized by low levels of precipitation and snowpack, increased temperatures, and more significant repercussions for agriculture, urban areas, and ecosystems. Strategies like water recycling, groundwater recharge, and conservation have helped mitigate these challenges, but gaps in forecasting and infrastructure remain. The chapter highlights the transition from viewing droughts as isolated emergencies to addressing them as recurring challenges shaped by long-term aridification.

Australia's drought policy has evolved over decades to address the challenges posed by its highly variable climate, marked by recurring droughts exacerbated by climate change. As discussed in Chap. 11, what began as ad hoc and reactive measures evolved significantly in 1992 with the implementation of the National Drought Policy. This policy recognized drought as a natural part of Australia's climate, encouraging self-reliance among farmers and focusing on sustainable practices. Over time, mechanisms like the Exceptional Circumstances (EC) provisions and Farm Management Deposits (FMD) scheme have supported financial resilience, yet the implementation has often reverted to emergency measures during severe droughts, undermining the policy's long-term goals. Recent evaluations stress the importance of a cohesive and forward-looking approach, integrating innovative solutions like Revenue Contingent Loans to balance financial aid with fiscal responsibility. A more comprehensive policy framework is needed, combining risk-sharing mechanisms, behavioral economic insights, and environmental sustainability.

1.4 Description of Chapters in Part III: Improving Drought Management in Chile: Policy Proposals and Adaptation Options

Public policy proposals for proactive drought management in Chile are formulated based on the assessment of its actual policy and the lessons learnt from the international comparative study. The proposed measures aim to integrate drought management into broader water governance frameworks, ensuring long-term resilience and alignment with international best practices.

As Chap. 12 shows, Chile's approach to drought management is primarily reactive, fragmented, and lacks a unified strategy. Efforts are concentrated on crisis response rather than preventative risk management. Although Chile acknowledges drought in its policies, the lack of a standardized definition and proactive measures limits effective management. The assessment of Chile's drought management identifies critical gaps in six policy dimensions: drought definition, evaluation and learning, institutional context, measures, resources, and coordination. While some progress has been made in defining drought and implementing measures like agricultural insurance, the fragmented institutional framework limits the effectiveness of these initiatives. Coordination between the multiple state entities remains weak, with overlapping responsibilities and insufficient resources impeding proactive actions. Measures are primarily reactive, focusing on mitigating impacts post-drought rather than preparing for future events. To enhance Chile's drought resilience, the chapter recommends a comprehensive shift towards proactive management.

The subsequent chapters of this section provide a comprehensive analysis of the main proposed drought management measures.

Chap. 13 highlights the importance of quantitative indices to enhance planning, management, and decision-making. A new platform is proposed to systematize drought tracking by integrating standardized indices like the Standardized Precipitation Index (SPI) and Standardized Streamflow Index (SSI). This platform should automatically calculate indices using diverse data sources, including national and international data, satellite imagery, and modeling. Designed for flexibility, the system aims to visualize drought severity across different regions and timeframes, enabling tailored interventions. Insights from international tools and local stakeholder consultations emphasize a user-friendly, public-access platform that incorporates a 30-year moving window for climate data, spatialized drought descriptions, and multiple severity levels. These features would enable Chile to transition from reactive to preventive drought management.

The use of remote sensing phenology as a tool to monitor the impact of drought on forest ecosystem productivity in South America's temperate forests is explored in Chap. 14. Through satellite-based vegetation indices like NDVI and EVI, the authors analyze phenological metrics such as the start, peak, and end of growing seasons, providing insights into vegetation dynamics and their responses to climate variability. These indices allow for the identification of productivity trends over time, highlighting differences in forest types and their adaptations to environmental conditions. For example, deciduous forests exhibit higher variability in productivity due to their seasonal leaf loss, while evergreen forests show more stable productivity levels year-round. This approach highlights the role of remote sensing in capturing large-scale ecosystem changes, offering a valuable tool for drought monitoring.

Chapter 15 provides a comprehensive examination of the role of hydrological processes, including fog, groundwater, and vegetation dynamics, as sentinel indicators of drought and climate change in Chile's Mediterranean and semi-arid regions. It provides insights into how these variables respond to prolonged drought and climatic stress. Coastal fog serves as a key supplementary water source, attenuating drought impacts in northern and central Chile through its role in reducing evaporation and providing moisture to ecosystems. Groundwater, essential for agriculture and human consumption, has sustained these activities during droughts but requires improved monitoring through indices like the Standardized Groundwater Storage Index (SGWI) to detect anomalies and guide resource management. Finally, vegetation phenology, studied using satellite data over two decades, highlights the impacts of drought and heatwaves on ecosystem productivity and underscores the need for sustainable land management and conservation. The findings stress the importance of integrating sentinel variables into adaptive management strategies for mitigating drought impacts.

A policy instrument involving pre-agreement contracts for temporary water transfer from agricultural sectors to urban utilities is proposed in Chap. 16, focusing on the Aconcagua River basin in Chile. The methodology includes scenario modeling, agreement activation based on drought indices, and economic evaluation of compensations. By analyzing both optimistic and pessimistic climate scenarios, the study identifies the feasibility and triggers for implementing such agreements and assesses their economic implications, considering opportunity costs for agriculture and consumption thresholds for human use. The study also addresses regulatory aspects, suggesting that the framework aligns with Chilean legal priorities for human consumption and domestic use.

Chapter 17 focuses on water reuse in Chile as a viable alternative to address the increasing challenges of drought and water scarcity. Water reuse, particularly of greywater and wastewater, is highlighted as a sustainable solution to alleviate the pressures on high-quality water resources. Despite its history, the concept of water reuse is relatively new in Chile, where the enactment of Law No. 21,075 in 2018 marked a significant step toward formalizing greywater reuse practices. This legal framework encourages urban and rural applications, recognizing the environmental and economic benefits of reducing the demand for potable water through recycled sources. However, implementing water reuse faces several barriers, including technological limitations, economic costs, and public perception issues. Additionally, public skepticism about the safety and quality of recycled water remains a significant obstacle, requiring educational campaigns to build trust.

The growing significance of desalination as a solution to address water scarcity and drought conditions is presented in Chap. 18. With a lengthy coastline, Chile has seen rapid growth in desalination facilities, with 31 plants currently operational, under construction, or planned. While desalination offers an alternative to freshwater, particularly for the mining and industrial sectors, its sustainable implementation faces numerous challenges. Authors conclude that the main gaps are in regulatory frameworks, fragmented institutional management, and the absence of a cohesive national water plan to integrate desalination effectively into broader water resource strategies. Additionally, there are significant environmental concerns, such as the impact of brine discharge on marine ecosystems, greenhouse gas emissions from energy-intensive operations, and vulnerabilities to seismic and tsunami events. Economic considerations, including the high costs of desalinated water and potential subsidies for sectors like agriculture, are pivotal for its broader application.

Finally, Chap. 19 evaluates the operational policies for multipurpose reservoirs, focusing on balancing hydropower generation and agricultural irrigation amidst drought-driven water scarcity. Using Laja Lake in Chile as a case study, the research integrates a hydrothermal scheduling model and a water resources simulation model to examine the impacts of modified water allocation policies under various hydrological scenarios. The findings reveal significant trade-offs between energy emissions and agricultural water reliability, particularly under dry conditions. By testing scenarios of water allocation changes, the study demonstrates that targeted policy adjustments can mitigate deficits in irrigation and maintain energy reliability while balancing environmental and economic considerations.

Chapter 2
Definition and Characterization of Drought

Jorge Gironás and Marialina Núñez

Abstract Drought is a multifaceted and unpredictable climate event with profound effects on ecosystems, communities, and economies. This chapter explores the nature of drought, its origins, and provides definitions tailored to various affected systems. Additionally, it presents related concepts while addressing potential misunderstandings and their consequences. The chapter also introduces statistical properties and indices for identifying and characterizing drought.

Keywords Drought identification · Drought characterization · Types of drought · Drought indices

J. Gironás (✉)
Departamento de Ingeniería Hidráulica y Ambiental, Pontificia Universidad Católica de Chile, Santiago, Chile

Centro de Desarrollo Urbano Sustentable, Santiago, Chile

Centro de Investigación para la Gestión Integrada del Riesgo de Desastres, Santiago, Chile

Centro de Cambio Global, Pontificia Universidad Católica de Chile, Santiago, Chile
e-mail: jgironas@uc.cl

M. Núñez
Departamento de Ingeniería Hidráulica y Ambiental, Pontificia Universidad Católica de Chile, Santiago, Chile

D. Rivera, G. Donoso (eds.), *Droughts in Chile*, Global Issues in Water Policy 31,
https://doi.org/10.1007/978-3-031-85040-0_2

2.1 Introduction

Drought corresponds to a natural climate phenomenon (Ding et al., 2011; Fernández & Gironás, 2021) of an accidental nature, whose properties cannot be totally anticipated. It is one of the natural hazards that affect the greatest number of people and generates the largest economic losses in the world (FAO, 2002; World Bank, 2003). It is a large regional phenomenon of slow development whose spatial limits as well as beginning and end are hard to identify (Mishra et al., 2009). Additionally, its non-violent impacts are silent and can accumulate over time. Hence, its impacts can manifest even years after the actual drought end. Finally, a drought can have different meanings for different people or social groups (Fernández & Gironás, 2017), which complicates its understanding and the way it is approached. This chapter describes drought and its causes, as well as some definitions of drought based on the affected systems. Furthermore, other related concepts typically confused with drought will be discussed, together with the possible implications of possible misconceptions. Finally, statistical properties and identification of drought are discussed, together with its quantitative characterization, particularly through indices.

2.2 Drought Basics

2.2.1 Definition of Drought

Managing water resources during drought conditions requires a clear and precise definition of the phenomenon, particularly when assigning responsibilities, allocating additional resources, or providing socio-economic assistance. The literature contains hundreds of interpretations of what constitutes drought. As highlighted by Dai (2011) and Mishra and Singh (2010), the wide variability in hydro-meteorological behavior complicates the establishment of a universally accepted definition. Economic factors, coupled with the stochastic nature of water supply and demand across different regions, further add to the difficulty of defining drought precisely. This lack of clarity often leads to indecision or the absence of proactive measures among decision-makers, regulators, and other stakeholders (Wilhite et al., 1986).

Several commonly used definitions come from international organizations and reflect their interests and areas of influence. For instance, the World Meteorological Organization (WMO) (1986) defines drought simply as a prolonged period of rainfall deficiency over a specific region. The Convention to Combat Desertification (UN Secretariat General, 1994) describes drought as a natural phenomenon occurring when precipitation is significantly below average, leading to serious hydrological imbalances that negatively affect systems reliant on natural resources.

A more general and operational definition is that of a temporary water shortage caused, at least in part, by unusual climatic conditions, leading to impact specific activities, social sectors, or the environment (Wilhite & Glantz, 1985). Salas (1978) define drought as a water deficit severe enough to adversely affect vegetation, livestock, or society and its activities in a given region. Rossi et al. (2003) suggests that a definition oriented toward water resource management should account for the water needs based on the biological, economic, and social characteristics of a region. Overall, and based on the previous and other definitions, the random water shortages characteristic of drought can be said to possess the following properties: (1) the shortage is severe in comparison to normal conditions, (2) it persists over an extended period, and (3) it affects a large geographical area. These properties are inherently subjective, as they depend on local water demands and the perceived negative impacts of such deficits. Thus, a definition that encompasses these properties is that provided by Yevjevich et al. (1983) which states, based on a general agreement, that droughts are unpredictable, temporary events of meteorological origin, which lead to a significant reduction in water availability relative to a region's normal conditions, and result in negative environmental and social consequences. From such a definition, drought can then be understood both as a natural hazard and a disaster (Pereira et al., 2002). It is a hazard due to the natural unpredictability with a measurable frequency, while it is also a disaster due to the disruption of the hydrological cycle, with implications on water availability for ecosystems, agriculture, and human activities.

2.2.2 Causes of Drought

The intricate temporal and spatial variability of the water cycle is one the main features of Earth's climate. This characteristic results in significant fluctuations in precipitation, air and soil moisture across different regions and time scales. As a result, drought can eventually occur at any time and in any location around the globe. Meteorological factors are generally the primary cause of drought, beginning with a lack of rainfall, which subsequently affects other components of the hydrological cycle (i.e. surface water, soil moisture, and groundwater), as well as socio-natural systems.

In mid-latitude regions, such as central Chile, a key characteristic of drought is the persistence of hot, dry air in the troposphere. This phenomenon is associated with high-pressure zones around 30 degrees latitude and near the poles, which are responsible for the formation of Earth's arid and semi-arid regions. The expansion or shift of these high-pressure zones toward higher latitudes is a typical feature of drought in these areas (Namias, 1983).

Another relevant triggering factor of drought is the occurrence of large-scale ocean-atmosphere phenomena (Cook et al., 2007), which have been found to be correlated with hydrological indices. Some of the indicators used to characterize these phenomena that are related to drought in Chile are the Southern Oscillation Index (SOI) (Ropelewski & Halper, 1996), the Multivariate ENSO Index (MEI) (Wolter & Timlin, 1993), and the Pacific Decadal Oscillation (PDO) index. For example, Fernández and Fernández (2002) identified areas in the southern cone of

South America where droughts are influenced by La Niña, whereas Oertel et al. (2020) describe the relationship between the ENSO Index and drought indices.

2.2.3 Aridity, Desertification and Water Shortage

There are other relevant concepts often used interchangeably that refer to distinct hydrological and environmental phenomena and are different from drought. These concepts are aridity, desertification and water shortage. Each concept involves different scales, processes, and impacts on ecosystems and human systems.

Aridity refers to the long-term climatic condition of a region characterized by low average precipitation, leading to a dry environment where evaporation exceeds precipitation (Tabaria & Aghajanloob, 2012), and where vegetation, if present, is highly specialized and adapted to dry conditions. Arid regions often have insufficient rainfall to support significant plant growth, although specialized ecosystems may still thrive. Aridity is a non-random, natural, long-term characteristic of certain regions.

On the other hand, desertification refers to the process of land degradation in arid, semi-arid, and dry sub-humid areas due to various factors, including climatic variations and human activities, that result in the transformation of fertile land into desert-like conditions (UNCCD, 1994). While climatic factors contribute, human activities are primary drivers of desertification, and thus they can slow or even reverse the process. Hence, desertification is not the same as drought, but it can be accelerated by it.

Finally, water shortage refers to a situation in which the demand for water exceeds the available supply in a given region. It can be caused by natural factors (climate, reduced rainfall) or human-induced factors (over-extraction, pollution, inefficient use). Thus, some human activities that can cause shortage are the overexploitation of surface water and groundwater, or the degradation of its quality due to bad land management and pollutant discharges to water bodies (Paulo & Pereira, 2006). Water shortage can be temporary (e.g., seasonal fluctuations) or chronic (e.g., in areas where demand consistently exceeds supply), and is exacerbated by poor water infrastructure, pollution, and mismanagement.

The analysis of these concepts shows that what might be understood as a long-term drought could eventually signify the occurrence of another distinct phenomenon, such as water scarcity or aridity. Regarding the concepts of scarcity and drought, both terms are often confused, leading to social discussions about whether it is one phenomenon or the other affecting a region or country, even though there are quite clear quantitative metrics for identifying them.

In the case of Chile, there are two relevant phenomena to consider. The first is that, while the historical reference period for administratively identifying droughts corresponds to the last 3 decades, the condition of scarcity is influenced by decisions regarding water rights that were granted in earlier times of greater abundance, with little knowledge of the long-term availability of surface and groundwater resources. This leads to situations of scarcity during times when there is not necessarily a drought, and to confusion among the population when a severe drought occurs, but only a scarcity issue is visualized.

The other relevant situation to consider is the potential transition of semi-arid and Mediterranean zones of the country to more arid conditions. Extended droughts in the context of climate change could be reflecting the early stages of a prolonged or permanent shift in the climate of these areas. This would mean that some dry periods would no longer correspond to droughts but would instead be manifestations of new normal climatic conditions. There should be continuous monitoring over time to detect whether this potential transition to more arid climates is indeed occurring. If this is the case, water resource management must address this new normality; only then can long-term preparedness be established to ensure water sustainability.

2.2.4 *Types of Droughts*

Drought can manifest in various forms depending on its underlying causes and impacts. Typically, the literature (e.g., Wilhite & Glantz, 1985; American Meteorological Society, 2004; Mishra & Singh, 2010; Fernández & Gironás, 2017) identifies four main types of drought: meteorological, agricultural, hydrological, and socioeconomic. Each type of drought has its own characteristics, affects different sectors, and often overlaps with others, creating a chain of impacts.

2.2.4.1 Meteorological Drought

Meteorological drought refers to a reduction in precipitation over a specific period, compared to the long-term average in a particular region. It is essentially the starting point for other types of droughts and reductions in water availability for ecosystems and urban areas. Its effects might become more serious when coinciding with critical crop growing stages.

2.2.4.2 Agricultural Drought

Agricultural drought occurs when there is insufficient soil moisture -or Soil Moisture Deficiency- to meet the needs of crops or other vegetation. This type of drought typically follows a meteorological drought but focuses specifically on agriculture's ability to sustain growth and yields. The impacts can be even more significant in regions with crops that are rain-fed and have shallow root systems, whereas crops in irrigated systems may be more resilient, depending on water supplies.

2.2.4.3 Hydrological Drought

Hydrological drought refers to deficiencies in water supply, such as reduced streamflow, lake levels, groundwater depletion, and reservoirs. It typically follows extended meteorological drought and reflects the longer-term impacts on water systems, with lag-times that depend on the hydrologic regime of the river basin (e.g., pluvial, nival). In general, hydrological drought can be longer but less intense and fewer in amount as compared to meteorological drought (Fernández & Gironás, 2015). Furthermore, the spatial dependence among the stream flows series in a region reflects the spatial dependence of the rainfall, and thus the regional extensions of meteorological and hydrological droughts are usually similar. Water Supply Shortages, ecosystem stress or reduction in hydropower generation are among the typical impacts of hydrological droughts. Finally, note that some authors (Van Lanen & Peters, 2000; Mishra & Singh, 2010) have restricted hydrological drought to mostly surface water, and have proposed the concept of groundwater drought when referring to the decrease of recharge and the subsequent long-term reductions in groundwater levels and discharges.

2.2.4.4 Socioeconomic Drought

Socioeconomic drought occurs when water demand surpasses supply due to a combination of meteorological, agricultural, or hydrological droughts, affecting economic activities and social well-being. It is driven by both natural and human factors (e.g., population growth, urbanization, and water mismanagement), and typically has a wide range of indirect effects such as disruptions in industries reliant on water, sanitation issues and waterborne diseases, food scarcity, energy shortages, and social conflicts. Note that while socioeconomic drought is a temporary phenomenon driven by natural drought events and affects water supply in a specific period, water scarcity is a long-term, structural issue that exists independently of short-term weather conditions and reflects a fundamental lack of available water to meet demand.

2.2.4.5 Ecological Drought

Although the four types of droughts previously described are the most mentioned in the literature, in recent years, a new type of drought known as *ecological drought* has been proposed. The Ecological Drought Working Group established by the Science for Nature and People Partnership (SNAPP) defined ecological drought as episodic deficit in available water induced by climate and human factors during the vegetation growth period, which ultimately affects other systems (Jiang et al., 2021). Furthermore Crausbay et al. (2017), provided a more detail definition for the concept, stating that an ecological drought is an episodic deficit in water availability that drives ecosystems beyond thresholds of vulnerability, impacts ecosystem services, and triggers feedbacks in natural and/or human systems. The definition of this new type of drought is expected to allow the assessment of drought on ecosystems and the natural capital they provide to human communities (Crausbay et al., 2017).

2.3 Run Theory in Drought Identification

Drought properties can be considered random variables, the values of which cannot be accurately predicted until the event occurs. Therefore, the statistical analysis of these characteristics is crucial, particularly for the planning and management of water resources. Various probabilistic and stochastic methods can be employed to analyze different aspects of droughts, including frequency analysis, the estimation of return periods, and spatial and temporal multivariate analyses (Mishra & Singh, 2010).

Run theory is a statistical tool applied to the identification and characterization of drought events, which are considered stochastic events. This method was first introduced by Yevjevich (1967) and has since become a fundamental approach in the analysis of drought. In this approach, two random variables are defined: water supply (or availability) and water demand. A new time series is then constructed based on the difference between these two variables. A drought is identified as a negative run or a sequence of negative values in this new series, indicating that water supply cannot meet demand.

Once a drought is identified using run theory, several important characteristics can be measured (Fig. 2.1). Given the random nature of the supply and demand series, these characteristics are also random variables, and eventually interrelated.

- Initiation Time: This is the moment when the series transitions from positive to negative values, indicating the onset of a drought.
- Termination Time: The point at which the series returns from negative to positive, marking the end of the drought.

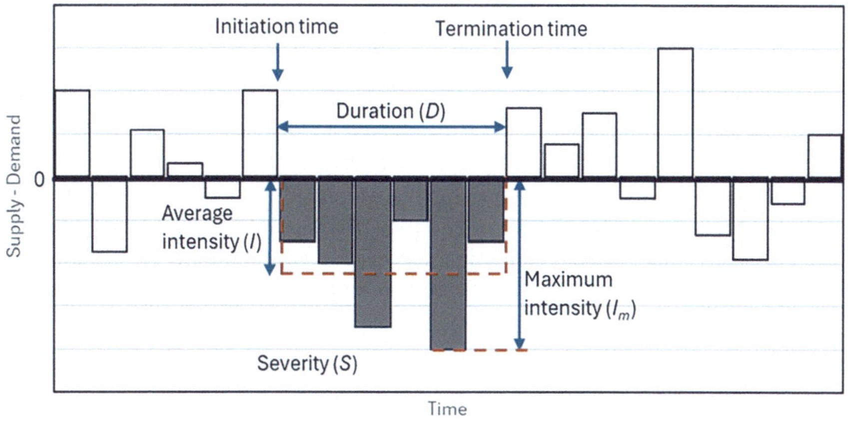

Fig. 2.1 Characteristics of local drought events using run theory

- Duration (D): The length of time between the initiation and termination of the drought.
- Severity (S): The cumulative deficit over the drought period, calculated as the sum of all deficits or negative values in the series.
- Average Intensity (I): Defined as the severity divided by the duration, providing insight into the average deficit over the drought period.
- Maximum Intensity (I_m): The largest individual deficit during the drought period, highlighting the peak intensity of water scarcity.

Although the general formulation of run theory refers to supply and demand, given the previously explained definition of drought, the demand can be considered a threshold or condition representative of historical availability, such as the average. In this way, a supply greater than the demand would indicate that the supply exceeds the average available water amount during the historical period also called *reference period*. Furthermore, the run theory can be applied not only to physical variables representing supply and demand but also to a drought index, like those that will be presented later. These indices are defined so that positive and negative values indicate abundance or deficit conditions, with zero representing the normal state. Finally, the application of run theory to multiple locations allows the characterization of regional drought features such as the extent of the affected area, the magnitude of the event, and the total volume of the water deficit across the region (Fernández & Gironás, 2017).

2.4 Drought Indices

Drought indices are essential tools for quantifying water supply and demand in a region. They condense multiple variables into a single value, providing a clear overview of drought conditions (Hayes, 2006). This simplifies both analysis and communication, especially when dealing with complex hydro-meteorological data (Zargar et al., 2011). However, the choice of indices can be controversial, as different indices may yield conflicting results depending on how drought is perceived and characterized (Paulo & Pereira, 2006). Consequently, no universal index exists, and a wide variety of indices have been developed to deal with different types of droughts, data availability, and temporal scales (Mishra & Singh, 2010). Over 150 drought indices have been proposed, with more being introduced as technology advances (Niemeyer, 2008; Zargar et al., 2011). These indices, combined with the run theory, facilitate better communication of drought conditions to stakeholders and the general public, making real-time monitoring, early detection, and severity assessments more straightforward (Zargar et al., 2011). The use of indices allows water managers to act proactively and implement mitigation measures (Tsakiris et al., 2007).

A drought index measures deviations from normal conditions, with the selection of this normal condition or threshold being a key aspect, particularly because of its relevance in identifying drought onset and conclusion (Tsakiris et al., 2007). While annual time scales are common for regional drought assessments, they are less suitable for detailed monitoring and decision-making. Monthly scales offer more precision for identifying drought onset and termination. However, using finer time steps introduces statistical complexities like non-stationarity and hydrological cycle phenomena (Panu & Sharma, 2002; Mishra & Singh, 2010).

2.4.1 Classification of Drought Indices

The choice of variables for drought indices depends on the type of drought and its impacts. For meteorological droughts, precipitation-based indices are most common, capturing early drought manifestations. Examples include the Rainfall Anomaly Index (RAI), the Drought Severity Index (DSI), and the widely used Standardized Precipitation Index (SPI) (McKee et al., 1993). On the other hand, to account for demand or the impacts of meteorological drought, other variables like temperature or evapotranspiration are integrated into indices like the Reconnaissance Drought Index (RDI) and the Standardized Precipitation Evapotranspiration Index (SPEI) (Vicente-Serrano et al., 2010).

For hydrological droughts, indices include streamflow data, reservoir levels, and snowpack measurements. Examples are the Surface Water Supply Index (SWSI), and the Reclamation Drought Index (RDI) (Shafer & Dezman, 1982; Weghorst, 1996). On the other hand, Agricultural drought indices focus on the water available for crops, combining factors like precipitation, temperature, and soil moisture. Common indices include the Crop Moisture Index (CMI) and the Crop Specific Drought Index (CSDI) (Meyer et al., 1993; Palmer, 1968).

The previous indices are just some of the indices available in the literature (e.g., Keyantash & Dracup, 2004; Hayes, 2006; Dai, 2011; Zargar et al., 2011; Mishra & Singh, 2010). New indices are constantly proposed based on available information (i.e. from in-situ measurements to remote sensing information), regional characteristics and a better understanding of drought impacts. Table 2.1 summarizes the drought indices identified by Fernández and Gironás (2017), along with their main properties (i.e. type of drought involved, input information, corresponding temporal and spatial scales, strengths and drawbacks).

Table 2.1 Drought indices commonly used for planning, managing and monitoring water resources

Indices	Properties	Strengths	Drawbacks
Percent of normal (Heim, 2002)	Type: Meteorological Data Needed: Precipitation Temporal scale: Month to year Spatial scale: Point	Very simple and easy to understand and communicate	Not applicable for non-normal variables. Cannot be used for comparisons across seasons and regions
Rainfall deciles (Gibbs & Maher, 1967; Mpelasoka et al., 2008)	Type: Meteorological Data Needed: Precipitation Temporal scale: Month to year Spatial scale: Point to regional	It delivers information about risks. It can be applied regionally	It does not consider other variables. Requires long records for thresholds definition
SPI, Standardized Precipitation Index. (McKee et al., 1993)	Type: Meteorological Data Needed: Long records of Precipitation Temporal scale: Month to year Spatial scale: Point to regional	Used for different time scales and related to probability. Symmetric for comparison	It requires long-term records. No consideration of evapotranspiration
SPEI, Standardized Precipitation Evapotranspiration Index, (Vicente-Serrano et al., 2010)	Type: Meteorological Data Needed: Precipitation, temperature, and evapotranspiration Temporal scale: Multiple, weekly or monthly Spatial scale: Regional	Sensitive to long-term trends in temperature change. It considers water balance and evapotranspiration	Calculation of PET can be complicated
PDSI, Palmer Drought Severity Index. (Palmer, 1965)	Type: Agricultural Data Needed: Precipitation, temperature and water supply Temporal scale: Month to year Spatial scale: Agricultural regions	More comprehensive than indices based only on precipitation. PET and soil moisture are also considered	Calibrated for US Great Plains' conditions. Limited applicability in locations with climatic extremes, mountainous terrain or snow-packs unless calibrated
CMI Crop Moisture Index (Palmer, 1968)	Type: Agricultural Data Needed: Precipitation, temperature Temporal scale: Weekly Spatial scale: Regional	It is very suitable for the prediction of short-term droughts	Low applicability for predicting long-term droughts. CMI can only be used in the growing season, it is not suitable for winter drought predictions

(continued)

Table 2.1 (continued)

Indices	Properties	Strengths	Drawbacks
NDVI, Normalized Difference Vegetation Index, (Ji & Peters, 2003)	Type: Ecological and Agricultural Data Needed: Multiband satellite images Temporal scale: Image dependent Spatial scale: Image dependent	Simple to calculate	Cloud, seasonal smoke, aerosols, etc. can contaminate the pixels. In wet conditions, the reflectance may not be equal in the two bands
TCI, Temperature Condition Index, (Kogan, 1995)	Type: Ecological and Agricultural Data Needed: Brightness and temperature Temporal scale: Weekly, monthly Spatial scale: Image dependent	Simple to calculate	It only depends on the temperature of the air and does not consider other variables
VCI, Vegetation Condition Index, (Kogan, 1990)	Type: Ecological and Agricultural Data Needed: NDVI Temporal scale: Weekly. Monthly Spatial scale: Point to regional	Simple to calculate. It considers the vegetation conditions of the entire series studied	Atmospheric interference, soil condition effects (wet), anisotropy
VHI, Vegetation Health Index, (Kogan, 1995)	Type: Ecological and Agricultural Data Needed: TCI and VCI Temporal scale: Weekly, monthly scale (Image dependent) Spatial scale: Point to regional	Simple to calculate. It considers the vegetation conditions of the entire series studied and the temperature applying a factor to each one of these factors	It depends on the atmospheric conditions
NDWI, Normalized Difference Water Index, (Gao, 1996)	Type: Ecological and Agricultural Data Needed: NIR, SWIR Temporal scale: Weekly, monthly Spatial scale: Image dependent, point to Region	Calculates the water content in the vegetation, which complements NDVI	It depends on the atmospheric conditions

(continued)

Table 2.1 (continued)

Indices	Properties	Strengths	Drawbacks
SWSI, Surface Water Supply Index (Shafer & Dezman, 1982)	Type: Hydrological Data needed: Precipitation, temperature, Stream flow, snowpack, remote sensing Temporal scale: Monthly Spatial scale: River basin.	Simple calculation, it includes water management. It considers snowpack and water storage.	It doesn't represent extreme events well. Interbasin comparisons are not possible (Hayes et al., 1999) or drought analysis on a global scale
RDI Reclamation Drought Index (Weghorst, 1996)	Type: Hydrological Data needed: Precipitation, temperature, Stream Flow, snowpack, remote sensing Temporal scale: Monthly Spatial scale: River basin	Created for the operational detection of drought events and for the triggering of relief	RDI behaves more slowly even compared to the PHDI
VegDRI, Vegetation Drought Response Index, (Brown et al., 2008)	Type: Hydrological Data needed: NDVI, SPI, and PDSI Temporal scale: Monthly to annual Spatial scale: Depends on the satellite imagery	Provides near-real-time maps of drought severity and spatial extent	It's necessary to combine NDVI, SPI and PDSI
RSDI, Regional Streamflow Deficiency Index (Stahl, 2001)	Type: Hydrological Data needed: Streamflow, remote sensing data Temporal scale: Daily Spatial scale: Region (cluster in homogeneous zones)	The RDI series includes the spatial dimension in the definition of severe dry periods. Relations and comparisons between regions are possible	Several homogenous basins affected by hydrological droughts are necessary
ADI, Aggregate Drought Index (Keyantash & Dracup, 2004)	Type: Hydrological Data needed: Precipitation, streamflow, reservoir storage, evapotranspiration, soil moisture and snow water content Temporal scale: Monthly Spatial scale: Regional	Considers all physical forms of drought. Principal component analysis is used to extract dominant hydrologic signals. Direct mathematical formulation	Much data required for calculation

(continued)

Table 2.1 (continued)

Indices	Properties	Strengths	Drawbacks
SMDI, Soil Moisture Deficit Index, and ETDI, Evapotranspiration Deficit Index (Narasimhan & Srinivasan, 2005)	Type: Hydrological Data needed: Soil moisture (originally Simulated using SWAT model)/ET Temporal scale: Weekly Spatial scale: High spatial resolution	SMDI and ETDI use a high-resolution comprehensive hydrologic model that incorporates a crop growth model. ETDI and SMDI could be good indicators of short-term agricultural droughts	It is difficult to calculate the historical soil moisture, NDVI or hydrological models can be used, but a great deal of data and calculated variables are required. The data supplied by the satellite imagery can be affected by clouds and other conditions

2.5 Conclusions

Drought is a complex, unpredictable climate phenomenon that significantly impacts ecosystems, societies, and economies. This chapter described drought and its causes and provided some definitions of drought based on the affected systems. Moreover, other related concepts related to drought were discussed, together with the possible implications of misconceptions. Finally, statistical properties and identification of drought were introduced, together with the quantitative characterization through indices. The main conclusions are as follows:

- Drought is a stochastic, slow-developing phenomenon with varying severity and duration, making it challenging to predict, prepare for, and manage. Its onset and conclusion are difficult to pinpoint due to its gradual impact on ecosystems and human systems.
- Different sectors experience drought in unique ways. Meteorological, agricultural, hydrological, socioeconomic, and ecological droughts emphasize that no single definition fits all, as each type affects different systems and requires different management approaches.
- The run theory provides a mathematical framework for analyzing droughts by defining key variables like drought duration, severity, and intensity. It also aids in identifying the onset and end of droughts based on deviations from average conditions.
- Drought indices consolidate complex data into manageable metrics, facilitating timely communication between scientists, policymakers, and the public. They support real-time monitoring and allow for early detection and proactive mitigation efforts.
- Continuous monitoring of droughts, especially in regions facing long-term climatic shifts, is crucial for distinguishing between temporary droughts and permanent aridification. This requires adaptive management strategies to ensure sustainable water use.

Acknowledgments The authors gratefully acknowledge funding from ANID project FSEQ210018. Jorge Gironás also acknowledges the support from CEDEUS (ANID FONDAP 1523A0004) and CIGIDEN (ANID FONDAP 1523A0009).

References

American Meteorological Society (AMS). (2004). Statement on meteorological drought. *Bulletin of the American Meteorological Society, 85*, 771–773.
Brown, J. F., Wardlow, B. D., Tadesse, T., Hayes, M. J., & Reed, B. C. (2008). The vegetation drought response index (VegDRI): A new integrated approach for monitoring drought stress in vegetation. *GIScience & Remote Sensing, 45*, 16–46.
Cook, E. R., Seager, R., Cane, M. A., & Stahle, D. W. (2007). North American drought: Reconstructions, causes, and consequences. *Earth Science Reviews, 81*, 93–134.
Crausbay, S. D., Ramirez, A. R., Carter, S. L., Cross, M. S., Hall, K. R., Bathke, D. J., et al. (2017). Defining ecological drought for the twenty-first century. *Bulletin of the American Meteorological Society, 98*(12), 2543–2550.
Dai, A. (2011). Drought under global warming: A review. *Wiley Interdisciplinary Reviews: Climate Change, 2*, 45–65.
Ding, Y., Hayes, M. J., & Widhalm, M. (2011). Measuring economic impacts of drought: A review and discussion. *Disaster Prevention and Management: An International Journal, 20*(4), 434–446. https://doi.org/10.1108/09653561111161752
FAO. (2002). *Report of FAO-CRIDA Expert Group consultation on farming system and best practices for drought-prone areas of Asia and the Pacific Region*. Food and Agricultural Organization of United Nations. Publishes by Central Research Institute for Dry land Agriculture, Hyderabad, India.
Fernández, H. W., & Fernández, B. (2002). The influence of ENSO in the precipitation regime in southern South America. *Hydraulic Engineering in Mexico (in Spanish), 13*(3), 5–16.
Fernández, B., & Gironás, J. (2015). Using standardized precipitation and runoff indices for the identification of extraordinary droughts events in Chile. In Andreu et al. (Eds.), *Droughts: Research and science-policy interfacing* (pp. 133–138). Taylor and Francis Group.
Fernández, B., & Gironás, J. (2017). Drought characteristics. Chapter 79. In V. P. Singh (Ed.), *Handbook of applied hydrology* (pp. 79-1–79-7). McGraw-Hill Education.
Fernández, B., & Gironás, J. (2021). Droughts. In B. Fernández & J. Gironás (Eds.), *Water resources of Chile*. World Water Resources, vol 8. Springer. https://doi.org/10.1007/978-3-030-56901-3_8
Gao, B. C. (1996). NDWI a normalized difference water index for remote sensing of vegetation liquid water from space. *Remote Sensing of Envorinment, 58*(3), 257–266.
Gibbs, W., & Maher, J. (1967). Rainfall deciles as drought indicators (Background discussion of drought in Australia and maps of annual rainfall for years 1885–1965).
Hayes, M. J. (2006). *Drought indices*. Wiley Online Library.
Hayes, M. J., Svodoba, M. D., Wilhite, D. A., & Vanyard, O. V. (1999). Monitoring the 1996 drought using the standardized precipitation index. *Bulletin of the American Meteorological Society, 80*(3), 429–438.
Heim, R. R. (2002). A review of twentieth-century drought indices used in the United States. *Bulletin of the American Meteorological Society, 83*(8), 1149–1165.
Ji, L., & Peters, A. (2003). Assessing vegetation response to drought in the northern Great Plains using vegetation and drought indices. *Remote Sensing Environment, 87*(1), 85–98.
Jiang, T., Su, X., Singh, V. P., & Zhang, G. (2021). A novel index for ecological drought monitoring based on ecological water deficit. *Ecological Indicators, 129*, 107804.
Keyantash, J. A., & Dracup, J. A. (2004). An aggregate drought index: Assessing drought severity based on fluctuations in the hydrologic cycle and surface water storage. *Water Resources Research, 40*, W09304.
Kogan, F. (1990). Remote sensing of weather impacts on vegetation in non-homogeneous areas. *International Journal of Remote Sensing, 11*, 1405–1419.
Kogan, F. (1995). Droughts of the late 1980s in the United States as derived from NOAA polar-orbiting satellite data. *Bulletin of the American Meteorological Society, 76*, 655–668.
McKee, T. B., Doesken, N. J., & Kleist, J. (1993). The relationship of drought frequency and duration to time scales. In *Proceedings of the 8th Conference on Applied Climatology* Vol. 17, No. 22, pp. 179–183.

Meyer, S. J., Hubbard, K. G., & Wilhite, D. A. (1993). A crop-specific drought index for corn: I. Model development and validation. *Agronomy Journal, 85*, 388–395.

Mishra, A. K., & Singh, V. P. (2010). A review of drought concepts. *Journal of Hydrology, 391*, 202–216.

Mishra, A. K., Singh, V. P., & Desai, V. R. (2009). Drought characterization: A probabilistic approach. *Stochastic Environmental Research and Risk Assessment, 23*(1), 41–55.

Mpelasoka, F., Hennessy, K., Jones, R., & Bates, B. (2008). Comparison of suitable drought indices for climate changes impact assessment over Australia towards resources management. *International Journal of Climatology, 28*(10), 1283–1292.

Namias, J. (1983). Some causes of United States drought. *Journal of Climate and Applied Meteorology, 22*, 30–39.

Narasimhan, B., & Srinivasan, R. (2005). Development and evaluation of soil moisture deficit index (SMDI) and evapotranspiration deficit index (ETDI) for agricultural drought monitoring. *Agricultural and Forest Meteorology, 133*, 69–88.

Niemeyer, S. (2008). New drought indices. *Options Méditerranéennes. Série A: Séminaires Méditerranéens, 80*, 267–274.

Oertel, M., Meza, F. J., & Gironás, J. (2020). Observed trends and relationships between ENSO and standardized hydrometeorological drought indices in Central Chile. *Hydrological Processes, 34*(2), 159–174.

Palmer, W. C. (1965). *Meteorological drought*. US Department of Commerce, Weather Bureau Washington, DC, USA.

Palmer, W. C. (1968). Keeping track of crop moisture conditions, nationwide: The new crop moisture index. *Weatherwise, 21*(4), 156–161. https://doi.org/10.1080/00431672.1968.9932814

Panu, U., & Sharma, T. (2002). Challenges in drought research: Some perspectives and future directions. *Hydrological Sciences Journal, 47*, S19–S30.

Paulo, A. A., & Pereira, L. S. (2006). Drought concepts and characterization: Comparing drought indices applied at local and regional scales. *Water International, 31*, 37–49.

Pereira, L. S., Cordery, I., & Iacovides, I. (2002). Coping with water scarcity. IHP-VI. In *Technical documents in hydrology, N 58*. UNESCO.

Ropelewski, C. F., & Halper, M. S. (1996). Quantifying southern oscillation—Precipitation relationships. *Journal of Climate, 9*, 1043–1059.

Rossi, G., Cacelliere, A., Pereira, L. S., Oweis, T., Shataniawi, M., & Zairi, A. (Eds.). (2003). *Tool for drought mitigation in Mediterranean regions*. Kluwer Academic Publishers.

Salas, J. D. (1978). Statistical properties of droughts. United States-Argentinian Workshop on Droughts. Mar del Plata, Argentina.

Shafer, B., & Dezman, L. (1982). Development of a Surface Water Supply index (SWSI) to assess the severity of drought conditions in snowpack runoff areas. *Proceeding of the Western Snow Conference*. pp 164–175.

Stahl, K. (2001). Hydrological drought. A study across Europe. Ph.D. thesis, Albert-Ludwigs-Universität, Freiburg, Germany.

Tabaria, H., & Aghajanloob, B. N. (2012). Temporal pattern of aridity index in Iran with considering precipitation and evapotranspiration trends. *International Journal of Climatology, 33*(2), 396–409.

Tsakiris, G., Loukas, A., Pangalou, D., Vangelis, H., Tigkas, D., Rossi, G., & Cancelliere, A. (2007). Drought characterization in drought management guidelines technical annex. *Cap, 7*, 85–102.

UN Secretariat General. (1994). *United Nations convention to combat drought and desertification in countries experiencing serious drought and/or desertification, particularly in Africa*. Paris.

UNCCD (United Nations Convention to Combat Desertification). (1994). *Elaboration of an International Convention to Combat Desertification in Countries Experiencing Serious Drought and/or Desertification, particularly in Africa*. U.N. Doc. A/AC.241/27, 33 I.L.M. 1328, United Nations, New York, USA.

Van Lanen, H. A. J., & Peters, E. (2000). Definition, effects and assessment of groundwater droughts. In J. V. Vogt & F. Somma (Eds.), *Drought and drought mitigation in Europe* (pp. 49–61). Kluwer Academic Publishers.

Vicente-Serrano, S. M., Beguería, S., & López-Moreno, J. I. (2010). A multiscalar drought index sensitive to global warming: The standardized precipitation evapotranspiration index. *Journal of Climate, 23*, 1696–1718.

Weghorst, K. M. (1996). *The reclamation drought index: Guidelines and practical applications* (p. 6). Bureau of Reclamation.

Wilhite, D. A., & Glantz, M. H. (1985). Understanding the drought phenomenon: The role of definitions. *Water International, 10*, 111–120.

Wilhite, D. A., Rosenberg, N. J., & Glantz, M. H. (1986). Improving federal response to drought. *Journal of Climate and Applied Meteorology, 25*, 332–342.

Wolter, K. and Timlin, M. S. (1993). Monitoring ENSO in COADS with a seasonally adjunted principal component index. In: Proc. *Seventh Annual Climate Diagnostic Workshop*, Norman, OK, NOAA, pp. 52–57.

World Bank. (2003). *Report on financing rapid onset natural disaster losses in India: a risk management approach*. Report no. 26844-IN, Washington, DC.

World Meteorological Organization (WMO). (1986). *Report on drought and countries affected by drought during 1974–1985* (p. 118). WMO.

Yevjevich, V. (1967). *An objective approach to definitions and investigations of continental hydrologic droughts*. Colorado State University Fort Collins.

Yevjevich, V., Da Cunha, L., & Vlachos, E. (1983). *Coping with drought*. Water Resources Publications.

Zargar, A., Sadiq, R., Naser, B., & Khan, F. I. (2011). A review of drought indices. *Env. Rev., 19*, 333–349.

Part I
Setting and Country Context: Drought Impacts in Chile

Chapter 3
The Chilean Economy: A General Profile

Felipe Morandé

Abstract This chapter provides a comprehensive overview of the Chilean economy, detailing its current state, historical development, institutional frameworks, and economic policies. It also discusses recent advancements, challenges, and future opportunities for growth and development. Chile's GDP in 2023 was nearly US$ 330 billion, with a per capita figure of over US$ 30,000 in PPP terms. The country's economy is highly integrated into international trade, with exports and imports comprising 75% of GDP. Chile exhibits high social indicators such as life expectancy (81.4 years) and literacy rate (96.4%), but income inequality remains a significant issue, as reflected by a Gini coefficient of 0.46. Since the military coup in 1973, Chile has undergone extensive economic reforms, including trade liberalization, privatization, and the establishment of a market economy. These reforms have significantly shaped the current economic landscape. Post-1990 democratic governments focused on "growth with equity," maintaining market-oriented policies while increasing social spending to address high poverty rates and improve social equity. The autonomous Central Bank of Chile has been instrumental in achieving and maintaining low and stable inflation rates, which is crucial for economic stability and growth. Chile's abundant natural resources, particularly in lithium and renewable energy, present significant opportunities for economic growth, especially in the context of global efforts to combat climate change.

Keywords Economy · Economic sectors · Macroeconomic indicators · Social indicators · Policy reforms · Chile

F. Morandé (✉)
Escuela de Gobierno, Pontificia Universidad Católica de Chile, Santiago, Chile
e-mail: felipe.morande@uc.cl

D. Rivera, G. Donoso (eds.), *Droughts in Chile*, Global Issues in Water Policy 31,
https://doi.org/10.1007/978-3-031-85040-0_3

3.1 Introduction

This chapter aims to briefly present the Chilean economy as it is today and the road it has followed in the last 50 years. More than giving a complete account of events, this chapter emphasizes data that describes the economy, and at the same time focuses on the institutions that characterize the economic side of Chile's society. Of course, the economy does not perform in a vacuum, so the text also touches—very briefly—on the political context and the political system.

Section 3.2 provides an overview of the current state of the Chilean economy. Section 3.3 discusses the institutional developments over the past fifty years that have shaped the country's present circumstances. Section 3.4 outlines the economic policies implemented by the democratic governments, while Sect. 3.5 covers the recent advancements in Chile's development. Section 3.6 brings the chapter to a conclusion.

3.2 Chile at a Glance

Inhabited by close to 20 million people, Chile's GDP reached almost US$ 330 billion in 2023 (close to Sweden or Belgium), while in per capita terms the figure is slightly over US$ 30,000 in Purchasing Power Parity (PPP) adjusted terms, according to the World Bank (2022).

As Table 3.1 presents, most of the population lives in urban areas (87.8%) and almost 36% live in Santiago, the capital city. Other important cities are Valparaíso, the main port at 100 km from Santiago, and Concepción, in the Center-South zone (Table 3.2).

The climatic diversity throughout the country, the presence of the Andes Cordillera, the geological characteristics of the mountainous areas, the long coast on the Pacific characterized by the cold Humboldt current, the characteristics of the soils and the availability or lack of fresh water, make Chile rich in a variety of

Table 3.1 Main indicators

Population (2023)	19,960,889
GDP, millions of US$ (2023)	327,694
GDP per capita, US$ (2022)	15,355
GDP per capita, PPP, US$ (2022)	30,208
Exports, millions of US$ (2022)	107,077
Imports, millions of US$ (2022)	118,094
Geographic area, km^2	756,626
Urban population rate (2017)	87.8%

Sources: Based on data from National Statistics Institute of Chile (INE), Central Bank of Chile, and the World Bank

Table 3.2 Main cities

City	Population	Geographical area	Maximum temperature (°C)	Minimum temperature (°C)	Average year rainfall (cm)
Santiago	7,116,706	Center	30	3	21.15
Valparaíso/Viña del Mar	1,054,253	Center, coast	20	7	18.44
Great Concepción	1,036,142	Center south	23	6	71.52
Antofagasta	441,211	North	25	13	0.31
Iquique	370,877	North	26	15	0.15
Coquimbo/La Serena	534,239	Center north	21	9	6.11
Talca	307,931	Center south	29	3	44.88
Temuco	392,794	South	24	4	81.86
Puerto Montt	328,924	South	20	4	96.61
Punta Arenas	144,938	South end	14	0	26.19

Source: Based on INE and Weather Spark (n.d.)

Table 3.3 Sectoral indicators

Economic sector	% GDP 2022
Industry	9.74
Fishing	0.74
Mining	14.24
Agriculture/forestry	2.80
Construction	5.88
Services	56.40
Product taxes	10.20

Source: Central Bank of Chile (2024)

natural resources. Mining is a relevant activity, representing almost 15% of the GDP, particularly copper, lithium, and iron, mainly localized in the northern part of the country (Table 3.3 and 3.4). Agriculture, particularly fruit, and agro-industry, take advantage of the Mediterranean climate of the central area of the country, as does the wine industry. Both in the north and in the south, there are good conditions for fishing and in the south, salmon farming is predominant.

Table 3.5 presents Chile's macroeconomic indicators. At a macroeconomic level, total investment reaches almost a quarter of GDP, while the proportion of exports plus imports over total GDP—75%—shows that Chile is a highly integrated economy to international trade. The availability of natural resources affects the composition of exports (Table 3.5) which are dominated by mining, salmon, wine, and food-related exports.

The financing of total investment counts on 16.4% of national savings, a variable that has declined in recent years along with a significant slowdown in economic growth (2.11% annual average in the last 6 years). However, Chile has broad access to external financing, which so far allows it to secure a significant total investment.

Table 3.4 Regional GDP weights

Region	Regional GDP/National GDP	Main productive sectors (non-services)
Arica Parinacota	0.9%	Construction and mining
Tarapacá	2.7%	Mining and construction
Antofagasta	8.8%	Mining and construction
Atacama	2.2%	Mining and construction
Coquimbo	3.4%	Mining and construction
Valparaíso	8.2%	Mining and construction
Metropolitana (Santiago)	46.6%	Industry and construction
O'Higgins	4.6%	Mining and agriculture/forestry
Maule	4.3%	Industry and agriculture/forestry
Ñuble	1.5%	Agriculture/forestry and construction
Bío Bío	6.9%	Industry and construction
Araucanía	3.2%	Industry and agriculture/forestry
Los Ríos	1.5%	Industry and agriculture/forestry
Los Lagos	3.8%	Industry and construction
Aysén	0.7%	Fishing and construction
Magallanes	1.0%	Industry and construction

Source: CORFO (2022)

Table 3.5 Macroeconomic indicators

Indicator		2022
GDP, billions of CLP		262,593
Private consumption (% of GDP)		63.9
Public consumption (% of GDP)		14.4
Gross Fixed Capital Formation (GFCF) (% of GDP)		24.8
Exports (% of GDP)		35.7
Imports (% of GDP)		39.3
Gross National Savings (% of GDP)		16.4
Inflation (end of 2023)		3.90%
Monetary policy rate (December 2023)		8.25%
Average GDP growth (2018–2023)		2.11%
Main exports 2022 (% of exports)		
	Copper	45.3%
	Lithium	8.2%
	Salmon	6.3%
	Wine	1.9%
	Fruits	5.9%
	Other agro-industrial	28.5%
	Other exports	3.8%

Source: Based on data from the Central Bank of Chile (2024)

Table 3.6 Social indicators

Life expectancy at birth (INE, 2023)	81.4 years
Poverty rate (MSDF, 2022)	6.5%
Extreme poverty rate (MSDF, 2022)	2%
UNDP Human Development Index (UNDP, 2022)	0.855 in a 0–1 scale
Gini coefficient (after government policies) (WB)	0.46 in a 0–1 scale
Literacy rate (MSDF, 2022)	96.4%
Tertiary enrollment rate (WB)	96%
Drinking water coverage (SISS, 2022)	99.94%
Sewer coverage (SISS, 2022)	97.48%
Homicides, per 100,000 people (PDI, 2023)	6.7

Sources: National institute of Statistics (INE), Ministry of Social Development and Family of Chile (MSDF, 2022), Superintendency of Health Services of Chile (SISS), UNDP, World Bank and Investigation Police of Chile (PDI, 2023)

Inflation, which has hit the entire world hard in the years following the Covid pandemic, is currently on the decline in Chile (3.9% last December) as a consequence of a contractionary monetary policy implemented since 2021—now receding—and an austere fiscal policy during 2022 and 2023.[1]

Chile is a middle-income country and an emerging economy. Many of its social indicators, however, rival those of developed countries. Such is the case of life expectancy (81.4 years), literacy rate (96.4%), and population under the poverty line (6.5%).[2] More generally, the UNDP Human Development Index places Chile ahead of all Latin American countries and at place 42 in the whole world.

One social indicator which is deficient is the Gini coefficient (Table 3.6), which at 0.46[3] shows that in Chile income distribution is still skewed. With this record, Chile ranks among the most unequal countries in the OECD and around average in Latin America. However, there has been a relevant improvement since 1990, as detailed in Sect. 3.3.

Chile grew very rapidly during the 90 s and the 2000s, so fast so that its per capita GDP overpassed those of the rest of Latin America by far.[4] Still today, after a decade of rather slow growth, Chile leads the region according to this indicator. In the world at large, Chile's position in this respect is 57, similar to Eastern European countries, still 2/3 of Spain and 40% of the US.

[1] See Table 3.5

[2] See Table 3.6

[3] The Gini indicator goes from 0—all income concentrated in one person—to 1—income is uniformly distributed among the population.

[4] See Sect. 3.3 for a concise account of Chile's development in the last 50 years.

Table 3.7 Demographic and labor market indicators

Indicator		
Population (thousands)	End 2023	19,961
	Men	9849
	Women	10,112
	0 to 14 years	18.9%
	15 to 29 years	22.6%
	30 to 44 years	19.6%
	45 to 59 years	19.2%
	60 +	19.7%
Working age population (thousands)	End 2023	16,281
Labor force (thousands)	End 2023	10,012
	Employed	91.27%
	Unemployed	8.73%
Total participation rate	End 2023	
	Men	70.2%
	Women	50.8%
Informal employment rate	End 2023	27.4%

Sources: Based on data from National Statistics Institute of Chile (INE), Central Bank of Chile (2024) and Ministry of Social Development and Family of Chile (MSDF, 2022)

Demographic data shows Chile has an aging population (cohorts of less than 14 years of age are smaller than older cohorts[5]) but still less so than more developed countries (Europe and Japan, for example). It is expected that by year 2050 people older than 60 will account for more than 25% of the population.

Out of the approximately 20 million people that is reported as Chile's population, close to 1.6 million are immigrants who have migrated mostly in the last 10 years. Venezuela, Perú, Colombia and Haiti are the main sources of migrants. They are normally in their 20 s or 30 s and contribute also with more children than the average Chilean family.

Two issues, among others, are worth noting in the labor market (Table 3.7). One is the rather low rate of female participation in the labor force (around 50%) and the other is the significant proportion of informal labor (27%). In both cases there were relevant improvements before 2020, but the Covid pandemic brought some setbacks.

As will be seen in Sect. 3.3, there has been an explosion of tertiary education students starting in the 2000s, phenomenon which is associated to the emergence of new universities and technical education institutions, as well as to the creation of a financial mechanism subsidized by the government that facilitates the financing of enrollment and tuition fees. As of today, around 45% of students graduating from secondary education pursue a degree or a diploma in either universities or technical education institutions (Cincel, 2002).

[5] See Table 3.7

Table 3.8 Infrastructure indicators

Indicator	2023
Broadband internet penetration (SUBTEL, 2022)	67.4%
Cell phones per inhabitant (World Bank, 2022)	1.35
Kilometers of road (2021)	88,150
% paved road (2021)	24%
Santiago metro network (2023)	149 km
Electric generation renewable energies (2023)	62%

Sources: World Bank (2022), Generadoras de Chile (2023), SUBTEL (2022), Gobierno de Chile (2023) and Observatorio logístico (2021)

In terms of public infrastructure, as can be seen in Table 3.8, Chile does well with more than 20,000 km of paved roads (24% of total roads), an underground transport system in Santiago that runs over 149 km and a private-public association that has allowed significant improvements in interurban and urban highways, public buildings, irrigation dams, transport infrastructure, hospitals, and jails. All urban population has access to potable water and more than 95% to a sewage system. On the other hand, 52.4% of rural population has access to potable water. Around two thirds of people have access to broad-band Internet connection, although just around half of that is through fiber-optic.

The political system in Chile is a representative democracy with a presidential scheme and a two-chamber Congress. The judiciary is a third autonomous power and there is also an independent national comptroller. The Central Bank is also autonomous. Although less than perfect, and in spite of some strong turbulences in the last 4 years, Chile's political system works reasonably well, the rule of law prevails, and free-speech and religious and association freedom is totally respected. Corruption exists but it is of low intensity and many indicators place Chile well-positioned world-wide and leading Latin America.

The economic system is one of a free-market, open economy, private sector dominated type. Still after many social-democrat changes applied during the last three decades to a very liberal economy at the end of the 80s, today Chile ranks high in terms of economic freedom (22 according to The Heritage Foundation (2023)). The State is present in the provision of public health and education, but private institutions also play a significant role in both services. The social security system is totally private and of an individual capital accumulation sort,[6] although a basic minimum pension guaranteed to all is financed with general tax money.

In any event, pension funds are now around 2/3 of GDP and are a basic pillar of capital and financial markets.[7] Even more, they have been instrumental in the

[6] Worker's contributions to pension funds are mandatory at a 10% rate. There is currently a bill being discussed in Congress to overhaul much of the system introducing a partial pay-as-you-go scheme applied to an extra contribution of 6% of wages. However, it is uncertain at this point in time whether this reform will be approved.

[7] See Sect. 3.3 for more details on the pension funds system.

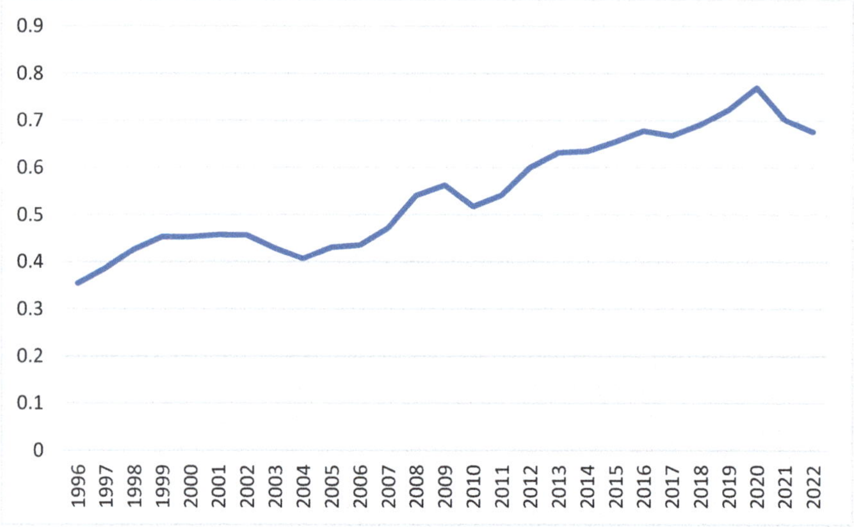

Fig. 3.1 Financial deepness—M2/GDP (Central Bank of Chile, 2024)

development of long-term domestic financial markets and more recently in the financing of public infrastructure.

Chile's financial markets are fairly developed and sophisticated. One indicator is the M2/GDP ratio, which in Chile hovers around 70%—see Fig. 3.1—comparable to France, Spain, UK and a number of other developed economies, and certainly higher than the 20% of Brazil or the 16% of Mexico.[8] The funding provided by the pension funds has been key to this development.

One problem that has emerged strongly in the last few years is that of crime and security. What is particularly worrisome is the emergence of gangs and organized crime around drug trafficking that have brought violent practices and firearms use from other countries in the Latin American region. Although homicides per 100.000 people is at a moderate level (6.7 in 2022, similar to the US, smaller than other LA countries but much higher than European or Asian countries), it is three times what it was not more than 5 years ago.

3.3 The Trajectory of the Chilean Economy in Recent Decades

Chile has had significant economic progress in the last 35 years. In particular, its GDP per capita has multiplied by 6.2 times since 1990 (Fig. 3.2), while poverty has been reduced from 68% to 6.5% in the same period (Fig. 3.3) and the Human

[8] See IMF (2023), Financial Soundness Indicators.

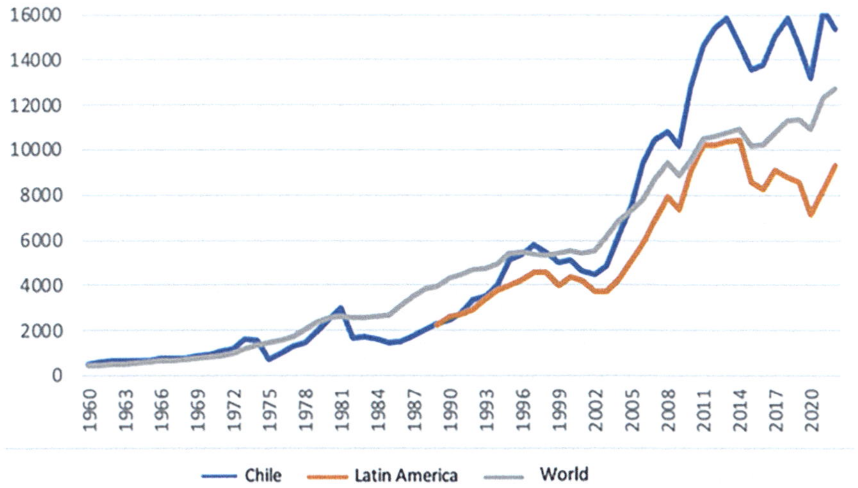

Fig. 3.2 GPD per capita (current US$) (World Bank, 2022)

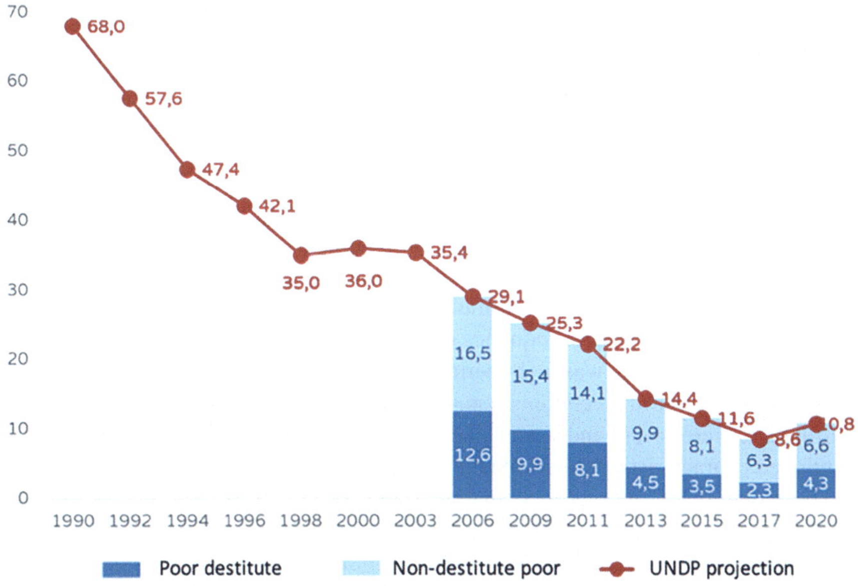

Fig. 3.3 Poverty (MSDF, 2022)

Development Index has improved from 0.706 in 1990 to 0.855 in 2022 (Fig. 3.4). In these three general indicators, Chile also occupies a leading position in Latin America, after having been a very lagging country in the region until before the 90s, certainly behind Argentina, Uruguay, Venezuela, and Mexico.

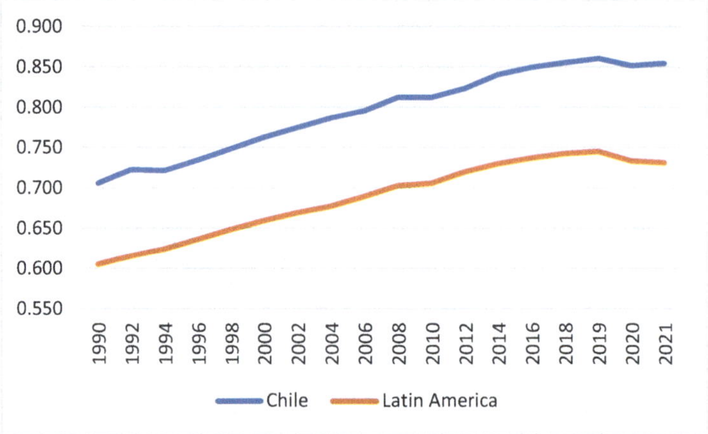

Fig. 3.4 Human Development Index (UNDP, 2022)

It is true that we are still far from developed countries and that in the last decade the economic and social performance has been quite mediocre but, in any event, what has been achieved so far is commendable. This has not been the result of chance but of public policies that, since the 70s, have been changing the institutions and the way the Chilean economy works in many key aspects.

3.3.1 Chile in 1973

Until 1973, Chile was a fairly poor economy (GDP per capita of US$ 1632 in today's currency adjusted by PPP), which presented four characteristics that largely explain this situation. First, the Chilean economy was practically isolated from international markets as a consequence of deliberate policies on the part of the government; second, there was a hypertrophied public sector; third, state control extended throughout the economy, causing great distortions in relative prices; and fourth, there were severe and unsustainable macroeconomic imbalances. The aforementioned characteristics were linked to each other and influenced each other.

The isolation of international markets was the result of an "inward-looking development" strategy that excessively protected the national industry from external competition through high import tariffs, a regime of multiple exchange rates and a wide range of non-tariff barriers. In September 1973, tariffs reached 750% in some cases and non-tariff barriers included import quotas, as well as extensive lists of prohibited imports, and prior deposits of 10,000% of the value of what was imported for 90 days. The gap between different exchange rates reached 52 times between the highest and the lowest. As a result, imports were concentrated in intermediate inputs and some basic goods—including oil and wheat—while exports were concentrated in more than 80% in copper, the production of which was in the hands of the State.

The State involvement in the economy was reflected in the almost 40% of GDP represented by state companies and public services, concentrating between 40% (manufacturing industry) and 100% (public services) of the corresponding activity in various sectors.[9] Likewise, government policy contemplated widespread price controls, at levels well below the effective costs of production, which caused the proliferation of black markets, credit rationing, and the setting of interest rates below inflation.

Meanwhile, fiscal policy was highly expansionary, and its financing was entirely monetary. Towards mid-1973 the fiscal deficit exceeded 20% of GDP, while official inflation rose above 100% annually—and even more so if the price repression and the existence of black markets with prices much higher than the regulated ones were considered.[10] By the same date, the current account deficit of the balance of payments exceeded 15% of GDP and international reserves had been exhausted.

3.3.2 The Economic Reforms of the Military Government

In September 1973, the Armed Forces staged a military coup that overthrew President Allende, democratically elected in 1970, and established a de facto government led by a Military Junta that suspended the rule of law and most citizen guarantees.[11]

This government was long—from 1973 to 1989—and re shaped the country from an institutional point of view, going through different stages in implementing very disruptive economic reforms with respect to the trajectory that the Chilean economy had until then. The substantive idea was to convert Chile into a market economy, open to the world and oriented to the private sector. Thus, the range of reforms was wide and deep, including full integration to international trade, a reshaping of the social security system, a reform to the fiscal regime, the liberalization of prices, structural reforms to the labor regime, financial markets, health, and education, and two waves of privatization of companies and banks.

The central body of these reforms largely persists to this day, although with some modifications. Thus, they will be presented in some detail and in chronological order, along with the context in which they were implemented and their main economic consequences.

[9] 85% in mining, 70% in transportation, 85% in the financial sector and 70% in telecommunications Hachette and Luders (1987).

[10] At the end of 1973, once repressed markets were free, inflation exceeded 500%, and some estimates of the inflation repressed by price controls place the annual figure for that year above 1000%. See Cortazar and Marshall (1980).

[11] Rather than focusing on the human rights abuses that followed, and which have been widely reported and condemned, we will focus on the economic management of the autocratic government.

3.3.3 The Restoration of Macroeconomic Balances and the Liberation of Prices

Very soon, the de facto government attacked the two most acute problems facing the Chilean economy in September 1973: the severe macroeconomic imbalances and the shortage of goods resulting from artificial price controls. Public spending was sharply reduced and in 1974 the deficit was only 4% of GDP—versus 25% the previous year—which significantly restricted monetary issuance. The shortage of goods was resolved by eliminating the vast majority of price controls. In addition, the system of multiple exchange rates was scrapped and an adjustable exchange rate with respect to past inflation—crawling peg—was applied.

During the following years, a conservative fiscal spending policy was maintained, the public payroll was reduced—both in quantity and in real remuneration, the tax structure was simplified and the Value Added Tax, VAT, was introduced. Previously confiscated companies were also returned to their owners and others were privatized, which led to the elimination of large subsidies, all of which together helped to achieve a surplus public balance by 1980, on the order of 6% of GDP.

The rapid decrease in the public deficit starting in 1974 was consistent with the decision to prohibit the Central Bank from granting credit to the Public Treasury—as had been the usual practice for many years and until 1973. This precept was explicitly incorporated into the 1980 Constitution and remains to this day.

The trajectory of inflation, however, was one of a very gradual decline. In fact, after 500% in 1973, inflation was still in double digits in 1976 (although declining year after year) and only in 1981 would it reach a single digit fig. (8%). This was the result of an explicit decision not to eliminate the high degree of indexation of many key prices to past inflation, such as some long-term interest rates and the exchange rate. The market itself also indexed the prices of homes and real estate assets to the inflation of previous periods, a practice that was the result of a long inflationary history facilitated by the establishment of an indexed unit of account parallel to the local currency, the *Unidad de Fomento*, UF.[12]

Despite the relatively slow reduction in inflation,[13] the principle of respecting "macroeconomic balances"—a euphemism to indicate a conduct of fiscal and monetary responsibility—was established as a necessary condition for the stability of the economy and economic progress, which that has lasted for almost 50 years.

Something similar happened with price freedom. After the hasty liberalization of the prices of most goods and services in 1973, interest rates were also liberalized around 1975, and a year later those of basic products such as bread. In 1982 salaries were also freed. Notwithstanding that some basic services provided by natural monopolies, such as electricity and drinking water, have had their prices governed

[12] The UF existed since the 60 s but in 1975 it was "perfected", being officially calculated per day according to the daily proportion of inflation from the previous month.

[13] And an even more gradual reduction in unemployment, after a strong increase in this variable in 1975.

since the mid-1980s by ad-hoc laws, which otherwise simulate the value that a perfectly competitive provider would charge the public, to date price freedom remains the norm in the functioning of markets.

3.3.4 Opening of the Economy to Foreign Trade of Goods, Services, and Capital

One of the most significant reforms from the point of view of resource allocation was the liberalization of foreign trade since it meant breaking with more than 35 years of application of an inward-looking industrialization strategy. Begun in 1974, the opening to the world economy included the complete and immediate abolition of non-tariff barriers and a plan to reduce tariffs in 4 years. The average tariff was reduced from 105% in 1973 to 10% in 1978, and a transition was made from a cascading structure to a flat rate for all products and services (10%).

A very relevant factor in this opening process between 1974 and 1976 was the management of the exchange rate by the Central Bank. After unifying the multiple exchange rates at the end of 1973, the government applied a crawling-peg exchange rate system to take into account local inflation but also with the purpose of increasing the real exchange rate, as a way to confront the fall in terms of trade in 1975–76, and also for the purpose of compensating for the reduction of tariffs and promoting exports. The real exchange rate depreciated 45% between 1974 and 1975. Starting in 1977, however, the exchange rate was increasingly used for anti-inflationary purposes, which was leveraged by the significant increase in capital inflows. This, in turn, corresponded to the greater international liquidity towards the end of the 70s—a consequence of the recycling of petrodollars—but also to the gradual removal that had been taking place of the restrictions on capital flows by the Central Bank. It is not surprising that in these circumstances the real exchange rate appreciated substantially.

The Chilean economy today remains fully integrated into international trade and investment flows, both financial and direct. What's more, this integration has deepened since 1990, as a result of government policies that the country has had since the return of democracy and which we will review later.

3.3.5 Pension, Health, and Education Reforms

These three reforms were also groundbreaking and with a marked interest in assigning a greater, and perhaps preponderant, role to the private sector in services traditionally provided by the State. All of them were implemented around 1981.

In the case of retirement provision, the change also involved moving from a traditional pay-as-you-go scheme to one of individual capitalization. The old system

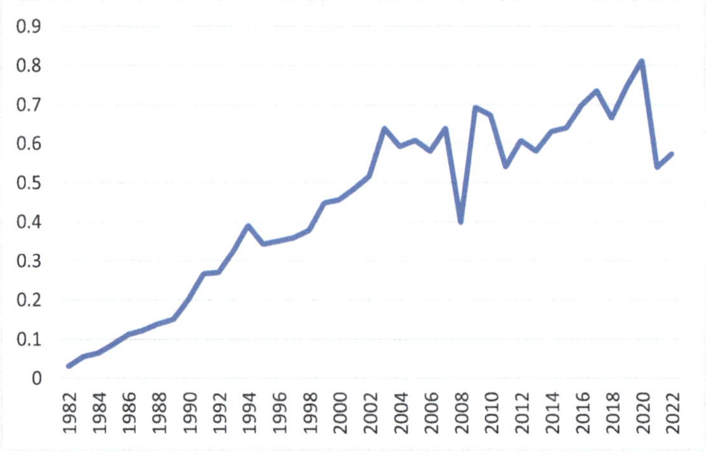

Fig. 3.5 Stock of savings (% of GDP) (CMF, 2022)

was actuarially broken, dispersed in more than 20 different regimes and implied a contribution rate of no less than, for the time, close to 20% of salaries. To carry out this reform, the government prepared the ground with a couple of years of strong fiscal surplus (1979 and 1980) in order to finance the transition starting in 1981. The scheme contemplated the creation of private fund managers[14] that had to compete among themselves to capture the savings of active workers and, over the years, return those resources, plus their profitability, at the time of retirement. As expected, the new scheme over time brought a substantial accumulation of private savings—partly at the expense of public savings, partly as a consequence of the economy's own growth.

Pension funds have reached around 80% of GDP at their best (2019) but the stagnation of the economy in recent years, the aging of the population itself, which has implied an increase in the number of retirees and, above all, the withdrawal of one fourth of these funds that was allowed during the COVID pandemic has reduced the participation to 58% of GDP in 2022 (Fig. 3.5). In any event, the accumulation of savings allowed with time the development of deep financial and capital markets, particularly real estate and long-term investments.

However, the design of the capitalization scheme with private managers has received various criticisms, the most important being that the resulting retirements have shown replacement rates close to 30 to 40%, greatly reduced compared to expectations 40 years ago.[15] It is not surprising then that it has undergone several

[14] The AFPs, or Pension Fund Administrators.

[15] After the implementation of the PGU, Universal Guaranteed Pension, in 2022, replacement rates have increased significantly for lower to middle-income pensioneers.

modifications and that an increase in the contribution rate is currently being considered, from the 10% of salaries that was initially implemented, to 16%. However, the individual capitalization system is now more than 4 decades old and, despite the reform proposal currently under discussion, it will continue to prevail in the foreseeable future.

The health reform, also in 1981, created insurance institutions that administer a mandatory contribution for each worker (7%) offering various coverage plans in a market of private providers that compete with each other. These institutions have the acronym ISAPRE. Benefited by State subsidies (during the 1980s) and based on a more attractive offer for middle and high-income workers than the state alternative—which has a greater solidarity component and named FONASA—the ISAPREs reached almost 20% of the total of workers in the country. Currently that percentage has fallen below 15% for various reasons and the ISAPREs are very stressed financially. It is very likely that they will not persist much longer, although other forms of private insurers will take their place.

The two most striking aspects of the school education reform were, first, the transfer of public establishments to the municipalities (decentralization); and second, the implementation of a voucher system per student in both public and subsidized private schools. The latter resulted in a very strong expansion of subsidized private education in practice, to the point that it currently covers 55% of the school population, while municipal education (currently in the process of institutional reform), takes 35%. The remaining 10% is paid private education. Meanwhile, the voucher system continues to prevail to this day.

At the level of higher education, in 1981 the possibility of creating new universities, mainly private, was opened, which resulted in the total number of university establishments (and other tertiary education institutions) to date reaching 140, much higher than the 8 that existed before this reform. Consequently, enrollment in higher education has increased 890% since the early 1980s up to the early 2020s, although an important part of this increase is linked to non-university technical education (Fig. 3.6).

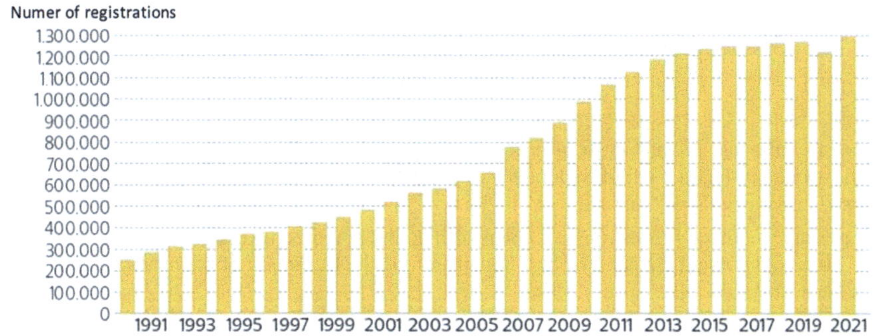

Fig. 3.6 Higher education enrollment (Gallardo, 2023)

3.3.6 Labor Reform

The new Labor Code implemented in 1979 established collective bargaining at the company level—and not at the branch or industry level—the right to strike and lockout and freedom of association. Likewise, the possibility was established for the employer to dismiss without giving cause but paying the affected worker financial compensation according to their years of service. Beyond the rights it granted to workers, the main objective of the law was certainly to guarantee flexibility in the labor market.

Although this legislation has undergone several changes over time, collective bargaining at the company level has been maintained until now, a key aspect that has contained the level of labor conflict in the private sector.[16]

3.3.7 A New Constitution

A commission of experts commissioned by the de facto government—led by General Pinochet since 1976 in place of the original Government Junta—drafted a new Constitution ratified in a referendum in 1980.[17]

The legal body sought to support the institutional structure that was being built through the reforms that we have been commenting on. Some of the most important economic aspects contained in this Magna Carta are the following:

(a) Subsidiarity of the State: the principle is established that all economic activity must be carried out by the private sector and only in a subsidiary manner by the State, and as long as it is authorized by an ad-hoc law. It was assumed that only a few companies considered strategic would remain in the hands of the State, such as the copper mining giant, CODELCO, the oil refineries, ENAP, and other smaller ones linked to the Armed Forces.[18]
(b) Freedom of choice: people can choose the providers of mandatory education and health services between the State and private companies.
(c) Equal treatment is established between the foreign investor and the national investor.
(d) The possibility of mining exploitation by foreign investors opens.

[16] The most massive and disruptive strikes or stoppages occur in Chile in the public sector, which has its own labor regime. Paradoxically, this regime does not allow strikes, so the strikes that are observed are illegal—but tolerated.

[17] According to official results, 66% of voters ratified the proposed text, although the referendum was carried out without electoral records and, therefore, the legitimacy of the process has been in question ever since.

[18] For many years, the Armed Forces maintained the right to receive 10% of CODELCO's annual sales and until not long ago they could make discretionary and reserved use of those funds for the purpose of purchasing weapons.

(e) The existing system of private water rights is validated, which is very relevant for the smooth functioning of agriculture, mining, and the provision of water itself for use by the population.
(f) The State has the right to expropriate private interests but in exchange for compensation at market value.
(g) The Central Bank is prohibited from buying public debt or financing the State.
(h) A series of laws are established with high quorum to make it difficult to change them by circumstantial majority, among them the future law of the Central Bank and the one that sanctions the pension system of individual savings accounts and administered by private parties.

The 1980 Constitution has undergone many modifications over time, perhaps the most relevant in 2005 under the government of President Ricardo Lagos. However, most of the modifications concentrated more on political than economic issues and have not greatly altered the eight aspects linked to the economy mentioned in the previous paragraph.[19]

3.3.8 The Great Recession of 1982/83 and New Reforms

In 1982, real GDP fell 14.2% (15.8% in per capita terms), while the unemployment rate rose to almost 30% in early 1983. At the beginning of 1983, almost all banks were intervened by the government to avoid a financial run-in view of the liquidity and solvency problems experienced by many of those banks, including several of the largest ones. Unlike the usual pattern of a crisis following a policy of irresponsible fiscal expansion, this time the crisis was triggered by the abrupt cessation of external financing—after years of easy external money towards the end of the 70s—to an unsustainable expansion of private spending (consumption and investment) financed with generous resources from abroad. Only in 1981 did external indebtedness reach 25% of GDP. But in 1982 that came to an abrupt halt, both as a result of a crisis in Mexico that scared foreign investors about Latin America and a recession in the US and developed countries. But there was also a clear internal responsibility, since the economic conglomerates that had emerged in the country had done so around a bank at the core, which recycled external funds towards the group's companies and their growth. This was possible due to the financial deregulation implemented in the 80 s, which was clearly done without the essential prudential safeguards. The previous strong real appreciation of the peso also helped—a fixed exchange rate since 1979 and internal inflation 3 times higher than international

[19] There have been two attempts, between 2019 and 2023, to completely change the constitution through two consecutive assemblies specifically elected by popular vote. For various reasons that are not relevant to analyze here, the proposals that emanated from these assemblies were rejected in two consecutive referendums, one in September 2022 and the other only in December 2023. As a consequence, the 1980/2005 Constitution remains in force, although without the figure of laws with a high quorum.

inflation—and a mechanism for automatic salary adjustment to past inflation that was included in the 1979 labor reform.

The Chilean economy took 4–5 years to fully recover, and the authorities implemented substantive modifications to their policies. The most relevant was undoubtedly a new, much more demanding banking law that limited banking activity and introduced much more severe prudential regulation. From a macro point of view, Chile returned to a crawling peg exchange rate scheme according to past inflation and the principle of a responsible fiscal policy was safeguarded, despite the fact that the bailout of the financial system cost the State between 10% and 15% of GDP between 1983 and 1990, resources that were contributed financially by the Central Bank as fiscal agent, and indirectly by the growing private savings in pension funds.

The process of privatization of companies in the hands of the State that had begun in the previous decade deepened in the second half of the 80s, including several public services (utilities), such as the generation of electrical energy and its transmission and distribution, and the telecommunication companies. For each of these services, legal frameworks existed before their privatization to establish their rates in order to protect consumers. Although at first it was mainly Chilean investors who gained ownership of these companies,[20] over time they have been acquired by large foreign investors with knowledge of the respective industries. It is worth noting that all of these companies remain in private hands to date and that technological advances have made pricing by the regulator redundant, as in the case of telecommunications.

The last reform implemented prior to the return to democracy was the one that transformed the Central Bank into an autonomous entity of the government and with the objective of achieving (first) and maintaining (later) price stability and the proper functioning of payment methods and the financial system.

3.4 The Economic Policies of Democratic Governments

After losing a referendum on his continuity in power in October 1988, General Pinochet left the position of President of the Republic in March 1990, giving way to the government of Patricio Aylwin, leader of the opposition to Pinochet and head of a center-left coalition.[21] His motto, from an economic point of view, was "growth with equity", which summarized very well what his government and that of his successor, Eduardo Frei, would be: maintaining a market economy oriented to the world, and its institutional bases, legacy of Pinochet and his government, with a new emphasis on equity and social justice. The latter was not a minor issue because

[20] Not without controversy because some of them were executives of the respective state companies placed there by the government and who could have had information advantages in the privatization processes.

[21] This coalition was named "Concertación de partidos por la democracia", and later simply "Concertación".

poverty reached 68% of the population by early 90 s, a left-over from the 1982/83 crisis and a result as well of insufficient public social spending in the years afterward.

To address this high poverty rate, the Aylwin government implemented a significant increase in public spending on social programs, which was financed through a tax reform that allowed an additional 2% of GDP to be collected per year. This also evidenced the strict adherence of the ruling coalition to the principle of fiscal prudence—the so-called "macroeconomic equilibrium"—a guarantee that there was no desire to return to a past of high inflation. On the contrary, the management of the autonomous Central Bank began with a contractionary adjustment in its monetary policy in 1990—to confront an inflation that had reached 21.4% in 1989—and subsequently the monetary and exchange policies were administered in order to achieve a gradual reduction of inflation, from 27.3% in 1990 to 4.7% in 1998.

The government and the opposition at that time also agreed on changes to labor legislation that involved the elimination of the business lockout, the increase in severance pay and greater facilities for forming unions. However, essential aspects of the labor legislation then in force were maintained, such as collective bargaining at the company level and freedom of union membership.

From an economic perspective, there are two important reforms that took place during the 1990s. The first is the deepening of the integration of the Chilean economy into international markets (mentioned in passing above) and the beginning of public-private partnerships for the financing and construction of infrastructure works.

3.4.1 The Deepening of Commercial Integration to Global Markets

The trade policy followed since 1990 has not only kept Chilean markets open to international trade but has also deepened said policy (Fig. 3.7). On the one hand, the general and single tariff was reduced from 15% to 11% in 1991 and then to 6% in 2003, a rate that is maintained to this day. On the other hand, free trade agreements have been negotiated covering 64 countries that cover almost 85% of the world's GDP, which led the effective and average tariff on imports being less than 1% already in the mid-2000s.[22] This type of approach has been called "reciprocal opening", in the sense that an additional reduction in the local customs tariff is negotiated in exchange for access—or tariff-free access—to the markets of the country with which it is negotiated. As part of this same strategy, Chile also became a member of APEC in 1994 and signed an association agreement with MERCOSUR in 1996.

It should be noted that Chile has state-of-the-art economic integration agreements with the European Union (recently dated) and with the United States, which

[22] See Ffrench-Davis (2018).

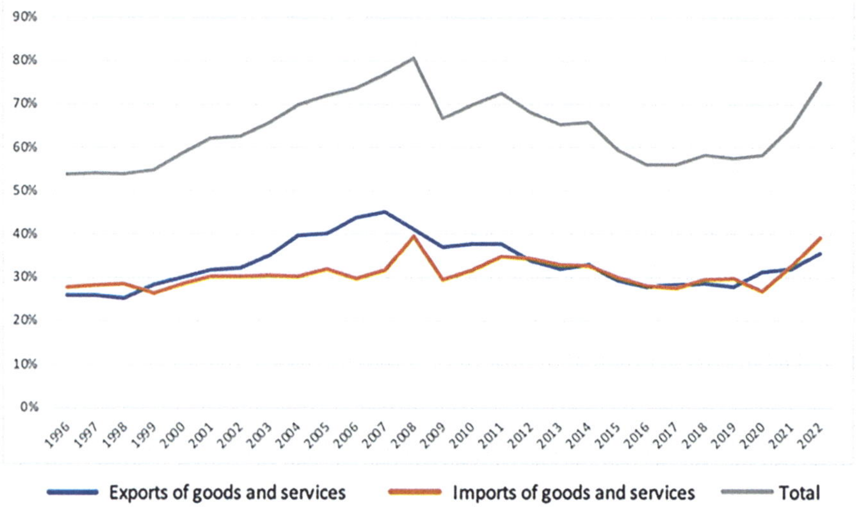

Fig. 3.7 Exports and imports (% of GDP) (World Bank, 2022)

include labor, environmental and gender aspects—in the European case. It also has a broad-spectrum agreement in force with China, Chile's main trading partner, constituting 44% of our exports and 35% of imports.

Although foreign direct investment was observed in the second half of the 1980s, especially in the case of privatized public service companies, it was not until it became clear that the new democratic governments would maintain the basic features of an open market economy that a substantial flow of direct investment was not observed. This arrived in masse in the mining sector, taking advantage of a pro-investment modification made in 1981 that had not achieved concrete effects until before 1990, partly due to the questioned political regime in Chile prior to that date. Mining investments and many others would come when Chile combined a market economy integrated with world markets with a democratic government regime. It is estimated that in 2022 the total foreign direct investment in Chile amounts to more than US$ 300,000 million.

Integration into the world economy also allowed Chilean companies to invest abroad, which initially—in the 90s—materialized more than anything in Latin America. Today, investments have diversified to the US, Europe, and other regions, reaching a figure of close to US$ 50,000 million.

Likewise, financial inflows grew strongly in the 90s until the Asian crisis in 1998–1999, during which they practically dried up, but resumed with vigor starting in 2001. Both banks and large companies directly have had a fluid access to external financing, which has been facilitated by the investment grade that the country obtained from the main international credit agencies between 1995 and 2005. This in turn was the result of the policy of fiscal prudence, the robustness of the financial system, the prevalence of the rule of law, the non-discriminatory tax treatment of foreign capital and Chile's own insertion into the global economy.

The risk premium on sovereign debt calculated by the JP Morgan bank has consistently shown Chile to be the lowest risk country in Latin America between 1990 and the present, currently observing values of this indicator similar to countries such as Poland and Portugal. This is despite the fact that in the last 8 years Chile has experienced reductions of one or two notches in its risk classification, although it is still far from losing the "investment grade".[23]

Another sign of Chilean integration into the world in general terms, and also of the positive evaluation made of its economic policies and institutional consolidation in international markets and fora, was the country's admission as a member of the OECD in 2010. This multilateral organization essentially brings together developed economies with democratic regimes—"like-minded"—and its mission is to gather and recommend best practices in economic policies based on empirical evidence.

3.4.2 The Era of Public Infrastructure Works Concessions

In 1993, a law was approved that established a public-private partnership formula that facilitated private investment (and its construction) in public infrastructure works in exchange for the right to exploit that infrastructure during a certain period of time, after which the infrastructure reverted to the State. In this, a BOT-type mechanism initially developed in the United Kingdom was being followed. After a somewhat hesitant start, the mechanism caught on very strongly towards the second half of the 90 s, accumulating until 2022 a total investment of US$ 21,000 million, or 7% of GDP.

With this type of mechanism, interurban roads, bridges, and tunnels were initially financed, later urban roads and public buildings, then hospitals and prisons. The business model has been adapted to the type of work—in the case of roads, it is the users who pay through a toll, in the case of prisons and hospitals, it is the State that assumes a per capita payment per year. Different auction allocation models and different forms of guarantees by the State have also been used to resolve some risks.

Although it has had some ups and downs over time, the concessions procedure is still fully in force and today it is also being used for water reservoirs for agriculture and infrastructure for public transportation.

3.4.3 The Consolidation of the Autonomous Central Bank

Between 1930 and 1990, the average annual inflation in Chile exceeded 30% annually. Although there were clear episodes of very high inflation in the 50s and especially in the 70s, the truth is that the rest of the time double-digit rates were not

[23] Standard&Poors and Moody's.

Fig. 3.8 General inflation (Central Bank of Chile, 2024)

unusual. The origin of the phenomenon was always fiscal deficits financed by issuing money, in a context in which the Central Bank was subordinated to the political authority. The 1980 Constitution cut off monetary financing of the fiscal deficit, an issue that was later institutionalized with the Central Bank autonomy law that came into force in 1989.

The autonomous Central Bank, mandated to achieve price stability, defined as "low and stable" inflation, managed, as already noted, to reduce inflation from 28% in 1990 to 2% in 1999 (Fig. 3.8). Establishing a system of inflation targets following the patterns of Australia, New Zealand, and other pioneering countries in the matter towards the end of the 90 s, inflation has since then remained around the self-imposed target of 3% annually, with a tolerance margin between 2% and 4% annually. Indeed, from 2000 to 2023, inflation has averaged 3.18% annually, and if we do the calculation until 2019 to avoid the anomalies created by the pandemic,[24] that average is almost the same figure, 3.19% annually.

The government of the Central Bank rests on a council of 5 members appointed each 2 years for alternating periods of 10 years each. Their names are proposed by the President of the Republic and their appointments are approved by the Senate, which has resulted in political representation in the Council similar to the citizen preferences represented in Congress. But more importantly, the appointments have generally gone to technicians of high reputation who have effectively honored the mandate of the Central Bank.[25]

[24] In 2022, as in most countries in the world, inflation accelerated to reach 14% at its peak, as a consequence of the enormous demand stimuli, both fiscal and monetary, implemented during the pandemic. Since the end of 2021, the Central Bank has followed an aggressive contractionary monetary policy that has managed to reduce inflation to 3.9% in December 2023 and with a downward trend. The 3% target is expected to be reached before the end of 2024, as anticipated by specialists and financial prices, and consequently the Central Bank is currently in a process of normalizing its monetary policy on the way to a neutral position.

[25] The person who presides over the Council is also President of the Central Bank for 5 years and is nominated by the President of the Republic from among the directors.

The good performance of the Central Bank in terms of keeping inflation at bay has been an unquestionable institutional basis and recognized by all political sectors as a necessary condition for economic progress.

3.4.4 The Fiscal Rule

In 2001, during the government of President Lagos, a fiscal rule called "structural fiscal balance", or "cycle-adjusted balance" was designed and implemented, which basically consisted of determining the growth of effective spending according to "structural" income instead of projected income and a target for the "structural" balance. Structural or cyclically adjusted income is calculated based on a potential or trend growth of the economy and a trend or long-term copper price.[26]

Thus, if the economy is in the high phase of the cycle or the price of copper is very high, the structural income will be less than the effective one and the structural balance will be less than the effective one. The opposite occurs when there is a recession, or the price of copper is greatly reduced[27] (Fig. 3.9). To close the equation, the government must determine at the beginning of its mandate what its balance target is for the 4 years that it lasts, without prejudice to the fact that this target can be changed for justified reasons within the 4 years. When the rule was created, the target was a surplus of 1% of GDP and the treasury was able to accumulate in the order of US$ 20 billion until 2007, based on a period of high copper prices. For the Great Recession of 2008–09, caused by a global financial crisis, the government used US$9 billion of those savings to support the economy and avoid a major recession. Since then, the structural balance target of the following governments has always been negative values, between −2 and −1% of GDP, while the effective balances have been mostly negative, especially in 2008–2009 and then for the pandemic (2020 and 2021).

It should also be noted that the public debt follows the fluctuations of the effective balances. In the 2000s, net public debt became negative for some years, while gross debt observed an increasing trend since the Great Recession to date. However, it remains below 40% of GDP by 2023.[28]

The relative worsening of public accounts revealed by the increase in debt has been partly the result of the low rates of economic growth that the economy has been showing since the middle of the last decade, which in turn can be attributed to

[26] This takes into account that the State's income from Codelco's profits and taxes and royalties from private mining companies are a significant item in the budget.

[27] The economic cycle in Chile is usually correlated with (if not caused by) the international economic cycle, which in turn influences the international price of copper.

[28] The Fiscal Advisory Council, created by law in 2017, and whose mission is to monitor the health of the State's financial accounts and to recommend policies and actions in this regard, proposed in 2022 that, in addition to the fiscal rule already in use, it be added a ceiling for public debt of 45% of GDP. A bill on the matter is being discussed in Congress.

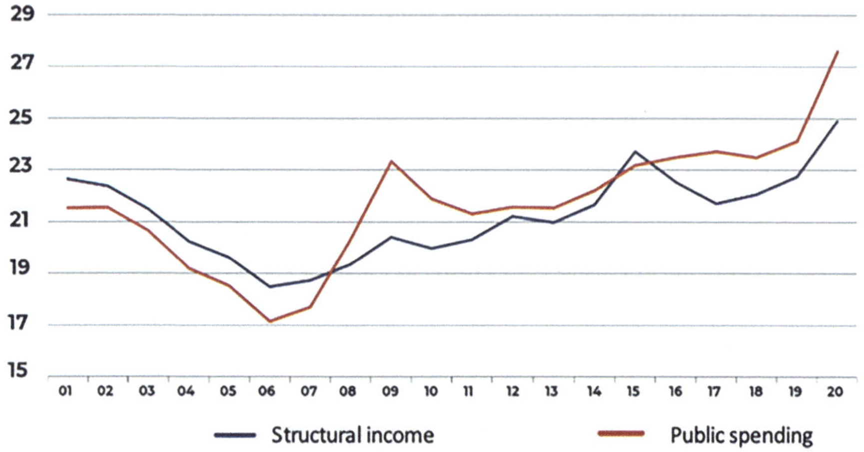

Fig. 3.9 Public expenditure and cyclically adjusted income (% of GDP) (Lagos & García, 2021)

a stagnation of productivity and a lower dynamism of private investment, issues in turn linked to inappropriate public policies.

3.5 Recent Developments and Challenges in the Economic Field

To date, the Chilean economy continues to be the one with the highest GDP per capita in Latin America and the best human development index, and it is also well managed from a macroeconomic point of view. Furthermore, it has a set of institutions that are strong enough to guarantee a well-functioning democratic regime and rules that maintain full integration into the global economy.

However, it is not yet a developed economy. In fact, its PPP-adjusted GDP per capita is less than half that of the US and the G-7 to date (Table 3.9). Despite the significant progress made in reducing poverty since 1990, income inequality remains high and an issue that neither higher per capita income nor greater resources allocated to social policies have been able to substantially improve.

3.5.1 The Problem of Inequality

The GINI coefficient reached in 2022 a value of 0.46, much higher than the average of OECD countries, although it decreased from 0.5 in 2010 and 0.549 in 2000.[29]

[29] The GINI is an index that goes from 1 (extremely unequal) to 0 (complete equality).

Table 3.9 Per capita GDP at PPP: Chile as a percentage of USA, G-7, and L.A

Year	USA	G-7	Latin America
1970	31.4	34.8	100.8
1973	27.8	30.2	86.8
1989	24.7	25.9	91.0
1997	34.3	37.0	127.0
2007	36.3	42.2	139.3
2013	42.4	49.4	147.1
2016	41.7	48.9	155.4
2019	39.3	59.9	152.5
2022	39.6	61.2	156.8

Source: Ffrench-Davis (2018) based on World Bank data

Table 3.10 Income inequality, real wages, and social spending in real terms

		Social spending per capita[a]			Real
Year	Gini	Education	Health	Total	Wages[a]
1989		100.0	100.0	100.0	100.0
1990	52.1	94.9	94.0	97.4	101.8
1996	52.2				
1997		183.4	174.4	154.5	137.8
1999		213.1	189.7	174.7	144.8
2000	54.9				
2003	52.8	261.4	236.1	198.0	153.7
2006	50.4	290.6	306.0	229.3	162.5
2009		433.5	451.1	319.3	174.7
2011	50.0	473.6	501.4	350.7	183.0
2013	48.8	562.5	584.3	382.1	196.2
2015	47.6	650.4	710.6	435.6	203.2
2016	44.4	714.9	704.1	462.5	206.1

Source: Ffrench-Davis (2018) with data from CASEN, INE and DIPRES
[a]Index 1989 = 100

It is clear that when it comes to reducing inequality, especially income inequality, there are no shortcuts that allow radical changes in the very short term. What's more, the efforts made by successive governments in this matter has not been minor: between 1989 and 2016, fiscal spending on education and health grew 7 times, and if we add housing, social security and others, the increase exceeds 4 times (Table 3.10).[30] However, the improvement in the GINI coefficient after considering social policies is much smaller than in the case of developed countries. This may be due to the fact that a significant component of social spending in developed countries goes to finance the pension deficit, benefiting pensioners with lower incomes, unlike in Chile, whose workers retire with what they have saved during their active

[30] Ffrench-Davis (2018).

life.[31] But it is also evident that much of the social spending in the richest countries is monetary transfers, which have proven to be more effective in compensating for income inequalities. On the tax side, furthermore, the proportion of revenues from income tax over total tax resources in Chile barely reaches 12%, while the norm in developed countries is for this percentage to exceed 20%. The progressivity of the income tax is another factor that helps equalize income inequality.

3.5.2 Productivity Stagnation

As mentioned above, since 2014 the economy has shown meager economic growth, of 2.25% on average until 2023, much lower than the 6.26% of the 90s and the 4.51% of the 2000s. The decline in growth is especially linked to a stagnation in total factor productivity, which has shown virtually no increase since 2013.[32] In the decade of the 90s, the decade of greatest economic expansion in the last 80 years, the contribution of TFP to growth was 3.7 percentage points per year (1990–1995).

There could be many reasons behind this productivity stagnation. On one side, the highly positive growth in the 90s could have been the result of allocative efficiency gains accrued to the liberalizing reforms of the 70s, 80s and 90s. These gains would have vanished as of lately as no new significant reforms have taken place. On other side, dynamic efficiency gains, accrued to investment in R&D and innovation in general is not present either: R&D spending in Chile is less than 0.5% of GDP, much less than the 2% in developed economies or the 3.5–4% in China and other Asian economies. Human capital improvements as represented by the massive increase in tertiary education enrollment, could be a helping factor, but on the other hand PISA results show that Chile's high-school students clearly rate well below OECD average. In any event, much needs to be done in the productivity area.

3.5.3 Some New Opportunities Ahead

The world is facing an existential challenge: the effects of climate change. Prospects do not look good for the future of the planet and the human race if the temperature continues rising beyond control. In order to keep the global warming within supportable levels—perhaps not beyond 2 degrees above the preindustrial levels—then there must be a decisive set of policies in place worldwide that at this point are very

[31] In early 2022, during the last days of President Piñera's tenure, the Pensión Garantizada Universal (Universal Guaranteed Pension) was introduced to secure 80% of the population a minimum pension regardless of what they have saved during their active lifes.

[32] Part of the explanation lies in the productivity decline in mining due in turn to lower copper content in the raw mineral in the main and larger mines. But starting in 2014 the stagnation is widespread across sectors.

well known, both for mitigating the sources of man-made emissions of warming gases and for adapting infrastructure and living conditions to a warmer planet and more frequent natural disasters.

This existential challenge is ironically an opportunity to Chile's economy.[33] First, Chile has huge reserves of lithium, a key component in the manufacturing of batteries for electric vehicles, the production of which is an essential part of the fight against climate change. Total production could be doubled before 2030 reporting a 10% increase in total exports. In the same vein, Chile is currently the main producer of copper in the world and has important reserves of this mineral for many years to come.[34] Copper is also key to electric vehicles. And second, Chile is blessed with clear and permanent sunshine in the north of the country and strong and persistent winds in the extreme south. This allows for a rich renewable energy generation activity which, in turn, is key to the production of green hydrogen. It is calculated that around US$ 60 billion could be invested in this industry before 2030.

3.6 Conclusions

Until a few years ago, being a developed economy seemed perfectly feasible in a matter of a decade or even less. And it was a political objective (and a dream) for many based on the impressive growth path in the 90 s and 2000s. However, the productivity and overall stagnation of the last decade, plus political turmoil, has put this goal on hold. To some, confronting inequalities must have priority over pure economic growth. So, today reaching the frontier of development—something like US$ 40,000 to 50,000 GDP per capita—seems farther away. Some experts claim that Chile fell in the "middle-income country trap".

It is true that the term "developed country" involves much more than a high per capita income, but normally there is a close association between income per capita and progress in most social aspects. And Chile is no exception. Public policies must improve in many areas, like health and education, if Chile wants to grow fast again and at the same time confront inequality. Everyone would agree with that but not on the way to shape those policies. Also, with no increase in R&D investment we are condemned to low growth in the long run, and private investment will not rebound if regulations are too many and too inefficient, as they are today.

Still, Chile has a shot. Focusing again on growth, stressing quality in education, investing in training for the new jobs of the digital era and taking advantage of the triad copper, lithium and green hydrogen, development can be a reachable dream again.

[33] Having a bright side does not preclude that Chile is one of the world most affected countries by climate change, according to experts.

[34] Although the extraction cost has been steadily increasing in the last two decades.

References

Central Bank of Chile. (2024). https://si3.bcentral.cl/siete

Cincel. (2002). Informe sobre la educación superior en chile: 1980–2003 https://www.cincel.cl/documentos/Recursos/informe_educacion_superior_chile_2003.pdf

CMF. (2022) CMF presenta radiografía de la actividad bancaria a nivel regional. https://www.cmfchile.cl/portal/prensa/615/w3-article-55221.html

CORFO. (2022). *Chile y sus regiones en datos económicos. Informe económico para la descentralización* https://wapp4.corfo.cl/archivos/WCSCONTI/IP/GRC/Chile%20y%20sus%20regiones%20en%20datos%20economicos%20Gerencia%20de%20Redes%20y%20Territorios%20CORFO%202022.pdf

Cortázar, R., & Marshall, J. (1980). Índice de precios al consumidor. *Colección Estudios CIEPLAN*, No. 4.

Ffrench-Davis, R. (2018). Reformas económicas en Chile, 1973–2017. *Taurus. Penguin Random House.*

Gallardo, R. (2023). *Cifra de matriculados en educación superior aumenta cinco veces en los últimos 30 años.* La tercera. https://www.latercera.com/la-tercera-sabado/noticia/cifra-de-matriculados-en-educacion-superior-aumenta-cinco-veces-en-los-ultimos-30-anos/P62CGUCBGZBZVOIU4ZDOGTE3MQ/#

Generadoras de Chile. (2023). *Generación Eléctrica en Chile.* https://generadoras.cl/generacion-electrica-en-chile#:~:text=A%20diciembre%20de%202022%20cuenta,9%2C8%25%20petróleo

Gobierno de Chile (2023). *Extensión de Línea 2 de Metro de Santiago comenzó su marcha blanca.* https://www.gob.cl/noticias/marcha-blanca-extension-metro-linea-2-san-bernardo-santiago/

Hachette, D. Y., & Luders, R. (1987). El proceso de privatización de empresas en Chile, 1974–1982. *Boletín Económico No. 22, Instituto de Economía, U. Católica de Chile.*

IMF (2023). International soundness indicators. Washington, DC.

INE. Instituto Nacional de Estadísticas. (2023). https://www.ine.gob.cl.

Lagos, L. F., & García, M. (2021). Propuesta de regla fiscal: credibilidad y simplicidad. Libertad y Desarrollo. https://lyd.org/wp-content/uploads/2021/04/SIE-291-1.pdf

MSDF. Ministry of Social Development and Family of Chile. (2022). *CASEN.* https://datasocial.ministeriodesarrollosocial.gob.cl/; https://observatorio.ministeriodesarrollosocial.gob.cl/encuesta-casen-2022

Observatorio logístico: Ministerio de Transporte y Telecomunicaciones. (2021). *Red vial nacional según categoría.* https://www.observatoriologistico.cl/infraestructura/red-vial/?id=5d719224d2c6f20029110412#:~:text=La%20red%20vial%20nacional%20(de,vial%20nacional%20en%20ese%20año.

PDI. (2023). *Primer Informe unificado sobre homicidios consumados.* https://www.pdichile.cl/centro-de-prensa/detalle-prensa/2023/07/17/primer-informe-unificado-sobre-homicidios-consumados

SISS. Superintendencia de servicios sanitarios. (2022) *Informe de Coberturas Sanitarias 2022.* https://www.siss.gob.cl/586/articles-22927_recurso_1.pdf

SUBTEL. Subsecretaría de Telecomunicaciones. (2022). *Hogares con acceso a Internet fijo alcanzan el 67% y usuarios aumentan preferencia por redes de alta velocidad.* https://www.subtel.gob.cl/hogares-con-acceso-a-internet-fijo-alcanzan-el-67-y-usuarios-aumentan-preferencia-por-redes-de-alta-velocidad/

The heritage foundation. (2023). *2023 Index of Economic Freedom* https://www.heritage.org/index/explore

UNDP. (2022). *Human development index* https://hdr.undp.org/data-center/human-development-index#/indicies/HDI

Weather Spark. (n.d). *Chile y tiempo promedio durante todo el año en Chile.* https://es.weatherspark.com/countries/CL

World Bank. (2022). https://datos.bancomundial.org

Chapter 4
Climatic Profile and Drought Characteristics in Chile

Francisco Javier Meza

Abstract From a climatic perspective, Chile is a country of extremes. From the driest desert on Earth to Patagonia, communities and ecosystems have adapted to different climates and to enduring periods of persisting drought. With mounting evidence of global change processes, it is still unclear how these regimes will evolve in the future, and what are the challenges that water managers will need to manage droughts. This chapter describes the major climatic regimes in the country, highlighting the variability and uncertainties pertaining to the occurrence of droughts. Climate change projections will be discussed and its impact on droughts.

Keywords Climate · Drought · Climatic trends · Chile

4.1 Climate of Chile

4.1.1 Major Atmospheric, Oceanic and Geographic Drivers

Spanning over 4200 km and with an average width of 230 km, Chile presents a wide variety of climatic regimes, which are the result of a combination of its latitudinal extension and position, the presence of the Pacific Ocean and the Humboldt current, the subtropical High-pressure system (Pacific Anticylone) and an irregular topography dominated by the Coastal and the Andes Cordillera. Along its territory, it is possible to observe a north-south transition from the driest desert in the world to a

F. J. Meza (✉)
Facultad de Agronomía y Sistemas Naturales, Pontificia Universidad Católica de Chile, Santiago, Chile

Instituto para el Desarrollo Sustentable, Pontificia Universidad Católica de Chile, Santiago, Chile

Centro de Cambio Global, Pontificia Universidad Católica de Chile, Santiago, Chile
e-mail: fmeza@uc.cl

D. Rivera, G. Donoso (eds.), *Droughts in Chile*, Global Issues in Water Policy 31,
https://doi.org/10.1007/978-3-031-85040-0_4

Mediterranean climate, which later gives way to a temperate and sub polar oceanic environments in the central and southern parts of the country, respectively. Along the West-East axis, there is a transition from ecosystems under oceanic influence and coastal fog towards Alpine-type climatic features.

Covering almost 38° latitudinal degrees, incoming solar radiation is the most important controller of the climate. Consequently, warm climates are in the northern part, whereas maritime polar climates are found in the extreme south. A remarkable feature, due to semipermanent clear skies, is that Chile exhibits the highest values of available solar radiation in the Atacama Desert (Rondanelli et al., 2015).

The influence of the Pacific Ocean, and specially the surgency of the cold current that runs parallel to the coast (Humbolt current) diminishes thermal amplitude along the entire territory and substantially reduces the occurrence of extreme temperature values. Temperatures in locations placed in the coastal regions do not show high degree of thermal amplitude and temperature range from north to south are not as high as one would have expected according to the latitudinal extension. From west to east, coastal zones are characterized by cloudy skies and higher relative humidities, this effect can be observed several kilometers into the continent.

In the central and southern part of Chile, the Coastal Cordillera acts as an obstacle for the prevailing westerly winds, forcing humid air to ascent which increases the frequency of isolated precipitation events on its west side. Adiabatic descent on the Eastern side, on the contrary, causes high atmospheric stability favoring dry conditions in the "Secano interior" (interior drylands). Precipitation decreases in the central valley, but then experience a substantial rise due to orographic effects associated to the presence of the Andes.

The descendent branch of the Hadley cell usually lands around latitude 30° (north and south). This quasi-permanent center of high pressure with counterclockwise rotation in the southern hemisphere is characterized by high temperature due to diabatic heating and low relative humidity and blocks the arrival of frontal systems originated at higher latitudes, yielding clear skies, high thermal amplitude and especially low to almost null precipitation. The Pacific Anticyclone shows cycles at different time scales (seasonal to interdecadal). At the seasonal scale, the Anticyclone SPSA is more intense around latitudes 35° south in the austral summer, while in austral winter it moves northward (Latitudes 28°–30°S), being responsible for the strong seasonal behavior in precipitation within those latitudes (Ancapichún & Garcés-Vargas, 2015).

4.1.2 Rainfall Variability

Temporal variability in rainfall can be associated with different climatic patterns. In the Altiplano region (Lat 17°-25° S), water vapor is transported from the Intertropical Convergence Zone during the austral summer creating a period of intense precipitation that falls commonly as snow and contributes to the recharge of aquifers. Changes in the prevailing easterly winds and the position of subtropical high-pressure systems introduce a high degree of variability between years (Valdés-Pineda et al., 2016; Garreaud, 2000).

The Atacama Desert (Lat 26° to 29°S) is characterized for being the driest with extensive periods of no precipitation at all.

Further south (Lat 29° to 37°S) there is an extensive area of Mediterranean climates that hosts not only major cities, but also as significant proportion of high value export-oriented agriculture, several important hydropower plants that are responsible for more than 25% of the electricity generation, major copper and other mining exploitations and a significant share of the plantations that support forest industry. Here Mid-latitude storms move eastward and north in latitudinal bands or storm tracks. This region shows a significant El Niño-Southern Oscillation footprint (ENSO), which is the interaction between the atmosphere and ocean in the tropical Pacific that results in a somewhat periodic variation between below-normal and above-normal sea surface temperatures and dry and wet conditions over the course of a few years.[1]

During the warm phase of ENSO (El Niño event), positive pressure anomalies are found over the Amundsen and Bellingshausen Seas. These anomalies block westerly winds and generate a displacement of the frontal systems to the north, bringing above normal precipitation over central Chile during the austral winter (Cai et al., 2020), this is enhanced as the Pacific Anticyclone moves northward and towards the ocean. The opposite occurs during the cold phase (La Niña years).

The ENSO phenomenon is monitored using an array of buoys (TOA/TRITON) that have been deployed in the equatorial pacific. Four different regions have been identified as they show the highest correlation with climatic features in the globe. For Chile, one of the most used corresponds to the region denominated as ENSO3–4 (5 N–5 S; 120–170 W). Table 4.1 shows the correlation of sea surface temperatures in the region ENSO3–4 and seasonal precipitation in several stations in Chile for the period 1990–2019.

A strong positive ENSO footprint (red colors) can be identified for winter (JJA) precipitation in almost all locations, but signal becomes weaker as we move southward. ENSO has an influence on springtime precipitation for locations in the south.

A warm or El Niño phase (cold, La Niña) can be declared when sea surface temperature anomalies (deviations from the long-term mean) surpass certain threshold, around 0.5 °C, for several months. There is a large variability within phases but still it can be asserted that one can expect above (below) normal precipitation during El Niño (La Niña) years. Figure 4.1 provides an example of wintertime precipitation in the station of Quinta Normal.

ENSO is not the only large scale oceanic atmospheric process that has an influence on the climatic regime of Chile, among them we can found the Pacific Decadal Oscillation (PDO), Antarctic Oscillation (AAO) and the Madden-Jullian Oscillation (MJO) (Valdés-Pineda et al., 2016), but due to its simplicity, ability to monitor and

[1] https://iridl.ldeo.columbia.edu/maproom/ENSO/ENSO_Info.html

Table 4.1 Heat map of Pearson correlation coefficient between sea surface temperatures in region ENSO 3–4 and seasonal precipitation for several locations of Chile MAM: March, April, May; JJA: June, July, August; SON: September, October, November; DJF: December, January, February

Location	Latitude (°S)	Mean Annual PP (mm)	DJF	MAM	JJA	SON
La Serena	29.9	76.1	-0.15	0.16	0.41	0.29
Illapel	31.6	155.6	0.13	0.28	0.56	0.50
Los Andes	32.9	214.4	0.08	0.43	0.36	0.13
Quinta Normal	33.4	260.9	0.25	0.35	0.44	0.30
San Fernando	34.6	587.2	0.11	0.29	0.30	0.57
Curico	35	551.4	0.02	0.36	0.29	0.65
Los Angeles	37.4	961.0	-0.18	0.15	-0.02	0.66
Temuco	38.7	1023.4	-0.08	0.12	0.17	0.53
Valdivia	39.9	1596.7	-0.09	0.09	0.20	0.46
Osorno	40.6	1143.7	0.08	0.23	0.35	0.48

Data source: Climate Explorer https://explorador.cr2.cl/ and NCAR climate data guide https://climatedataguide.ucar.edu/climate-data/nino-sst-indices-nino-12-3-34-4-oni-and-tni

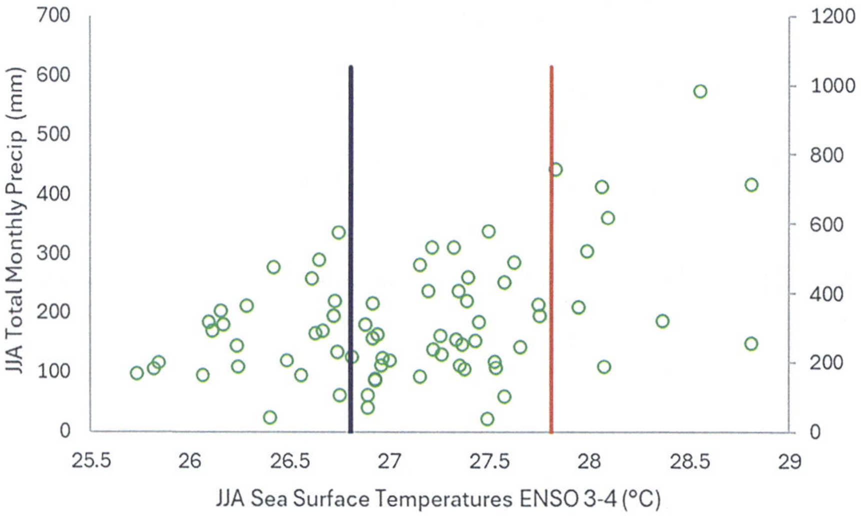

Fig. 4.1 Seasonal precipitation in winter (JJA) for station Quinta Normal as function of sea surface temperature values in region ENSO 3–4. (Source: own elaboration)

forecast and the time scale in which it operates, it has been the most commonly used index involved in climate forecast and early warning systems.

4.2 Climatic Trends and Future Projections

As described above, temperatures show little variation along the coast mainly due to the presence of the Humbolt current. In terms of long-term trends, Falvey & Garreaud, 2009 have identified a cooling trend in coastal stations that can reach up to −0.2 °C/decade. This effect has apparently ended and can be associated with a shift in the phase of the Tripole Index for the Interdecadal Pacific Oscillation (IPO). Inland stations show a positive trend whose magnitude depend on latitude and elevation (Burger et al., 2018). As temperature has a direct effect on the rate of evapotranspiration it is likely that atmospheric water demands tend to increase marginally contributing to exacerbate the incidence and magnitudes of droughts.

Precipitation in central Chile has shown a discernible trend, but it is not possible to isolate this feature of the effect of a long-term drought that has lasted for almost a decade (Garreaud et al., 2020). The degree of variability is quite considerable at annual and seasonal scales. Several authors have described this effect in the late part of the twentieth century with possible attribution to anthropogenic climate change (Boisier et al., 2016, 2018).

Figure 4.2 shows the time series of annual precipitation since 1976 (when a documented shift in PDO occurred) for a sample of stations located between latitudes 29 and 39 S.

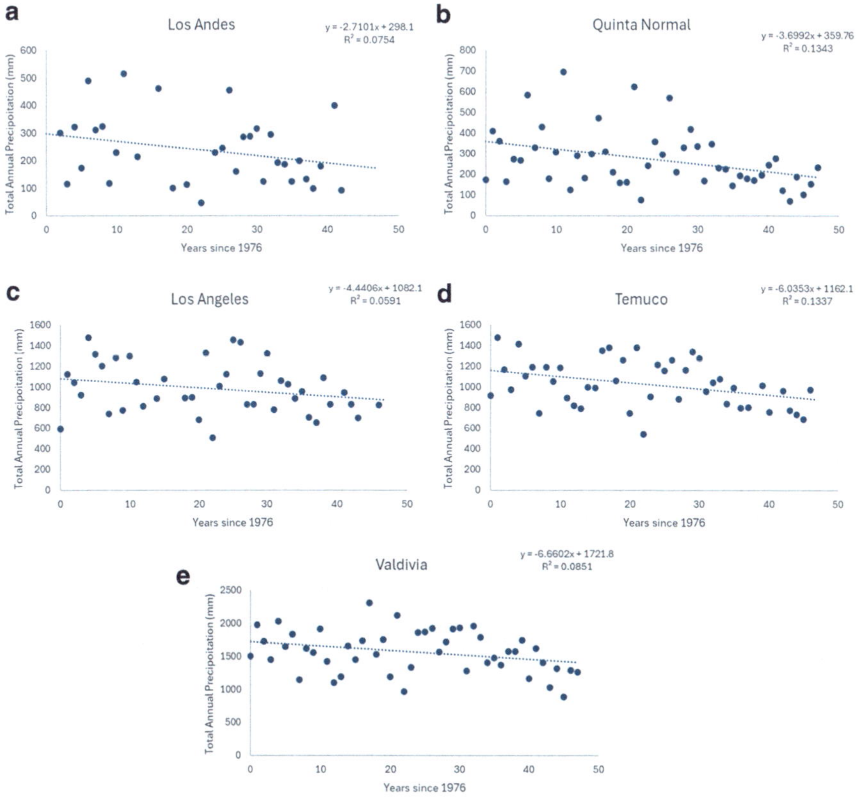

Fig. 4.2 Time series of annual total precipitation of selected stations in Central Chile (Source: own elaboration)

Although other hydrological variables such as streamflows and depth of water table are the result of multiple nonlinear interactions and depend on land use, water use and available infrastructure, it is important to highlight that the effects of climate can percolate down into a basin and contribute to aggravate drought conditions. Sangüesa et al. (2023) analyzed trends in monthly flows in 31 basins in central Chile, finding that trends were predominantly negative for semiarid and subhumid climatic zones. Consistent results have been documented by Fuentes et al. (2021) when describing diminishing trends in rain gauges and monitoring wells.

With respect to future climate conditions, Global Circulation Models of the Coupled Model Intercomparison Project version 6 (CMIP6) run under shared socioeconomic pathways (SSP) show a reduction in mean annual precipitation between latitudes 20 ° and 40 °S, the level of agreement between models is quite considerable (90%; Salazar et al., 2024). Previous assessments developed under CMIP5 are also consistent and show a decreasing trend in precipitation whose intensity depends on the scenario used (3%–30%) and a significant warming that varies between 1.2 and 3.5 °C at the end of the present century (Bozkurt et al., 2019).

4.3 Droughts in Chile: General Characteristics

As droughts are extreme meteorological events that have consequences in the hydrological cycle and affect agricultural and ecosystem performance, a single and universally accepted definition of drought is not available. The World Meteorological Organization defines droughts a sustained and persistent decrease in precipitation. Others such as the UN Convention to Combat Desertification (UNCCC) indicates that drought occur when precipitation falls below normal levels causing impact on the hydrological system and land productivity, whereas the Food and Agricultural Organization (FAO) highlights the impacts on crop productivity and even crop failure because of sustained soil moisture deficits. Regardless of the definition and the socioeconomic or socioecological sector that needs to be analyzed, a drought can be conceptualized as a substantial departure of the difference between water supply and demand and treated a s a time series of a variable that accounts for those differences (see an example in Oertel et al., 2018). Thus, a succession of negative values (supply unable to meet demand) defines a drought period, and the magnitude of the difference relates to the severity of it (Fernández & Gironás, 2021).

In Chile, droughts are a recurrent phenomenon that carry significant losses in agriculture and tourism (Aldunce et al., 2016) being able to impact water provision for human consumption, electricity generation and other relevant economic sectors (Fernández et al., 2023). The occurrence of dry periods and droughts can be traced back to the mid sixteenth century with estimates of the occurrence of these types of phenomena that vary between 51 to 79 events until the twentieth century (Fernández & Gironás, 2021).

More recently, the adoption of drought indices such as the Standardized Precipitation index (SPI, McKee et al., 1993), the Standardized Precipitation

Evaporation index (SPEI, Vicente-Serrano et al., 2010), Standardized Streamflow Index (SSI, Modarres, 2007) has been the focus of research for the characterization, monitoring and the adoption of policy and disaster mitigation responses.

Fernández and Gironás (2021) present an analysis of droughts characteristics in central Chile using both meteorological (SPI) and hydrological indices (Standardized Runoff index, SRI) concluding that meteorological droughts are more frequent but are usually shorter, more intense and have shorter duration times. Perhaps the one event that does not follow this general rule is the recent drought that has lasted more than one decade (starting in 2010), being regarded as a Megadrought (Garreaud et al., 2020).

As we can see in Fig. 4.3 (Fernández & Gironás, 2021), there is a marked trend towards decreasing SPI values with a strong influence of the recent Megadrought. It is also noticeable that this feature is present in a significant proportion of the territory indicating that droughts require national level responses to avoid major loses in the country.

Aside from the more recent drought that has a strong footprint on the evolution of drought indices along the territory, it is important to highlight that large scale atmospheric oceanic patterns such as the one associated with ENSO phenomenon, have also a direct influence on the occurrence and duration of meteorological and hydrological droughts, being explained by effects of ENSO that brings higher probabilities of more intense precipitation during El Niño years and higher precipitation deficits during La Niña years (Oertel et al., 2020).

Human induced climate change as a consequence of greater accumulation of greenhouse gases in the atmosphere have a direct effect on the occurrence and

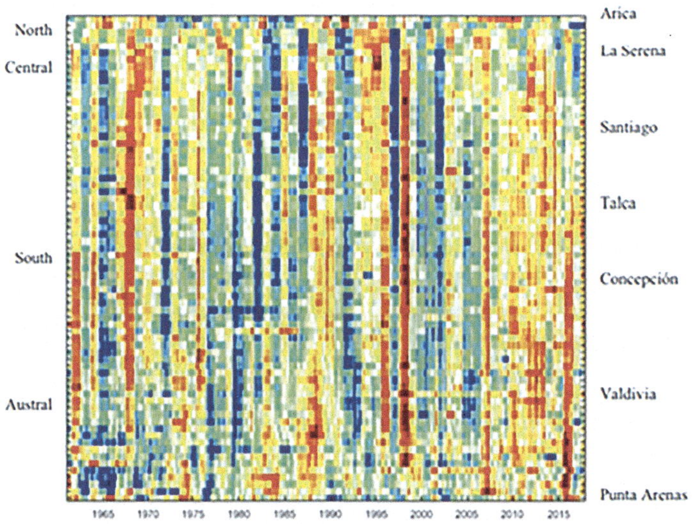

Fig. 4.3 Evolution of SPI values over 2016–2018 from north (Arica) to south (Punta Arenas) (Source Fernández & Gironás, 2021; reproduced from DMC, 2019)

severity of droughts. As temperatures increase and precipitation decreases (high consensus in the direction of the trend among GCM models), rainfall and streamflow values tend to decrease, and drought indices react in a consequently manner. Bobadilla et al. (2024) confirm this assessment showing that meteorological droughts increase in magnitude and duration (especially in the Andes Mountain Range), and that hydrological droughts show increases in magnitude and duration.

4.4 Research Needs in Drought Monitoring and Management

Droughts in Chile have always been present and, although represent an extreme meteorological (hydrological) event, they cannot be regarded as single or isolated events. In addition, there is strong evidence that these types of events will become more frequent in the future as climate change intensifies its effect over the country. Therefore, drought characterization will become increasingly important, being necessary to expand the scope to incorporate other type of impacts (e.g. effects on water quality as shown by Peña-Guerrero et al., 2020), and incorporate ground-based indices that are able to capture the effects on agriculture and natural vegetation (Zambrano et al., 2018).

There are also institutional challenges for drought management and the development of effective and coordinated responses, as many institutions are involved, and the current legal framework needs to be improved to facilitate better anticipatory management responses (Nuñez Cobo et al., 2018).

There is also a yet to be explored field of the role of seasonal climate forecasting for drought management as droughts have a significant ENSO footprint that can be monitored and projected over seasonal time scales. Issues of operational forecast products, combination with ground information about soil moisture and vegetation conditions, current stored capacity in reservoirs, forecast lead time to ensure timely responses, ex-ante evaluation of management responses, communication and raising public awareness are a sample of strategies that can be analyzed individually or as part of larger and more ambitious drought management portfolios that increase coping capacity and build resilience in an era of climate change.

Acknowledgements This work was partially supported by Fondecyt Grant 1210526 and by ANID Anillo de Ciencia y Tecnología ATE 220055.

References

Aldunce, P., Bórquez, R., Adler, C., Blanco, G., & Garreaud, R. (2016). Unpacking resilience for adaptation: Incorporating practitioners' experiences through a transdisciplinary approach to the case of drought in Chile. *Sustainability, 8*, 905.

Ancapichún, S., & Garcés-Vargas, J. (2015). Variability of the Southeast Pacific subtropical anticyclone and its impact on sea surface temperature off north-central Chile. *Ciencias Marinas, 41*(1), 1–20.

Bobadilla, A., Stehr, A., & Toro, N. (2024). Evaluation of climate-change impacts on the temporal and spatial behaviour of drought in South-Central Chile. *Hydrological Sciences Journal, 69*(2), 165–184.

Boisier, J. P., Rondanelli, R., Garreaud, R. D., & Muñoz, F. (2016). Anthropogenic and natural contributions to the Southeast Pacific precipitation decline and recent megadrought in Central Chile. *Geophysical Research Letters, 43*(1), 413–421.

Boisier, J. P., Alvarez-Garreton, C., Cordero, R. R., Damiani, A., Gallardo, L., Garreaud, R. D., et al. (2018). Anthropogenic drying in Central-Southern Chile evidenced by long-term observations and climate model simulations. *Elementa: Science of the Anthropocene, 6*, 74.

Bozkurt, D., Rojas, M., Boisier, J. P., Rondanelli, R., Garreaud, R., & Gallardo, L. (2019). Dynamical downscaling over the complex terrain of Southwest South America: Present climate conditions and added value analysis. *Climate Dynamics, 53*, 6745–6767.

Burger, F., Brock, B., & Montecinos, A. (2018). Seasonal and elevational contrasts in temperature trends in Central Chile between 1979 and 2015. *Global and Planetary Change, 162*, 136–147.

Cai, W., McPhaden, M. J., Grimm, A. M., Rodrigues, R. R., Taschetto, A. S., Garreaud, R. D., et al. (2020). Climate impacts of the El Niño–southern oscillation on South America. *Nature Reviews Earth & Environment, 1*(4), 215–231.

DMC, Dirección Meteorológica de Chile (2019) *Reporte Anual de la Evolución del Clima en Chile 2018* (in Spanish).

Falvey, M., & Garreaud, R. D. (2009). Regional cooling in a warming world: Recent temperature trends in the Southeast Pacific and along the west coast of subtropical South America (1979–2006). *Journal of Geophysical Research: Atmospheres, 114*, D4.

Fernández, B., & Gironás, J. (2021). Droughts. In B. Fernández & J. Gironás (Eds.), *Water resources of Chile. World water resources* (Vol. 8). Springer. https://doi.org/10.1007/978-3-030-56901-3_8

Fernández, F. J., Vásquez-Lavín, F., Ponce, R. D., Garreaud, R., Hernández, F., Link, O., et al. (2023). The economics impacts of long-run droughts: Challenges, gaps, and way forward. *Journal of Environmental Management, 344*, 118726.

Fuentes, I., Fuster, R., Avilés, D., & Vervoort, W. (2021). Water scarcity in Central Chile: The effect of climate and land cover changes on hydrologic resources. *Hydrological Sciences Journal, 66*(6), 1028–1044.

Garreaud, R. (2000). Intraseasonal variability of moisture and rainfall over the South American Altiplano. *Monthly Weather Review, 128*(9), 3337–3346.

Garreaud, R. D., Boisier, J. P., Rondanelli, R., Montecinos, A., Sepúlveda, H. H., & Veloso-Aguila, D. (2020). The Central Chile mega drought (2010–2018): A climate dynamics perspective. *International Journal of Climatology, 40*(1), 421–439.

McKee, T. B., Doesken, N. J., & Kleist, J. (1993). The relationship of drought frequency and duration to time scales. In *Proceedings of the 8th Conference on Applied Climatology* (Vol. 17, No. 22, pp. 179–183). to time scales. Proceedings of the 8th Conference on Applied Climatology, American Meteorological Society, Boston: 179–184.

Modarres, R. (2007). Streamflow drought time series forecasting. *Stochastic Environmental Research and Risk Assessment, 21*(3), 223–233. https://doi.org/10.1007/s00477-006-0058-1

Nuñez Cobo, J., Garraud, R., Verbist, K. (2018) Chile, En Atlas de Sequias de America Latina y El Caribe. UNESCO-CAZALAC. (in Spanish).

Oertel, M., Meza, F. J., Gironás, J., Scott, A., Rojas, C. F., & Pineda-Pablos, N. (2018). Drought propagation in semi-arid river basins in Latin America: Lessons from Mexico to the Southern Cone. *Water, 10*(11), 1564.

Oertel, M., Meza, F. J., & Gironás, J. (2020). Observed trends and relationships between ENSO and standardized hydrometeorological drought indices in Central Chile. *Hydrological Processes, 34*(2), 159–174.

Peña-Guerrero, M. D., Nauditt, A., Muñoz-Robles, C., Ribbe, L., & Meza, F. (2020). Drought impacts on water quality and potential implications for agricultural production in the Maipo River Basin, Central Chile. *Hydrological Sciences Journal, 65*(6), 1005–1021.

Rondanelli, R., Molina, A., & Falvey, M. (2015). The Atacama surface solar maximum. *Bulletin of the American Meteorological Society, 96*, 405–418. https://doi.org/10.1175/BAMS-D-13-00175.1

Salazar, Á., Thatcher, M., Goubanova, K., Bernal, P., Gutiérrez, J., & Squeo, F. (2024). CMIP6 precipitation and temperature projections for Chile. *Climate Dynamics, 62*(3), 2475–2498.

Sangüesa, C., Pizarro, R., Ingram, B., Balocchi, F., García-Chevesich, P., Pino, J., et al. (2023). Streamflow trends in Central Chile. *Hydrology, 10*(7), 144.

Valdés-Pineda, R., Valdes, J. B., Diaz, H. F., & Pizarro-Tapia, R. (2016). Analysis of spatio-temporal changes in annual and seasonal precipitation variability in South America-Chile and related ocean–atmosphere circulation patterns. *International Journal of Climatology, 36*(8), 2979–3001.

Vicente-Serrano, S. M., Beguería, S., & López-Moreno, J. I. (2010). A multiscalar drought index sensitive to global warming: The standardized precipitation evapotranspiration index. *Journal of Climate, 23*(7), 1696–1718.

Zambrano, F., Vrieling, A., Nelson, A., Meroni, M., & Tadesse, T. (2018). Prediction of drought-induced reduction of agricultural productivity in Chile from MODIS, rainfall estimates, and climate oscillation indices. *Remote Sensing of Environment, 219*, 15–30.

Chapter 5
Economic Impacts of Droughts on Agriculture and Urban Water Supply Sector

Francisco J. Fernández and Felipe Vásquez-Lavín

Abstract This chapter explores the economic effects of long-term droughts on agriculture and urban water supply sectors in Chile, with a particular focus on the ongoing megadrought that started in 2010. Using a comprehensive framework that incorporates Infrastructure Capacity (IC), the Amount of Water in Storage (AWS), and adaptation strategies, we offer a detailed analysis of resilience mechanisms and the economic impacts of drought conditions. Our findings underscore the critical balance between water availability and economic sustainability, revealing both the challenges and successes in the sectors' adaptive responses. The agricultural sector's evolution towards more sustainable practices and the urban water supply sector's strategic investments highlights significant progress in drought resilience. However, the potential for future tipping points in water scarcity underscores the urgent need for integrated water resource management, sustainable agricultural practices, and the strategic allocation of resources to prevent critical shortages. This chapter provides recommendations for managing the economic impacts of long-term droughts in Central Chile, both present and future.

Keywords Drought impacts · Economic impacts · Agriculture · Urban water supply · Chile

F. J. Fernández (✉)
Faculty of Economics and Government, Center of Economics for Sustainable Development (CEDES), Universidad San Sebastián, Santiago, Chile

Center for Climate Change and Resilience (CR2), Santiago, Chile

Center of Applied Ecology and Sustainability (CAPES), Santiago, Chile
e-mail: francisco.fernandez@uss.cl

F. Vásquez-Lavín
Center for Climate Change and Resilience (CR2), Santiago, Chile

Center of Applied Ecology and Sustainability (CAPES), Santiago, Chile

School of Business and Economics, Universidad del Desarrollo, Concepción, Chile

Instituto Milenio en Socio-ecología Costera (SECOS), Santiago, Chile

D. Rivera, G. Donoso (eds.), *Droughts in Chile*, Global Issues in Water Policy 31,
https://doi.org/10.1007/978-3-031-85040-0_5

5.1 Introduction

In Chile, droughts occur regularly and are a natural phenomenon (Fernández & Gironás, 2021). Garreaud et al. (2017) have reported that droughts (defined as years when the regional precipitation index (RPI) is below 75%) were observed for 24 years between 1915 and 2009 (about a quarter of the time). During the last century, the country experienced several severe drought episodes, notably in 1924, 1968, and 1998. Since 2010, Chile has been facing a continuous and unprecedented drought, ongoing up to 2023. Within this period, the years 2019 and 2021 have been among the most severe drought episodes on record, following the significant droughts of 1924 and 1998.

According to historical data on rainfall in Santiago, Chile, it rained only 66 mm in 1924, making it the driest year in nearly two centuries (Camus & Jaksic, 2020). The second driest year recorded in the twentieth century and part of this twenty-first century was 1998. Chile experienced a record-setting drought this year, with an average of 89 mm of rainfall. The third driest year between 1900 and the current year was recorded in 1968 when some regions had a rainfall deficit of 80%. Today, the current drought, which began in 2010, has been labeled the Central Chile megadrought due to its extraordinary duration and spatial pattern (MD; Garreaud et al., 2017). This drought has become the longest drought since the twentieth century. Its severity has not reached the levels of the 1924 or 1998 droughts. In 2019, precipitation in Chile reached an average deficit of 39%, while in 2021, it reached an average deficit of 43%, making it the second driest year since 1968 (DGA, 2022). Precipitation in the country has been decreasing at a rate of 4% per decade over the last 61 years, a trend that increases to −8% per decade when considering the previous 41 years (DGA, 2022).

Economic agents have learned significantly since 1924. Governments and private water users have become more capable of adapting to droughts. Increasing water storage infrastructure and groundwater extraction have been key responses to these events. While private actors have primarily focused on adapting technologies and practices to enhance resilience, public actors have been instrumental in making policies to prepare for and mitigate the economic impact of future droughts in the long run. Although these actions have increased Chilean resilience and water users' reaction capabilities, climate change has worsened droughts' characteristics, making them more frequent, intense, and longer-lasting.

This chapter describes the economic cost of the current megadrought in Chile, considering the changes to those factors that increase resilience to the region and the climate-driven changes in the characteristics of the droughts. We present a brief literature review of the past economic impact assessment of droughts in Chile, most related to those described in the literature as the country's most intense drought on record. Then, we present a new approach to assess the economic impacts of long-term droughts. This approach responds to the new challenges to evaluate the economic impact due to the new characteristics of the droughts driven by climate and population changes. A particular focus will be on the effects on agriculture and other highly water-dependent economic sectors, such as water supply in urban areas.

We build upon the foundation of our previous article (Fernández et al., 2023). This chapter extends the analysis presented in this paper. While the paper provided a comprehensive overview of the methodological approach and initial findings, here we delve deeper into the nuances of the economic impacts. We expand upon the initial framework to offer a more detailed examination of the effects on agriculture and other water-dependent sectors, providing richer insights into the adaptation strategies and resilience factors at play. We explore the broader implications of our findings and engage in a more thorough discussion of policy recommendations.

5.2 An Overview of Previous Economic Impacts of Droughts in Chile

Since the drought of 1924 until now, there are few records about their economic impacts. Camus and Jaksic (2020) address this issue by analyzing the literature referring to the 1924 Chilean political crisis, which was developed parallel to the harshest drought recorded over two centuries. Camus and Jaksic (2020) explore the 1924 crisis from a socio-ecological perspective and suggest that the severe drought may have contributed to the rise of military power. The worst drought in two hundred years exacerbated Chile's existing economic, political, and social crisis caused by the disappearance of the nitrate industry and the scarcity of foreign currency. The authors examine how the drought affected agriculture, livestock, and water availability, leading to famine and migration.

From an economic point of view, there is not enough information on the economic cost of that drought. Camus and Jaksic (2020) cited Urrutia and Lanza (1993) in their compilation of the "Catástrofes de Chile." Urrutia and Lanza (1993) indicate that wheat production decreased by at least one million quintals (nearly 14% of total national production) during the drought experienced in the country's central zone (between Coquimbo and Maule). The wheat losses reached one hundred and thirty quintals in the Coquimbo province (less than 1% of regional production).[1] In contrast, in the rest of the affected area, losses were estimated to be seventy percent of average production in dry lands and twenty percent on irrigated land. Furthermore, also in the province of Coquimbo, two hundred thousand sheep, two hundred fifty thousand goats, and twenty thousand cattle disappeared due to water scarcity. Railways agreed to reduce the cost of animal freight from the north to the south of Curicó by forty percent as long as the destination did not fall below one hundred and fifty kilometers from the place of departure. In the north, fallow lands could not be planted, natural grass did not develop, and branches did not grow, so there was no animal pasture.

In summary, the 1924 drought caused crop failures and limited access to water and forage, threatening food security and leading to a significant decline in

[1] Percentage losses are approximations based on Urrutia and Lanza (1993) and Mellado (1998).

agricultural production and livestock productivity. The water scarcity caused by the drought exacerbated the already vulnerable water management system, leading to a prolonged water crisis that affected the economy and people's livelihoods, particularly in rural areas.

Fernández and Gironás (2021) compiled records from several institutions about past economic impacts due to droughts. For instance, the authors indicate that the 68's drought affected six million hectares and spread over 11 provinces, from the current Atacama region in the north to the Bío-Bío region in the south. Cereal and vegetable crops and livestock were severely damaged, and it is estimated that between 225,000 and 450,000 people were left unemployed at an estimated cost of almost USD 1000 million[2] (Donoso, 2022; FAO, 2010; UNESCO, 2012). In 1998, the losses of the agronomic sector were estimated at 250 million USD, according to the National Agriculture Society (SNA). A more detailed assessment is presented by the General Directorate of Water Resources, DGA (1997) and Fernández et al. (1999). Both studies presented a detailed assessment of drought events in Chile in the 90 s and assessed the impacts on specific productive sectors based on quantitative criteria. They assessed a series of monthly deficits for 48 years to evaluate further the economic impact on the agriculture, water supply, mining-industrial and hydroelectric sectors. They estimated that the annual average losses of these sectors were 202 MM $CLP (USD 438 thousand), equivalent to 0.005% of the total national GDP in 1998.

5.3 The Economic Impact of the Current Megadrought: A New Conceptual Framework

The methodological framework depicted in Fig. 5.1, as developed by Fernández et al. (2023), represents a comprehensive approach to understanding the economic impacts of the current megadrought in Chile. This framework extends beyond traditional assessments by incorporating key factors that influence the long-term economic effects of droughts. Specifically, it evaluates the interplay between Infrastructure Capacity (IC), the Amount of Water in Storage (AWS), and the range of adaptation responses from water users. By focusing on these key factors, this approach aims to provide a comprehensive understanding of the economic impact across several economic sectors, emphasizing the effects of prolonged drought conditions on them.

The framework begins by delineating the spatial and temporal boundaries of the assessment. The spatial framework identifies the geographic scope of the evaluation, recognizing the heterogeneous impacts of droughts and emphasizing the

[2] Although there are no details on estimating this value, the book's authors specify that this value is assuming a drought equivalent to the 67-68 drought with the water demand level of 1999 (Norero & Bonilla, 1999).

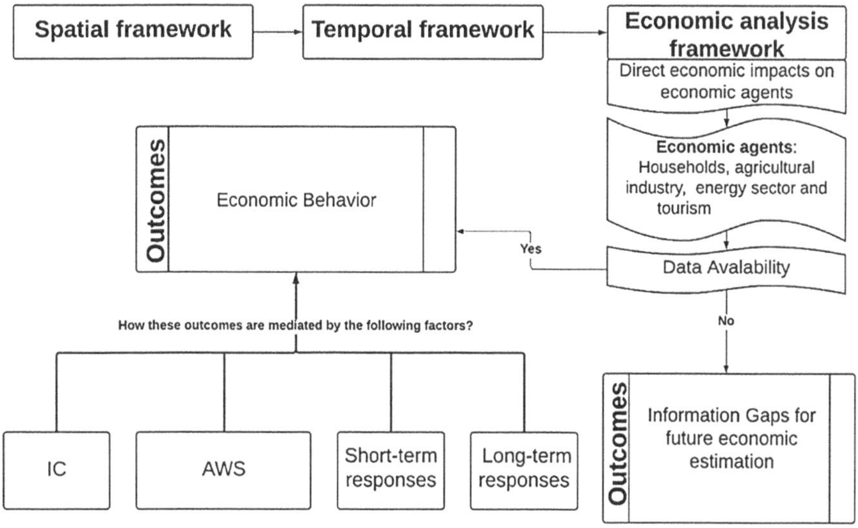

Fig. 5.1 Methodological framework. Source: Fernández et al. (2023)

importance of identifying zero-sum transfers between impacted entities, which occur when the gain of resources by one entity directly results in an equivalent loss by another, thus leaving the overall resource balance unchanged (Ding et al., 2011). The temporal aspect considers the duration of the drought and its long-term effects, acknowledging that the economic impacts may unfold over an extended period and may not be immediately apparent. Then, the framework specifies the economic sectors under assessment, which are selected based on their water dependency and significance to the national economy. The choice of methodology for economic impact assessment on each sector is guided by data availability and the necessity to reflect the economic behaviors under continuous water scarcity.

By applying this framework, the analysis seeks to estimate the drought's economic impact on each sector through diverse methodological approaches based on data availability. These estimates are expected to capture the adaptive behaviors of water users in response to sustained water deficits. This approach, mediated by IC, AWS, and both short- and long-term adaptations, offers a comprehensive perspective on the economic impact of the current megadrought. Infrastructure Capacity assesses the ability of the existing water infrastructure to mitigate the effects of drought. It includes the evaluation of dams, reservoirs, and other water management systems that contribute to the region's resilience. The AWS serves as a measure of the total volume of accessible water resources. This includes surface water bodies, like rivers and lakes, and subsurface reservoirs, known as groundwater. Distinctly, AWS is considered separately from 'Infrastructure Capacity' to underscore the unique physical presence and economic value of stored water. Unlike IC, which evaluates the infrastructure's capacity to manage water distribution and storage, AWS quantifies the actual water availability. This distinction is key for assessing the sustainability and efficiency of water use, as it allows us to address the inherent differences in how surface and groundwater are replenished, stored, and extracted. Finally, adaptation responses capture water users' short- and long-term responses to drought conditions. It includes changes in water usage patterns, shifts in agricultural practices, and other behavioral adaptations that can influence economic outcomes.

5.4 Case Study

The case study focuses on the central region of Chile, located between the Andes Cordillera and the Pacific coast, extending from 30° to 38°S. This region is particularly interesting due to being the most populous area and has faced increasing environmental and social impacts due to the MD. The temporal framework selected for the analysis spans from 2010 to 2020. This period was chosen because it marks the duration of the central Chile megadrought and data availability. The region has experienced a rainfall deficit ranging from 25% to 45% annually during this time, with significant variability in the magnitude of the deficit. The year 2019 was particularly dry, with only a quarter of the average annual precipitation.

The economic sectors selected for the study were agriculture and the urban water sector, chosen based on their water requirements, national economic importance, and the data availability for their economic assessment. With this, we covered most of the country's water demand, representing approximately 95% (FCH, 2019).

5.4.1 Agricultural Sector

For the agricultural sector, the economic impact of the drought was analyzed using a mathematical programming modeling approach, specifically Positive Mathematical Programming (PMP), calibrated with national statistics. In this case, the effects of IC, the AWS, and both short-term and long-term responses are implicitly captured by data from the sector (baseline path) during the evaluation period (from 2010 to 2020). Two scenario paths were constructed to simulate the agricultural structure from the temporal framework under conditions "without" water scarcity, isolating the effects of prices, yields, and costs and providing an economical range of the megadrought's impact. By modeling the sector's baseline conditions and simulating scenarios "with" and "without" water scarcity, we isolated the impacts of drought on agricultural economics. This comparison allowed us to estimate the economic range of drought impacts, providing insights into potential income losses and highlighting the sector's resilience and adaptive capacities in water scarcity. Table 5.1 shows a description of the baseline and scenario paths used.

Table 5.1 Scenario description for the Agricultural sector

Paths	Description	Assumptions	Focus
Baseline path	Reflects the actual agricultural activities and outcomes from 2010 to 2020	Incorporates actual data on crop yields, water availability, market prices, and production costs	To establish a reference point for evaluating the impacts of drought conditions by comparing with simulated scenarios
Scenario 1 (Scen1)	Simulates agricultural activities under conditions "without" water scarcity, isolating the effects of changes in crop prices, and production costs	Assumes no change in water availability compared to baseline levels, allowing for the isolation of economic variables	To assess the economic impact of drought by analyzing how changes in prices, and costs would have affected agriculture in the absence of water scarcity
Scenario 2 (Scen2)	Similar to Scen1 but includes changes on yields from 2010 and 2020	Assumes no change in water availability compared to baseline levels, allowing for the isolation of economic and productivity variables (yields)	To explore the economic impact of drought by analyzing how changes in prices, yields and costs would have affected agriculture in the absence of water scarcity

5.4.2 *Urban Drinking Water Sector*

For the urban drinking water sector, our economic impact assessment focused on the supply side, represented by sanitation companies, and the demand side, represented by household water consumption. On the supply side, we estimated the megadrought-induced costs by analyzing sanitation companies' investments in response to the drought.[3] These investments encompassed various measures, including developing new water sources, enhancing water treatment capacities, and expanding the distribution network. By quantifying these investments, we assessed the financial burden placed on sanitation companies due to the drought and its implications for the urban water supply infrastructure.

On the demand side, we conducted simulations of household water usage incorporating climatic factors such as temperature, rainfall, and the standardized deviation of the accumulated Normalized Difference Vegetation Index (zcNDVI). Our study included data from 215 communes in Central Chile from 2010–2020. We applied an Ordinary Least Squares (OLS) methodology and a fixed-effects (FE) panel data approach in our analysis. To refine our estimates, we utilized data on drinking water bills from sanitation companies from 2010 to 2020. The developed model for water demand was employed to forecast average household water use, determine the consumer surplus, and assess the impact of water rationing scenarios on household well-being, which is in line with the "Water Rationing Protocol for

[3] Data used for calculations is unavailable for public use due to a confidentiality agreement with the participating firms.

Greater Santiago. By evaluating the consumer surplus and the effects of potential water rationing scenarios on household welfare, we were able to gauge the economic impact of drought on urban residents. This dual-focus approach on both supply and demand sides allowed us to assess the economic implications of drought on the urban drinking water sector, highlighting the sector's vulnerability to drought conditions and the economic costs associated with adapting to decreased water availability. The study assumes an initial equilibrium between water demand and supply in 2020, including the responses of sanitation companies to the megadrought through investments related to IC and AWS.

5.5 Results

5.5.1 Impact on the Agricultural Sector

5.5.1.1 Baseline Change from 2010 to 2020

According to our model, when comparing 2010 and 2020, we observed a decrease in net water use of about 15.3% and a decrease in total agricultural land of 13.5% (Table 5.2). At a more detailed level, our model shows that dryland and irrigated land decreased by 3.1% and 15.7%, respectively. Although we cannot establish causality, drought is probably a significant factor in decreasing irrigated land, where surface irrigation areas such as laying and furrows have presented significant losses.

Table 5.2 Agricultural changes between 2010 and 2020

Description	2010	2020	% change
Net water use (10^9 m^3)	**67.05**	**10.27 reduction**	**−15.3%**
Total land (ha)	**472,762**	**63,919 loss**	**−13.5%**
By system			
Irrigated land (ha)	388,189	61,258 loss	−15.7%
Dryland (ha)	84,572	2661 loss	−3.1%
By subsector			
Cereals (ha)	246,326	47,169 loss	−19.1%
Fruits (ha)	175,656	21,144 increase	+ 12.0%
Legumes and tubers (ha)	30,916	25,351 loss	−81.9%
Industrial (ha)	19,863	12,543 loss	−63.1%
Income (million USD)	**3.59**	**6.66 increase**	**+ 185%**
By subsector			
Cereals (million USD)	1.58	0.62 increase	+ 37.1%
Fruits (million USD)	1.88	4.9 increase	+ 326.2%
Legumes and tubers (million USD)	0.09	0.06 loss	−69.1%
Industrial (million USD)	0.04	0.004 loss	−10.1%
Labor Demand (Working days/year)	**20,587,618**	**1,309,189 increase**	**+6.3%**

Source: Fernández et al. (2023)

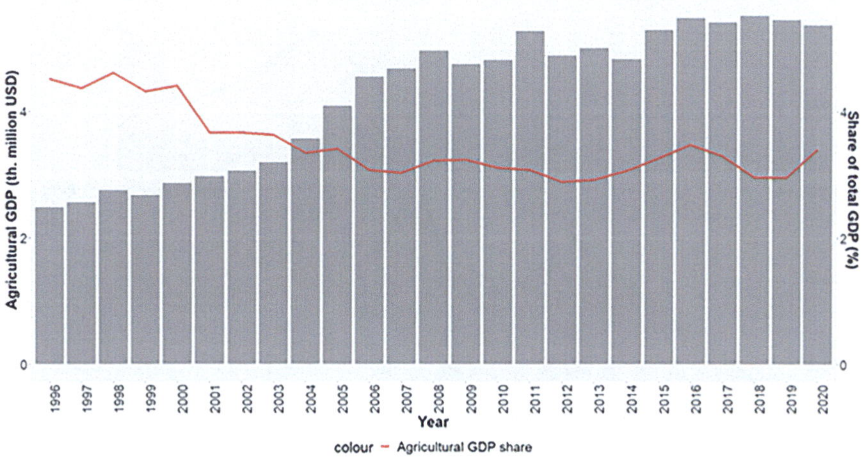

Fig. 5.2 Agricultural, livestock, and forestry chain-linked GDP at 2013 prices (bars) and Sectorial proportion in GDP (red line). Source: https://si3.bcentral.cl/siete

However, sprinkler irrigation and micro-irrigation have increased their surface. Our model also indicates substantial decreases in the surface of cereals (19%), legumes (81.9%), and Industrial crops (63%). On the other hand, fruit trees increased by more than 21 thousand hectares.

Based on agricultural statistics from 2010 to 2020, our current path is in line with the trend. The latest Agricultural and Forestry Census (INE, 2022) shows that the agricultural area in the country has decreased by 8.1% (4,165,209 ha) since 2007, and the irrigated area has decreased by almost 19% since 2007. Despite this, Chile's agricultural GDP increased from 4810 to 5361 (million USD) between 2010 and 2020, maintaining its sectoral contribution to the country's GDP at around 3% throughout the current MD (Banco Central, 2021) (Fig. 5.2).

5.5.1.2 Scenario Analysis

To determine the economic impact of MD, we tested the isolated and combined effects of crop prices, costs, and yield evolutions from 2010 to 2020. Figure 5.3 illustrates the income losses when comparing the agricultural income of the baseline path, which is the actual activity between 2010 and 2020, with two scenarios. These scenarios assume the activity levels of the 2010s and the evolution of prices, costs, and yields from 2011 to 2020. We assume that the difference between the baseline path and the scenarios' paths represents the income losses of the agricultural sector due to factors like MD and water scarcity without considering the impact of income, prices, and yield evolution.

Figure 5.3 illustrates that the gap between the baseline path and both Scen1 and Scen2 has been widening over time since 2016. This can be attributed to the

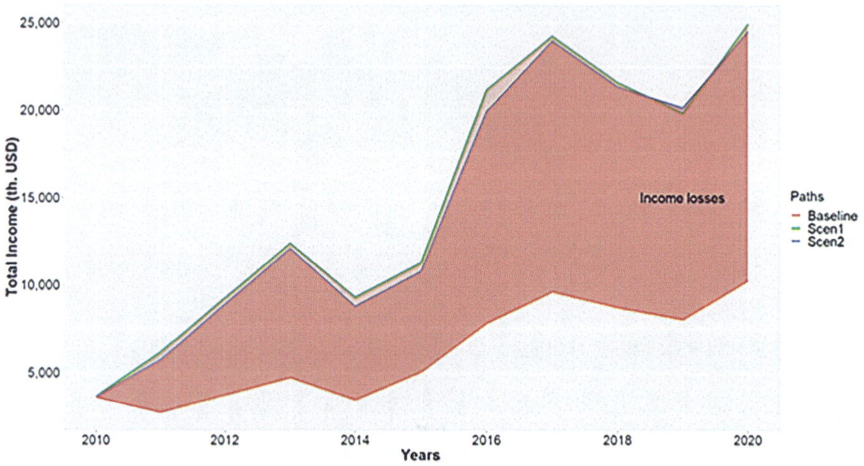

Fig. 5.3 Total agricultural income 2010–2020 for the baseline path and both scenarios (The area between the baseline path and scenarios indicates income losses). Source: Fernández et al. (2023)

Infrastructure Capacity (IC) the set of responses available to cope with the megadrought (MD). In the initial year of the MD, the loss of income seems to be insignificant, which could be due to the water systems, such as groundwater or dams, having sufficient reserves (AWS). However, with time, the ability to achieve the potential income becomes more challenging, as water reserves continue to decrease. The cumulative income loss has been estimated to be between 90.46 million USD (Scen 1) and 86.76 million USD (Scen 2).

It is estimated that Chile spent around 160 million USD on agricultural emergency response between 2008 and 2017. Of this amount, 64% was allocated to public emergency responses for drought, according to FAO's report in 2021 (FAO, 2021). Based on this estimate, the figure could potentially rise to 196.45 million USD, which amounts to 0.036% of the Agricultural GDP from 2020.

5.5.1.3 Regional Impact

At regional level, Fig. 5.4 shows that in the northernmost regions represented by Coquimbo, Valparaíso and Metropolitan region, the agricultural structure shows a predominant focus on fruit cultivation. In contrast, the regions of O'Higgins, Maule and Bio Bio display a more diverse agricultural structure with significant areas dedicated to cereals and fruits. Figure 5.4 suggests that during the megadrought period, there might have been a transition in agricultural structure, possibly as an adaptive response to the prolonged dry conditions.

Within the northern regions of central Chile, for instance, in the Coquimbo and Valparaíso regions, there was a decrease across all crop groups between 2010 and

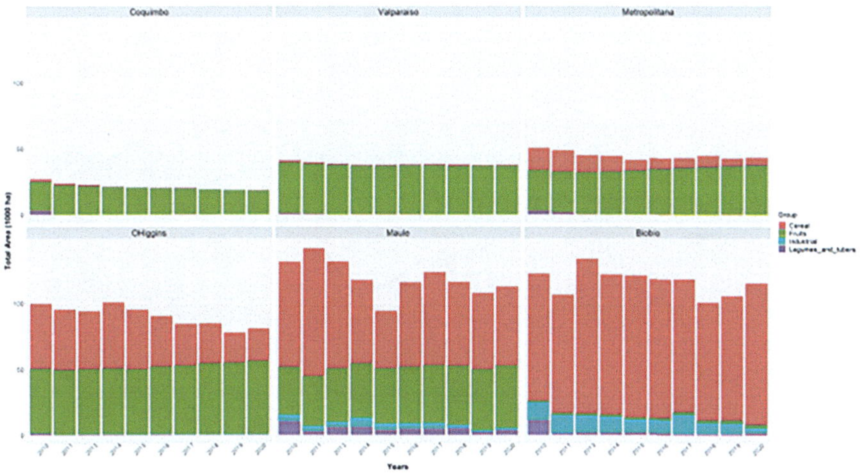

Fig. 5.4 Agricultural structure by region and year

2020. This decrease for all groups leads to an overall decrease in the total agricultural land in Coquimbo and Valparaíso of 32.4% (8790 ha approx.) and 8.2% (3390 ha approx.), respectively. Further south, within the Metropolitan region, similar trends are observed except for fruits that experienced a 23.0% increase, indicating a shift towards fruit cultivation, with an overall decrease in total agricultural land of 14%.

Within the southern regions, the O'Higgins and Maule regions presented a mixed picture, with a decrease in cereals, legumes and tubers, and industrial crops but increasing in fruits by 13.8% and 31.5%, respectively. The Biobio region is the only one where land for cereals increased (11.8%), together with a 52.8% increase in fruits. However, legumes and tubers decreased by 66.1%, and industrial crops by 85.8%, contributing to a total decrease in land use of 5.6%.

In economic terms, there is also a spatial variability in economic impacts attributed to MD. Figure 5.5 shows this variability among regions. In regions such as Coquimbo and Valparaíso, the income in the baseline path and the scenarios move closely together until around 2016, when scenarios begin to diverge slightly from the baseline, suggesting increased economic impact from MD. However, the gap between both regions' baseline and scenarios paths indicates a steady income loss due to MD. On the other hand, O'Higgins and Maule present scenarios that show a sharp deviation from the baseline, indicating a considerable economic impact induced by MD. Conversely, regions like Biobio manifest a more contained impact compared with the nearest regions (Maule). However, since 2015–2016 the divergence between the baseline path and Scen1 and Scen2 has been increasing.

Table 5.3 shows the tabulated income losses by region to enhance the interpretation of Fig. 5.5. The total income losses in Coquimbo stand at −3.88 million

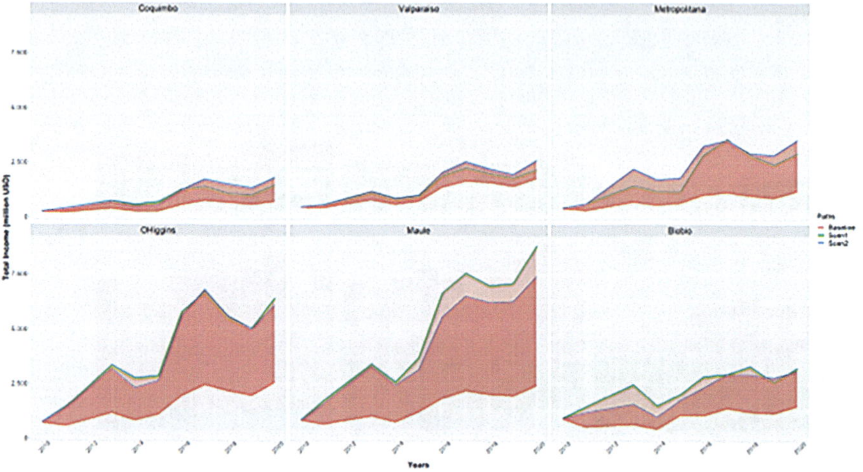

Fig. 5.5 Income losses by region

Table 5.3 Tabulated income losses by region and between scenarios

Region	Total Income loss Baseline—Scen 1 (million USD)	Total Income loss Baseline—Scen 2 (million USD)
Coquimbo	−3.88	−5.13
Valparaiso	−2.99	−4.49
Metropolitana	−11.27	−14.73
O'Higgins	−24.79	−23.68
Maule	−34.24	−28.21
BioBio	−13.26	−10.51
Total	−90.46	−86.76

USD for Scen1 and −5.13 million USD for Scen2. In Valparaiso, losses are quantified at −2.99 million USD for Scen1 and −4.49 million USD for Scen2. Notably, Metropolitan, O'Higgins, and Maule regions exhibit more pronounced financial repercussions. Metropolitan region faces −11.27 million USD and −14.73 million USD losses under Scen1 and Scen2 respectively, while O'Higgins encounters the highest discrepancy between the scenarios with losses at −24.79 million USD for Scen1, slightly higher than −23.68 million USD for Scen2. The Maule region exhibits the most considerable losses at −34.24 million USD for Scen1 and −28.21 million USD for Scen2, signaling the region's heightened vulnerability to MD. Finally, BioBio's losses are noted at −13.26 million USD for Scen1 and −10.51 million USD for Scen2, indicating a lesser yet significant economic impact.

5.5.2 Impact on Urban Water Supply

5.5.2.1 Induced Costs of Sanitation Companies

Sanitation companies in Chile have experienced an increase in capital and operational costs due to MD between 2010 and 2020. These costs were classified based on their stage within the process as an investment for production, distribution, or "others" (like studies or public awareness campaigns for rational use). At a more detailed level, the production costs include expenses related to acquiring new water sources, such as desalination, storage facilities, increased water volumes through water treatment, or purchasing and leasing water rights. While the distribution costs include pipe extension, pumping capacity, and other water transport activities. Figure 5.6 shows that the weight given to each category by companies varies depending on their location.

It is observed that production constitutes most of the expenditure across all zones, with particularly high prevalence in North and Central zones, with 97.2% and 84.1%, respectively. New water sources (desalination plants) and groundwater treatment capacity for companies in the northern zone increase production costs. In the central zone, surface storage, surface water treatment capacity, and well deepening and drilling account for a significant portion of the expenditure. The South zone exhibits a relatively balanced distribution between production (58.9%) and distribution (41.1%), their major transport, water rights, and wastewater treatment costs.

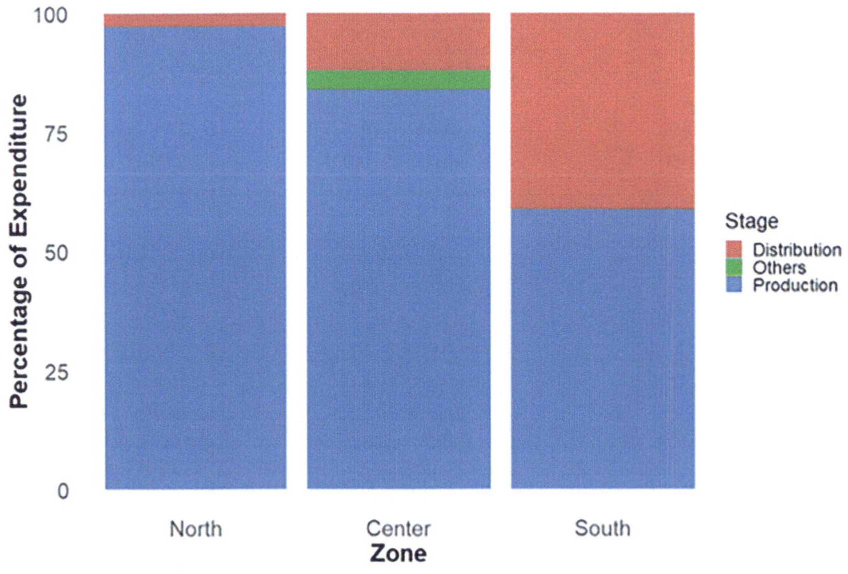

Fig. 5.6 Expenditure comparison by stage and zone

The capital expenditure between 2010 and 2020 was 1119 million USD. However, when operational costs for the drought are included, the amount comes to 318 million USD, for a total of 1437 million USD—considering only the facilities in Chile's Central area, which had a capital expenditure of 660 million USD. Operational expenses were 155 million USD, implying total expenditures of 815 million USD.

5.5.2.2 Urban Water Consumption—Avoided Costs

Table 5.4 shows the results of the OLS and RE model concerning water demand. The direction of the price effect on residential water use is as expected in all models. According to previous research (Acuña et al., 2020; Fercovic et al., 2019; Vásquez Lavín et al., 2017), all elasticities are inelastic (from −0.16 to −0.35).

Table 5.4 shows that income positively and significantly affects water use. The climatic variables that were examined had the expected signs. Precipitation showed a negative effect, while temperature increases positively impacted residential water consumption. Specifically, temperature increases led to a 9% to 15% rise in water demand compared to average temperatures between 0° and 10° Celsius. On the contrary, a positive precipitation trend reduced water consumption by 2–4%. The findings suggest that accumulated rainfall between 100 and 200 mm decreases water demand by about 4% compared to the base category (i.e., between 0 and

Table 5.4 Estimation results. Standard errors are in parentheses

	Spatial framework (30°–38°S)	
Variable	OLS	Panel RE
Price	−0.35*** (0.01)	−0.16*** (0.00)
Income	0.35*** (0.01)	0.08*** (0.01)
Population	0.01*** (0.00)	0.09*** (0.01)
Rainfall		
Between 100 and 200	−0.09*** (0.01)	−0.04*** (0.00)
Between 200 and 300	−0.04*** (0.01)	−0.02*** (0.01)
Greater than 300	−0.03** (0.01)	−0.03** (0.01)
Temperature		
Between 10 and 20	0.06*** (0.01)	0.09*** (0.00)
Greater than 20	0.07*** (0.01)	0.15*** (0.01)
zcNDVI		
Less than −1	0.03*** (0.01)	−0.01** (0.00)
Between −1 and −0.1	0.06*** (0.01)	−0.01 (0.00)
Between 0.1 and 1	0.03*** (0.01)	0.00 (0.00)
Greater than 1	0.01** (0.01)	0.02*** (0.01)
Constant	0.39*** (0.01)	1.58*** (0.08)
Observations	23,538	23,538
N		215

Source: Own elaboration

Statistical significance is indicated by asterisks: ***$p < 0.01$, **$p < 0.05$, *$p < 0.1$

100 mm of accumulated rainfall). Similarly, accumulated rainfall between 200 and 300 mm decreased demand by 2% concerning the base category. Furthermore, accumulated rainfall above 300 mm reduced water demand by approximately 3% compared to low levels of rainfall (i.e., between 0 and 100 mm of accumulated rainfall).

The zcNDVI variable indicates water consumption behavior during a vegetational drought. It is an indirect measure of plant water status, where higher zcNDVI values indicate higher water consumption. In this case, we use it to indicate human water consumption related to the irrigation of yards, gardens, and green areas in urban spaces. If the vegetation index zcNDVI is greater than 1, then residential water consumption increases by around 2%. However, if it is less than 1, then water consumption drops by around 1%.

Figure 5.7 shows the impact of potential rationing on welfare loss. The variable W_0 represents the current household water demand, and W_{MD} represents the water supply constraint under a 20% rationing scenario. The variable P_0 is the fixed water price due to institutional constraints. W_{c0} represents the consumption of households in commune c in the initial scenario, while W_{cMD} represents the water consumption of households in commune c under the rationing scenario. Thus, households will use less water under a 20% rationing scenario, and the water consumption in this scenario is W_{MD}. Since P_0 is fixed, a virtual water price P_v is necessary to calculate consumer surplus loss in commune c.

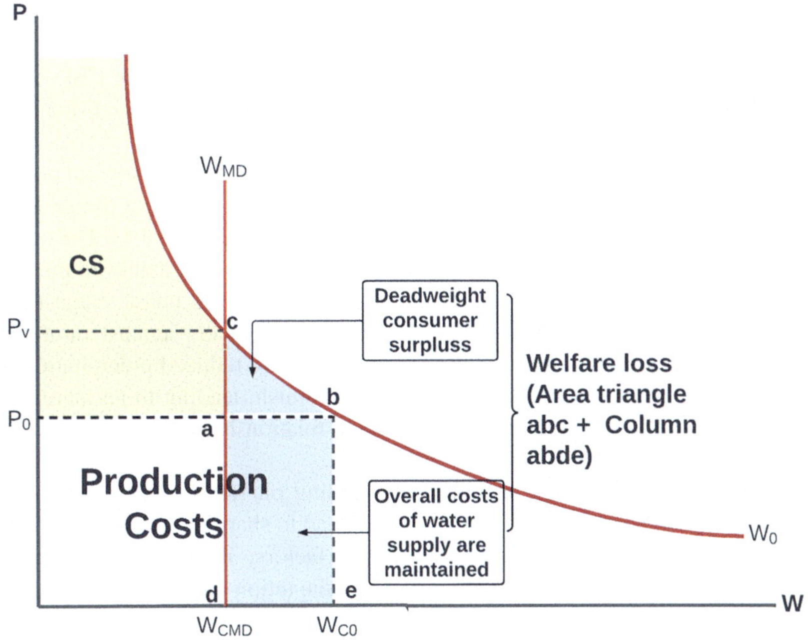

Fig. 5.7 Household-level water demand and consumer surplus

The triangle *abc* represents household welfare loss under a resource rationing scenario. Household welfare decreases by 122 USD per year. Additionally, it is important to mention that rectangle *abed* is generally not a saving in production costs. This is because variable costs comprise only a small portion of water supply costs. Thus, the conventional assumption that when less water is delivered, the cost of water supply operations falls in approximately the same proportion is unlikely to be held in this case. The rectangle *abed* amounts to a welfare loss of 94 USD per year, bringing the total to 217 USD of annual welfare loss per household. Considering the spatial framework of interest, the avoided cost per year, represented by the total loss of surplus, corresponds to 542.40 million USD per year.

We compare the company's investments in the study area's most populous region with the expected annual economic impact under the 20% rationing scenario. Considering the 217 USD/household/year welfare loss under the 20% restriction, our results show that a dollar invested in drought preparedness can prevent approximately 8,3 dollars in economic losses.

5.6 Total Economic Impact—Resilience and Adaptation Strategies

Our research findings indicate that central Chile has demonstrated a high level of resilience against the negative economic impacts of a drought that has persisted for over 10 years. When we combined the economic impact of the agricultural sector with the investments made by sanitary companies to tackle the drought in the region, we discovered an accumulated economic impact of $1011.45 million USD within the defined analysis timeframe. This figure represents less than 0.5% of the country's GDP.[4]

Despite this marginal economic effect, a closer examination of historical precipitation patterns, the depletion of water in storage, and economic cost trends across different sectors reveals a significant negative correlation (Fig. 5.8). During the early stages of the drought, economic costs were relatively low, thanks to the near-full capacity of the Amount of Water in Storage (AWS). This phase suggests that sufficient water reserves could protect the economic sectors against immediate drought impacts, maintaining economic stability despite reduced precipitation. As the drought persisted, water reserves began to diminish, leading to an increase in economic costs. This trend highlights the critical relationship between water availability and economic resilience.

The analysis points to the necessity of focusing on future "tipping points"—moments when substantial water scarcity could lead to sharp increases in economic costs. These tipping points hinge on several factors, including Infrastructure Capacity (IC), AWS, and the effectiveness of adaptation strategies in mitigating

[4] GDP (2020) 252.9 billion USD.

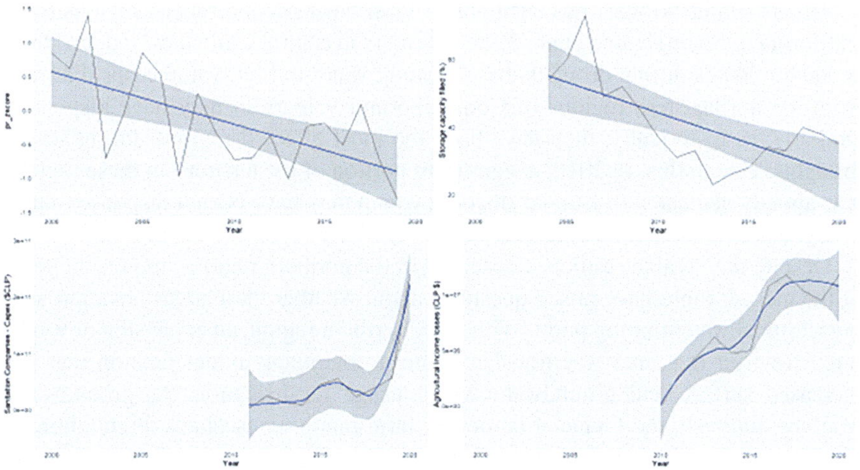

Fig. 5.8 Water supply (precipitation and stored water) and economic impacts (sanitation companies' investments and Agricultural income losses). pr_zscore: precipitation z-scores calculated as $\left({pp_{y} - \overline{pp}_{1979-2019}} \big/ {s_{1979-2019}} \right)$

drought impacts. However, the current gaps in hydrological data and information hinder the precise prediction of these tipping points. This limitation affects the accuracy of economic cost projections and challenges the development of robust adaptation strategies.

After conducting a detailed analysis of the drought's impact at the sectoral level and disaggregating the data by region, we identified effects that were not apparent at the aggregate level. In the agriculture sector, we observed that the economic impacts in the northern regions were less severe in economic terms despite being the regions most affected by the lack of precipitation and decrease of water in storage in reservoirs during the droughts. As we move towards the south, we observe an increase in the economic impact until the Bio Bío region, where the economic impacts decrease again. The IC, AWS, and the effectiveness of adaptation strategies are key factors that explain this phenomenon.

From the perspective of IC, the surface water storage capacity is concentrated in the northern zone of central Chile. In the last hundred years, more than 17 reservoirs have been built north of Santiago, adding 112,420 hectares of irrigation (Haindl, 2022). Additionally, from the perspective of the AWS, the agricultural sector, has significantly buffered the MD economic impacts through groundwater use (Donoso, 2021). Groundwater has helped farmers adjust to surface water imbalance in the short term. From the effectiveness of adaptation strategies perspective, irrigation efficiency has increased its average efficiency (Martin & Saavedra, 2018), and resources have been allocated more efficiently economically to higher-value crops. This means that the productivity structure has changed from annual crops such as cereals, legumes, and tubers to perennial activities such as fruits, which are commonly associated with significant profitability.

This is similar to what has been observed in other Mediterranean regions (e.g., California or Spain) (Lund et al., 2018). Farmers in central Chile have reduced their economic losses during droughts by allocating water to high-value crops. Farmers from the northeastern regions of Coquimbo and Valparaíso may have responded similarly, as historically, they have been the most affected regions during severe droughts. Due to this, in 2010, a significant portion of the territory in these regions was already dedicated to fruit production, which may have been a response to past droughts and an action that increased their resilience to the current drought.

Despite the notable resilience observed in the northern regions, the recent years of the megadrought have raised questions about whether these areas are approaching a critical water tipping point. Within the northern region, an expansion of water-intensive fruit tree area was translated into a continuous intensification that has increased surface and groundwater water usage for irrigation. As groundwater reserves diminish, the financial burden of farm pumping escalates, highlighting a direct correlation between decreasing water availability and increasing operational costs (Rinaudo & Donoso, 2018; Suárez et al., 2014). Moreover, this overexploitation of surface and groundwater resources has led to reservoirs being close to the dead pool and lower groundwater tables, putting more pressure on pumping costs over time. Within the last years of the period of analysis, farmers have been forced to cut down their trees or leave crops unharvested due to the high costs. This has been observed within the Limarí Valley in the Coquimbo region during the last years of the analyzed period (Nauditt et al., 2021).

We observed substantial economic impacts from the MD among the central regions, primarily Maule and O'Higgins. Unlike the northern regions, which have historically experienced the most severe impacts of drought and consequently adapted their agricultural structures towards less water-dependent crops, Maule and O'Higgins have relied significantly on annual crops. This reliance has rendered these regions particularly vulnerable to the adverse effects of droughts. The historical agricultural structure of Maule and O'Higgins, characterized by a diverse mix of cereals, legumes, and other annual crops, has thus faced significant challenges in the wake of increasing drought frequency and severity.

We observe that the regions of Maule and O'Higgins are embarking on a similar path of structural change as witnessed in the northern regions, transitioning towards perennial crops and adopting more water-efficient agricultural practices. However, the urgency and complexity of recent droughts highlight the need to refine these strategies, ensuring that Maule and O'Higgins do not replicate the north's challenges. In this context, it becomes crucial to prioritize sustainable practices in expanding high-value crop areas and enhancing irrigation efficiency. Careful management of water resources, including the avoidance of aquifer overexploitation and the efficient utilization of the AWS, is imperative to sustain the agricultural industry's long-term viability.

Within the urban water supply sector, we highlight three major findings. First, the regional differences related with the induced costs associated with drought mitigation efforts, reflect the context-specific characteristics of resilience and adaptation. Secondly, a comparison of avoided costs against induced costs reveals the economic efficacy of short- and long-term measures taken by water supply companies. The investments made in enhancing the water supply infrastructure—notably in production and distribution—have been key in mitigating the economic impact of the drought. These investments can be seen as preemptive actions against potential water scarcity, showing how current strategies have significantly saved resources by preempting more severe economic impacts. Finally, despite the industry's resilience, a threshold exists beyond which the cost of drought mitigation could increase exponentially. This threshold is closely tied to the eventuality of water restrictions—a scenario that Santiago was perilously close to in April 2022. Such a scenario underscores the critical balance between current adaptation strategies and the potential for future water scarcity.

5.7 Policy Recommendations and Future Directions

The adaptation strategies observed in the northern regions, now mirrored in the Maule and O'Higgins regions, emphasize the necessity of sustainable agricultural practices. The transition to higher-value crops should be complemented by educational programs for farmers on the benefits of sustainable practices. It is vital to ensure that this shift considers the long-term viability of water resources, avoiding the pitfalls experienced due to overexploitation in the north.

Developing and implementing an Integrated Water Resource Management plan that takes a holistic view of water resources, including both surface and groundwater, is crucial. This approach requires a collaborative effort from all stakeholders, such as governments, communities, industries, and agricultural sectors, to work towards shared water use objectives. It is essential to ensure that water allocation decisions are made transparently and sustainably. Moreover, the effectiveness of adaptation strategies, particularly in the urban water supply sector, illustrates the potential of economic incentives in promoting water-saving technologies and practices. Implementing a framework that offers financial benefits for water efficiency can significantly impact both agricultural and urban sectors.

The looming threat of reaching critical water tipping points, accentuated by the near-crisis scenario in Santiago in April 2022, highlights the importance of robust hydrological monitoring and forecasting. Investing in advanced systems for real-time water availability assessment and enhancing predictive models for drought impacts can significantly improve resilience. Strengthening this area involves technological upgrades and a commitment to open data policies that allow for the free flow of information. This enables more accurate planning and response strategies, ensuring that adaptation measures are informed by the most current and comprehensive data available.

5.8 Conclusion

This chapter has explored the economic impacts of droughts on agriculture and the urban water supply sector in Chile, underscored by the megadrought from 2010 to the present. We have identified key insights into resilience and adaptation mechanisms by analyzing historical data, the adaptation strategies of various sectors, and the implementation of an original conceptual framework for the analysis of long-term droughts. Our study reveals that, despite the substantial challenges posed by prolonged drought conditions, there have been significant steps in building resilience across the agricultural and urban water supply sectors.

Central to our findings is the interplay between Infrastructure Capacity, the Amount of Water in Storage, and adaptation strategies that have collectively underpinned the economic sectors' response to drought. The agricultural sector's gradual transition towards more drought-resilient crops and efficient irrigation practices mirrors a broader shift towards sustainability, informed by the hard lessons learned from past drought events. Similarly, the urban water supply sector's proactive investments and demand management strategies underscore a strategic approach to ensuring water security in the face of increasing scarcity.

Yet, our study also highlights the looming threat of reaching critical water scarcity tipping points, which could precipitate sharp increases in economic costs and test the limits of current resilience and adaptation strategies. Therefore, the need for robust, forward-looking water management policies is more pressing than ever, emphasizing sustainable practices, integrated water resource management, and the critical role of hydrological data and forecasting in informed decision-making.

As Chile continues to navigate the challenges posed by climate change and its implications for water availability, the lessons drawn from this megadrought provide invaluable insights for future drought resilience and adaptation efforts. The path forward necessitates a joint effort from all stakeholders to prioritize sustainable water use, invest in adaptive infrastructure, and foster a culture of innovation and collaboration in drought response strategies.

Acknowledgments We gratefully acknowledge support from the Center of Applied Ecology and Sustainability (CAPES), the ANID PIA/BASAL AFB240003, and Fondecyt Iniciación 11230334.

References

Acuña, G. I., Echeverría, C., Godoy, A., & Vásquez, F. (2020). The role of climate variability in convergence of residential water consumption across Chilean localities. *Environmental Economics and Policy Studies, 22*(1), 89–108. https://doi.org/10.1007/S10018-019-00249-3/TABLES/7

Banco Central. (2021). *Producto interno bruto por actividad económica, volumen a precios año anterior encadenado, series empalmadas, referencia 2013 (miles de millones de pesos encadenados).* Cuentas Nacionales. https://si3.bcentral.cl/Siete/ES/Siete/Cuadro/CAP_

CCNN/MN_CCNN76/CCNN2013_P2/CCNN2013_P2?cbFechaInicio=1996&cbFechaTermino=2021&cbFrecuencia=ANNUAL&cbCalculo=NONE&cbFechaBase=.
Camus, P., & Jaksic, F. (2020). La extraordinaria sequía de 1924: Crisis socio-ecológica e irrupción del poder militar en Chile. *Revista de geografía Norte Grande, 2020*(77), 397–416. https://doi.org/10.4067/S0718-34022020000300397
DGA. (1997). Impacto Económico de las Sequías. In *Análisis Impacto Económico Originado por Sequías Hidrológicas* (p. 152). Ministerio de Obras Públicas.
DGA. (2022). *Reporte Anual de la Evolución del Clima en Chile*, p. 46. https://cambioclimatico.mma.gob.cl/wp-content/uploads/2023/05/reporteEvolucionClima2022.pdf
Ding, Y., Hayes, M. J., & Widhalm, M. (2011). Measuring economic impacts of drought: A review and discussion. *Disaster Prevention and Management, 20*(4), 434–446. https://doi.org/10.1108/09653561111161752
Donoso, G. (2021). Management of water resources in agriculture in Chile and its challenges. *International Journal of Agriculture and Natural Resources, 48*(3), 171–185. https://doi.org/10.7764/ijanr.v48i3.2328
Donoso, G. (2022). Gestión de los recursos hídricos en la agricultura y sus desafíos. In A. Valdés, F. William, J. Ortega, R. Pérez, & G. Vargas (Eds.), *Perspectivas y desafíos de la agricultura y desarrollo rural en Chile* (pp. 194–231).
FAO. (2010). *Gestión Del Riesgo De Sequía Y Otros Eventos Climáticos Extremos En Chile*. https://bibliotecadigital.ciren.cl/items/364eec8a-0dc4-4d24-a06c-17c90c8cabbf
FAO. (2021). *Chile—Building agricultural resilience to climate risks. Case study.* pp. 60–60. https://doi.org/10.4060/cb7062en.
FCH. (2019). *Radiografía del Agua. Brecha y Riesgo Hídrico en Chile.* pp. 144. https://escenarioshidricos.cl/wp-content/uploads/2020/06/radiografia-del-agua-1.pdf
Fercovic, J., Foster, W., & Melo, O. (2019). Economic development and residential water consumption in Chile. *Environment and Development Economics, 24*(1), 23–46. https://doi.org/10.1017/S1355770X18000463
Fernández, B., & Gironás, J. (2021). Droughts. In B. Fernández & J. Gironás (Eds.), *Water resources of Chile* (Vol. 8, pp. 173–187). Springer. https://doi.org/10.1007/978-3-030-56901-3_8
Fernández, B., Donoso, G., Luraschi, M., Orphanópoulos, D., & Salazar, C. (1999). Estimacion Del Impacto Economico Asociado A Sequias Hidrologicas. *VI Jornadas del CONAPHI-CHILE*, 16.
Fernández, F. J., Vásquez-Lavín, F., Ponce, R. D., Garreaud, R., Hernández, F., Link, O., Zambrano, F., & Hanemann, M. (2023). The economics impacts of long-run droughts: Challenges, gaps, and way forward. *Journal of Environmental Management, 344*, 118726. https://doi.org/10.1016/j.jenvman.2023.118726
Garreaud, R. D., Alvarez-Garreton, C., Barichivich, J., Pablo Boisier, J., Christie, D., Galleguillos, M., LeQuesne, C., McPhee, J., & Zambrano-Bigiarini, M. (2017). The 2010-2015 megadrought in Central Chile: Impacts on regional hydroclimate and vegetation. *Hydrology and Earth System Sciences, 21*(12), 6307–6327. https://doi.org/10.5194/HESS-21-6307-2017
Haindl, E. (2022). Los embalses y el desarrollo agrícola de Chile. In *Perspectivas y desafíos de la agricultura y desarrollo rural en Chile*. Ediciones UC.
INE. (2022). *VIII Censo Agropecuario y Forestal*. Resultados preliminares enero 2022. https://www.ine.cl/censoagropecuario
Lund, J., Medellin-Azuara, J., Durand, J., & Stone, K. (2018). Lessons from California's 2012–2016 drought. *Journal of Water Resources Planning and Management, 144*(10), –04018067. https://doi.org/10.1061/(asce)wr.1943-5452.0000984
Martin, F., & Saavedra, F. (2018). Irrigated agriculture. In G. Donoso (Ed.), *Water policy in Chile* (pp. 165–178). Springer. https://doi.org/10.1007/978-3-319-76702-4
Mellado, M. (1998). Análisis del Cultivo del Trigo en Chile durante el siglo Veinte. *Agricultura Técnica, 58*(3), 230–240.

Nauditt, A., Sycz, J., & Ribbe, L. (2021). The Limarí River basin. En J. Schmandt, A. Kibaroglu, R. Buono, & S. Thomas (Eds.), Sustainability of engineered Rivers in arid lands: Challenge and response (pp. 152-163). Cambridge University Press.

Norero, A., & Bonilla, C. (1999). *Las Sequías en Chile: Causas, Consecuencias y Mitigación*. Colección en agricultura, Facultad de Agronomía e Ingeniería Forestal, Universidad Católica de Chile. 128 pp.

Rinaudo, J. D., & Donoso, G. (2018). State, market or community failure? Untangling the determinants of groundwater depletion in Copiapó (Chile), *35*(2), 283–304. https://doi.org/10.1080/07900627.2017.1417116

Suárez, F., Muñoz, J. F., Fernández, B., Dorsaz, J. M., Hunter, C. K., Karavitis, C. A., & Gironás, J. (2014). Integrated water resource management and energy requirements for water supply in the Copiapó River basin, Chile. *Water, 6*(9), 2590–2613. https://doi.org/10.3390/W6092590

UNESCO. (2012). *Análisis de riesgos de desastres en Chile*. https://unesdoc.unesco.org/ark:/48223/pf0000232991

Urrutia, R., & Lanza, C. (1993). *Catástrofes de Chile 1541—1992* (Primera). La Noria. https://bibliotecadigital.ciren.cl/items/7d3f3df0-0659-430b-8ce0-a4084e249865

Vásquez Lavín, F. A., Hernandez, J. I., Ponce, R. D., & Orrego, S. A. (2017). Functional forms and price elasticities in a discrete continuous choice model of the residential water demand. *Water Resources Research, 53*(7), 6296–6311. https://doi.org/10.1002/2016WR020250

Chapter 6
Impacts of Droughts on Water Quality: Processes and Monitoring

José A. Díaz, Mauricio Montecinos, Alejandra Vega, Jorge Gironás, María Molinos, and Pablo Pastén

Abstract Water uses may be threatened by hydrochemical shifts that are triggered by droughts. Reduced rainfall, streamflow, and temperature fluctuations may induce distinct changes in salinity, suspended solids, turbidity, dissolved oxygen, organic matter, nutrient cycling, pH, stratification, and hydraulic residence times. The extent of these variations is determined by the drought stage, location and altitude within the watershed, and land use. The processes leading to these changes are described and the implications for water quality monitoring and socioeconomic impacts in Chilean Andean watersheds are discussed. A coordinated monitoring approach is needed to further reveal the intricate nexus between changing climatic patterns, drought events, and the ensuing consequences for water quality within Chilean Andean watersheds.

Keywords Water quality monitoring · Water quality consequences · Hydrochemical shifts · Chilean Andean watersheds.

6.1 Introduction: Processes and Impacts of Droughts

This chapter discusses the implications of droughts on water quality, focusing on aspects that could be relevant to Chilean watersheds. The range of human activities and ecosystems that water can support is determined by its quantity and quality. Water quality is defined by a set of chemical, physical, and biological properties of water necessary for each use. Droughts are complex natural phenomena that involve

---

J. A. Díaz · M. Montecinos · A. Vega · M. Molinos
Center for Sustainable Urban Development, CEDEUS, Santiago, Chile

J. Gironás · P. Pastén (✉)
Center for Sustainable Urban Development, CEDEUS, Santiago, Chile

Departamento de Ingeniería Hidráulica y Ambiental, Pontificia Universidad Católica de Chile, Santiago, Chile
e-mail: ppasten@uc.cl

D. Rivera, G. Donoso (eds.), *Droughts in Chile*, Global Issues in Water Policy 31,
https://doi.org/10.1007/978-3-031-85040-0_6

the perturbation of natural meteorological/climatic and hydrological factors, affecting the hydrogeochemical controls of water quality (Mosley, 2015). In Chile, water quality is characterized by diverse hydrochemical environments (Pastén et al., 2021), which could be distinctively influenced by shifts in hydrometeorological drivers and variables (e.g., precipitation, maximum and minimum temperatures, streamflow) in different regions. Hydrometeorological variables exhibit significant climate-related variability in different regions of Chile and undergo profound fluctuations during extreme weather and climate-related events (Castro & Gironás, 2021), including droughts. The changes in patterns and amounts of rainfall have implications in variables like snowpack, runoff, streamflows, lake and reservoir levels, groundwater storage, and wetland coverage (Mishra & Singh, 2010), which are known to play a key role in sustaining the populations and diversity in aquatic ecosystems, as life cycles are tied to levels of water in streams, rivers, lakes, and wetlands (Poff et al., 2002). While the effects of droughts on water quantity in Chile have received much attention in recent research (Garreaud et al., 2017; Oertel et al., 2020; Fernández & Gironás, 2021; Fuentealba et al., 2021), the consequences for water quality have remained overlooked. Droughts are one of the most relevant and imminent climate extreme conditions (Fabian et al., 2023), thus the urgency to better understand the linkages between droughts and water quality.

6.1.1 Drought Perspectives and Drivers on Water Quality

Drought is a natural random phenomenon that affects the greatest amount of people and generates significant economic losses in the world (FAO, 2002; World Bank, 2003). It is a large regional phenomenon of slow development whose beginning, end and spatial limits are hard to identify. Furthermore, drought's silent and non-violent impacts can manifest even years after its end, with severe impacts not only on water availability, but on different ecosystems (Bambach et al., 2013; Meza et al., 2015). Drought is part of the normal climate, rather than a departure from it (Glantz, 2003) and must be actively considered by water planning and management, a complex activity facing large uncertainties associated with climate change, growing population, and land-use modifications. Moreover, in addition to ensuring access to water for social and productive uses, planning and management must cope with the protection of the environment, other natural resources and ecological values (Gleick, 2000).

Spanning over 4200 km, Chile covers a wide range of climatic zones affected by a combination of climate patterns and natural features such as the Pacific Ocean, the Andes, and the Atacama Desert (Oertel et al., 2020). There is a north-south transition from the driest desert in the world to a Mediterranean climate, which gives way to temperate and sub polar oceanic environments. Along a West-East axis, there is a transition from ecosystems under oceanic influence and coastal fog towards mountain-type climatic features. Precipitation is strongly influenced by the Pacific Subtropical High, whose intensity and proximity explains rainfall seasonality

(Quintana & Aceituno, 2012; Barrett & Hameed, 2017). Furthermore, there is a strong interannual variability, particularly in the Mediterranean and semiarid areas (Pittock, 1980; Aceituno, 1988; Montecinos & Aceituno, 2003). Chile is highly vulnerable to climate change, which is likely to affect drought frequency, duration and magnitude, particularly in Mediterranean and semiarid regions by the end of the twenty-first century (Dai, 2011; Chadwick et al., 2018). Temperatures in spring and summer have already increased while overall precipitation has been decreasing, and more frequent extreme drought events have occurred (Núnez et al., 2011; Vicuña et al., 2013; Burger et al., 2018; Urdiales et al., 2018). In fact, the ongoing Megadrought in Chile that started 2010 cannot be explained only considering the natural variability of the climate (Boisier et al., 2016). This drought has affected the country as no other drought during the last century (e.g., Boisier et al., 2016; Aldunce et al., 2017; Garreaud et al., 2017, 2020). Overall, because climate change driven by anthropic activities promotes meteorological and hydrological changes, it is expected that modifications in temperature and water availability and timing will have a variety of effects, including those over the main forcing factors determining hydrochemistry and water quality.

6.1.1.1 Drought Perspectives

Analyzing the relationship between droughts and water quality requires categorizing droughts. Meteorological droughts refer to a prolonged deficit of precipitation (Bates et al., 2008), or a dryness in a region in terms of the amount of precipitation in comparison to a reference value, or a sustained period time without significant rainfalls (Wilhite & Glantz, 1985). Hydrologic droughts refer to streamflow, lake and groundwater below normal levels (Bates et al., 2008). A persistent deficit in precipitation results in reduced soil moisture, diminished surface runoff into streams, and decreased groundwater recharge, leading to a higher severity of a hydrological drought (Giri et al., 2021). As a result, hydrologic droughts are intricately linked to meteorological droughts.

Droughts have also been linked to human related triggers or conditions. Increased water extraction for consumptive purposes contributes to the incidence of hydrological droughts, independent of changes in meteorological factors (Mosley, 2015). Agricultural droughts are linked to the deficiency of subsoil moisture in croplands (Bates et al., 2008), or gaps between plant water demands and available soil water in croplands. Socioeconomic droughts occur when a drought results in available water falling short of meeting the demands of established human activities, or when this demand exceeds the natural or normal water availability (Wilhite & Glantz, 1985). This chapter discusses how hydrochemical parameters and processes respond to changes in precipitation and snowfall (meteorological droughts), and streamflow and runoff (hydrologic droughts).

6.1.1.2 Drought-Related Drivers of Hydrochemical Changes

Hydrochemical parameters are modified by droughts through three primary drivers or forcing factors:

(a) **changes in temperature patterns:** droughts entail air and soil warming and consequent increasing waterbodies temperatures. Temperatures control lake stratification, nutrients dynamics, and phytoplankton composition (Shimoda et al., 2011), and modify chemical reaction rates, such as sulfide oxidation (Schoonen et al., 2000). Furthermore, air temperatures affect snowfall and snowmelt, thus changing streamflows, a driver discussed below.

(b) **changes in streamflows:** changes in rainfall/snowfall and snowmelt patterns associated with droughts impact the distribution and magnitude of flows, consequently affecting the water quality of streams. It is well-known that significant changes in water quality of rivers and lakes arise from low flow conditions in temperate, semi-arid and arid zones (Van Vliet & Zwolsman, 2008; Mosley et al., 2012). Some outcomes of low flows include diminished volumes and extended residence times in water bodies, leading to elevated water temperatures due to reduced thermal capacity and higher sensitivity to atmospheric heat input (Mosley et al., 2012), along with alterations in various other biogeochemical attributes, involving mass, concentration, or chemical speciation. For instance, concentrations of suspended solids in streams can either increase or decrease as function of streamflows and velocity profiles. The deposition of suspended solids along the flow or the release and suspension of riverbed sediments depend on factors such as particle weight, buoyancy, flow shear stress, and turbulence. A threshold exists between erosion and deposition, tied to the equilibrium of particle-laden flow (Dorrell et al., 2018), which can respond to alterations in stream hydrodynamics triggered by droughts.

(c) **changes in watershed heterogeneity:** the spatial distribution of key hydrological processes in the watersheds are particularly relevant in the way droughts impact water quality of waterbodies (Mishra et al., 2021). In the headwaters of snow-dominated regime watersheds, warmer air temperatures accelerate snowmelt potentially decreasing the temperatures of surface waters in the upper sections (Yevenes et al., 2018; Peña-Guerrero et al., 2020). Conversely, in streams of the lower sections of a watershed with pluvial-dominated or mixed regime, droughts may increase water temperatures (Yevenes et al., 2018; Peña-Guerrero et al., 2020).

6.1.2 Drought Stages and Hydrochemical Impacts

Droughts encompass various stages with distinct impacts on water quality (Bonsal et al., 2011; Parry et al., 2016a, b; Ahmadi & Moradkhani, 2019), depending on the variations in meteorological and hydrological processes. Droughts impact water quality during the drought itself and after the drought ends (Crusberg & Eslamian, 2017), referred to as the "drought" and the "post-drought", stages (Ylla et al., 2010; Mosley, 2015; Aldunce et al., 2017; Li et al., 2018). Soil quality and moisture, vegetation cover, and the accumulation of contaminants in soil and water sinks undergo changes over the course of a drought as well. When hydrochemical impacts are analyzed, it is critical to identify at which stage of drought water quality samples or measurements were taken and incorporated in the analyses. Four distinct periods are delineated for drought water quality studies:

- **Pre-drought:** it configures the baseline conditions before meteorological (precipitation) and hydrological (streamflow and lake levels) droughts ensue. Water quality data from the drought period is directly compared with data from the pre-drought period enabling the assessment of the effects of drought on water quality (Ylla et al., 2010; Mosley et al., 2014; Aldunce et al., 2017; Yevenes et al., 2018; Méndez-Freire et al., 2022).
- **Drought:** a period characterized by diminished precipitation, streamflow and/or available water in the watershed in comparison to pre-drought period levels. Its starting point is referred to as "drought onset" and its conclusion is marked as "drought termination". Within a hydrological framework, three sub-stages are defined; growth, persistence and retreat, (Ahmadi & Moradkhani, 2019). As water levels decline across the watershed, hydrological, ecological, and chemical alterations unfold across land and waterbodies during a drought. On land, droughts induce vegetation decrease (Garreaud et al., 2017), the accumulation of salts and pollutants, and soil drying, among others.
- **Initial rewetting and drought recovery:** a brief stage when precipitation and stream flows begin to return to normal conditions, causing an increase in available water in the watershed, such as increasing rainfall, streamflow or lake levels (Mosley et al., 2014). This period is often considered as a transitional stage between the development of drought and the post-drought period, it extends beyond "drought termination" (Ahmadi & Moradkhani, 2019). Rewetting and recovery trigger water quality impacts in streams and lakes (Ylla et al., 2010; Mosley et al., 2014; Fabian et al., 2023), dissolving salts and mobilizing pollutants stored in different environmental compartments during droughts. It can also emerge as a significant and abrupt change, particularly in the context of a drought-flood cascading event (an extreme flood following a drought), causing a sharp rise in stream flows and substantial effects on water quality indicators (Mishra et al., 2021).
- **Post-drought:** this stage ensues when precipitation and stream flows ultimately return to normal conditions or pre-drought levels in terms of meteorology and/or hydrology. In certain water quality studies, the description of outcomes during this period is occasionally not distinctly separated from drought recovery (e.g. consider flooding, which is not a typical condition) (Mosley, 2015). During this stage, changes in catchments and waterbodies can lead to shifts in water quality. Furthermore, in certain studies, water quality data from the post-drought period is compared with data from the pre-drought and/or drought periods to analyze

the effects of drought, rewetting, and recovery on water quality (Ylla et al., 2010; Aldunce et al., 2017; Li et al., 2018).

The reduced water availability in watersheds due to droughts, including diminished stream flows and lake volumes, is reversed to normal levels during the recovery/rewetting and post-drought periods. Ahmadi and Moradkhani (2019) discovered that the recovery time for water quality can be longer than the time required for streamflow recovery. This is due to a lag time between the recovery of water quantity and the restoration of water quality. They demonstrated that specific water quality parameters can return to normal levels during drought termination, but with varying recovery times among them. As an illustration, an analysis of water quality recovery in Californian watersheds during hydrological drought indicates that, on average, temperature and dissolved oxygen in streams needed more time than turbidity to reach threshold levels (Ahmadi & Moradkhani, 2019). However, they did not determine the fundamental drivers underlying these differences.

Drought recovery can trigger short-lived and/or long-term adverse impacts on water quality (Caruso, 2001). In fact, the quality of water in streams might deteriorate during drought recovery, sometimes surpassing the degradation observed during the drought itself (Fabian et al., 2023). Alterations in the landscape of catchments, such as vegetation stress, wildfires, pollutants accumulation, or shifts in soil properties, may not always revert to pre-drought conditions immediately or at all. Similarly, within lake systems, extensive refilling may not restore the water quality to pre-drought levels (Li et al., 2018). Nutrients and other material could be retained in the hypolimnion and other poorly flushed zones, such as lakes bottoms (Reynolds et al., 2012). Such systems are particularly susceptible to the impacts of droughts.

To comprehensively understand how the propagation of a drought alters pre-drought conditions and subsequently affects water quality, as well as how rewetting and drought recovery influence water quality, a comprehensive analysis of meteorological, hydrometeorological, hydrological and (bio)chemical processes occurring within a watershed is needed.

6.1.3 Processes Affected by Droughts

While water quality outcomes during droughts can vary due to site-specific characteristics and particularities of the drought stage, common insights exist regarding the integration and identification of key processes that underlie the impacts of drought on water quality. According to Mosley (2015), three primary factors emerge: (a) hydrological drivers and mass balances; (b) role of temperature and stratification; and (c) increased influence of internal processes and resuspension. These drivers are engaged in various watershed processes that modify hydrochemical environments. The main processes and factors will be briefly discussed below and are snowmelt, erosion, transport, groundwater recharge, mixing zones and dilution, chemical reactions, evapoconcentration, residence times, and nutrients and carbon cycle.

6.1.3.1 Snowmelt

Elevated temperatures contribute to the gradual elevation of the 0 °C isotherm and the altitude of the snowline (Mardones & Garreaud, 2020) as well as glacier retreat in high mountain regions (Ayala et al., 2020). These factors alter the catchment areas exposed to the mobilization of eroded material (Méndez-Freire et al., 2022). These areas, which have not been previously exposed, may contribute distinctively towards sediment and pollutant loads. Further, it has been observed that shifts in water tables of snowmelt-recharged hillslopes alter weathering zones, thereby altering the flux of chemical species exported from the hillslope, like reactive nitrogen (Tokunaga et al., 2022).

Premature snowmelt can increase contaminant concentrations in surface streams during the periods following low flows (Crouch et al., 2013). This is attributed to the reduced dilution capacity resulting from natural contaminant loads. Disrupted snowmelt also modifies the temperatures of rivers, which can subsequently influence the temperature profile and nutrient dynamics (Pizarro et al., 2016) of receiving water bodies.

6.1.3.2 Erosion and Wildfires

Droughts are intricately tied to shifts in rainfall patterns and streamflow quantities. Changes in the total annual rainfall (meteorological drought) within a watershed impact aridity or dryness, influencing erosion processes in soils. Every rainfall event encompasses various low-scale processes that collectively configure alterations in the larger-scale water cycle of the watershed, as described in Fig. 6.1. Raindrops impact the land surface and continue their course controlled by soil properties (e.g. soil texture, vegetation cover, organic matter content), slope, and soil moisture (Léonard & Richard, 2004; Kinnell, 2005; Fernández-Raga et al., 2017). Soil erosion is driven by raindrop impact and the detachment of soil particles, resulting in the disruption of soil structure (Kinnell, 2005). Alternatively, raindrops might directly fall into a waterbody. Raindrops fall and run off through rills and gullies (Kinnell, 2005), releasing and mobilizing chemical constituents such as metals and metalloids, nutrients, non-reactive solids, anthropogenic pollutants, and other components in sediments and soils (Walling & Collins, 2008; Rickson, 2014). Plant-covered soils mitigate the release of contaminants through interception, depending on the degree of coverage (Fernández-Raga et al., 2017; Vaezi et al., 2017). Conversely, arable soils and bare grounds increase erosion during runoff, facilitating the transport of detached soil particles when rain falls (Cerdan et al., 2010).

Reduced rainfall due to drought leads to a decline in the non-point source contaminants (Caruso, 2001). Drought periods disrupt catchment processes, including

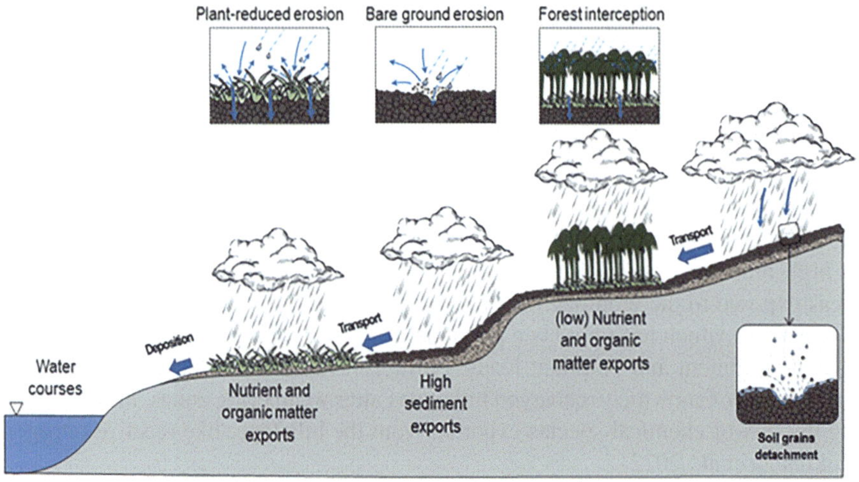

Fig. 6.1 Representation of a raindrop impact on soils, and its consequences in forest, bare ground and vegetated areas. Decreased rainfall associated with drought leads to lower raindrop erosion. Source: author's own work, conceptualized from literature review

soil erosion in the landscape due to diminished rainfall (Giri et al., 2021). However, dryness leads to increased vegetation stress (Kogan, 1997; Zambrano et al., 2016), resulting in changes in carbon and nutrient pools and their associated fluxes (Deng et al., 2021). Such transformations imply that later transport is dominated by accumulation during the drought propagation. The dead vegetation or accumulated manure can be mobilized into watercourses, and eventually enter lakes and reservoirs during the first flush rainfall or floods (Crusberg & Eslamian, 2017), a process that manifests during drought recovery or post-droughts periods, where rainfall becomes a driver of the introduction of suspended solids and nutrients into streams.

These processes exhibit disparities between different land uses (Fig. 6.1). In alpine mountains or headwaters characterized by steep slopes, sparse vegetation or the presence of loess and bare soils, the propensity for erosion during rainfalls is high (Descroix & Mathys, 2003; Cerdan et al., 2010). Periods of reduced rainfalls lead to a decrease in the release of minerals and sediment from these areas. In croplands, non-point emissions of nutrients and toxic compounds from agrochemicals decrease when irrigation is restricted (Barros et al., 1995). However, if agrochemical use persists during drought, contaminants could accumulate in the soil. The absence of cropland irrigation during drought may lead to the flushing of soil contaminants upon rewetting. In grasslands, rainfall reduction limits non-point emissions of nutrients, along with the accumulations of inorganic forms in dry soils. Nonetheless, this can also result in spikes in nitrate concentrations during rewetting (Schaeffer et al., 2017; Deng et al., 2021), which may subsequently run off into streams. Similarly, in grasslands and shrublands, litter input and decomposition decrease during drought, causing an increase in dissolved organic carbon and nitrate concentrations in soils, respectively (Deng et al., 2021), potentially yielding rewetting pulses. In forested areas, droughts induce analogous impacts on litter but result in a significant rise up ammonia concentrations (Deng et al., 2021), and may lead to the loss of soil-soluble organic phosphorous (Sardans & Peñuelas, 2007). As a result, the runoff during the drought recovery can mobilize these substances through leaching or erosion. Furthermore, droughts can exert substantial impacts on water quality through other alterations of forestlands, for instance, through the increase in the frequency of wildfires (Westerling et al., 2006).

Droughts can increase the likelihood of wildfires, which, in turn, modify soil properties, nutrient cycle, and decrease vegetation. Subsequently, this condition enhances erosion and facilitates the movement of sediment, nutrients, and metals into surface waters during rainfall events following wildfire (Loiselle et al., 2020; Mishra et al., 2021). Notably, the impact of wildfires on water quality is more prevalent following the drought (drought recovery) than during the drought itself (Crusberg & Eslamian, 2017).

A key factor found in fire-affected areas is residual ash, which contains soluble forms of nutrients and can lead to substantial effects on receiving waterbodies when rainfall returns (Crusberg & Eslamian, 2017). An example of this is the induction of nitrogen-fixing cyanobacterial blooms, whose growth is typically limited by the availability of phosphate, the soluble form of phosphorous.

6.1.3.3 Transport in Catchments and Rivers

During drought, the transport of water constituents and pollutants experiences interruptions, leading to their accumulation in various aquatic and terrestrial compartments. In semiarid and arid zones, low flow periods occurring during drought propagation result in minimal catchment runoff, effectively halting the transport of sediments, nutrients and salts (Mosley et al., 2012). Consequently, these materials accumulate across the landscape during this phase. Moreover, drought periods alter stream processes, including the reduction of stream bank and streambed erosion attributed to low flows and decreased turbulent stream velocity (Giri et al., 2021). Additionally, low-flow conditions can promote the sedimentation of organic and inorganic particulate matter within the rivers (Mosley et al., 2012).

During drought recovery, moderate flows mobilize substantial quantities of the accumulated sediments, nutrients and salts, flushing them from the catchment, riverbed, riverbanks, and floodplains (Mosley et al., 2012). In the aftermath of wildfires, rainfall triggers the same process (Crusberg & Eslamian, 2017). Therefore, the duration of the drought or dry period not only significantly influences surface hydrology and fire activity along Central Chile (Garreaud et al., 2017), but also the accumulation of pollutants and constituents, subsequently affecting the magnitude of their mobilization within the watershed (Mosley et al., 2012). This is a relevant issue, as drought duration and frequency are expected to increase due to climate change (Dai, 2011; Chadwick et al., 2018).

6.1.3.4 Groundwater Interactions

Lower water tables resulting from reduced groundwater recharge lead to elevated oxygen levels, which contribute to the occurrence of acid rock drainage when sulfides are present, and the release of contaminants and acidic waters (Todd et al., 2012; Crouch et al., 2013). As groundwater levels decline, a larger subsurface volume of rock becomes exposed to oxygen (Crouch et al., 2013), encompassing more reactive rocks that were not previously exposed. This exposure potentially increases the overall extent of sulfide oxidation in groundwater discharging to surface streams (Todd et al., 2012).

In the aftermath of wildfires, precipitation leaches the soil of burnt areas, leading to soil acidification (Crusberg & Eslamian, 2017), which releases metals from bedrock, further influencing water quality. Along coastal zones, the decline in groundwater levels may disrupt the equilibrium between freshwater and salted water interfaces, leading to salination (Djabri et al., 2014).

6.1.3.5 Mixing Zones, Confluences and Dilution

Alterations in surface hydrochemistry generally arise when mixing ratios of end-members change in river confluences. Simple dilution attenuates concentrations of pollutants (Anawar, 2013), beyond the chemical reactions occurring within a mixing zone. Natural environments have a range of mixing zones, including rivers confluences, inflows of streams into lakes and reservoirs, and surface-groundwater interaction zones within rivers and lakes. In these areas, shifts in water flow or changes in hydrochemical parameters from one source can modify the resultant water composition, implying poorer water quality if the flow of cleaner water decreases.

Additionally, mixing zones manifest in anthropogenic discharges into waterbodies, such as effluents from wastewater treatment plants, industrial wastewater, urban stormwater, etc. During droughts, the deterioration of water quality in the receiving surface waters can be more pronounced due to lower water flows and reduced dilution capacity (Meza et al., 2021). Conversely, during the post-drought stage, increased river flows lead to substantial dilution when accumulated constituents in catchment have already been flushed out by the initial rainfall events of drought recovery (Mosley et al., 2012). Groundwater inputs into surface waters are another example of mixing zones affected by drought. When catchment water inputs decrease due to lower rainfall, the influence of saline groundwater inputs increases (Mosley et al., 2012).

A relevant type of mixing zones impacting water hydrochemistry are reactive confluences. Reactive confluences are complex river domains where intricate chemical reactions take place, inducing significant changes in hydrochemical conditions (Guerra et al., 2016; Montecinos et al., 2020; Díaz et al., 2023; Pasten et al., 2023). Reactive confluences typically occur when rivers with vastly different water chemistries merge. For example, the mixing of natural or anthropogenic acidic waters with alkaline or neutral waters creates a mixing zone containing a range of chemical gradients, including pH and electric conductivity (Guerra et al., 2016). Within this zone, water constituents may undergo chemical transformations that impact their transport or toxicity (Balistrieri et al., 2007; Abarca et al., 2017). Notably, the chemical precipitation of certain metals and metalloids lead to the removal of other trace pollutants through sorption and/or coprecipitation within the transition zone between the acidic and near-neutral pH waters of the confluence (Casiot et al., 2009). The parameters dictating water quality in river confluences are governed by chemical and hydrodynamic environments (Abarca et al., 2017; Montecinos et al., 2020), that may be influenced by droughts due to shifts in river flow patterns (e.g. mixing ratio, turbulent velocity).

6.1.3.6 Chemical Reactions

Drought-induced warm temperatures have significant consequences on water quality by influencing chemical reactions. Temperature controls the thermodynamic feasibility and kinetics of precipitation–dissolution, sorption-desorption, acid–base, and redox reactions (Schoonen et al., 2000). For instance, the equilibrium between oxygen solubility in water and oxygen partial pressure is temperature-dependent (Tromans, 1998).

Higher temperatures impact reaeration rates of rivers, resulting in lower concentrations of dissolved oxygen saturation (Mosley, 2015). In some surface waters, increasing summer temperatures promote the development of anoxic or suboxic conditions in sediments, boosting bacterial activity, and intensifying the reduction of metal-rich oxides and oxyhydroxides, thereby mobilizing metals and metalloids within the sediments (Masson et al., 2007).

Drought impacts specific processes tied to chemical reactions, including acid rock drainage and acid mine drainage, as well as lake acidification.

Acid drainage: rising temperatures and shifts in rainfall patterns alter the generation of acid drainage and its impacts on surface and groundwater. However, investigations into the effects on transport mechanisms remain scarce and warrant further exploration (Anawar, 2013). Changes in hydrologic regimes may induce shifts in biogeochemical processes within aquatic ecosystems associated with acid mine drainage (AMD). Higher metal concentrations in streams have been noted during summer low-flow conditions due to elevated temperatures and altered snowmelt by a few weeks (Todd et al., 2012; Crouch et al., 2013). These shifts are tied to reduced instream flows and diel pH variability, introducing greater uncertainty into site-specific aquatic risk assessments of ionizable contaminants from acid mine drainage (Anawar, 2013). Decreased rainfall diminishes the oxidation of sulfide minerals and the generation of AMD, while warmer temperatures increase the reaction rates and kinetics in these processes (Schoonen et al., 2000; Anawar, 2013).

Lake acidification: severe droughts within rivers can lead to substantial reductions in lake depths and volumes, exposing extensive areas of sediments to the atmosphere along lake margins (Mosley et al., 2014; Aldunce et al., 2017). In the Murray-Darling Basin of Australia, sediment drying and exposure during drought trigger pyrite oxidation in sediments. When rewetting occurs with initial rainfalls and lake refilling, acidic surface waters emerge in these zones, resulting in high dissolved concentrations of metals (Mosley et al., 2014) and heightened salinity (Aldunce et al., 2017). In certain locations, natural neutralization ensues after rewetting, leading to neutral pH levels and the precipitation of metals. Conversely, other areas required limestone dosing to counteract the chemical consequences brought about by drought.

6.1.3.7 Evapoconcentration

Elevated temperatures combined with reduced flows and diminished water levels contribute to evapoconcentration, leading to the accumulation of salinity, major ions and metals within rivers, streams, lakes and reservoirs (Mosley et al., 2012; Jirsa et al., 2013; Mosley, 2015). Evapotranspiration in croplands from arid zones lead to salt build-up in soils, particularly when utilizing sub-surface water irrigation (Crusberg & Eslamian, 2017). Consequently, drought conditions intensify the deposition of salts on soil surfaces. The process of drought recovery disrupts this accumulation and washes off salts and ions, resulting in a concentrated runoff.

In some instances, lakes profoundly affected by extreme droughts struggle to receive sufficient flushing from rivers during rewetting, leading to higher nutrient concentrations (Aldunce et al., 2017). As a result, evapoconcentration may contribute to heightened eutrophication within these lakes.

6.1.3.8 Residence Times

Reduced volumes and extended residence times, resulting from low flows, increase water temperatures (Mosley et al., 2012). This effect is attributed to the reduced thermal capacity of waterbodies and the heightened sensitivity to atmospheric heat input (Mosley & Fleming, 2010).

Extended residence times and stagnant flow conditions in rivers induce changes in geochemical processes that govern nutrient concentrations, including nitrification, uptake by algae, and adsorption by suspended solids and sediments (Van Vliet & Zwolsman, 2008). Prolonged residence times lead to increased cell growth and biomass accumulation, that trigger phytoplankton blooms when coupled with increased nitrate inflow (Whitehead et al., 2009; Glibert et al., 2014).

6.1.3.9 Thermal Stratification

Drought-related high air temperatures have consequential effects on the stratification of lake systems. These changes influence nutrient concentrations, algal production, the occurrence of cyanobacterial blooms and the distribution of dissolved oxygen throughout the water column (Mosley, 2015).

Variations in snowmelt timing also affect lake stratification (Pizarro et al., 2016). For stratifying lakes, declining tributary water temperatures cool the hypolimnion, while warmer air temperature raises the epilimnion temperatures, extending the stratification (Shimoda et al., 2011). The prolonged stratification period hinders vertical mixing, leading to decreased dissolved oxygen at maximum depths. This change results in detrimental hypoxic or anoxic conditions for aquatic organisms and further release of nutrients into the water column (Williamson et al., 2009).

6.1.3.10 Nutrient Cycle and Carbon Cycle

Droughts can exert a significant influence on the nutrient and carbon cycle within various ecosystems. The productivity of photosynthetic organisms, which involves the fixation of carbon mass, can be amplified by reduced water levels and elevated temperatures (Mosley, 2015). In lakes and reservoirs, multiple factors associated with droughts lead to increased primary productivity, impacting chlorophyll-a concentrations, dissolved oxygen levels, and dissolved organic carbon (Van Vliet & Zwolsman, 2008; Jirsa et al., 2013).

External inputs: drought-induced lack of precipitation reduces the flushing of nutrients and their exports from the land (Caruso, 2002; Qiu et al., 2023), leading to decreased riverine input during drought periods. Additionally, drought can trigger shifts in nutrients and carbon cycling within vegetated soils, altering nutrient availability for plant uptake and potential release during rewetting (Johnson et al., 2008; Deng et al., 2021). Available substrates such as dissolved organic carbon and nitrogen accumulate during droughts, resulting in higher losses of these elements (Deng et al., 2021). Furthermore, soil-soluble organic phosphorous, which is not a readily available form for plant uptake, can accumulate in soils during drought and be exported through leaching or erosion during subsequent rainfalls (Sardans & Peñuelas, 2007). Phosphorous concentration frequently limits nutrient availability and drives eutrophication in freshwater ecosystems (Fabian et al., 2023), making its export from land a significant factor impact in water quality (Lisboa et al., 2020).

Moreover, the instream biological processes occurring in rivers during low flow conditions—shallow and stagnant flows—such as denitrification and sulfate reduction, can lead to reduced concentrations of certain nutrients (Andersen et al., 2004). Furthermore, in addition to their impacts on biogeochemical cycling on land, droughts can yield significant consequences for the internal nutrient cycling and productivity of lakes and reservoirs. These effects can ultimately alter the trophic state of these waterbodies, often leading to an increase in eutrophication (Mosley, 2015; Aldunce et al., 2017).

Internal processing: changes in internal processing within lakes and reservoirs have been identified as contributing to nutrient cycling during drought. Water bodies with stratification and reduced oxygen levels experience increases in ammonia, organic nitrogen, and total phosphorus concentrations (Baldwin et al., 2008). Shallower lakes lead to increased resuspension (Mosley et al., 2012) and periodic nutrients pulses from enhanced turnover of nutrient-laden bottom waters (Baldwin et al., 2008).

Eutrophication: eutrophication and cyanobacterial blooms are promoted by various factors at different stages: (a) elevated water temperatures during drought (Paerl & Huisman, 2009), (b) increased internal nutrient inputs during drought (Baldwin et al., 2008; Mosley et al., 2012), (c) higher external inputs of nutrients during rewetting or post-wildfires rainfalls due to soil flushing (Li et al., 2018; Lisboa et al., 2020), and (d) nutrient evapoconcentration resulting from reduced river flushing (Aldunce et al., 2017). These conditions lead to elevated concentrations of cyanobacterial toxins and/or other changes in phytoplankton community structure. The depletion of dissolved oxygen due to algae mortality induces anoxic conditions in the water column, triggering the release of inorganic phosphate from bottom sediments (Crusberg & Eslamian, 2017), further fueling cyanobacterial growth.

6.1.4 Anthropic Factors: Pollution Sources and Hydraulic Structures

The complexity of droughts-water quality connections is compounded by the impact of human activities on Earth's surface, affecting water quantity in various ways. Water extraction, changes in surface permeability, construction of hydraulic infrastructure, and human-driven climate change modify the water cycle, which governs water quantity in waterbodies and continental aquatic ecosystems. Land use changes, agricultural practices leading to soil erosion and other non-point sources of pollutants, as well as the discharge of industrial and sewage waste into waterbodies (point sources), contribute to the deterioration of water quality. The consequences of drought on water quality differ depending on land uses, whether rural or urban (Elsdon et al., 2009).

Urban environments exert significant impacts on water quality of waterbodies through three major pathways: stormwater runoff, sewage wastewater, and industrial effluents (Walsh, 2000). Numerous parameters are affected by urban pollution, including concentrations of dissolved nutrients due to elevated atmospheric nitrogen levels and inadequate sewage infrastructure (Walsh, 2000). During precipitation events and dry periods, atmospheric deposition aids in the transport of urban air pollutants to the land surface (Müller et al., 2020). Stormwater runoff in urban regions is characterized by heightened concentrations of suspended sediments, metals and toxic organic pollutants (Walsh, 2000; Müller et al., 2020). Dry periods in urban areas lead to the accumulation of pollutants, and the intensity and duration of rainfall play a differential role in pollutants wash-off, influenced by the composition and chemical speciation of pollutants (Murphy et al., 2015). Consequently, meteorological droughts impact the water quality of natural watercourses that receive urban stormwater runoff, with discernible effects during the drought period (accumulation of contaminants) and when rainfall resumes (contaminants wash-off).

Low flows during droughts reduce the dilution capacity of surface waters towards industrial or municipal wastewater effluents (Andersen et al., 2004; Crusberg & Eslamian, 2017). Water conservation measures implemented by water agencies to curtail water demand can lead to a decline in the quality of treated municipal wastewater. This, coupled with an increased proportion of wastewater in the resulting surface waters, restricts the potential reuse of these waters downstream (Tran et al., 2017).

In agricultural catchments, reduced non-point source pollution is a likely outcome of decreased runoff due to diminished rainfall (Caruso, 2002). Agricultural pollution can lessen during periods of extreme low flow when nutrient drainage to rivers decreases owing to restricted irrigation following severe water allocations limitations (Mosley & Fleming, 2010). Drought-induced reductions in irrigation of croplands also decreases agrochemicals runoff and groundwater pollution (Barros et al., 1995). Converting forested or shrublands areas to croplands and grazing lands enhances water quality impacts during drought (Oyarzun et al., 1997). However, agricultural irrigation practices and other water extraction demands contribute to causing drought, which in turn deteriorate surface water quality (Mosley et al., 2012; Aldunce et al., 2017).

Similar to urban regions, domestic wastewater impacts water quality in rural areas. During drought periods, nutrient emissions from domestic septic wastewater systems may emerge with less dilution affecting the quality and potential use of groundwater base flows or local springs (Crusberg & Eslamian, 2017).

Notably, managed or regulated river flows have exhibited resilience to droughts (Meyer et al., 1999). Unregulated river systems typically experience greater flow and water levels variations than those controlled by structures like dams (Meyer et al., 1999). Reservoir management can increase the proportion of natural flows over wastewater effluents, enhancing the dilution capacity and thus improving water quality. However, this practice may pose challenges in meeting year-round water demand due to diminished water storage (Momblanch et al., 2015).

Even in hydraulic structures, water quality can be compromised by droughts. Reduced water levels lead to pollutant concentration, while waterbodies can also experience eutrophication, similar to lakes (as previously mentioned). Drinking water reservoirs may be impacted by elevated toxin concentrations resulting from cyanobacterial blooms triggered by higher temperatures and nutrient inputs (Crusberg & Eslamian, 2017).

Droughts can influence the hydrochemistry of mining environments, including mining sites, tailings, and untreated mining waste, although the impacts of geochemical reactions on mining wastes have not been extensively studied or reviewed (Anawar, 2013). The water quality of reservoirs that receive inflows of acid mine drainage may be impaired when alterations in site-specific hydrology affect physical, chemical, and biological processes (Brooks et al., 2011).

Therefore, several impacts on water quality may occur at different stages of drought, which may be conceptualized as shown in Fig. 6.2. The processes represented in Fig. 6.2 include non-point and point sources of pollution, as described above. Other components of the water cycle, such as snowpack and lakes were not considered to simplify this conceptual model.

Note: This conceptual model provides insights into the cascading effects of drought on multiple water quality processes, underlining the dynamic interplay between drought stages and various human activities that influence water quality. It serves as a framework for understanding how drought-induced changes influence the transport, accumulation, and transformation of contaminants within water bodies.

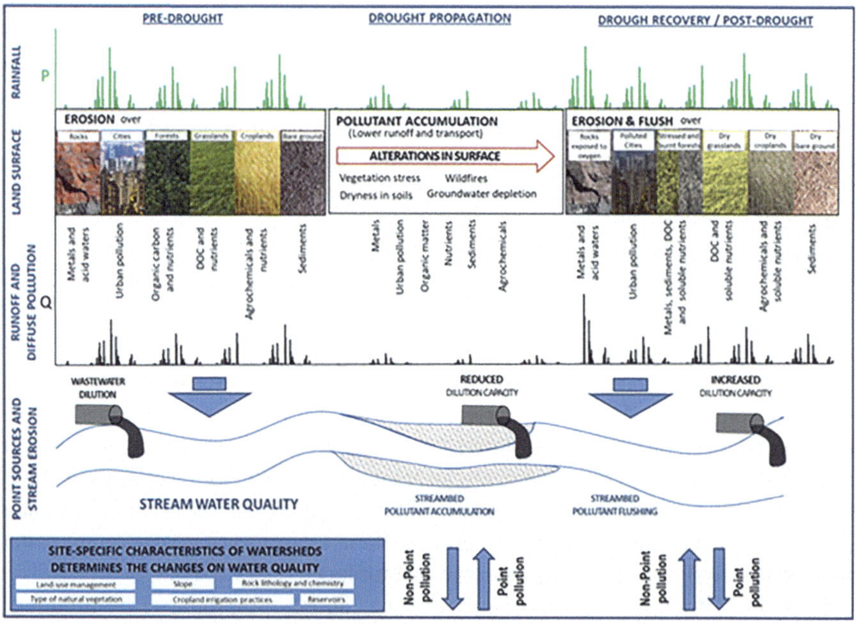

Fig. 6.2 Representation of consequences of drought propagation and drought recovery on non-point and point impacts of contaminants into streams. Source: author's own work, conceptualized from literature review

6.2 Droughts Impact on Water Quality in Andean Watersheds

The distribution and impacts of droughts on water quality in mountain regions depend on hydrological, geomorphological, and geological features. Human activities can lead to shifts in soil properties, vegetation cover, and the mobilization of pollutants. A significant proportion of research about the relationship between drought and water quality has been predominantly conducted in North America, Europe and Australia (Mosley, 2015). Conversely, regions such as Africa, Asia and South America need more comprehensive research into this phenomenon. Notably, Fabian et al. (2023) recently examined the impact of droughts (and floods) on water quality in Asia. In the case of Chile, despite experiencing a mega-drought in its Central zone since 2010 (Arroyo et al., 2020), the consideration of drought impacts on water quality has only gained prominence recently, with limited studies emerging since 2017. Through the identification of the influencing factors for each parameter and a comparative analysis with studies from other regions, it becomes feasible to understand the hydrochemical effects of drought within Andean watersheds.

Figure 6.3 illustrates the distribution of slopes and land uses in four model Andean watersheds. The mineralogy, erosion patterns, soil uses, and human

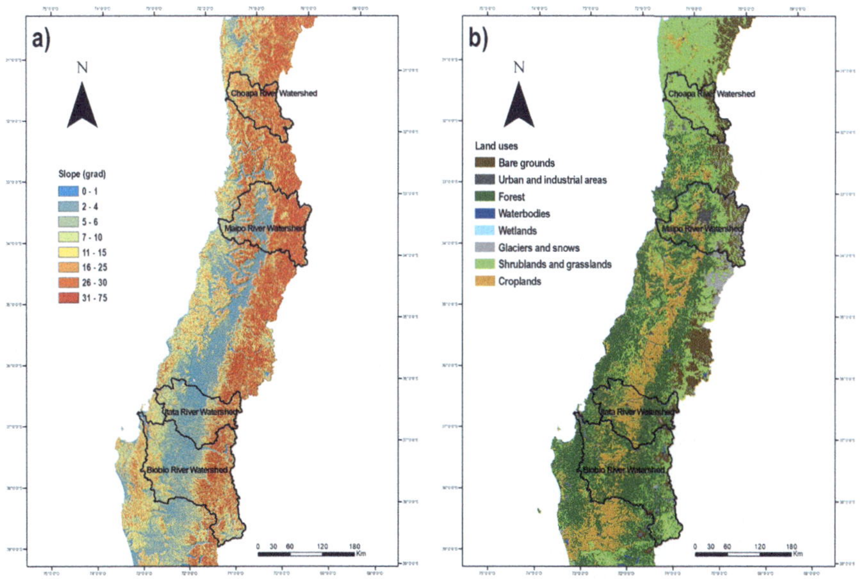

Fig. 6.3 Distribution of land uses and geological characteristics in the Choapa, Maipo, Itata and Biobio rivers watersheds, Chile. (**a**) slope map and (**b**) land uses. Andean watersheds present a marked distribution of geological characteristics associated to the geomorphological units present from west to east. This results in specific land uses along watershed, according to the slope and altitude. Source: author's own work based on Chile's publicly available geospatial datasets

activities vary depending on altitude, resulting in distinct distributions and exposures to forcing factors that ultimately determine the water quality. River morphology and regime evolve with changing altitudes, resulting in different responses to drought for the upper and lower portions of the watershed.

6.2.1 Relevant Characteristics of Andean Watersheds

Mishra et al. (2021) highlight the importance of watershed characteristics in comprehending the effects of droughts on the water quality of surface and groundwater. Alnahit et al. (2020) present a conceptual framework and mathematical modeling that identified the key variables responsible for the spatial heterogeneity of water quality across watersheds in the USA. Their study concludes water quality variations are influenced not only by anthropogenic factors but also basin characteristics (e.g. soil properties).

Quantitatively linking water quality to droughts in the context of Chile without extended datasets from a spatially dense water quality monitoring network is a difficult endeavor. Chile is characterized for its wide range of hydrogeochemical environments, posing challenges in the realms of water quality management and monitoring (Pastén et al., 2021). The term Andean watersheds refers to basins that possess the following attributes:

- Localized within Chilean mountain zones (Cordillera de los Andes), or transverse valleys and intermediate depressions.
- Main streams of short length (<500 km).
- Steep slopes that foster turbulent streamflow.
- Subject to seasonal climates, marked by a predominant rainfall/snowfall season during winter and occasional summer storm events in high mountains.
- Experiencing arid, semiarid, or temperate climates.
- Exhibiting snowmelt-dominated regimes that shift to rainfall-dominated regimes in the lower reaches (Peña-Guerrero et al., 2020), resulting in complex confluences in terms of hydrology and hydrochemistry. In Chile, the central Andes snowpack (30°–37°S) is the predominant source (>85%) of streamflow variability (Masiokas et al., 2006).

Andean watersheds are also exposed to anthropogenic impacts, including:

- Mining activities in the headwaters or valleys, mainly in the Northern and Central Macrozones.
- Legacy and large area deposits of mining residues.
- Agriculture and croplands in the valleys.
- Agricultural dams or reservoirs for hydropower and drinking water supply.

The occurrence of acid rock drainage poses a recurrent and challenging issue in these watersheds, stemming from the oxidation of sulfide minerals, particularly pyrite (Duhalde et al., 2021). This phenomenon is exacerbated by mining operations that involve the extraction and accumulation of sulfur-laden materials, such as the disposal of mining residues like tailings and dumps (Anawar, 2013). Climate change can potentially exacerbate this impact. However, the arid climate prevalent in the north-central Chilean watersheds is intrinsically linked to natural alkalinity, which, mitigates the impacts of acid drainage to a certain extent (Duhalde et al., 2021). The enrichment of metals may be observed in streams impacted by acid drainage across several watersheds in northern and Central Chile.

6.2.2 *Drought Impacts on Water Quality Parameters*

This section outlines the effects of droughts on specific water quality parameters within the context of Andean watersheds in Chile, considering findings found in recent work. Figure 6.4 presents on a model of an Andean watershed the forcing factors of droughts (temperature, rainfall and streamflows), which lead to shifts in

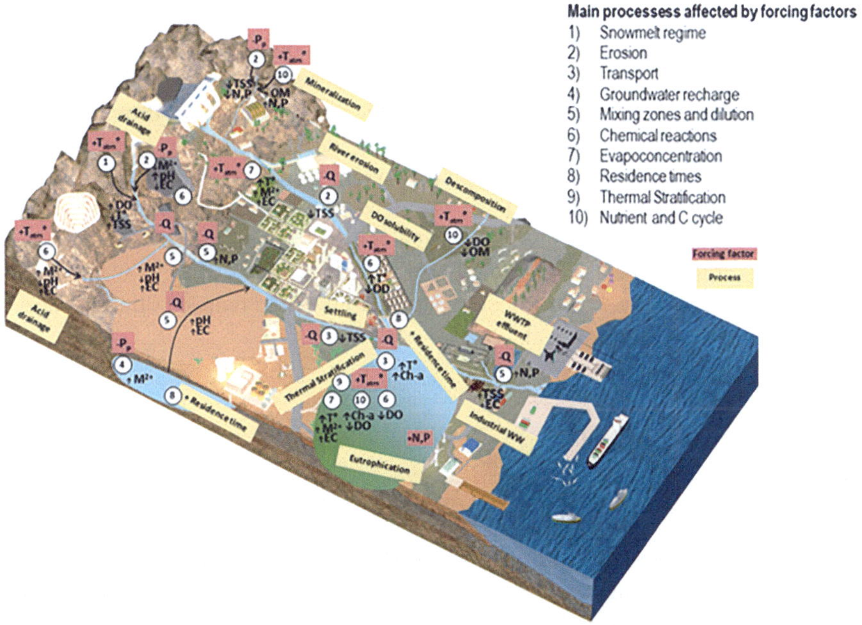

Fig. 6.4 Representation of watershed processes influenced by droughts and their consequences on water quality parameters in surface and groundwater. Yellow rectangles represent the driving forces behind hydrochemical changes during droughts, including rising temperatures and reduced flows. Red rectangles depict the processes affected by these factors, such as snowmelt regimes, erosion, transport, groundwater recharge, mixing zones and dilution of waters, chemical reactions (e.g. precipitation-dissolution), evapoconcentration, residence times, thermal stratification, and nutrients and carbon cycles. Source: author's own work using the CEDEUS Andean watershed model

hydrometeorological, hydrological, chemical, and biological characteristics. These changes in turn impact water quality parameters, showcasing the interplay between drought-related factors and water quality dynamics within Andean watersheds.

6.2.2.1 Salinity and Electric Conductivity

Water availability plays a critical role in determining salinity and conductivity levels in rivers. Reduced water availability corresponds to decreased dilution capacity, leading to elevated concentrations of major ions. This consequence of droughts has been reported in different sites. In the Meuse River (western Europe), Van Vliet and Zwolsman (2008) revealed that drought conditions led to increased concentration of chloride, fluoride and sulfate. The concentration of these anions that are considered non-reactive tend to increase due to the limited dilution of anthropogenic sources containing such anions, like those coming from diffuse or point industrial, urban, and mining sources. Likewise, droughts promote increased salinity and specific conductance in streams and forested areas. This is a consequence of sustained evaporation of surface water and reduced dilution from saline groundwater sources (Sprague, 2005; Mishra et al., 2021).

In Andean river systems, an inverse concentration-flow relationship (C-Q) concerning electric conductivity and the concentration of major ions (Cl, Na, SO_4, Mg, Ca, K) may be observed. During periods of low-flow induced by drought, major dissolved ions exhibit an increase in their values. This behavior has been observed in watersheds in northern, central and southern Chile, including the Choapa (Hernandez et al., 2023), Maipo (Peña-Guerrero et al., 2020) and Biobio (Yevenes et al., 2018) basins, respectively. These high values often surpassed the regulatory limits set by the Chilean standards for agricultural water use (NCh1333/78), and where attributed to increased salts evapoconcentration and reduced dilution capacity of rivers (Peña-Guerrero et al., 2020).

6.2.2.2 Nutrients and Algae

The trends in nutrient concentrations and algae development within river ecosystems exhibit close correlations with human activities, erosion processes and location of the geomorphological units (i.e., altitude of the watershed). In periods of drought, the reduced dilution capacity of wastewater treatments plants effluents can result in elevated concentrations of NH_4^+, NO_3^- and NO_2^- (Sprague, 2005; Van Vliet & Zwolsman, 2008). During droughts, wastewater effluents may become a larger proportion of a river's flow (Sprague, 2005). Reduced dilution and elevated evaporation rates result in increased concentrations of NO_2^- and PO_4^{3-} derived from highly urbanized areas (Sprague, 2005; Van Vliet & Zwolsman, 2008; Hrdinka et al., 2012; Mishra et al., 2021), and concentration of NO_3^- and K^+ from agricultural lands due to fertilizer leaching (Van Vliet & Zwolsman, 2008; Mishra et al., 2021). The combination of nutrient inputs with higher temperatures and residence times, create favorable conditions for algal blooms, leading to high concentrations of chlorophyll-a in rivers during summer periods (Van Vliet & Zwolsman, 2008; Mosley, 2015; Mishra et al., 2021).

In Andean watersheds, a similar response to drought can be identified, with variations linked to different altitudes. Yevenes et al. (2018) studied the effect of drought on the water quality of the Biobío River by comparing data collected during the drought period (2010–2015) with data from pre-drought periods (2000–2009), at various locations within the watershed. Their findings revealed that in the upper reach of the watershed, no significant differences in the concentrations of NO_3^-, NO_2^- or NH_4^+ were found between the two periods. In contrast, in the lower reach of the watershed, the authors observed a decrease in the concentrations of total phosphorus (TP) and NO_3^-. This decline could be attributed to a reduction in the drainage and nutrient export caused by the drought conditions. Peña-Guerrero et al. (2020) yielded different results in their study of the Maipo River, where they identified negative correlations between daily discharge and the concentrations of NO_3^- and PO_4^{3-} in the upper and medium part of the watershed. This behavior was related to agricultural zones in the Central valley and Coastal cordillera delivering consistent nutrient fluxes to the rivers (Jorquera et al., 2015). Therefore, future trends of concentration of nutrients in rivers will be highly dependent on the human uses of land and in the occurrence of runoff which mobilizes these contaminants to water sources.

6.2.2.3 Metals and Metalloids

In the case of the Meuse River, Van Vliet and Zwolsman (2008) found that the response of metal and metalloid concentration depended on their adsorption capacities to suspended solids. Elements with low partition coefficients, such as Se, Ba and Ni, which exhibit low affinity to suspended solids, are transported in the dissolved phase. Consequently, their concentrations display a negative relation with discharge since they are controlled by the dilution capacity. Conversely, elements with high partition coefficient, such as Pb, Cr, Hg and Cd, exhibited a positive correlation with discharge, akin to suspended solid concentration. Elements with moderately high partition coefficient, such as Cu, Zn and As, did not exhibit consistent changes with drought.

In Andean watersheds the concentration of metals and metalloids, such as Cu, As, Fe, Al and Zn, is closely linked to the eroded material stemming from acid rock and mine drainage (Guerra et al., 2016; Montecinos et al., 2020). In this type of watersheds, a reduction in the fluxes of suspended sediments driven by drought could lead to a reduction in the total concentration of metals and metalloids. The response of metals and metalloids concentration during drought varies depending on their source interactions with other rivers. For instance, in the Maipo River watershed, Peña-Guerrero et al. (2020) identified for most monitoring stations significant positive relationships ($p < 0.05$) between daily discharge rates and the concentrations of Al, As, Cu, Mn, Fe and Zn. Consequently, metal and metalloid concentrations decreased due to the retention of contaminants during the drought period, subsequently concentrating and delivering them downstream during high flows. Conversely, in the Elqui River watershed, Hernandez et al. (2023) identified a favorable forecast in Fe concentration in the upper part of the watershed. Based on a preliminary qualitative assessment, these authors anticipated that a future reduction in flows would result in lower concentrations of Fe. In contrast, Yevenes et al. (2018) observed in the Biobio River watershed an increase in the concentrations of Cr, Fe and Co during low flows promoted by drought. This increase was attributed to intensified evaporation from rivers and groundwater, coupled to prolonged residence times and increased interactions between water and sediment in the groundwater (Caruso, 2002; Arumí et al., 2012). Consequently, different responses can be expected depending on the characteristics of watersheds. Drought may limit the fluxes of metals, by restricting the transport of enriched sediments or increasing their concentration, reducing the dilution, and enhancing the interaction with enriched sediments.

6.2.2.4 Organic Matter

Few studies have addressed the behavior of organic matter in Andean watersheds affected by droughts. In contrast, in Chile, Peña-Guerrero et al. (2020) suggested that elevated temperatures resulting from drought in the Maipo River watershed leaded to increased mineralization and organic matter release. These processes are expected to release nitrogen, phosphorus and carbon, which may subsequently be transported by storm events following the drought period (Delpla et al., 2009).

6.2.2.5 Suspended Solids

The concentration of total suspended solids (TSS) in rivers is influenced by sediment generation and transport capacity, both severely impacted during droughts. Many studies have reported a decrease in river turbidity during drought due to reduced runoff and soil erosion resulting from decreased precipitation (Giri et al., 2021; Mishra et al., 2021). Lower sediment inputs are associated with lower water flow velocities during drought, which limits the streambank and streambed erosion and the transport capacity of suspended solids (Van Vliet & Zwolsman, 2008; Hrdinka et al., 2012; Giri et al., 2021). However, it is important to note that under certain conditions, such as point sources with high concentrations of suspended solids, lower dilution capacity can lead to an increased turbidity (Mishra et al., 2021).

In Andean watersheds, the impacts of drought on sediments fluxes vary depending on the location within the watershed (upper or lower) and the natural conditions of each basin. Yevenes et al. (2018) found that in the upper part of the watershed, drought increased the concentration of TSS in the river. This increase was attributed to earlier snow melting driven by rising temperatures (Bozkurt et al., 2017), which led to increased catchment runoff (Van Heemst et al., 2013). However, in the lower part of the watershed, TSS concentrations showed a strong decrease, due to the reduced transport capacity of the river under lower flow conditions. Lower flow conditions in this region promoted the settling of TSS along the Biobio River (Tolorza et al., 2014). Similar results were found by Méndez-Freire et al. (2022) for the Itata River watershed, where drought led to decreased sediment fluxes between 28% and 67% of their regular values due to reduced transport capacity under lower water discharge. The season of maximum sediment transport also shifted from May–June to August, influenced by changes in the precipitation regime and conditions required to mobilize sediments. These shifts in the sediment transport patterns can be expected in other Andean watersheds, due to higher reduction in surface runoff in the upper parts of the catchments, where sediments are primarily sourced (Martínez-Retureta et al., 2021).

The response of sediment loads in rivers in the context of drought depends on the watershed's ability to transport eroded material. In central Chile, the transport of sediments downstream will depend on natural conditions, such as slope, precipitation and vegetation cover (Pepin et al., 2010), but also on the human activities and land use. Dams play a significant role in trapping sediments, leading to substantial reductions in sediment fluxes downstream (Méndez-Freire et al., 2022). These hydraulic infrastructures introduce additional complexity and uncertainty in concentration-flow response studies (Hernandez et al., 2023), and can further constrain sediment transfer to the ocean during drought (Méndez-Freire et al., 2022), and post-drought.

6.2.2.6 Temperature of Water and Dissolved Oxygen

Various studies conducted in rivers in Europe and the USA have reported an increase in water temperature during drought events, with values exceeding baseline temperatures by up to 2 °C (Sprague, 2005; Van Vliet & Zwolsman, 2008; Hrdinka et al., 2012; Todd et al., 2012). This temperature rises primarily attributed to elevated air temperatures associated to drought, and the strong relationship between water and air temperatures (Van Vliet & Zwolsman, 2008). Higher water temperatures lead to reduced oxygen solubility, resulting in lower concentration of dissolved oxygen in rivers (Sprague, 2005; Hrdinka et al., 2012; Todd et al., 2012). The decreased levels of dissolved oxygen, coupled to reduced re-aeration rates due to low flow conditions, and increased organic matter decomposition rates (Van Vliet & Zwolsman, 2008; Hrdinka et al., 2012), pose a significant risk to aquatic organisms. In some cases, the growth of phytoplankton and zooplankton exacerbates oxygen gradients (Hrdinka et al., 2012). Nevertheless, in situations where water residence times are prolonged, nutrients facilitate the development of algal blooms, which trigger decomposition reactions that result in anoxic conditions (Mosley, 2015; Mishra et al., 2021).

In Andean watersheds impacted by drought, water temperature responses can vary depending on the river's regime and the location in the watershed, influencing dissolved oxygen concentrations. For instance, in Peña-Guerrero et al. (2020), river transects located in the lower part of the Maipo River watershed exhibited a significative negative relationship between daily discharge and temperature. These transects, characterized by a mixed and pluvial regime, were situated in irrigation districts where reduced flows led to increased water residence times. This change increased the sensitivity of rivers to atmospheric heat, resulting in higher water temperatures. Similar findings were reported for the Biobio River basin by Yevenes et al. (2018), where surface waters in the lower part of the watershed experienced a temperature increase of 1.9 °C during the drought period. Conversely, the upper part of the watershed showed opposite responses. In the Maipo River watershed, river transects located in higher zones, characterized by a snow regime and colder waters, showed a significative positive relationship between daily discharge and water temperature (Peña-Guerrero et al., 2020). This, in turn, was associated with a significant negative relationship with dissolved oxygen concentrations. However, in the upper Biobio River watershed, surface waters experienced a temperature increase of 0.9 °C during drought, which did not significantly impact the dissolved oxygen concentration in the river transects (Yevenes et al., 2018).

6.2.2.7 pH

The response of pH to drought is contingent on site-specific conditions and can be enhanced by the reduced dilution capacity promoted by drought. For instance, in the South Platte River Basin, Sprague (2005) evidenced that drought led to an increase in pH at forested sites. This was attributed to decreased dilution of calcium bicarbonate in groundwater during drought conditions. However, in Andean watersheds characterized by naturally high mineral concentration the outcome can be different. These watersheds often exhibit low pH values from acid rock and mine drainage, which can be neutralized to different extents by freshwater inputs (Guerra et al., 2016; Montecinos et al., 2020).

During drought, the reduction in precipitation and snowpack can diminish the mobilization of enriched material, resulting in localized increases in pH in impacted rivers. However, lower flows and limited dilution capacity can reduce the neutralization of freshwater, ultimately leading to an overall decrease in pH observed throughout the watersheds.

The abovementioned consequences of drought on pH of Andean watershed can interact with the outcomes in the processes and parameters analyzed earlier in this chapter, as depicted by Fig. 6.5.

6.3 Water Quality Challenges from Droughts in Chile

The previous section highlighted key connections between droughts, hydrometeorological drivers, and biogeochemical-biophysical processes that trigger shifts in hydrochemical parameters. These hydrochemical changes imply two challenges

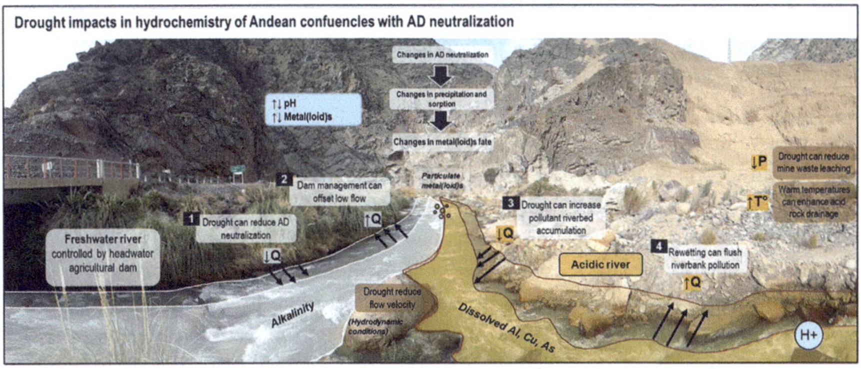

Fig. 6.5 Drought impacts on water quality in Andean confluences with acid drainage neutralization, using the Upper Elqui River watershed as an example (Díaz et al., 2023). This figure illustrates representative confluences that involve acid drainage neutralization, with their dynamics, equilibrium, and resulting water quality affected by drought drivers (precipitation and temperature) and various processes. These processes encompass snowmelt timing, sediment and metal export, as well as metal precipitation and deposition. Source: author's own work

that will be discussed here. First, a significant data gap in water quality and quantity monitoring networks prevents an adequate assessment and prediction of site-specific drought impacts. Second, the extent of the impact from changes in hydrochemical parameters largely depends on the water intended uses, as distinct limits affect its suitability for different activities. Thus, a careful analysis linking specific parameters to the impacts on the societal and ecological uses of water is necessary.

6.3.1 *Monitoring of Water Quality in the Drought Context*

The analysis of drought impacts relies on data related to hydrochemistry, meteorology and/or hydrology. In the Chilean studies mentioned above, various sources of data were utilized, with a significant portion obtained from the National River Monitoring Network of the National Water Direction/Chilean National Water Agency (Dirección General de Aguas, DGA). Meteorological, hydrological, and hydrochemical datasets are publicly available and are collected by the DGA network of monitoring stations. These stations include water quality sampling, stream gauging, suspended sediments, groundwater sampling, groundwater levels, lake/lagoon sampling, and lake/lagoon levels. Precipitation and temperature records are also obtained from monitoring stations managed by the Chilean Weather Service Agency (Dirección Meteorológica de Chile, DMC).

The length of the records from both institutions is not uniform and often contains a significant percentage of missing data. Only 82 and 343 thermometric stations and precipitation stations across Chile have records spanning 30 years or more (Pastén et al., 2021). Given the scarcity of meteorological stations, spatial interpolation methods often involve high levels of uncertainty, but this challenge has been addressed through gridded products generated from the "National Water Balance Update" by DGA in 2017 (DGA, 2017). Nonetheless, one of the primary challenges in studies related to water quality and droughts is to establish a comprehensive and continuous dataset for stream discharge and water quality simultaneously, encompassing numerous watersheds across Chile.

The water quality monitoring programs conducted by DGA typically encompass a minimum of four sampling campaigns per year, covering the seasons of summer, fall, winter, spring for surface waters. Groundwater monitoring involves two campaigns annually. These monitoring programs incorporate a wide range of parameters, with the more comprehensive programs focused on secondary ambient water quality standards (NSCAs). Parameters that are occasionally measured include chlorophyll "a", turbidity, pesticides, phosphate, dissolved oxygen, and bicarbonate. However, the lack of systematic measurements across all the entire network, especially for critical parameters such as BOD, alkalinity, turbidity, suspended solids, and dissolved metals, limits the potential for analyzing water quality outcomes resulting from both natural and anthropic impacts in certain watersheds (Pastén et al., 2021). Addressing this challenge requires efforts to overcome current limitations in budget, infrastructure, and logistics to enhance the number and frequency of measured parameters.

Given the relatively limited literature on the impact of droughts on water quality in South America (Mosley, 2015), and particularly in Chile, it is crucial to improve the river discharge monitoring network and sustain long-term water quality monitoring programs. In some studies, the availability of data was relatively low due to the quarterly and/or bimonthly monitoring frequency conducted by the DGA (Hernandez et al., 2023). To achieve more in-depth analysis of water quality during different seasonal droughts and understand the role of natural conditions, more frequent and comprehensive monitoring is required (Peña-Guerrero et al., 2020). Furthermore, newly relevant parameters such as sediment contaminants and emerging contaminants should be systematically included in the monitoring network, with a focus on prioritizing specific watersheds (Pastén et al., 2021). This expansion aims to enhance our understanding of the fluxes, dynamics, and impacts of contaminants from wastewater sources, including industrial effluents, urban and rural sewage, especially in the context of drought and climate change. These factors pose challenges to water availability and necessitate measures in water consumption and allocation management.

Data related to these pollutants may exist from sampling campaigns conducted by DGA and other agencies (e.g., Ministry of Environment-MMA, maritime water agency-DIRECTEMAR, Superintendence of Sanitary Services-SISS, etc.), albeit with limited spatial and temporal coverage within the Observation Network in surveillance programs of NSCA (referred to as PMCCA). Expanding data collection and analysis in this area could provide valuable insights into the impacts on healthy ecosystems and the potential (re)use of waters from rivers receiving wastewater, all within the context of drought and climate change. These efforts are essential to address the challenges posed by decreasing water availability and the need for effective water resource management.

6.3.2 Socioeconomic Implications of the Drought in Chile on Water Quality

Water quality for various purposes in Chile is regulated by the technical standard NCh1333/87. This standard delineates specific thresholds for physicochemical parameters for irrigation and mandates that potable water adheres to the technical norm NCh409/1. Agriculture, accounting for approximately 73% of Chile's total water consumption (ODEPA, 2021), is significantly affected when drought degrades water quality. This degradation, in turn, adversely impacts irrigated farming, leading to socioeconomic setbacks (Peña-Guerrero et al., 2020). Even though NCh1333/87 prescribes standards for 27 chemical parameters, salinity emerges as the principal determinant in evaluating water quality for irrigation (El Sabagh et al., 2020). A noticeable inverse correlation exists between periods of low water inflow and heightened salinity. Soil afflicted by salinization displays diminished productivity and impedes efficient water uptake by plants. Highlighting the repercussions of increased water salinity due to drought in Chile, studies by Arroyo et al. (2020) and Ortiz et al. (2023) have documented its adverse effects on seed set and grass respectively.

Droughts have also taken a toll on animal populations. As stated by the Chilean Ministry of Agriculture, 2019 witnessed the death of over 10,000 animals, predominantly goats, cattle, and sheep, primarily due to water scarcity and insufficient fodder. Moreover, wild carnivores, adjusting to the drought's effects, have altered their hunting preferences from wild prey to domesticated animals (Acosta-Jamett et al., 2016). Fish are not immune to these adversities either; they face threats from harmful algal blooms (HAB). The prolonged drought in Chile has led to elevated salt and nutrient levels in water systems, conditions that favor the proliferation of HAB (Chen et al., 2021). Supporting this, Quiñones et al. (2019) have highlighted the detrimental effects of HAB, coupled with other environmental challenges, on both shellfish and salmon aquaculture.

Focusing on potable water, according to SISS (2022), 56.4% of the drinking water supplied in Chile in urban areas is groundwater, 42.1% is surface water and 1.5% is desalinated water. Fluctuations in the quality of raw water require adjustments in the physico-chemical processes at drinking water treatment plants to comply with the standards outlined in the technical norm NCh409/1. Addressing this, Chilean water service providers invested approximately US$317 million in 2020 to guarantee the supply of potable water, both in terms of volume and quality (ANDESS, 2020). However, in rural regions of Chile, the situation remains concerning. It is estimated that about 47.2% of the rural populace, encompassing roughly 383,294 households or approximately one million individuals, lack consistent access to drinking water that meets standards for volume and quality. Of these individuals, a staggering 84.7% rely on untreated groundwater or surface water, which is neither monitored for quality nor treated (Amulén, 2020). Consequently, shifts in water quality, exacerbated by drought, pose significant health risks to these communities.

6.4 Conclusions

The interconnections between water quality and drought phenomena have been investigated by several groups in North America, Europe, and Asia. In contrast, the research on the impact of drought-induced changes in water quality in South America, particularly in Chilean Andean watersheds, remains comparatively limited. Given the diverse range of climatic, meteorological, and hydrological conditions present in Chile, it becomes imperative to develop an overarching conceptual model that elucidates the influences of drought on site-specific water quality dynamics. This approach is further warranted by the distinctive hydrochemical environments prevalent across the country.

This work underscores the necessity for a comprehensive conceptual framework that accounts for the variations across altitudes within a watershed. Pronounced disparities encompassing climatic, meteorological, hydrological, topographical, and vegetation attributes, alongside a range of anthropogenic factors that differ between distinct altitudes. Notably, the examination of just four research papers pertaining to Chilean Andean watersheds accentuates the need to expand the understanding of local hydrochemical processes that govern the hydrochemical dynamics.

Among the primary factors intensifying water quality perturbations during drought events are high temperatures, altered stream flows, and their spatial distribution. These drivers set a cascading effect of processes within the watershed, which encompass changes in snowmelt, erosion, wildfires, transport, groundwater recharge, mixing zones, chemical reactions, evapoconcentration phenomena, residence times, and the intricate interplay of nutrient and carbon cycles. For instance, drought conditions result in a significant rise in salinity levels and specific conductance in streams and forested areas, due to prolonged evaporation and limited dilution from saline groundwater sources. Moreover, reduced water availability diminishes dilution capacity of rivers and streams, leading to elevated concentrations of major ions, nutrients, metals, and toxic compounds delivered from anthropogenic wastewater discharges. Other hydrochemical alterations in natural systems associated to changes in rainfall and temperatures include shifts in seasonal patterns of parameters levels, and shifts in the nutrient cycle, including mineral phases, soil-plant interfaces, as well as storage and dynamics within lakes and reservoirs. While these processes are inherently operative within Andean watersheds, their relative significance exhibits marked variation. For instance, in river systems governed by snow-dominated regimes, the timing of snowmelt becomes critical, whereas corresponding effects on lagoons and reservoirs are more unclear and necessitate more in-depth investigation.

The intricate dynamics of chemical processes and the phenomena of mixing zones emerge as key determinants in the generation and neutralization of acid drainage, ultimately dictating the fate of metals and metalloids. This aspect assumes heightened importance within Andean watersheds due to the pronounced incidence of natural acid drainage generation and extensive mining activities. Given the nascent level of exploration in this domain, it is crucial to enhance and sustain our monitoring systems to acquire a more comprehensive dataset for decision and public policy making.

Chile faces considerable water quality challenges for specific purposes, which are further compounded by drought conditions and are governed by technical standards and regulations. Agriculture, as the primary water user, faces setbacks during drought-induced quality degradation, especially due to increased salinity. Livestock and wildlife are affected by water scarcity and impaired quality; for example, wild and farmed fish contend with harmful algal blooms linked to elevated salt and nutrient levels. Urban areas in Chile rely on a combination of groundwater, surface water, and desalination for drinking water, prompting treatment plants to make necessary adjustments due to prolonged drought conditions. Conversely, rural regions in Chile face significant hurdles, lacking access to safe drinking water due to low levels of treatment and monitoring.

As the impacts of climate change continue to manifest across Chile, there emerges a compelling need to refine our comprehension of how drought events influence hydrochemical dynamics. In this context, the harmonization of streamflow gaging and water quality monitoring stations, alongside the alignment of monitoring schedules, is relevant. This coordinated approach will enable us to unravel the intricate nexus between changing climatic patterns, drought events, and the ensuing consequences for water quality within Chilean Andean watersheds.

Acknowledgements The authors gratefully acknowledge the research support provided by ANID FSEQ210018, ANID FONDECYT 1200984, and CEDEUS, ANID FONDAP 1523A0004.

References

Abarca, M., Guerra, P., Arce, G., Montecinos, M., Escauriaza, C., Coquery, M., & Pastén, P. (2017). Response of suspended sediment particle size distributions to changes in water chemistry at an Andean mountain stream confluence receiving arsenic rich acid drainage. *Hydrological Processes, 31*, 296–307.

Aceituno, P. (1988). On the functioning of the southern oscillation in the South American sector. Part I: Surface climate. *Monthly Weather Review, 116*, 505–524.

Acosta-Jamett, G., Gutiérrez, J. R., Kelt, D., Meserve, P., & Previtali, M. A. (2016). El Niño southern oscillation drives conflict between wild carnivores and livestock farmers in a semiarid area in Chile. *Journal of Arid Environments, 126*, 76–80.

Ahmadi, B., & Moradkhani, H. (2019). Revisiting hydrological drought propagation and recovery considering water quantity and quality. *Hydrological Processes, 33*, 1492–1505.

Aldunce, P., Araya, D., Sapiain, R., Ramos, I., Lillo, G., Urquiza, A., & Garreaud, R. (2017). Local perception of drought impacts in a changing climate: The mega-drought in Central Chile. *Sustainability, 9*, 2053.

Alnahit, A. O., Mishra, A. K., & Khan, A. A. (2020). Quantifying climate, streamflow, and watershed control on water quality across Southeastern US watersheds. *Science of the Total Environment, 739*, 139945.

Amulén (2020). *Pobres de Agua. Radiografía del agua rural de Chile: Visualización de un problema oculto*. https://www.fundacionamulen.cl/wp-content/uploads/2020/07/Informe_Amulen.pdf

Anawar, H. M. (2013). Impact of climate change on acid mine drainage generation and contaminant transport in water ecosystems of semi-arid and arid mining areas. *Physics and Chemistry of the Earth, Parts a/b/c, 58*, 13–21.

Andersen, C. B., Lewis, G. P., & Sargent, K. A. (2004). Influence of wastewater-treatment effluent on concentrations and fluxes of solutes in the Bush River, South Carolina, during extreme drought conditions. *Environmental Geosciences, 11*, 28–41.

ANDESS. (2020). *Empresas de agua y saneamiento. Usos del Agua en Chile.*

Arroyo, M. T., Robles, V., Tamburrino, Í., Martínez-Harms, J., Garreaud, R. D., Jara-Arancio, P., Pliscoff, P., Copier, A., Arenas, J., & Keymer, J. (2020). Extreme drought affects visitation and seed set in a plant species in the central Chilean Andes heavily dependent on hummingbird pollination. *Plants, 9*, 1553.

Arumí, J. L., Rivera, D., Muñoz, E., & Billib, M. (2012). Interacciones entre el agua superficial y subterránea en la región del Bío Bío de Chile. *Obras y proyectos*, 4–13.

Ayala, Á., Farías-Barahona, D., Huss, M., Pellicciotti, F., McPhee, J., & Farinotti, D. (2020). Glacier runoff variations since 1955 in the Maipo River basin, in the semiarid Andes of Central Chile. *The Cryosphere, 14*(6), 2005–2027.

Baldwin, D. S., Gigney, H., Wilson, J. S., Watson, G., & Boulding, A. N. (2008). Drivers of water quality in a large water storage reservoir during a period of extreme drawdown. *Water Research, 42*, 4711–4724.

Balistrieri, L. S., Seal, R. R., II, Piatak, N. M., & Paul, B. (2007). Assessing the concentration, speciation, and toxicity of dissolved metals during mixing of acid-mine drainage and ambient river water downstream of the Elizabeth Copper Mine, Vermont, USA. *Applied Geochemistry, 22*, 930–952.

Bambach, N., Meza, F. J., Gilabert, H., & Miranda, M. (2013). Impacts of climate change on the distribution of species and communities in the Chilean Mediterranean ecosystem. *Regional Environmental Change, 13*, 1245–1257.

Barrett, B. S., & Hameed, S. (2017). Seasonal variability in precipitation in central and southern Chile: Modulation by the South Pacific high. *Journal of Climate, 30*, 55–69.

Barros, M. C., Mendo, M. J. M., & Negrao, F. C. (1995). Surface water quality in Portugal during a drought period. *Science of the Total Environment, 171*, 69–76.

Bates, B., Kundzewicz, Z., & Wu, S. (2008). *Climate change and water*. Intergovernmental Panel on Climate Change Secretariat.

Boisier, J. P., Rondanelli, R., Garreaud, R. D., & Muñoz, F. (2016). Anthropogenic and natural contributions to the Southeast Pacific precipitation decline and recent megadrought in Central Chile. *Geophysical Research Letters, 43*, 413–421.

Bonsal, B. R., Wheaton, E. E., Meinert, A., & Siemens, E. (2011). Characterizing the surface features of the 1999–2005 Canadian prairie drought in relation to previous severe twentieth century events. *Atmosphere-Ocean, 49*, 320–338.

Bozkurt, D., Rojas, M., Boisier, J. P., & Valdivieso, J. (2017). Climate change impacts on hydroclimatic regimes and extremes over Andean basins in Central Chile. *Hydrology and Earth System Sciences Discussions, 2017*, 1–29.

Brooks, B. W., Valenti, T. W., Cook-Lindsay, B., Forbes, M. G., Doyle, R. D., Scott, J. T., & Stanley, J. K. (2011). *Influence of climate change on reservoir water quality assessment and management: Effects of reduced inflows on diel pH and site-specific contaminant hazards*. Springer.

Burger, F., Brock, B., & Montecinos, A. (2018). Seasonal and elevational contrasts in temperature trends in Central Chile between 1979 and 2015. *Global and Planetary Change, 162*, 136–147.

Caruso, B. S. (2001). Regional river flow, water quality, aquatic ecological impacts and recovery from drought. *Hydrological Sciences Journal, 46*, 677–699.

Caruso, B. (2002). Temporal and spatial patterns of extreme low flows and effects on stream ecosystems in Otago, New Zealand. *Journal of Hydrology, 257*, 115–133.

Casiot, C., Egal, M., Elbaz-Poulichet, F., Bruneel, O., Bancon-Montigny, C., Cordier, M.-A., Gomez, E., & Aliaume, C. (2009). Hydrological and geochemical control of metals and arsenic in a Mediterranean river contaminated by acid mine drainage (the Amous River, France); preliminary assessment of impacts on fish (Leuciscus cephalus). *Applied Geochemistry, 24*, 787–799.

Castro, L., & Gironás, J. (2021). Precipitation, temperature and evaporation. *Water Resources of Chile*, 31–60.

Cerdan, O., Govers, G., Le Bissonnais, Y., Van Oost, K., Poesen, J., Saby, N., Gobin, A., Vacca, A., Quinton, J., & Auerswald, K. (2010). Rates and spatial variations of soil erosion in Europe: A study based on erosion plot data. *Geomorphology, 122*, 167–177.

Chadwick, C., Gironás, J., Vicuña, S., Meza, F., & McPhee, J. (2018). Using a statistical preanalysis approach as an ensemble technique for the unbiased mapping of GCM changes to local stations. *Journal of Hydrometeorology, 19*, 1447–1465.

Chen, J., Wang, J., Wang, Q., Lv, J., Liu, X., Chen, J., & Li, N. (2021). Common fate of sister lakes in Hulunbuir grassland: Long-term harmful algal bloom crisis from multi-source remote sensing insights. *Journal of Hydrology, 594*, 125970.

Crouch, C. M., McKnight, D. M., & Todd, A. S. (2013). Quantifying sources of increasing zinc from acid rock drainage in an alpine catchment under a changing hydrologic regime. *Hydrological Processes, 27*, 721–733.

Crusberg, T. C., & Eslamian, S. (2017). *Drought and water quality. Handbook of drought and water scarcity* (pp. 205–218). CRC Press.

Dai, A. (2011). Drought under global warming: A review. *Wiley Interdisciplinary Reviews: Climate Change, 2*, 45–65.

Delpla, I., Jung, A.-V., Baures, E., Clement, M., & Thomas, O. (2009). Impacts of climate change on surface water quality in relation to drinking water production. *Environment International, 35*, 1225–1233.

Deng, L., Peng, C., Kim, D.-G., Li, J., Liu, Y., Hai, X., Liu, Q., Huang, C., Shangguan, Z., & Kuzyakov, Y. (2021). Drought effects on soil carbon and nitrogen dynamics in global natural ecosystems. *Earth-Science Reviews, 214*, 103501.

Descroix, L., & Mathys, N. (2003). Processes, spatio-temporal factors and measurements of current erosion in the French Southern Alps: A review. *Earth Surface Processes and Landforms: The Journal of the British Geomorphological Research Group, 28*, 993–1011.

DGA. (2017). *Actualización del Balance Hídrico Nacional.* Dirección General de Aguas, División de Estudios y Planificación, Santiago, Chile. https://repositoriodirplan.mop.gob.cl/biblioteca/items/fc5370ee-1498-41b7-8ff9-58d529168e0a

Díaz, J., Castillo, D., Oyarzún, R., Briso, A., Montecinos, M., & Pastén, P. (2023). The dynamics of arsenic and copper in solid and aqueous phases in reactive confluences receiving acid drainage: The role of turbidity and particle size. *Environmental Pollution, 327*, 121449.

Djabri, L., Bouhsina, S., Hani, A., Bosch, A., Mudry, J., & Cerif Djouamaa, M. (2014). Impacts of drought on water quality: The case of aquifers in eastern Algeria. *Proceedings of IAHS, 364*, 357–362.

Dorrell, R., Amy, L., Peakall, J., & McCaffrey, W. (2018). Particle size distribution controls the threshold between net sediment erosion and deposition in suspended load dominated flows. *Geophysical Research Letters, 45*, 1443–1452.

Duhalde, D., Castillo, D., Oyarzún, R., Oyarzún, J., & Arumí, J. L. (2021). Mining and industrial uses. *Water Resources of Chile*, 273–284.

El Sabagh, A., Hossain, A., Barutçular, C., Iqbal, M. A., Islam, M. S., Fahad, S., Sytar, O., Çiğ, F., Meena, R. S., & Erman, M. (2020). Consequences of salinity stress on the quality of crops and its mitigation strategies for sustainable crop production: An outlook of arid and semi-arid regions. *Environment, Climate, Plant and Vegetation Growth*, 503–533.

Elsdon, T. S., De Bruin, M. B., Diepen, N. J., & Gillanders, B. M. (2009). Extensive drought negates human influence on nutrients and water quality in estuaries. *Science of the Total Environment, 407*, 3033–3043.

Fabian, P. S., Kwon, H.-H., Vithanage, M., & Lee, J.-H. (2023). Modeling, challenges, and strategies for understanding impacts of climate extremes (droughts and floods) on water quality in Asia: A review. *Environmental Research, 115617*.

FAO. (2002). *Report of FAO-CRIDA expert group consultation on farming system and best practices for drought-prone areas of Asia and the Pacific Region*. Central Research Institute for Dryland Agriculture.

Fernández, B., Gironás, J. (2021). Droughts. *Water Resources of Chile*, 173–187.

Fernández-Raga, M., Palencia, C., Keesstra, S., Jordán, A., Fraile, R., Angulo-Martínez, M., & Cerdà, A. (2017). Splash erosion: A review with unanswered questions. *Earth-Science Reviews, 171*, 463–477.

Fuentealba, M., Bahamóndez, C., Sarricolea, P., Meseguer-Ruiz, O., & Latorre, C. (2021). The 2010–2020 'megadrought' drives reduction in lake surface area in the Andes of Central Chile (32°–36°S). *Journal of Hydrology: Regional Studies, 38*, 100952.

Garreaud, R. D., Alvarez-Garreton, C., Barichivich, J., Boisier, J. P., Christie, D., Galleguillos, M., LeQuesne, C., McPhee, J., & Zambrano-Bigiarini, M. (2017). The 2010–2015 megadrought in Central Chile: Impacts on regional hydroclimate and vegetation. *Hydrology and Earth System Sciences, 21*, 6307–6327.

Garreaud, R. D., Boisier, J. P., Rondanelli, R., Montecinos, A., Sepúlveda, H. H., & Veloso-Aguila, D. (2020). The Central Chile mega drought (2010–2018): A climate dynamics perspective. *International Journal of Climatology, 40*, 421–439.

Giri, S., Mishra, A., Zhang, Z., Lathrop, R. G., & Alnahit, A. O. (2021). Meteorological and hydrological drought analysis and its impact on water quality and stream integrity. *Sustainability, 13*, 8175.

Glantz, M. (2003). *Climate affairs: A primer*. Island Press.

Gleick, P. H. (2000). A look at twenty-first century water resources development. *Water International, 25*, 127–138.

Glibert, P. M., Dugdale, R. C., Wilkerson, F., Parker, A. E., Alexander, J., Antell, E., Blaser, S., Johnson, A., Lee, J., & Lee, T. (2014). Major–but rare–spring blooms in 2014 in San Francisco Bay Delta, California, a result of the long-term drought, increased residence time, and altered nutrient loads and forms. *Journal of Experimental Marine Biology and Ecology, 460*, 8–18.

Guerra, P., Gonzalez, C., Escauriaza, C., Pizarro, G., & Pasten, P. (2016). Incomplete mixing in the fate and transport of arsenic at a river affected by acid drainage. *Water, Air, & Soil Pollution, 227*, 1–20.

Hernandez, V., Arumí, J. L., Boll, J., Duhalde, D., MacDonell, S., & Oyarzún, R. (2023). Streamflow–concentration relationships of surface water in the Choapa basin: Historical analysis and projections under climate change. *Hydrological Sciences Journal, 68*(9), 1250–1263.

Hrdinka, T., Novický, O., Hanslík, E., & Rieder, M. (2012). Possible impacts of floods and droughts on water quality. *Journal of Hydro-Environment Research, 6*, 145–150.

Jirsa, F., Gruber, M., Stojanovic, A., Omondi, S. O., Mader, D., Körner, W., & Schagerl, M. (2013). Major and trace element geochemistry of Lake Bogoria and Lake Nakuru, Kenya, during extreme draught. *Geochemistry, 73*, 275–282.

Johnson, D. W., Todd, D. E., Jr., & Hanson, P. J. (2008). Effects of throughfall manipulation on soil nutrient status: Results of 12 years of sustained wet and dry treatments. *Global Change Biology, 14*, 1661–1675.

Jorquera, C. O., Oates, C. J., Plant, J. A., Kyser, K., Ihlenfeld, C., & Voulvoulis, N. (2015). Regional hydrogeochemical mapping in Central Chile: Natural and anthropogenic sources of elements and compounds. *Geochemistry: Exploration, Environment, Analysis, 15*(1), 72–96.

Kinnell, P. (2005). Raindrop-impact-induced erosion processes and prediction: A review. *Hydrological Processes: An International Journal, 19*, 2815–2844.

Kogan, F. N. (1997). Global drought watch from space. *Bulletin of the American Meteorological Society, 78*, 621–636.

Léonard, J., & Richard, G. (2004). Estimation of runoff critical shear stress for soil erosion from soil shear strength. *Catena, 57*, 233–249.

Li, T., Li, S., Liang, C., Bush, R. T., Xiong, L., & Jiang, Y. (2018). A comparative assessment of Australia's lower lakes water quality under extreme drought and post-drought conditions using multivariate statistical techniques. *Journal of Cleaner Production, 190*, 1–11.

Lisboa, M. S., Schneider, R. L., Sullivan, P. J., & Walter, M. T. (2020). Drought and post-drought rain effect on stream phosphorus and other nutrient losses in the Northeastern USA. *Journal of Hydrology: Regional Studies, 28*, 100672.

Loiselle, D., Du, X., Alessi, D. S., Bladon, K. D., & Faramarzi, M. (2020). Projecting impacts of wildfire and climate change on streamflow, sediment, and organic carbon yields in a forested watershed. *Journal of Hydrology, 590*, 125403.

Mardones, P., & Garreaud, R. D. (2020). Future changes in the free tropospheric freezing level and rain–snow limit: The case of Central Chile. *Atmosphere, 11*, 1259.

Martínez-Retureta, R., Aguayo, M., Abreu, N. J., Stehr, A., Duran-Llacer, I., Rodríguez-López, L., Sauvage, S., & Sánchez-Pérez, J.-M. (2021). Estimation of the climate change impact on the hydrological balance in basins of south-Central Chile. *Water, 13*, 794.

Masiokas, M. H., Villalba, R., Luckman, B. H., Le Quesne, C., & Aravena, J. C. (2006). Snowpack variations in the Central Andes of Argentina and Chile, 1951–2005: Large-scale atmospheric influences and implications for water resources in the region. *Journal of Climate, 19*, 6334–6352.

Masson, M., Schäfer, J., Blanc, G., & Pierre, A. (2007). Seasonal variations and annual fluxes of arsenic in the Garonne, Dordogne and Isle Rivers, France. *Science of the Total Environment, 373*, 196–207.

Méndez-Freire, V., Villaseñor, T., & Mellado, C. (2022). Spatial and temporal changes in suspended sediment fluxes in Central Chile induced by the mega drought: The case of the Itata River Basin (36°–37°S). *Journal of South American Earth Sciences, 118*, 103930.

Meyer, J. L., Sale, M. J., Mulholland, P. J., & Poff, N. L. (1999). Impacts of climate change on aquatic ecosystem functioning and health 1. *JAWRA Journal of the American Water Resources Association, 35*, 1373–1386.

Meza, F. J., Vicuña, S., Gironás, J., Poblete, D., Suárez, F., & Oertel, M. (2015). Water–food–energy nexus in Chile: The challenges due to global change in different regional contexts. *Water International, 40*, 839–855.

Meza, F., Gil, P., Melo, O. (2021). Agricultural uses. *Water Resources of Chile*, 243–258.

Mishra, A. K., & Singh, V. P. (2010). A review of drought concepts. *Journal of Hydrology, 391*, 202–216.

Mishra, A., Alnahit, A., & Campbell, B. (2021). Impact of land uses, drought, flood, wildfire, and cascading events on water quality and microbial communities: A review and analysis. *Journal of Hydrology, 596*, 125707.

Momblanch, A., Paredes-Arquiola, J., Munné, A., Manzano, A., Arnau, J., & Andreu, J. (2015). Managing water quality under drought conditions in the Llobregat River Basin. *Science of the Total Environment, 503*, 300–318.

Montecinos, A., & Aceituno, P. (2003). Seasonality of the ENSO-related rainfall variability in Central Chile and associated circulation anomalies. *Journal of Climate, 16*, 281–296.

Montecinos, M., Coquery, M., Alsina, M. A., Bretier, M., Gaillard, J.-F., Dabrin, A., & Pastén, P. (2020). Partitioning of copper at the confluences of Andean rivers. *Chemosphere, 259*, 127318.

Mosley, L. M. (2015). Drought impacts on the water quality of freshwater systems; review and integration. *Earth-Science Reviews, 140*, 203–214.
Mosley, L. M., & Fleming, N. (2010). Pollutant loads returned to the lower Murray River from flood-irrigated agriculture. *Water, Air, & Soil Pollution, 211*, 475–487.
Mosley, L. M., Zammit, B., Leyden, E., Heneker, T. M., Hipsey, M. R., Skinner, D., & Aldridge, K. T. (2012). The impact of extreme low flows on the water quality of the Lower Murray River and Lakes (South Australia). *Water Resources Management, 26*, 3923–3946.
Mosley, L., Zammit, B., Jolley, A., & Barnett, L. (2014). Acidification of lake water due to drought. *Journal of Hydrology, 511*, 484–493.
Müller, A., Österlund, H., Marsalek, J., & Viklander, M. (2020). The pollution conveyed by urban runoff: A review of sources. *Science of the Total Environment, 709*, 136125.
Murphy, L. U., Cochrane, T. A., & O'Sullivan, A. (2015). Build-up and wash-off dynamics of atmospherically derived Cu, Pb, Zn and TSS in stormwater runoff as a function of meteorological characteristics. *Science of the Total Environment, 508*, 206–213.
Núnez, J. H., Verbist, K., Wallis, J. R., Schaefer, M. G., Morales, L., & Cornelis, W. (2011). Regional frequency analysis for mapping drought events in north-Central Chile. *Journal of Hydrology, 405*, 352–366.
ODEPA. (2021). Oficina de Estudios y Políticas Agrarias (ODEPA). *Agua para la agricultura.*
Oertel, M., Meza, F. J., & Gironas, J. (2020). Observed trends and relationships between ENSO and standardized hydrometeorological drought indices in Central Chile. *Hydrological Processes, 34*, 159–174.
Ortiz, J., Hernández-Fuentes, C., Sáez, P. L., Fernández Del Saz, N., & Cavieres, L. A. (2023). Compound and simple leaf woody species of the Chilean matorral are equally affected by extreme drought. *Plant Ecology, 224*, 33–45.
Oyarzun, C., Campos, H., & Huber, A. (1997). Nutrient export from watersheds with different land uses in southern Chile (Lake Rupanco, X region). *Revista Chilena de Historia Natural, 70*, 507–519.
Paerl, H. W., & Huisman, J. (2009). Climate change: A catalyst for global expansion of harmful cyanobacterial blooms. *Environmental Microbiology Reports, 1*, 27–37.
Parry, S., Prudhomme, C., Wilby, R. L., & Wood, P. J. (2016a). Drought termination: Concept and characterisation. *Progress in Physical Geography, 40*, 743–767.
Parry, S., Wilby, R. L., Prudhomme, C., & Wood, P. J. (2016b). A systematic assessment of drought termination in the United Kingdom. *Hydrology and Earth System Sciences, 20*, 4265–4281.
Pastén, P., Vega, A., Lizama, K., Guerra, P., Pizarro, J. (2021). Water quality. *Water Resources of Chile*, 209–228.
Pasten, P.A., Montecinos, M., Díaz, J.A., Turri, A.M., Briso, A.E., Vega, A.S. (2023). *Arsenic and copper in arid and semiarid Andean rivers: controls of suspended particle size distribution by aluminum and iron geochemistry.* Goldschmidt 2023 Conference. GOLDSCHMIDT.
Peña-Guerrero, M. D., Nauditt, A., Muñoz-Robles, C., Ribbe, L., & Meza, F. (2020). Drought impacts on water quality and potential implications for agricultural production in the Maipo River Basin, Central Chile. *Hydrological Sciences Journal, 65*, 1005–1021.
Pepin, E., Carretier, S., Guyot, J.-L., & Escobar, F. (2010). Specific suspended sediment yields of the Andean rivers of Chile and their relationship to climate, slope and vegetation. *Hydrological Sciences Journal–Journal des Sciences Hydrologiques, 55*, 1190–1205.
Pittock, A. (1980). Patterns of climatic variation in Argentina and Chile—I precipitation, 1931–60. *Monthly Weather Review, 108*, 1347–1361.
Pizarro, J., Vergara, P. M., Cerda, S., & Briones, D. (2016). Cooling and eutrophication of southern Chilean lakes. *Science of the Total Environment, 541*, 683–691.
Poff, N., Brinson, M., & Day, J. (2002). Potential impacts on inland freshwater and coastal wetland ecosystems in the United States. *Aquatic Ecosystems & Global Climate Change*, 1–44.
Qiu, J., Shen, Z., & Xie, H. (2023). Drought impacts on hydrology and water quality under climate change. *Science of the Total Environment, 858*, 159854.

Quinones, R. A., Fuentes, M., Montes, R. M., Soto, D., & León-Muñoz, J. (2019). Environmental issues in Chilean salmon farming: A review. *Reviews in Aquaculture, 11*, 375–402.
Quintana, J., & Aceituno, P. (2012). Changes in the rainfall regime along the extratropical west coast of South America (Chile): 30-43° S. *Atmosfera, 25*, 1–22.
Reynolds, C. S., Maberly, S. C., Parker, J. E., & De Ville, M. M. (2012). Forty years of monitoring water quality in Grasmere (English Lake District): Separating the effects of enrichment by treated sewage and hydraulic flushing on phytoplankton ecology. *Freshwater Biology, 57*, 384–399.
Rickson, R. (2014). Can control of soil erosion mitigate water pollution by sediments? *Science of the Total Environment, 468*, 1187–1197.
Sardans, J., & Peñuelas, J. (2007). Drought changes phosphorus and potassium accumulation patterns in an evergreen Mediterranean forest. *Functional Ecology, 21*, 191–201.
Schaeffer, S. M., Homyak, P. M., Boot, C. M., Roux-Michollet, D., & Schimel, J. P. (2017). Soil carbon and nitrogen dynamics throughout the summer drought in a California annual grassland. *Soil Biology and Biochemistry, 115*, 54–62.
Schoonen, M., Elsetinow, A., Borda, M., & Strongin, D. (2000). Effect of temperature and illumination on pyrite oxidation between pH 2 and 6. *Geochemical Transactions, 1*, 23–33.
Shimoda, Y., Azim, M. E., Perhar, G., Ramin, M., Kenney, M. A., Sadraddini, S., Gudimov, A., & Arhonditsis, G. B. (2011). Our current understanding of lake ecosystem response to climate change: What have we really learned from the north temperate deep lakes? *Journal of Great Lakes Research, 37*, 173–193.
SISS. (2022). *Informe de gestion de empresas sanitarias en Chile*. https://www.siss.gob.cl/586/articles-22969_recurso_1.pdf
Sprague, L. A. (2005). Drought effects on water quality in the South Platte river basin, Colorado 1. *JAWRA Journal of the American Water Resources Association, 41*, 11–24.
Todd, A. S., Manning, A. H., Verplanck, P. L., Crouch, C., McKnight, D. M., & Dunham, R. (2012). Climate-change-driven deterioration of water quality in a mineralized watershed. *Environmental Science & Technology, 46*, 9324–9332.
Tokunaga, T. K., Tran, A. P., Wan, J. M., Dong, W. M., Newman, A. W., Beutler, C. A., Brown, W., Henderson, A. N., & Williams, K. H. (2022). Quantifying subsurface flow and solute transport in a snowmelt-recharged hillslope with multiyear water balance. *Water Resources Research*, 58.
Tolorza, V., Carretier, S., Andermann, C., Ortega-Culaciati, F., Pinto, L., & Mardones, M. (2014). Contrasting mountain and piedmont dynamics of sediment discharge associated with groundwater storage variation in the Biobío River. *Journal of Geophysical Research: Earth Surface, 119*, 2730–2753.
Tran, Q. K., Jassby, D., & Schwabe, K. A. (2017). The implications of drought and water conservation on the reuse of municipal wastewater: Recognizing impacts and identifying mitigation possibilities. *Water Research, 124*, 472–481.
Tromans, D. (1998). Temperature and pressure dependent solubility of oxygen in water: A thermodynamic analysis. *Hydrometallurgy, 48*, 327–342.
Urdiales, D., Meza, F., Gironás, J., & Gilabert, H. (2018). Improving stochastic modelling of daily rainfall using the ENSO index: Model development and application in Chile. *Water, 10*, 145.
Vaezi, A. R., Ahmadi, M., & Cerdà, A. (2017). Contribution of raindrop impact to the change of soil physical properties and water erosion under semi-arid rainfalls. *Science of the Total Environment, 583*, 382–392.
Van Heemst, C., Willems, J., Weller, A., Van Verseveld, H., Caamaño, D., & Aránguiz, R. (2013). Flood defence alternatives for the lower Bío Bío River, Chile. *Obras y Proyectos, 14*, 22–33.
Van Vliet, M., & Zwolsman, J. (2008). Impact of summer droughts on the water quality of the Meuse river. *Journal of Hydrology, 353*, 1–17.
Vicuña, S., Gironás, J., Meza, F. J., Cruzat, M. L., Jelinek, M., Bustos, E., Poblete, D., & Bambach, N. (2013). Exploring possible connections between hydrological extreme events and climate change in central South Chile. *Hydrological Sciences Journal, 58*, 1598–1619.

Walling, D. E., & Collins, A. L. (2008). The catchment sediment budget as a management tool. *Environmental Science & Policy, 11*(2), 136–143.

Walsh, C. J. (2000). Urban impacts on the ecology of receiving waters: A framework for assessment, conservation and restoration. *Hydrobiologia, 431*, 107–114.

Westerling, A. L., Hidalgo, H. G., Cayan, D. R., & Swetnam, T. W. (2006). Warming and earlier spring increase western US forest wildfire activity. *Science, 313*, 940–943.

Whitehead, P. G., Wilby, R. L., Battarbee, R. W., Kernan, M., & Wade, A. J. (2009). A review of the potential impacts of climate change on surface water quality. *Hydrological Sciences Journal, 54*, 101–123.

Wilhite, D. A., & Glantz, M. H. (1985). Understanding: The drought phenomenon: The role of definitions. *Water International, 10*, 111–120.

Williamson, C. E., Saros, J. E., Vincent, W. F., & Smol, J. P. (2009). Lakes and reservoirs as sentinels, integrators, and regulators of climate change. *Limnology and Oceanography, 54*, 2273–2282.

World Bank. (2003). *Report on financing rapid onset natural disaster losses in India: A risk management approach*. Washington, DC.

Yevenes, M. A., Figueroa, R., & Parra, O. (2018). Seasonal drought effects on the water quality of the Biobío River, Central Chile. *Environmental Science and Pollution Research, 25*, 13844–13856.

Ylla, I., Sanpera-Calbet, I., Vázquez, E., Romaní, A. M., Munoz, I., Butturini, A., Sabater, S. (2010). Organic matter availability during pre-and post-drought periods in a Mediterranean stream. Global Change and River Ecosystems—Implications for Structure. *Function and Ecosystem Services*, 217–232.

Zambrano, F., Lillo-Saavedra, M., Verbist, K., & Lagos, O. (2016). Sixteen years of agricultural drought assessment of the BioBío region in Chile using a 250 m resolution vegetation condition index (VCI). *Remote Sensing, 8*, 530.

Chapter 7
Governance and Drought Management Policies in Chile

Daniela Rivera and Guillermo Donoso

Abstract This chapter analyzes the current institutional, governance and drought management situation in Chile, identifying the main authorities and agencies that are legally entrusted with competences in this area, and specifying the main characteristics of these functions. The most pertinent and applicable tools are subsequently reviewed, conducting a thorough conceptual and practical analysis of the declarations of water scarcity zones due to severe drought and agricultural emergency resulting from water deficit. Finally, it examines the existing gaps and challenges in addressing extreme hydrological events that are generally occurring and continue to be managed as emergencies or disasters.

Keywords Drought · Emergency · Water management · Institutional framework · Chile

7.1 Introduction

Although droughts occur frequently, the global approach tends to treat them as atypical and urgent events, addressing them primarily with reactive measures typically used for disaster or catastrophe management (Wilhite et al., 2000, 2005; Neuman, 2003; Wilhite, 2009; Brufao, 2012; Courtenay & Hayes, 2012; Fu et al., 2013; Paneque, 2015; Stein et al., 2016; Embid, 2018; Setuaín, 2018; Berbel & Esteban, 2019; Hervás-Gámez & Delgado-Ramos, 2019; Iglesias et al., 2019; Martín, 2019–2020, 2022, 2024; Arana & Miranzo, 2021). This approach has generally

D. Rivera (✉)
Faculty of Law and the Center for Water Law and Management, Pontificia Universidad Católica de Chile, Santiago, Chile
e-mail: dirivera@uc.cl

G. Donoso
Faculty of Agronomy and Natural Systems and the Center for Water Law and Management, Pontificia Universidad Católica de Chile, Santiago, Chile

D. Rivera, G. Donoso (eds.), *Droughts in Chile*, Global Issues in Water Policy 31,
https://doi.org/10.1007/978-3-031-85040-0_7

lacked effectiveness, efficiency, coordination, and timeliness, with limited progress made in mitigating drought-related risks (Wilhite et al., 2000, 2005; Neuman, 2003; Paneque, 2015; Martín, 2022). However, some countries have begun shifting from crisis or emergency management to risk management, as noted by various authors (Wilhite et al., 2000, 2005; Neuman, 2003; Wilhite, 2009; Brufao, 2012; Courtenay & Hayes, 2012; Fu et al., 2013; Paneque, 2015; Stein et al., 2016; Embid, 2018; Setuaín, 2018; Berbel & Esteban, 2019; Hervás-Gámez & Delgado-Ramos, 2019; Iglesias et al., 2019; Martín, 2019–2020, 2022, 2024; Arana & Miranzo, 2021).

Chile has faced numerous drought episodes throughout its history, some of which have been particularly critical due to their duration -exceeding 14 years in some areas- and their spatial impact, affecting a significant portion of the national territory beyond the typically driest regions (Center for Climate and Resilience Research (CR)2, 2015; Fernández & Gironás, 2021). Considering the extensive knowledge and experience on the subject, one might expect that the country has developed an institutional framework, governance structure, and water management system conducive to proactive drought management. Furthermore, this suggests that society has enhanced its resilience and adaptive capacity in this domain.

This chapter aims to evaluate the current institutional, governance, and management situation of drought issues in Chile through a methodology that includes an examination of the existing legal framework, specialized literature, jurisprudential pronouncements, and data related to key drought management tools. First, it identifies the main organizations, and their functions involved in this area; subsequently, it reviews the most relevant and commonly implemented instruments in drought management. Following this, the state of drought governance is addressed, and the chapter concludes with pertinent findings and recommendations.

7.2 Institutionality and Competencies Related to Drought Management

7.2.1 *State Administration*

According to a 2013 World Bank report, Chile has 43 institutions -public, private, and autonomous- with responsibilities related to water management. These institutions collectively perform 102 distinct functions in a fragmented and dispersed manner (Banco Mundial, 2013). This information warrants review, as changes have occurred that make it necessary to update or re-evaluate these assertions. This chapter outlines the principal agencies within the State Administration that are responsible for drought management, as stipulated by relevant legal provisions governing their activities. We focus on administrative agencies, since water is a national public good and it is these entities that hold (or should hold) the main powers in this field (Embid, 2018). Except for the cases we will highlight, there are no explicit references to drought within the regulatory framework of most of these authorities.

Instead, terms such as emergencies, disasters, and catastrophes, which encompass the aforementioned extreme hydrological phenomenon, are utilized in a more general context.

7.2.1.1 President of the Republic

The President of the Republic, at the head of the State Administration, possesses three general powers that can be invoked in drought situations: the declaration of a state of constitutional exception due to catastrophe caused by public calamity (articles 32 N°5 and 41 of the Political Constitution); the declaration of a catastrophe zone as per the authority established in Law N°16,282, concerning earthquakes or catastrophes; and the declaration of a sanitary alert or emergency following the provisions of Article 36 of the Sanitary Code.

The constitutional state of emergency for a catastrophe encompasses broad grounds and authorizes the President of the Republic to appoint officials responsible for overseeing and implementing recovery programs in the affected regions (Tapia, 2021).

Law No. 16,282 addresses emergencies with an emphasis on natural disasters (Soto et al., 2022). In the event of earthquakes or other catastrophes that cause significant damage to people or property, the President is authorized to issue a supreme decree identifying the affected areas. This decree grants the President the power to reallocate funds from the national budget to cover expenses related to the disaster, especially when institutions are unable to operate due to the depletion of their own resources (articles 1 and 10 Decree 104, which sets the consolidated, coordinated and systematized text of Title I of Law No. 16,282). The General Directorate of Water (Dirección General de Aguas, DGA) may, at the request of water users' organizations such as supervisory boards, canal associations, and water communities, undertake the financing of these entities' functions in the affected areas. The expenses incurred will be reimbursed by the beneficiaries, although, in certain qualified cases, they may be wholly or partially covered by the State (article 42 Decree 104).

In the event of an emergency, the two regulatory exceptions could be enacted, each governed by distinct statutes, objectives, and authorities. Concerning droughts, the predominant declaration, as indicated by the study of Soto et al. (2022), has been the declaration of a disaster under Law No. 16,282. There were ten such decrees issued between 2010 and 2019.

Regarding the declaration of a sanitary alert or emergency as outlined in Article 36 of the Sanitary Code, this applies in cases of epidemics, diseases, or emergencies that present a serious risk to public health or life. A situation of water shortage could be included in these circumstances (opinion No. E31,873, of 2020).

Furthermore, the President of the Republic holds the specific authority to declare water scarcity zones due to severe drought conditions, as stipulated in article 314 of the Water Code. In practice, this responsibility is delegated to the Minister of Ministry of Public Works (Ministerio de Obras Públicas, MOP).

7.2.1.2 MOP–DGA

The DGA, under the MOP, is the main administrative authority on water resources, tasked with functions related to droughts. It must declare threat warnings on water resources, including droughts and floods (article 299 b) N°4 of the Water Code), and communicate these to the Servicio Nacional de Prevención y Respuesta ante Desastres (SENAPRED).

From a more specific perspective, the General Directorate of Water (DGA) plays a crucial role throughout the entire process of issuing and applying decrees related to water scarcity zones, as outlined in Articles 314 and 315 of the Water Code. Through Resolution No. 1331 of 2022, later amended by Resolution No. 579 of 2023, the DGA established the criteria for classifying a drought as severe. It then requests and informs the President to declare a water scarcity zone (delegated to the Minister of MOP).

The DGA also mandates redistribution agreements from supervisory boards and managers of state works for resource development, when its administration is not handled by the State. Furthermore, it approves and enforces these redistribution agreements and has the authority to suspend the functions of supervisory boards and the sectioning of natural sources, thereby managing the redistribution of surface and underground waters directly.

In addition, the DGA may authorize water extractions and the execution of works in watercourses during the period that the water scarcity decree is active, prioritizing human consumption, sanitation, or domestic subsistence use. In the absence of legally established users' organizations, the DGA will manage the redistribution efforts to ensure equitable distribution among the respective users. The DGA also reports any extensions of the scarcity zone decrees.

On the other hand, when imposing fines for infractions of the Water Code, the DGA is authorized to increase the penalty amount by up to 75% if the violations occur in watercourses that have been intervened following a declaration of a water scarcity zone (article 173 bis N°2 a) of the Water Code).

7.2.1.3 MOP—Directorate of Hydraulic Works (Dirección de Obras Hidráulicas, DOH)

Also, under the jurisdiction of MOP, the Department of Hydraulics (DOH) addresses various phases of the design, construction, and operation of hydraulic works and infrastructure. This includes reservoirs, desalination plants, and the construction or deepening of wells, as outlined in article 17(e) of DFL 850.

7.2.1.4 Ministry of Agriculture (Ministerio de Agricultura, MINAGRI)—Agricultural Development Institute (Instituto de Desarrollo Agropecuario, INDAP)—National Irrigation Commission (Comisión Nacional de Riego, CNR)

Moving to the scope of action of another Ministry, the main instrument utilized by the Ministry of Agriculture (MINAGRI) in this context is the declaration of agricultural emergency due to water deficit (Ordinary No. 989, of 2011, from the Undersecretary of Agriculture (Subsecretaría de Agricultura); Resolutions by the Undersecretary of Agriculture No. 309, of 2021, and No. 769, of 2022). This declaration is independent and is not issued based on the same criteria as the decree of water scarcity zone due to severe drought, which is managed by the MOP.

Furthermore, two agencies associated with the Ministry of Agriculture (MINAGRI), namely the National Institute for Agricultural Development (INDAP) and the National Irrigation Commission (CNR), are vested with promotional powers. These include the provision of loans by INDAP to small farmers and financial support for agricultural irrigation, which may be relevant in the context of droughts. Such support is often realized through competitions and programs aimed at infrastructure development or contributing to water security and climate change adaptation (articles 1 and 3 of Law No. 18,910; articles 1, 3, and 3 ter of Law No. 18,450; Resolution (Exempt) No. 309 of 2021 from the Undersecretariat of Agriculture).

Additionally, INDAP has the authority to declare an emergency and/or disaster situation, allowing it to reallocate resources from its budget to assist small farmers affected by such conditions (Resolution (Exempt) No. 309 of 2021 from the Undersecretariat of Agriculture). This includes a special line of financing or incentives specifically designed for emergencies (Resolution (Exempt) No. 1049 of 2010 from INDAP).

7.2.1.5 Ministry of Energy (Ministerio de Energía, MINEN)—National Energy Commission (Comisión Nacional de Energía, CNE)

The Ministry of Energy (MINEN), following a prior report from the National Energy Commission (CNE), is authorized to issue rationing decrees due to drought conditions as outlined in article 163 DFL 4 of the General Law of Electric Services, and Decree 51 of 2021. The legal regulation does not define what constitutes a drought nor specify any connection with the water scarcity decree issued by the MOP or the declaration of an agricultural emergency due to water deficit by MINAGRI. This rationing measure was implemented during 1998 and 1999. While the severe drought at that time was the primary cause, other factors related to management have also been identified (Bernstein, 1999; Díaz et al., 2000). According to information provided by the Undersecretary of Energy, no further rationing decrees have been issued since then. However, measures to prevent, reduce, and manage generation deficits have been undertaken in accordance with article 163 of DFL 4.[1]

7.2.1.6 National Disaster Prevention and Response Service (Servicio Nacional de Prevención y Respuesta ante Desastres, SENAPRED)

In 2021, under the authority of the Ministry of the Interior and Public Safety (Ministerio del Interior y Seguridad Pública), and as the successor to the former National Emergency Office, the National Disaster Prevention and Response Service (SENAPRED) was established to provide guidance, organize, plan, and oversee disaster risk management. Given that drought is traditionally classified as an emergency, it is important to understand the concept of emergency within the scope of this agency's activities: "an event, or the imminent threat thereof, which disrupts the functioning of a community due to the manifestation of a natural, biological, or anthropogenic threat. This threat interacts with conditions of exposure, vulnerability, and coping capacity, resulting in one or more of the following effects: losses, human impacts, material damage, economic disruptions, or environmental impacts" (Law No. 21,364, article 2, letter b).

7.2.1.7 Regional Presidential Delegates (Delegados Presidenciales Regionales) and Provincial Presidential Delegates (Delegados Presidenciales Provinciales)

The regional and provincial presidential delegates, who are also connected to the President of the Republic through the Ministry of the Interior and Public Security, have responsibilities related to the prevention and management of emergencies and disasters in general (articles 2 ñ) and 4 e) DFL 1–19,175). Specifically, the regional presidential delegates are administratively tasked with requesting the DGA to declare a water scarcity zone, which is then evaluated and forwarded to the MOP if deemed appropriate (No. 3 DGA Resolution No. 1331, of 2022).

7.2.1.8 Regional Governments

Their responsibilities include addressing, preventing, and safeguarding against emergencies or catastrophes (article 16 j) DFL 1–19,175). Additionally, they are tasked with preparing and approving, in collaboration with the Regional Committee for Climate Change, the instruments for climate change management within the

[1] Along these lines are Supreme Decrees N°26, N°38, N°39 and N°58, of 2011; N°51 and N°87, of 2021; N°1, N°29, N°66, N°69 and N°74, of 2022; N°12 and N°37, of 2022.

region (article 16 o) DFL 1–19,175), which presumably encompass drought management.

7.2.1.9 Municipalities

At the communal level, municipalities have similar functions to those of regional governments. These entities must manage disaster risk in the commune's territory and prepare the communal plan for disaster risk reduction and the communal emergency plan, in accordance with articles 4(i) and 26 bis DFL 1 of 2006.

7.2.1.10 Ministry of National Defense (Ministerio de Defensa Nacional, MINDEF)—Armed Forces (Fuerzas Armadas)

In connection with the disaster prevention and response system, MINDEF plays a role in coordinating and implementing several provisions of Law N°21,364: the National Policy for Disaster Risk Reduction, the National Strategic Plan for Disaster Risk Reduction, and the National Emergency Plan (articles 16 to 18 of Law N°21,364). Additionally, the Armed Forces participate at various stages of the disaster management cycle established by Law N°21,364 (articles 17 and 18 of Law N°21.364).

7.2.1.11 Regional Ministerial Health Secretariats (Secretarías Regionales Ministeriales (SEREMI) de Salud)

Within the Ministry of Health (Ministerio de Salud), the SEREMI of Health play a critical role in the authorization, surveillance, oversight, control, and sanctioning of water supply systems via water trucks, as outlined in articles 1, 11, and 16 of Decree 41, 2018. Tank truck water supply, which initially gained prominence during drought conditions, has expanded beyond such scenarios for various reasons (Tapia, 2021).

7.2.1.12 Summary

Table 7.1 provides a summary of the principal administrative authorities and agencies with drought-related functions, categorized by their respective sectors within the State Administration's organizational framework. The table identifies 17 entities dispersed across seven sectors, reflecting a notable degree of dispersion. The largest number is concentrated in the sector led by the Ministry of the Interior and Public Security, but the main tools belong to another area; precisely that led by the MOP and MINAGRI.

Table 7.1 Main sectors, authorities and administrative agencies involved in drought management

Presidency of the republic	Interior and public safety	Public works	Agriculture	Energy	Defense	Health
– President of the republic	– SENAPRED – Regional Presidential Delegations – Provincial Presidential Delegations – Regional Governments – Municipalities	– MOP – DGA – DOH	– MINAGRI – CNR – INDAP	– MINEN – CNE	– MINDEF – Armed Forces	– SEREMI of Health

Source: Own elaboration

Table 7.2 Main functions related to drought management

Drought management functions	Authority/agency involved
– Declaration of a state of constitutional exception of catastrophe due to public calamity (drought). – Declaration of drought catastrophe zone. – Declaration of sanitary alert or emergency due to water deficit.	– President of the republic
– Declaration of water scarcity zone due to severe drought.	– President of the republic – MOP – DGA – Regional Presidential Delegate
– Declaration of threat alert associated with water resources. – Qualification of criteria that determine the character of severe drought. – Management of scarcity zone declaration (redistribution agreements, suspension of supervisory boards and sectioning, direct redistribution, authorization of abstractions and works, extension of declaration, etc.) – Application of increases in fines for infractions in watercourses intervened by a declaration of scarcity zone.	– DGA
– Design, construction and operation of water works and infrastructure.	– DOH
– Declaration of agricultural emergency due to water deficit.	– MINAGRI
– Bonus and subsidies related to the agricultural sector and irrigation works – Granting of loans and subsidies to small farmers and peasants	– CNR – INDAP
– Electrical rewiring due to drought situations	– MINEN – CNE
– Disaster risk management in the country and its associated instruments	– SENAPRED – MINDEF – Armed Forces – Regional Presidential Delegate – Provincial Presidential Delegate – Regional Governments – Municipalities
– Climate change management at the regional level	– Regional Governments
– Sanitary authorization, supervision, control and sanction of drinking water supply systems through water trucks	– SEREMI of Health

Source: Own elaboration

Table 7.2 outlines the primary functions associated with droughts. It is evident that general functions are more prevalent than those specifically focused on drought. This indicates that drought is addressed as one of the potential catastrophic or emergency situations requiring intervention by authorities and administrative agencies. The

General Directorate of Water (DGA) has the most functions related to droughts, specifically and explicitly referencing this phenomenon. With few exceptions due to recent regulatory changes (Law No. 21,364 of 2021), such as the declaration of threats associated with water resources (DGA) and disaster risk management and its instruments (SENAPRED), these functions are predominantly reactive, emerging in response to already occurring events. Additionally, there are several instances of shared functions involving multiple administrative authorities or agencies, highlighting the need for better coordination among them to prevent functional overlap or interference.

7.2.2 *Private Entities*

Notwithstanding the various measures adopted by individuals to enhance water use efficiency during drought episodes, such as those implemented by holders of water rights, the role of supervisory boards is particularly significant. These boards, which are one of the water users' organizations regulated by the Water Code, exercise their authority over natural water sources. The legally mandated function of the boards of directors of these entities includes declaring water scarcity in the natural sources they manage and determining and implementing extraordinary distribution measures as necessary (article 274 No. 2 of the Water Code).

This declaration of scarcity differs from the declaration of a water scarcity zone due to severe drought, which primarily involves the Ministry of Public Works (MOP) and the General Directorate of Water (DGA). The Water Code does not specify the criteria to be used, nor does it establish any other guidelines for its issuance. This forms part of the broader task of adopting water distribution agreements assigned to the supervisory boards, which proceed based on the characteristics of the water rights of their members and the availability of water in the natural source at a given time. Some rules may be contained in the bylaws of these user organizations. These decisions—such as the declaration of scarcity and the definition and suspension of extraordinary distribution measures—are made autonomously and on a well-founded basis by the board of directors, relying on technical background.

There is a notable difference in terminology within the Water Code. In the instance regulates by article 274 No. 2, it refers to a "declaration of scarcity" and "extraordinary distribution measures", whereas in the context governed by article 314, it mentions a "declaration of scarcity zone due to severe drought" (the term "severe drought" does not appear in relation to actions taken by supervisory boards) and "redistribution agreements".

7.3 Drought Management Policies and Instruments

From a normative and regulatory perspective, drought has historically been treated as an abnormal event or "emergency," primarily managed with reactive tools typical of emergency management. This approach lacks integration with other extreme hydrological phenomena, such as floods, which, despite inherent differences, share a need for comprehensive treatment, due to their intensification resulting from climate change (Embid, 2018).

Among the current instruments, this discussion will concentrate on the two most used in drought situations: the declaration of a water scarcity zone due to severe drought, and the declaration of an agricultural emergency due to water deficit.

7.3.1 *Declaration of a Water Scarcity zone Due to Severe Drought*

Articles 314 and 315 of the Water Code regulate it as a tool for managing hydrological droughts. It is a reactive measure that addresses droughts as extraordinary events with an emergency approach, rather than viewing them as structural and normal occurrences in the country (Tapia, 2021).

Since this is the primary tool employed by the Water Code to address drought, we will examine its key aspects separately:

7.3.1.1 Procedure, Authorities Involved and Scope of Application

According to the recently cited provisions, it is the President of the Republic who has the power to decree a scarcity zone due to severe drought. This responsibility is one of the matters for which the MOP is authorized to sign the respective supreme decrees "by order of the President of the Republic"(article 1, IV, No. 9 Decree 19 of 2001). Consequently, in practice, such declarations are executed by the Minister of Public Works.

Given the limited regulatory framework on this matter, and despite the supplementary application of Law No. 19,880 of 2003, which establishes the bases for administrative procedures of State Administration bodies, there is no explicit legal or regulatory procedure encompassing all phases, actions, and requirements necessary to issue the final administrative act—a declaration of a scarcity zone. This absence represents a significant weakness and ambiguity in the current legal regime addressing droughts, particularly considering the exponential increase in such declarations in recent years.

The procedural details regarding the qualification, evaluation, and follow-up of severe droughts and the subsequent declaration of scarcity zones have been included through administrative acts by the DGA (Resolution No. 1331 of 2022, modified by Resolution No. 579 of 2023). These details have not been formalized in any legal or regulatory norm. This Resolution specifies that the respective Regional Presidential Delegation must request this measure from the DGA for evaluation, which, if deemed appropriate, should then be forwarded to the MOP. In other words, an administrative act has conferred power upon an administrative body without a legal basis, contravening the principle of legality, as it should have been regulated by law rather than by a lower hierarchical source.

As of 2022, under Law No. 21,435 which amends the Water Code, the declaration of scarcity zone is limited to a maximum duration of one year, renewable successively based on a report from the DGA.[2] Practically, this one-year term is consistently applied to all declarations without variation. The validity of instituting a fixed term for such measures is debatable and problematic, as drought conditions persist irrespective of legislative stipulations (Martín, 2019–2020). Nonetheless, it is important to note that these declarations activate a set of prerogatives and exceptional regulations typical of emergency responses, which cannot remain indefinitely and must be time-bound. Therefore, despite how seemingly arbitrary and unrealistic it may appear, having a definitive extension period for the declaration is preferable (Boettiger, 2019).

The declaration of water scarcity due to severe drought is territorially limited, applying to all holders of water rights within the specified area (Boettiger, 2023). According to DGA Resolution No. 1331, the territorial scope for such declarations may encompass a basin, region, province, or commune, despite lacking a definitive legal foundation. This range of options for defining the spatial implementation of this measure contradicts the principles of integrated water resources management, leading to inconsistent applications across different units (region, province, or commune). While a basin or its nearest unit is considered the most appropriate territorial scope for effective water management (Embid, 2021), declarations of water scarcity zones have predominantly been based on the political-administrative divisions of the country—namely communes, provinces, and regions—in descending order of practical usage. This trend has been particularly evident since 2015, with declarations based on the concept of a basin occurring only once in both 2015 and 2019.[3]

7.3.1.2 Dictation Requirements

The DGA is required to request the issuance of a water scarcity zone declaration and notify the President accordingly. This action must be supported by a technical report demonstrating the necessity of such a declaration (Boettiger, 2023).

The DGA must qualify "severe" droughts, a term used since the 2022 reform of the Water Code (previously referred to as "extraordinary" droughts). This qualification is a prerequisite for issuing a declaration of a scarcity zone. The characterization of a drought as severe requires specific attributes of severity. The Water Code assigns the DGA the responsibility of defining these attributes.

The DGA has established criteria through successive resolutions: No. 39 of 1984 (repealed), No. 1674 of 2012 (repealed), and No. 1331 of 2022, amended by No.

---

[2] Prior to the 2022 reform of the Water Code, the declaration of a scarcity zone was for a maximum period of 6 months and could not be extended.

[3] Background information available at: https://dga.mop.gob.cl/administracionrecursoshidricos/decretosZonasEscasez/Paginas/default.aspx

579 of 2023 (currently in force). According to the latest resolution, severe droughts are identified based on hydrometeorological conditions across four macro-zones. These conditions include average monthly flows, rainfall, reservoir statuses, and groundwater levels.

If hydrometeorological conditions cannot be established, the DGA may classify a drought as severe based on information submitted by a competent public agency to the Regional Presidential Delegation. A drought is considered severe if there is either a deficit of 50% or more of the monthly value within a given month or an excess probability exceeding 85% for that same period. Furthermore, these circumstances must lead to limitations on human consumption and sanitation or result in substantial losses in plantations, crops, or livestock, thereby negatively affecting the local economy.

DGA Resolution No. 1331 of 2022 must be reviewed and evaluated within a maximum period of 10 years from the date of its issuance. Additionally, the reference period for calculating the indexes is 30 years, concluding before the final year of the decade preceding said issuance.

7.3.1.3 Powers of the DGA

In addition to those previously mentioned, during the term of a water scarcity zone declaration, the DGA has the following powers:

1. Require supervisory boards and administrators of state works for resource development (excluding private reservoirs), when their administration does not correspond to the State, to submit a redistribution agreement within 15 calendar days of the declaration of a water scarcity zone. This agreement must be approved if applicable. The agreement should adhere to and respect the following critical aspects:

   (a) Include minimum technical conditions, obligations and limitations that guarantee that, in the redistribution of water among all users of the basin and the beneficiaries of the state works, human consumption, sanitation or subsistence domestic use will prevail.

   (b) Prevent serious misconduct or abuse in redistribution. As stated in DGA Circular No. 1 of 2020, it should be noted that:

      (i) The responsibility for water redistribution primarily falls to the supervisory boards. The DGA is authorized to undertake this task only when users, via the relevant supervisory boards, are unable to formulate or adhere to a mutually adopted redistribution agreement.

      (ii) The supervisory boards possess two legally recognized powers to address water scarcity. The first is articulated in article 274, N°2 of the Water Code, serving as an internal regulatory measure. The second power is delineated in article 314 of the Water Code, which applies under specific conditions—currently classified as severe drought (previously referred to as extraordinary drought)—and is explicitly identified in the DGA Circular No. 1 of 2020 as an emergency.

A water shortage zone must be declared by the MOP, necessitating agreements for the redistribution of water from natural surface and underground sources. The boards of directors of the supervisory boards are specifically authorized to enter into such agreements, covering waters within their own jurisdiction and forming agreements with neighboring supervisory boards if dealing with segmented streams. This mechanism aims to minimize the overall impact of drought, highlighting its role as a reactive or mitigating strategy.

(iii) Redistribution agreements must be adopted at a meeting specifically convened for this purpose, approved by an absolute majority of the existing directors unless the bylaws specify a different quorum, and recorded in the minutes. Additionally, article 315 of the Water Code explicitly stipulates that, in the case of natural streams or artificial watercourses where no water users' organization has been formed in accordance with the provisions of the Water Code, the DGA may, either ex officio or at the request of a party, mandate the users to carry out the redistribution of water or assume responsibility for it. The individuals appointed by this authority will possess the same powers as the directors or administrators of the said users' organizations, and their decisions may be subject to challenge.

(c) If the redistribution agreement is not submitted in a timely manner, or if measures related to the prioritization of human consumption, sanitation, or domestic subsistence use are not adhered to, the DGA may enforce compliance. This may include suspending the powers of the supervisory boards and sectioning off natural sources within the water scarcity zone and carrying out redistribution directly. The costs incurred will be charged to the supervisory boards and to the administrators of state works for resource development when the administration does not fall under the State's responsibility. In such cases, the DGA may liquidate and charge monthly the costs associated with direct redistribution.

2. While the declaration of a water scarcity zone is in effect, temporary authorization shall be granted for the extraction of surface water or groundwater, prioritizing human consumption, sanitation or domestic subsistence uses, and the execution of necessary works in watercourses. All the above without constituting water rights, without being subject to the general procedure of the Water Code and without the limitation of the minimum ecological flow.

Consequently, there is a comprehensive relaxation of the concession and authorization regulations that apply during periods of water "normality" or when there is no severe drought leading to the declaration of a water scarcity zone. In this context, even the concept of minimum ecological flow is impacted, as it will not be a requirement for temporary extractions that are authorized. This has been criticized for rendering the concept ineffective and creating unfair and poten-

tially conflicting disparities (Boettiger, 2019). The ecosystemic preservation function, which was strengthened following the 2022 reform of the Water Code, thus loses its efficacy in the event of severe drought. Essentially, environmental safeguards are diminished, reflecting a common characteristic of traditional global approaches to drought management (Iglesias et al., 2019). This is contradictory to the very purpose of declaring a water scarcity emergency (to minimize overall drought damage), as the environmental impacts resulting from severe drought could increase or become more complex to mitigate.
3. The administrative acts that adopt the measures must be enforced immediately. This does not exempt compliance with the process of acknowledgment by the Office of the Comptroller General of the Republic, which in this instance will occur subsequently, rather than a priori, as typically required. Furthermore, additional methods of control and avenues for challenge may be implemented (Boettiger, 2019).

However, we are observing a new example of the lowering of requirements and demands that derive from the treatment of droughts as an emergency, which is a common practice in various countries. This approach raises concerns primarily because it increases the discretionary powers of the administrative authority and public expenditure while implementing minimal forms of supervision (Brufao, 2012; Martín & Pinto, 2015; Martín, 2019–2020; 2022, 2024). It is further argued that droughts are often defined and utilized with political motives rather than scientific evidence to manage water crises outside the legal framework (Martín, 2022).

7.3.1.4 Obligations of Other Stakeholders

As a counterpart to the powers conferred upon the DGA and the guidelines established by the Water Code, the supervisory boards and the administrators of state works for the development of the resource, whose administration does not correspond to the State, have the following fundamental responsibilities:

1. Draft the redistribution agreement with provisions that prioritize human consumption, sanitation, or domestic subsistence use, while also preventing any serious misconduct or abuse.
2. Comply with the agreement approved by the DGA, within five calendar days of such approval.

The administrators of state works aimed at resource development, whose management does not fall under the State's purview, as well as associations of canal owners or water communities that include sanitation service providers among their members, must implement measures to ensure these resources can be used for human consumption, sanitation, or domestic subsistence purposes.

7.3.1.5 Criminal Consequences of Non-compliance with Redistribution Rules

In addition to any potentially applicable administrative sanctions, the amendment to the Penal Code enacted by Law No. 21,595 of 2023, classifies as a criminal offense the extraction of surface or groundwater in violation of redistribution and use regulations within a declared water scarcity zone. This offense may result in penalties including imprisonment, minor imprisonment, or fines (articles 305, 307, and 311 of the Penal Code). The significance of this revised instrument highlights the serious consequences that may arise from non-compliance with the decreed measures.

7.3.1.6 Right to Sue for Damages Regarding the Redistribution of Water

Article 314 of the Water Code states that any holder of a water right who receives less water than their entitled volume during redistribution has the right to seek compensation from the responsible party (the entity overseeing the redistribution). Fiscal compensation will only be applicable when the redistribution is conducted by the DGA due to the absence of a redistribution agreement by the supervisory boards or the administrators of state resource development projects. In most cases, where redistribution is managed by the supervisory boards, these entities would be the liable parties for the compensation action. However, the Water Code specifies that compensation is not applicable if the reduced volume of water received by a water right holder results from the prioritization of human consumption, sanitation, and subsistence domestic use.

To address this key issue, it is crucial to clearly define and precisely delineate the concept of "water for human consumption, sanitation, and domestic subsistence use," and to specify the volume allocated for this priority. This matter lacks clarity and is of paramount importance because using water for this purpose constitutes an exception to the right to compensation (and therefore impacts a water right holder), which cannot be established in ambiguous terms. Moreover, it is essential to note that the prioritization of subsistence usage is grounded in the human right to water, which has a specific and limited scope: water for personal and domestic use. Based on recommendations from the World Health Organization and judicial rulings by Courts of Justice in Chile, this amount is quantified at 100 liters of water per person per day. Therefore, this metric should be considered when assessing the appropriateness of the referred compensation. Additionally, it is important to remember that article 314 of the Water Code mandates the prevention of serious misconduct or abuse, ensuring that, notwithstanding the prioritization, all uses or functions of water are considered.

Establishing this exception to compensation does not imply that, in a drought context, private negotiations between holders of water rights involving the transfer of certain water volumes to water supply and sanitation providers must be free of charge. This situation pertains to a different matter, which may be encouraged, facilitated, or even mandated by prioritizing subsistence function. However, the implementation of such priority does not equate to actions or agreements without any compensation mechanisms for those transferring water volumes to water supply and sanitation providers.

7.3.1.7 Proliferation of Internal Administrative Acts of the DGA

Due to the limited existing legal framework regarding the declaration of a water scarcity zone (articles 314 and 315 of the Water Code) and the absence of comprehensive regulations within the Water Code, particularly concerning droughts, there has been an increase in directives issued by the DGA. These directives specify and detail requirements, procedures, and other relevant aspects, as well as interpret norms, such as those mentioned below:

1. Resolution No. 1482, on requirements, procedure and form of operation of temporary water extraction authorizations in water scarcity zones, of 2022, as amended by Resolution No. 265, of 2023.
2. Resolution No. 414, on the procedure for liquidation and collection of costs related to water redistribution from the oversight boards, of 2023.
3. Resolution No. 1104, on minimum technical conditions, obligations and limitations to be met by water redistribution agreements and their revision procedure, of 2022.
4. Circular No. 1, regarding the authority of the board of directors of the supervisory boards to promote and sign water redistribution agreements in water scarcity zones, of 2020.

If this administrative practice is not exercised with due care, it may incur legal deficiencies, as highlighted by the Office of the Comptroller General of the Republic in pronouncement E529150/2024. The pronouncement clarifies that while the DGA may issue instructions on matters within its jurisdiction through circulars, these circulars are not regulatory instruments per se and that "the DGA lacks interpretative powers of general effect" (opinion E529150/2024, p. 2). Consequently, "in exercising the aforementioned powers, it must limit itself to establishing how to apply the legal and regulatory standards in practice, without altering this regulatory framework"(opinion E529150/2024, p. 3). Specifically, regarding Resolution No. 1482, of 2022, the Comptroller's Office asserts that it contains provisions exceeding the legal framework established by article 314 of the Water Code, as it imposes requirements and operational modes for temporary extraction authorizations in scarcity areas that are not covered by the Code (opinion E529150/2024). In relation to Resolution 414, of 2023, the Comptroller's Office does not concur with the legality objections raised by the petitioner (opinion E529150/2024).

The other resolutions mentioned in the previous paragraph were not examined by the Comptroller's Office in the reviewed opinion, as they were not included in the respective request. However, Resolution No. 1104 was subject to a protection action. In its decision, the Court of Appeals of Santiago stated that the issuance of this resolution is based on the necessity to provide users with certainty regarding technical aspects that influence the approval or rejection of redistribution agreements. It falls within the legal authority of the DGA, thereby ruling out any illegality or arbitrariness by the agency in this matter (Junta de Vigilancia del Río Mataquito y otras vs. DGA, 2024).

7.3.1.8 Reactive Focus and Prioritization of the Livelihood Function

Article 314 of the Water Code explicitly states that the measures outlined are intended to minimize the overall damage caused by drought. This demonstrates a reactive or ex post approach in declaring a water scarcity zone. If additional tools were available to complement this approach with a preparedness or prevention perspective, there would be no significant issue. However, given the absence of such tools, the limited scope of the Water Code in this context is subject to criticism.

The main purpose of declaring a water scarcity zone is to ensure human consumption, sanitation, and domestic use. However, the Santiago Court of Appeals recently ruled that ecosystem preservation should also be prioritized in water redistribution during such declarations (Junta de Vigilancia del Río Mataquito y otras vs. DGA, 2024). In view of this, it should be remembered that the figure of minimum ecological flow has been made more flexible in the case of temporary authorizations for water use in areas of water scarcity, which leaves this environmental safeguard in a diffuse situation.

7.3.1.9 Assessment of the Application of the Declaration of Water Scarcity Zones

To illustrate the implementation of water scarcity zone declarations, we provide a systematic overview of the number and regional distribution of decrees issued from 2008 to October 2024 (as shown in Figs. 7.1 and 7.2):

The number of water scarcity zone decrees increased by 450%, clearly reflecting the prolonged drought that affected Chile. The highest number of decrees was recorded in 2021 (38), followed closely by 2022 (35) and 2020 (33). There has been a significant reduction in 2023 and 2024, which coincides with years experiencing higher rainfall.

We are not encountering an anomaly in application but rather a widespread presence that contrasts with the perception of droughts as extraordinary events (Boettiger, 2019; Tapia, 2021). As has been demonstrated in other regions such as Spain, California, and Mendoza, supposedly exceptional or emergency regulations and provisions often become general norms, leading to the indefinite relaxation of standard legal rules or standards. This practice is inadmissible from various perspectives (Brufao, 2012; Paneque, 2015; Urquijo et al., 2015; Boettiger, 2019; Martín & Pinto, 2015; Martín, 2019–2020, 2022).

The Metropolitana de Santiago Region experienced the highest number of water scarcity decrees between 2008 and 2024, reaching nearly 80 units. This is partly due

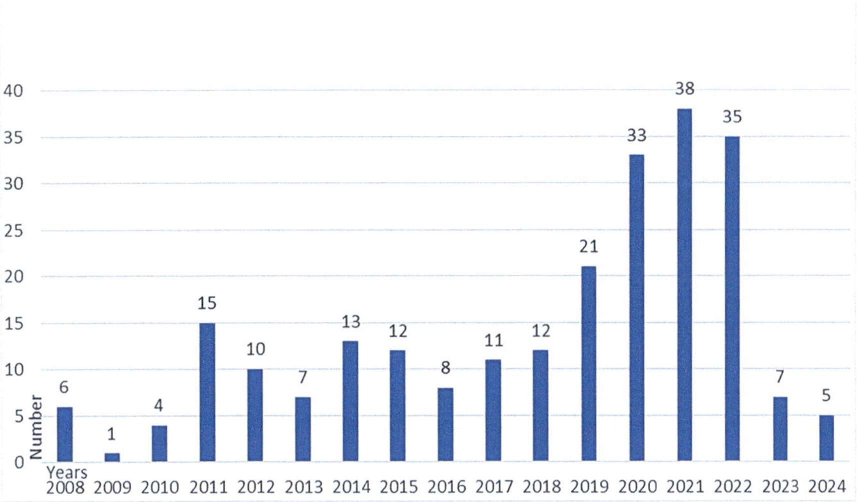

Fig. 7.1 Number of decrees of water scarcity zones issued between 2008 and October 2024 (Source: Prepared by the authors, based on information available on the DGA website: https://dga.mop.gob.cl/administracionrecursoshidricos/decretosZonasEscasez/Paginas/default.aspx)

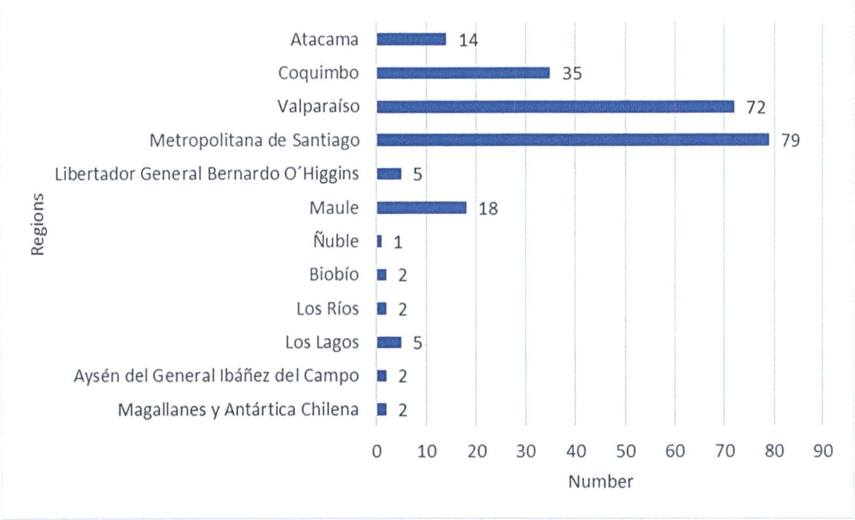

Fig. 7.2 Regional distribution of water scarcity zone decrees, 2008–October 2024 (Source: Prepared by the authors, based on information available on the DGA website: https://dga.mop.gob.cl/administracionrecursoshidricos/decretosZonasEscasez/Paginas/default.aspx)

to the significant impact of droughts, as this region comprises around the 50% of Chile's population. Valparaíso follows as the second highest, with 72 units, reflecting a notable presence in the dataset. Coquimbo presents 35 units, Maule 18 and Atacama 14. Other regions, including Libertador General Bernardo O'Higgins and Los Lagos, exhibit lower values, suggesting lesser involvement or impact. Finally, regions such as Ñuble, Biobío, Los Ríos, Aysén del General Ibáñez del Campo, and Magallanes y Antártica Chilena display very low or near-zero values, indicating minimal representation or impact within the data.

7.3.1.10 Relevant Jurisprudential Pronouncements

To complement the ongoing review, Table 7.3 provides a systematization of criteria established by the Ordinary Courts of Justice and the Office of the Comptroller General of the Republic regarding the content and implications of declarations of water scarcity zones:

7.3.2 Declaration of Agricultural emergency Due to Water Deficit

Agriculture is the sector with the highest water consumption in Chile, accounting for 82% of total consumptive water use (Meza et al., 2021). Geographically, this activity is predominantly concentrated in the central region, which has a Mediterranean climate. This area has a relatively lower water supply compared to the southern region (Barría et al., 2021; Donoso, 2021) and faces an imbalance between surface water availability and total agricultural water consumption (Vicuña et al., 2014).

Given these conditions, the Ministry of Agriculture (MINAGRI) provides various support programs for farmers. Historically, during droughts, this support has been implemented once the phenomenon is underway. One of the most significant and practically applicable tools is the declaration of an agricultural emergency due to water deficit. The key aspects of this measure are outlined below.

7.3.2.1 Concept

The declaration of agricultural emergency is a mechanism administered by the MINAGRI designed to assist and support regions affected by climatic events, disasters, or other occurrences that damage agricultural systems. This declaration facilitates the reallocation and earmarking of resources from the current budget year to finance exceptional measures aimed at mitigating such damages.

Table 7.3 Jurisprudential criteria on declarations of water scarcity zones

General Criteria	Jurisprudential pronouncements
Focus of the 2022 reform of the Water Code	
The reform to the Water Code by Law No. 21,435, of 2022, changed the circumstances of the regulation of the declaration of water scarcity zone, introducing more legal mechanisms to minimize the overall damage of drought, particularly to ensure human consumption, sanitation or domestic subsistence use.	– Junta de Vigilancia de la Primera Sección del Río Aconcagua vs. DGA, 2023, 2024
Reasons for declaring a water scarcity zone	
The administrative act that declares a water scarcity zone must be motivated, and must be accompanied, in its legality control, by all the background information that serves as the basis for the technical report that supports it	– Opinions N°7105, 2017; N°14,818, 2017; N°3142, 2018, among others
Powers of the DGA	
The powers of the DGA in the event of a declaration of a scarcity zone are not limited to the scope of competence of a specific supervisory board, but must be exercised throughout the affected area, and include all users, respecting equal treatment	– Opinion No. 7449, 1991. – Junta de Vigilancia de la Primera Sección del Río Aconcagua y otros vs. MOP y otros, 2018. – Sociedad Contractual Minera Los Pelambres vs. DGA, 2023.
To exercise the restrictive powers contained in article 314 of the Water Code, the DGA is not required to generate, after the declaration of a scarcity zone, a formal pronouncement from the users as to their lack of agreement on the redistribution of water. The DGA may otherwise note this lack of agreement	– Junta de Vigilancia de la Primera Sección del Río Aconcagua y otros vs. MOP y otros, 2018
Among the powers of a supervisory board that may be suspended by the DGA is the distribution of water in natural watercourses. Since this function corresponds to the boards of directors of the supervisory boards, the powers of these boards in this area are also suspended, which does not constitute an encroachment of functions	– Junta de Vigilancia de la Primera Sección del Río Aconcagua vs. DGA, 2023, 2024
The DGA's power to suspend sectioning may be partial, not being obligatory to include the whole of the respective stream, if the objective of minimizing the general damage of the drought is met, particularly to ensure human consumption, sanitation or domestic subsistence use	– Junta de Vigilancia de la Primera Sección del Río Aconcagua vs. DGA, 2023, 2024
The suspension of the powers of the supervisory boards and of the sectioning are not public burdens. They are measures of intervention, control and administration legally contemplated to minimize the general damages of drought, particularly to ensure human consumption, sanitation or domestic subsistence use	– Junta de Vigilancia de la Primera Sección del Río Aconcagua vs. DGA, 2023
Water redistribution	

(continued)

Table 7.3 (continued)

General Criteria	Jurisprudential pronouncements
Redistribution is the measure that the Water Code establishes as suitable to face a shortage event	– Junta de Vigilancia de la Primera Sección del Río Aconcagua vs. DGA, 2024
The purpose of administrative redistribution is to minimize the damage caused by drought and to guarantee human consumption, sanitation and subsistence domestic use, considering geographic and climatic diversity, the effective availability of water resources and the situation of each river basin	– Opinion No. E410,284, of 2023 – Opinion No. E522,141, of 2024
The redistribution mandated by the DGA does not necessarily have to be carried out pro rata to the existing use rights, but may, based on sound technical criteria, be carried out in a different proportion, provided that it complies with the purposes of drought damage reduction and prioritization of human consumption, subsistence domestic use and sanitation	– Opinion No. E410,284, of 2023 – Opinion No. E522,141, of 2024
In accordance with the principle of the unity of the stream, the redistribution carried out by the DGA is an administrative measure that must be adopted based on the total availability of the watercourse and of the water rights of all its users (not by section), regardless of whether the water source is sectioned	– Opinion N°E210,030, of 2022 – Opinion No. E273,030, of 2022 – Opinion No. E410,284, of 2023
When the DGA carries out the redistribution directly, it may issue instructions to the corresponding water distributors or river judge	– Junta de Vigilancia de la Primera Sección del Río Aconcagua vs. DGA, 2023
When the DGA orders the redistribution to a supervisory board, it is not obliged to address or inform each user of the basin, since that task corresponds to the supervisory board	– Sociedad Contractual Minera Los Pelambres vs. DGA, 2023
Redistribution implies a limitation on the exercise of water rights, which generates damage to the users of a basin, and is based on the public or general interest	– Sociedad Contractual Minera Los Pelambres vs. DGA, 2023
The obligations and limitations derived from redistribution apply to all users, including surface water and groundwater	– Sociedad Contractual Minera Los Pelambres vs. DGA, 2023
The redistribution agreement must be shared by all users in the area declared under water scarcity, including groundwater rights holders. (However, the DGA has stated the difficulty of including groundwater in redistribution models, which is due to the progressive or gradual implementation of monitoring systems for effective abstractions and transmission and the lack of groundwater communities. See Opinion N°E273,030 of 2022)	– Opinion No. 52,947, 2020 – Opinion N°E210,030, of 2022 – Opinion No. E273,030, of 2022
The redistribution agreement must be signed by all users and be complete, covering all aspects. If any aspect is missing, the DGA may supplement it by redistributing the water in all areas not covered by the agreement	– Junta de Vigilancia de la Primera Sección del Río Aconcagua vs. DGA, 2021.

(continued)

Table 7.3 (continued)

General Criteria	Jurisprudential pronouncements
Redistribution must prioritize human consumption, subsistence domestic use and sanitation. Water rights holders who receive a lower proportion of water than they would be entitled to are entitled to compensation, unless the lower proportion is the result of that prioritization.	– Opinion No. E273,030, of 2022 – Opinion No. E410,284, of 2023
The concept of "users" used by the Water Code in relation to redistribution agreements refers only to holders of water rights, since the measure in question must be taken based on such rights However, rural sanitation services that are not holders of water rights may request temporary authorizations to extract water for human consumption and sanitation during the period of validity of the water scarcity decree	– Opinion No. E346,302, of 2023
The DGA has full authority to review redistribution agreements or directly order such redistribution, including surface water and groundwater	– Sociedad Contractual Minera Los Pelambres vs. DGA, 2023
In the absence of agreement by the users or determination by the DGA regarding redistribution, the supervisory boards must continue to distribute the water according to the rights of use and the availability of the water source	– Junta de Vigilancia de la Tercera Sección del Río Aconcagua vs. Junta de Vigilancia de la Primera Sección del Río Aconcagua, 2014b – Junta de Vigilancia de la Tercera Sección del Río Aconcagua vs. Junta de Vigilancia de la Primera Sección del Río Aconcagua, 2014a
Emergency catchment works	
In the framework of a water scarcity zone declaration, municipal authorization is not required for the construction of emergency wells in a riverbed (national public property) by the DOH.	– Opinion No. 55,520, 2014

Source: own elaboration

While it is not inherently a preventive management tool, it operates under the principles of emergency (crisis) management. The declaration of agricultural emergency permits the release and/or reallocation of financial resources to address the effects of drought and other adverse conditions.

7.3.2.2 Procedure, Scope of Application and Authorities Involved

Unlike the declaration of water scarcity due to severe drought managed by the MOP-DGA, the process of declaring an agricultural emergency is not explicitly regulated in a legal or regulatory norm. It is mainly addressed in Resolution (exempt)

No. 309, of 2021 (modified by Resolution (exempt) No. 769, of 2022), which approves the National Operational Emergency Plan of MINAGRI, which refers to the Protocol for Agricultural Emergencies, approved by Ordinary No. 989, of 2011, of the Undersecretary of Agriculture.

The agricultural emergency applies to various threats to the agricultural, forestry and livestock sector, including both climatic and non-climatic events. According to Resolution No. 309 of 2021, the emergency plan can be activated in the event of drought, frost, snowfall, excessive and/or unseasonal rainfall, floods, hail, alluvium and/or landslides, volcanic eruptions, earthquakes, chemical emergencies, quarantine pests, pandemics, among others.

The agricultural emergency is declared through a resolution that identifies the phenomenon causing it and the affected communes. The process is led by the respective Regional Ministerial Secretary (Secretaría Regional Ministerial, SEREMI) of Agriculture. The Regional Presidential Delegate evaluates the technical background and the action plan provided by the SEREMI of Agriculture, determining whether to issue the declaration of emergency and indicating the applicable sources of financing. This is forwarded to the Minister of Agriculture, who issues the administrative act that declares the agricultural emergency, approves the plan and its financing, and defines the institutions responsible for assisting those affected (Resolution (exempt) No. 309, of 2021). Coordination at the level of the Ministry and regional authorities is required for this process.

In this regard, it has been argued that the issuing of the agricultural emergency declaration is quick and that coordination at the level of the Ministry and regional authorities is effective (Meza et al., 2010), while its implementation has resulted in high fiscal expenditure (Tapia, 2021).

Declarations of agricultural emergencies are contingent upon the availability of fiscal resources and therefore are effective during the respective budget year. This includes a specific budgetary item for financing unforeseen expenses arising from climatic phenomena, natural disasters, and emergency situations or damage affecting agricultural producers and rural inhabitants. Consequently, these declarations have a maximum duration of one year, necessitating a new administrative act to extend their validity if required.

Resolution No. 309 of 2021 provides a framework for action in emergencies and outlines a declaration procedure with decision-making responsibilities at each level of the process. However, it does not establish indicators, thresholds, or triggers that specify an agricultural emergency (such as drought) for a particular territory or crop.

The Agricultural Emergency Protocol states that the Agricultural Research Institute (Instituto de Investigaciones Agropecuarias, INIA), via its regional centers, prepares a report analyzing meteorological data and various agrometeorological indicators. The condition of the primary crops is also reviewed, providing recommendations to address the emergency. Additionally, information provided by the Agroclimatic Observatory is considered. Currently, there are no established objective criteria or indicators that determine when a territory qualifies for an agricultural emergency declaration. Nevertheless, information demonstrating that an abnormal situation has caused or could cause significant damage to production systems and/

or rural inhabitants, thereby necessitating assistance from MINAGRI must be presented (Ordinary No. 989, 2011, Undersecretariat of Agriculture). Indicators to consider include precipitation percentage, average monthly flow, significant anomalies compared to normal average precipitation data series, temperatures, and forecasts of climatic variables confirming the persistence of adverse phenomena, among others (Ordinary No. 989, 2011, Undersecretariat of Agriculture).

The use of additional indicators available from other institutions, such as the DGA, which would provide a more comprehensive overview of the agro-climatic situation, particularly concerning hydrological droughts affecting irrigated agriculture, is not currently incorporated. Additionally, there is no differentiation between information and indicators for declaring meteorological droughts, which affect rainfed agriculture, and hydrological droughts, which impact irrigated agriculture.

Therefore, the criteria for declaring an agricultural emergency require improvement. Current standards are not entirely suitable to timely address the effects of drought, lack homogeneous standards for assessing the level of impact, and leave significant room for subjectivity and the influence of social and political factors (Meza et al., 2010).

Regarding the territorial unit subject to the declaration of agricultural emergency due to water deficit, evidence shows a prevalence of declarations based on communes, though there are also some references to provinces and regions. The concept of a river basin, despite being the basic and optimal unit for water resource management, has limited application in this instrument.

7.3.2.3 Assessment of the Application of the Declaration of Agricultural emergency Due to Water Deficit

To illustrate the practical operation of the declaration of agricultural emergency due to water deficit, a concise systematization of the number and regional distribution of the decrees issued between 2008 and October 2024 is provided below (see Figs. 7.3 and 7.4):

Figure 7.3 shows that during the period studied, 2008 was the year with the highest number of these declarations (13), followed by 2012 and 2015 (9 declarations each). It is evident that there is no correlation in the number or annual distribution of the decrees for water scarcity zones (MOP-DGA) and those for agricultural emergency due to water deficit (MINAGRI), reaffirming the premise that they are based on different assumptions and serve distinct purposes. However, it would be beneficial to have some degree of coordination between these tools to promote a more unified response or procedure in addressing droughts. Currently, the response is sectoral and isolated, divided according to the type of drought (hydrological or agricultural) and the administrative agency involved.

Figure 7.4 shows that the O'Higgins region has the most declarations (19), followed by Valparaíso and Metropolitana de Santiago (17 each). During the analyzed period, there is no alignment in terms of the territorial concentration of the decrees of water scarcity zones and agricultural emergency decrees due to water deficit,

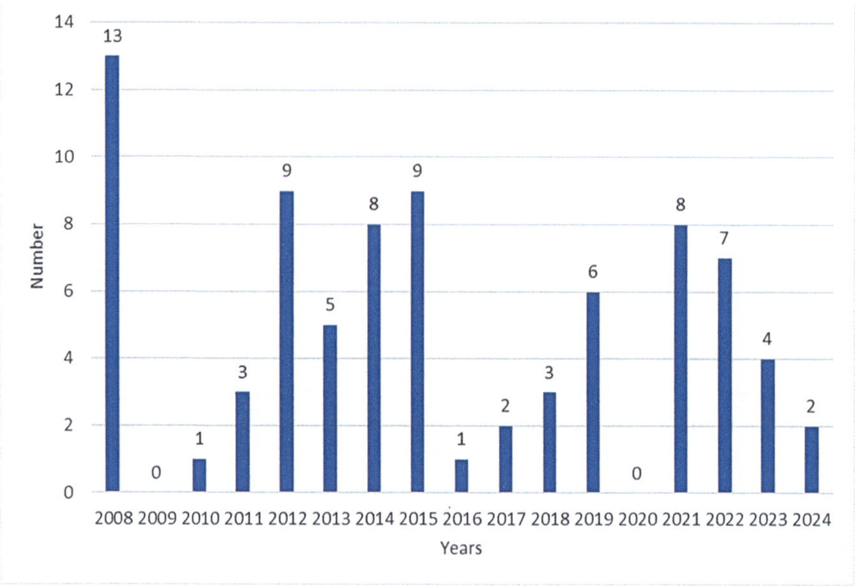

Fig. 7.3 Number of agricultural emergency decrees due to water deficit issued between 2008 and October 2024 (Source: Compilation based on information requested via Law No. 20,285 from MINAGRI and information available on the MINAGRI website: https://www.portaltransparencia.cl/PortalPdT/directorio-de-organismos-regulados/?org=AR001)

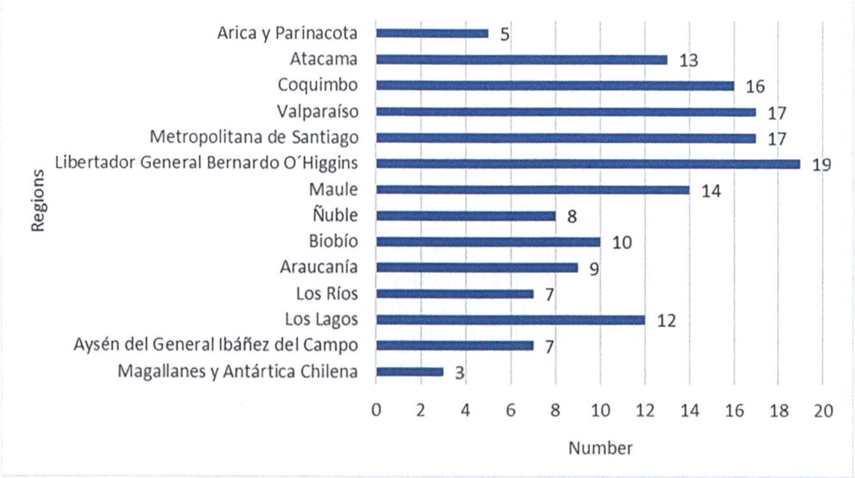

Fig. 7.4 Regional distribution of agricultural emergencies due to water deficit issued between 2008 and October 2024 (Source: Compilation based on information requested via Law No. 20,285 from MINAGRI and information available on the MINAGRI website: https://www.portaltransparencia.cl/PortalPdT/directorio-de-organismos-regulados/?org=AR001)

even though the Metropolitana de Santiago and Valparaíso regions have a strong presence of both tools.

7.4 Drought Governance

According to Glückler et al. (2019), governance refers to models and practices of coordinating collective action in situations involving diverse interests, distribution of resources, and the pursuit of legitimate and accepted common goals. Governance transcends the actions of an individual authority, necessitating negotiation between various interests and the coordination of independent stakeholders towards consensual ends (Glückler et al., 2019). Similarly, governance is seen as the interaction between public and private actors to address social problems or create social opportunities within an institutional context with an established normative basis (Bressers et al., 2016a).

Thus, the core concept of governance is the articulation or unity of action, which coincides with one of the particular legal bases of administrative organization and is imperative for the agencies that compose the State Administration (article 5 DFL 1–19,653, sets the consolidated, coordinated, and systematized text of Law No. 18,575, Constitutional Organic Law of General Bases of the State Administration). Water governance is crucial due to its complexity and interconnected nature, covering numerous areas (Bressers et al., 2016a). Its role is particularly critical during drought scenarios, where coordinated actions are required at multiple levels and time scales (Bressers et al., 2016b).

Theoretically, the 17 entities of the State Administration identified in this chapter as those exercising primary competences in drought management in Chile should act in a coordinated manner. However, in practice, there is a series of disconnected and disperse actions, reflecting weak institutional cooperation and fragmentation of governmental roles, a situation also observed in other countries such as those in the European Mediterranean (Iglesias et al., 2019).

This lack of cohesion is not specific to droughts but is a common feature of water management in Chile and has been repeatedly recognized as a significant gap in this area (Banco Mundial, 2013; Mesa Nacional del Agua, 2022). Notwithstanding this, droughts, given the emergency treatment they have traditionally received, generate an even more propitious space for the intensification of this gap, as many rules and controls are relaxed, supposedly to provide a more appropriate and timely response to this emergency.

Additionally, the regulatory framework, consistent with the shortage or deficit implied by drought, contrary to the importance of this issue, does not promote a systematic and integrated approach. The term "drought" is expressly mentioned only four times in the Water Code, whereas "scarcity" appears twelve times. For legislators, drought is an abnormal and exceptional occurrence, thus briefly regulated (Crocco, 2021; Tapia, 2021), specifically in only two articles of the Code (314 and 315), leaving many open questions and including indeterminate concepts of difficult precision. One fundamental weakness of the legal framework on droughts is the absence of clear institutional responsibility for decision-making, approval, execution, and monitoring actions, which is logically a consequence of the limited political importance assigned to this issue and the general lack of attention it receives in regulations or legislation (Iglesias et al., 2019).

As a result, there is a state of "ungovernance" concerning drought in Chile, marked by multiple actors, tools, resources, and purposes lacking planning or effective coordination. This situation has persisted despite ongoing and prolonged drought periods in the country. Despite a 14-year mega-drought still affecting some areas, no major regulatory or institutional changes have been made to emergency management. For instance, despite modifications introduced to articles 314 and 315 of the Water Code in 2022 (Law No. 21,435), the main tool continues to be the declaration of a water scarcity zone due to severe droughts, with no distinction being made between different water deficit scenarios, each of which should be linked to different management measures (including preparedness and adaptation), without having to wait for the drought to become severe in order to take action.

This contrasts with what is normally observed in comparative cases, where it is common for severe hydrological extremes to define and drive alterations in the legal system to include a preventive and risk management focus, as has occurred in Spain, California (Embid, 2018; Berbel & Esteban, 2019), and Australia (Courtenay & Hayes, 2012; Berbel & Esteban, 2019). In this sense, it has been argued that it is key to normalize drought and its consequences, integrating it into regular water resources planning and management (Neuman, 2003; Courtenay & Hayes, 2012; Urquijo et al., 2015; Setuaín, 2018).

It is evident that, to date, we have not effectively utilized the opportunities presented by these events to introduce regulatory adjustments and improvements in public and private practices or actions. Addressing cross-cutting and essential objectives to enhance resilience and reduce vulnerability to a phenomenon that remains part of our reality is crucial. Drought is not inherently a disaster or catastrophe but can become one depending on its impacts on the population, economy, and environment and our societal capacity to respond and recover (Wilhite, 2009; Brufao, 2012).

7.5 Conclusions

Reflecting the existing institutional fragmentation in water management in Chile, there are numerous actors, both public and private, operating at various scales and with limited coordination regarding drought issues. Consequently, multiple public policies and actions with differing approaches and objectives can be observed, lacking a coherent and unified strategy. This situation likely involves substantial public expenditure, as adequate funding is essential for public entities to effectively address such phenomena (Jiménez, 2018; Iglesias et al., 2019). The dispersion of efforts and resources results in the effects of these actions related to droughts being diluted or not readily apparent.

Key tools in this area include declarations of water scarcity zones due to severe drought and agricultural emergencies due to water deficit. These involve different agencies, procedures, and purposes. Both are governed by concise regulations: water scarcity has a legal basis for central aspects, while agricultural emergencies rely on administrative acts. They both treat droughts as emergencies with a reactive approach.

In this context, despite the frequent occurrence of such phenomena in Chile, addressing drought governance remains challenging. Currently, there is fragmentation, dispersion, and disorder evident in the system. While not the focus of this chapter but rather a topic for subsequent discussion,[4] it can be inferred that the measures implemented have shown limited effectiveness. These outcomes are contrary to the principles of good governance.

Although it can be estimated that the peak of the recent drought affecting the country (which continues in some areas) is ending, it is highly probable that another drought will occur in the future. Therefore, institutional and policy reordering is necessary, focusing on risk management and a planned and integrated combination of various preventive and reactive strategies and instruments (Stein et al., 2016; Hurlbert et al., 2019). Risk management alone is insufficient to mitigate the effects of extreme events; it is essential to address equitably the social drivers of their impact (Tellman & Eakin, 2022). Emphasis should be placed on water security, developing and strengthening resilience, and reducing vulnerability to this and other extreme hydrological phenomena, such as floods, considering both their differences and similarities.

Acknowledgments This work was carried out within the framework of the ANID FSEQ210018 project, for which we are grateful for the support and funding.

References

Arana, E., & Miranzo, J. (2021). *La gestión de la escasez de agua y de las sequías por parte de las entidades locales. Guía para una gestión eficiente* (p. 156). Editorial Aranzadi.

Banco Mundial. (2013). *Chile. Estudio para el mejoramiento del marco institucional para la gestión del agua* (p. 220). Banco Mundial.

Barría, P., Chadwick, C., Ocampo-Melgar, A., Galleguillos, M., Garreaud, R., Díaz-Vasconcellos, R., Poblete, D., Rubio-Álvarez, E., & Poblete-Caballero, D. (2021). Water management or megadrought: What caused the Chilean Aculeo Lake drying? *Regional Environmental Change, 21*(1), 19. https://doi.org/10.1007/s10113-021-01750-w

Berbel, J., & Esteban, E. (2019). Droughts as a catalyst for water policy change. Analysis of Spain, Australia (MDB), and California. *Global Environmental Change, 58*, 101969.

Bernstein, J. S. (1999). Racionamiento eléctrico: causas y posibles soluciones. *Puntos de Referencia, 209*, 1–11.

[4] See Chap. 12, entitled 'General Assessment of Drought Management and Proposals'.

Boettiger, C. (2019). Análisis crítico de la declaración de escasez. In E. Costa & S. Montenegro (Eds.), (Dir.) *La regulación jurídica de las aguas: nuevos desafíos del siglo XXI. Actas de las II Jornadas del régimen jurídico de las aguas* (pp. 345–376). Der Ediciones.

Boettiger, C. (2023). Limitaciones o restricciones cuantitativas al ejercicio de los derechos de aprovechamiento de aguas. *Actualidad Jurídica, 47*, 193–218.

Bressers, N., Bressers, H., & Larrue, C. (2016a). Introduction. In H. Bressers, N. Bresssers, & C. Larrue (Eds.), *Governance for drought resilience* (pp. 1–16). Springer.

Bressers, H., Bressers, N., Kuks, S., & Larrue, C. (2016b). The governance assessment tool and its use. In H. Bressers, N. Bresssers, & C. Larrue (Eds.), *Governance for drought resilience* (pp. 45–65). Springer.

Brufao, P. (2012). El régimen jurídico de las sequías: crítica a la regulación extraordinaria y urgente de un fenómeno natural y cíclico propio del clima. *Revista de Administración Pública, 187*, 199–239.

Center for Climate and Resilience Research (CR)2. (2015). *Report to the nation. The 2010–2015 megadrought: a lesson for the future* (Santiago, (CR)2). Available at: https://www.cr2.cl/eng/wp-content/uploads/2020/07/Mega-drought-Report.pdf

Courtenay, L., & Hayes, M. (2012). Drought triggers and declarations: Science and policy considerations for drought risk management. *Natural Hazards, 64*, 139–151. https://doi.org/10.1007/s11069-012-0231-4

Crocco, J. J. (2021). Megasequía: diagnóstico, impactos y propuestas. *Puntos de Referencia (Centro de Estudios Públicos), 559*, 24.

Díaz, C., Galetovic, A., & Soto, R. (2000). La crisis eléctrica de 1998-1999: causas consecuencias y lecciones. *Estudios Públicos, 80*, 149–192.

Donoso, G. (2021). Economics of water resources. In B. Fernández & J. Gironás (Eds.), *Water resources of Chile* (pp. 335–346). Springer.

Embid, A. (2018). Elementos para la construcción de una teoría general de las sequías e inundaciones como fenómenos hidrológicos extremos. In A. Embid (Ed.), (Dir.) *Sequía e inundación como fenómenos hidrológicos extremos* (pp. 25–84). Editorial Aranzadi.

Embid, A. (2021). El concepto de seguridad hídrica: contenido y funcionalidad. In A. Embid (Ed.), (Dir.) *La seguridad hídrica: desafíos y contenido* (pp. 27–70). Editorial Aranzadi.

Fernández, B., & Gironás, J. (2021). Droughts. In B. Fernández & J. Gironás (Eds.), *Water resources of Chile* (pp. 173–187). Springer.

Fu, X., Svoboda, M., Tang, Z., Dai, Z., & Wu, J. (2013). An overview of US state drought plans: Crisis or risk management? *Natural Hazards, 69*, 1607–1627. https://doi.org/10.10007/s11069-013-0766-z

Glückler, J., Rehner, J., & Handke, M. (2019). Gobernanza, redes y territorio. *Revista de Geografía Norte Grande, 74*, 5–20.

Hervás-Gámez, C., & Delgado-Ramos, F. (2019). Drought management planning policy: From Europe to Spain. *Sustainability, 11*, 1862. https://doi.org/10.3390/su11071862

Hurlbert, M., Gupta, J., & Verrest, H. (2019). A comparison of drought instruments and livelihood capitals. *Climate and Development, 11*(10), 863–872. https://doi.org/10.1080/17565529.2019.1585318

Iglesias, A., Garrote, L., & Granados, A. (2019). On the institutional framework for drought planning and early action. In A. Iglesias, D. Assimacopoulos, & H. Van Lanen (Eds.), *Drought. Science and policy*. Wiley Blackwell.

Jiménez, I. (2018). Gasto público e instrumentos económico-financieros alrededor de los fenómenos hidrológicos extremos: el tributo al servicio de la gestión de sequías e inundaciones. In A. Embid (Ed.), (Dir.) *Sequía e inundación como fenómenos hidrológicos extremos* (pp. 85–147). Editorial Aranzadi.

Martín, L., & Pinto, M. (2015). Escasez extraordinaria y derecho de aguas. *Revista de Derecho Administrativo Económico, 20*, 147–164.

Martín, L. (2019–2020). Emergencia hídrica, ambiental y climática. Elementos para su configuración y relación con el Estado de Derecho. *Revista Derechos en Acción*, Año 5, 5, 224–253.
Martín, L. (2022). From scarcity to emergency. A comparative study on water and drought management in the Americas. In C. Nicolas-Artero, S. Velut, G. Schneier-Madanes, F. Poupeau, & C. Chavarochette (Eds.), (Dir) *Luttes pour l'eau dans les Amériques* (pp. 43–57). Éditions de l'IHEAL.
Martín, L. (2024). Emergencia climática. Desafíos jurídicos y políticos. In B. Setuaín (Ed.), (Dir.) *Perspectivas jurídicas sobre clima, agua y energía. Estudios en reconocimiento al magisterio del profesor Antonio Embid Irujo* (pp. 29–55). Editorial Aranzadi.
Mesa Nacional del Agua. (2022). *Informe final* (p. 110). Mesa Nacional del Agua. Available at: https://derechoygestionaguas.uc.cl/es/documentos/new/507-2022-mesa-nacional-del-agua-informe-final/file
Meza, L., Corso, S., & Soza, S. (2010). *Gestión del riesgo de sequía y otros eventos climáticos extremos en Chile. Estudio piloto sobre la vulnerabilidad y la gestión local del riesgo* (p. 47). Organización de las Naciones Unidas para la Agricultura y la Alimentación, FAO.
Meza, F., Gil, P., & Melo, O. (2021). Agricultural uses. In B. Fernández & J. Gironás (Eds.), *Water resources of Chile* (pp. 243–258). Springer.
Neuman, J. (2003). Drought proofing water law. *University of Denver Water Law Review, 7*(1), 92–110.
Paneque, P. (2015). Drought management strategies in Spain. *Water, 7*, 6689–6701. https://doi.org/10.3390/w7126655
Setuaín, B. (2018). Respuestas jurídicas a la reciente situación de sequía en España. In A. Embid (Ed.), (Dir.) *Sequía e inundación como fenómenos hidrológicos extremos* (pp. 149–193). Editorial Aranzadi.
Soto, F., Ejsmentewicz, D., & Riveros, D. (2022). El tratamiento constitucional de los desastres socio-naturales: antecedentes y desafíos ante el debate constituyente. *Estudios Constitucionales, 20*(1), 81–109.
Stein, U., Özerol, G., Tröltzsch, J., Landgrebe, R., Szendrenyi, A., & Vidaurre, R. (2016). European drought and water scarcity policies. In H. Bressers, N. Bresssers, & C. Larrue (Eds.), *Governance for drought resilience* (pp. 17–43). Springer.
Tapia, F. (2021). Revisión a los instrumentos regulatorios de respuesta a la escasez hídrica en Chile: ¿emergencia o normalidad? In S. Montenegro, E. Costa, & T. Celume (Eds.), (Dir) *Repensando la regulación de las aguas: crisis socioambiental y proceso constituyente. Actas de las III Jornadas del régimen jurídico de las aguas* (pp. 389–418). Der Ediciones.
Tellman, B., & Eakin, H. (2022). Risk management alone fails to limit hazard impact. *Nature, 68*, 41–43.
Urquijo, J., De Stefano, J., & La Calle, L. (2015). Drought and exceptional laws in Spain: The official water discourse. *International Environmental Agreements: Politic, Law and Economics, 15*, 273–292. https://doi.org/10.1007/s10784-015-9275-8
Vicuña, S., Álvarez, P., Melo, O., Dale, L., & Meza, F. (2014). Irrigation infrastructure development in the Limarí Basin in Central Chile: Implications for adaptation to climate variability and climate change. *Water International, 39*(5), 620–634. https://doi.org/10.1080/02508060.2014.945068
Wilhite, D., Hayes, M. J., & Knutson, C. (2000). Planning for drought: Moving from crisis to risk management. *Journal of the American Water Resources Association, 36*(4), 697–710.
Wilhite, D., Hayes, M. J., & Knutson, C. (2005). Drought preparedness planning: Building institutional capacity. In D. Wilhite (Ed.), *Drought and water crises. Science, technology, and management* (pp. 93–135). CRC Press.
Wilhite, D. (2009). Drought monitoring as a component of drought preparedness plan. In A. Iglesias, L. Garrote, A. Cancelliere, F. Cubillo, & D. Wihite (Eds.), *Coping with drought risk in agriculture and water supply systems* (pp. 3–19). Springer.

Jurisprudence

Administrative

Dictamen (Opinion) N°7.449 (1991). Contraloría General de la República, 22 de marzo de 1991.
Dictamen (Opinion) N°55.520 (2014). Contraloría General de la República, 21 de julio de 2014.
Dictamen (Opinion) N°7.105 (2017). Contraloría General de la República, 1 de marzo de 2017.
Dictamen (Opinion) N°14.818 (2017). Contraloría General de la República, 26 de abril de 2017.
Dictamen (Opinion) N°3.142 (2018). Contraloría General de la República, 25 de enero de 2018.
Dictamen (Opinion) N°E31.873 (2020). Contraloría General de la República, 1 de septiembre de 2020.
Dictamen (Opinion) N°52.947 (2020). Contraloría General de la República, 18 de noviembre de 2020.
Dictamen (Opinion) N°E210.030 (2022). Contraloría General de la República, 4 de mayo de 2022.
Dictamen (Opinion) N°E273.030 (2022). Contraloría General de la República, 3 de noviembre de 2022.
Dictamen (Opinion) N°E346-302 (2023). Contraloría General de la República, 17 de mayo de 2023.
Dictamen (Opinion) N°E410.284 (2023). Contraloría General de la República, 30 de octubre de 2023.
Dictamen (Opinion) N°E522.141 (2024). Contraloría General de la República, 1 de agosto de 2024.
Dictamen (Opinion) N°E529.150 (2024). Contraloría General de la República, 20 de agosto de 2024.

Judicial

Junta de Vigilancia de la Tercera Sección del Río Aconcagua vs. Junta de Vigilancia de la Primera Sección del Río Aconcagua (2014a). Corte de Apelaciones de Valparaíso, 9 January 2014, Rol 6.845–2014.
Junta de Vigilancia de la Tercera Sección del Río Aconcagua vs. Junta de Vigilancia de la Primera Sección del Río Aconcagua (2014b). Corte Suprema, 25 March 2014, Rol 2.014–2014.
Junta de Vigilancia de la Primera Sección del Río Aconcagua y otros vs. MOP y otros (2018). Corte Suprema, 6 September 2018, Rol 8.811–2018.
Junta de Vigilancia de la Primera Sección del Río Aconcagua vs. DGA (2021). Corte de Apelaciones de Valparaíso, 2 December 2021, Rol 44755–2021. Judgment confirmed, without modifications, by the Supreme Court dated 17 June 2022, Rol 95.117–2021.
Junta de Vigilancia de la Primera Sección del Río Aconcagua vs. DGA (2023). Corte de Apelaciones de Santiago, 2 November 2023, Rol 150–2023. Appeal in cassation on the merits dismissed.
Sociedad Contractual Minera Los Pelambres vs. DGA (2023). Corte de Apelaciones de Santiago, 12 December 2023, Rol 73–2023. Executed sentence.
Junta de Vigilancia del Río Mataquito y otras vs. DGA (2024). Corte de Apelaciones de Santiago, 18 April 2024, Rol 515–2023. Executed sentence.
Junta de Vigilancia de la Primera Sección del Río Aconcagua vs. DGA (2024). Corte Suprema, 4 September 2024, Rol 247867–2023.

Standards and Administrative Acts Cited

Código Penal, 1874.
DFL 725, Código Sanitario, 1968.
Decreto 104, fija el texto refundido, coordinado y sistematizado del título I de la Ley 16.282, 1977.
DFL 1.122, fija texto del Código de Aguas, 1981.
Ley 18.450, aprueba normas para el fomento de la inversión privada en obras de riego y drenaje, 1985.
Ley 18.910, sustituye Ley Orgánica del Instituto de Desarrollo Agropecuario, 1990.
DFL 850, fija el texto refundido, coordinado y sistematizado de la Ley N°15.840, de 1964, y del DFL N°206, de 1960, 1998.
Decreto 19, faculta a los Ministros de Estado para firmar "por orden del Presidente de la República", 2001.
DFL 1-19.653, fija texto refundido, coordinado y sistematizado de la Ley N°18.575, Orgánica Constitucional de Bases Generales de la Administración del Estado, 2001.
Ley 19.880, establece bases de los procedimientos administrativos que rigen los actos de los órganos de la Administración del Estado, 2003.
Decreto 100, fija el texto refundido, coordinado y sistematizado de la Constitución Política de la República de Chile, 2005.
DFL 1-19.175, fija el texto refundido, coordinado, sistematizado y actualizado de la Ley N°19.175, Orgánica Constitucional sobre Gobierno y Administración Regional, 2005.
DFL 1, fija el texto refundido, coordinado y sistematizado de la Ley N°18.695, Orgánica Constitucional de Municipalidades, 2006.
DFL 4, fija texto refundido, coordinado y sistematizado del Decreto con Fuerza de Ley N°1, de Minería, de 1982, Ley General de Servicios Eléctricos, en materia de energía eléctrica, 2007.
Resolución 1.049 (exenta), establece el procedimiento operativo para el otorgamiento de incentivos de emergencia, Instituto de Desarrollo Agropecuario, 2010.
Ordinario N°989, protocolo de emergencia agrícola, Subsecretaría de Agricultura, 2011.
Decreto 41, Reglamento sobre condiciones sanitarias para la provisión de agua potable mediante el uso de camiones aljibe, 2018.
Circular N°1, aclara atribuciones del directorio de las juntas de vigilancia para promover y suscribir acuerdos de redistribución de aguas en zonas declaradas de escasez, en virtud del artículo 314 del Código de Aguas, Dirección General de Aguas, 2020.
Decreto 51, decreta medidas preventivas que indica de acuerdo a lo dispuesto en el artículo 163 de la Ley General de Servicios Eléctricos, 2021.
Ley 21.364, establece el sistema nacional de prevención y respuesta ante desastres, sustituye la Oficina Nacional de Emergencias por el Servicio Nacional de Prevención y Respuesta ante Desastres, y adecúa normas que indica, 2021.
Resolución 309 (exenta), aprueba plan nacional operativo de emergencia del Ministerio de Agricultura, Subsecretaría de Agricultura, 2021.
Ley 21.435, reforma el Código de Aguas, 2022.
Resolución 769 (exenta), modifica resolución exenta N°309, de 2021, que aprueba plan nacional operativo de emergencia del Ministerio de Agricultura y establece Comisión de Riesgos y Emergencias Silvoagropecuarias, Subsecretaría de Agricultura, 2022.
Resolución 1.104 (exenta), establece condiciones técnicas mínimas, obligaciones y limitaciones que deben cumplir los acuerdos de redistribución de aguas, y establece procedimiento de revisión, conforme al artículo 314 del Código de Aguas, Dirección General de Aguas, 2022.
Resolución 1.331 (exenta), deja sin efecto la Resolución DGA N°1.674 (exenta), de 12 de junio de 2012, y establece criterios que determinan el carácter de severa sequía, de conformidad a lo dispuesto en el artículo 314 del Código de Aguas, Dirección General de Aguas, 2022.
Resolución 1.482 (exenta), establece requisitos, procedimiento y forma de operación de autorizaciones temporales de extracción de aguas en zonas declaradas de escasez hídrica, de con-

formidad a lo dispuesto en el artículo 314 inciso 7° del Código de Aguas, Dirección General de Aguas, 2022.
Ley 21.595, ley de delitos económicos, 2023.
Resolución 265 (exenta), modifica Resolución DGA (exenta) N°1.482, de 23 de junio de 2022, establece requisitos, procedimiento y forma de operación en materia de autorizaciones temporales de extracción de aguas en zonas declaradas de escasez hídrica, de conformidad a lo dispuesto en el artículo 314 inciso 7° del Código de Aguas, en los términos que indica, Dirección General de Aguas, 2023.
Resolución 414 (exenta), establece procedimiento de liquidación y cobro a las juntas de vigilancia de los costos asociados a la redistribución de aguas, conforme al artículo 314 del Código de Aguas, Dirección General de Aguas, 2023.
Resolución 579 (exenta), modifica Resolución DGA N°1.331 (exenta), de 7 de junio de 2022, en los términos que indica, Dirección General de Aguas, 2023.

Part II
Comparative Experiences in Drought Management

Chapter 8
Drought Policy Overview: Insights from South America Region

Julia Urquijo-Reguera, Jairo Paizano-Potoy, Jesús López-Santiago, and David Pereira

Abstract Drought is one of the hazards that affects an important number of people and generates significative economic losses in the world. Drought patterns and impacts may worsen due to climate change; thus, an urgent need arises to address drought events under a proactive approach, which allows tackling underlying causes of drought and vulnerability factors in order to prevent, reduce, mitigate or adapt to drought impacts. However, drought is traditionally managed under a crisis-based approach rather than on a risk-based approach. Moving towards a proactive approach is key to improve drought policy responses under an adaptive and risk-based perspective, but limited literature is available on the progress towards a risk-based approach to drought management adoption. Thus, there is a need to contribute to this area of study by enhancing our understanding on how we are responding to drought worldwide, particularly for the South America region. The aim of this chapter is to analyze the characteristics of drought policy at a national level by identifying the existing regulatory and planning tools and their advances towards a proactive approach adoption to deal with drought events. The results will provide a regional overview on the progress, gaps and opportunities of improvement and outline recommendations for decision-makers to improve drought management at national level.

Keywords Drought policy · Proactive approach · Disaster risk reduction · Vulnerability and impact assessment · South America region

J. Urquijo-Reguera (✉)
Departamento de Ingeniería Agroforestal, ETSIAAB, Universidad Politécnica de Madrid, Madrid, Spain

CEIGRAM—Research Centre for the Management of Agricultural and Environmental Risks, Madrid, Spain
e-mail: julia.urquijo@upm.es

J. Paizano-Potoy · J. López-Santiago · D. Pereira
Departamento de Ingeniería Agroforestal, ETSIAAB, Universidad Politécnica de Madrid, Madrid, Spain

D. Rivera, G. Donoso (eds.), *Droughts in Chile*, Global Issues in Water Policy 31,
https://doi.org/10.1007/978-3-031-85040-0_8

8.1 Introduction

Drought is a complex natural hazard due to its slow onset and creeping development, which entails difficulties in determining the start and end, both at temporal and territorial scale. Moreover, drought affects all regions and countries of the world causing severe impacts (UNCCD, 2023; IPCC, 2022; UNDDR, 2021). It causes multiple and cascading impacts for all sectors of society, and they are context dependent as the same drought event characteristics may provoke different types and levels of impacts on the territory. They may be direct or indirect and take place on the short, medium, or long term and their causality is not well-known, particularly in relation to its interaction with people (Van Loon et al., 2016; Savelli et al., 2022) as limited systematic studies or databases on drought impacts exist (e.g. EM-DAT,[1] EDO,[2] EDII,[3] US Drought Monitor[4]). However, it is well-known that, drought is the natural hazard that affects a large proportion of the world's territory and population (CRED, 2021), which a recent study estimates in 1.84 billion people stricken by drought, especially in low- and middle-income countries, and women and children (UNCCD, 2023). Additionally, drought is expected to increase in magnitude, duration and intensity as well as its consequences as an increase in the population exposed to drought from 3% to 8% by 2100 is expected according to the IPCC (UNCCD, 2023).

Even though droughts generate high impacts on societies and hinder development, its silent evolution and limited media attention entails delayed and insufficient response and misperception of the problem. There is an urgent need to effectively deal with drought to prevent, mitigate and recover from its impacts, as well as to be better prepared to cope with future drought events. According to the literature on drought management, a proactive drought management approach (Hayes et al., 2004; Wilhite et al., 2000, 2014) allows addressing these needs, in contrast with the traditional reactive approach that mainly deals with the emergency and recovery situations triggered by drought events.

The proactive approach implies considering, at least, hazard characterization and drought event monitoring, vulnerability assessment and risk management including a set of measures and actions to address drought effects. In the last decade, drought planning has moved from a reactive crisis management approach towards a proactive risk-based approach, as seen as the best way to effectively mitigate drought impacts (UNISDR, 2021; Hagenlocher et al., 2019; Wilhite et al., 2014; Kampragou et al., 2011). When investing in a proactive approach a triple dividend is identified

---

[1] The International Disaster Database. Centre for Research on the Epidemiology of Disasters (EM-DAT), Brussels, Belgium, Available at https://www.emdat.be/

[2] European Drought Observatory (EDO). Available at https://edo.jrc.ec.europa.eu/edov2/php/index.php?id=1000

[3] European Drought Impact Report Inventory (EDII), Brussels, Belgium (Stahl et al., 2016).

[4] U.S. Drought Monitor, Lincoln, NE, Available at https://droughtmonitor.unl.edu/

avoiding future losses by reducing risk, increasing productivity and driving innovation, and fostering multiple social and environmental co-benefits (UNCCD, 2023).

Drought policy and management is an underrepresented topic and usually uncovered in policy documents (FAO, 2019). While numerous scholarly works have analyzed drought as a natural hazard, the extant literature about drought management is much more limited (Kallis, 2008), while research on drought and society is receiving increasing attention in the last decade (Savelli et al., 2022). In the first decade of the 2000 few guidelines to orient on how to develop drought policy or plans under this approach were developed (e.g. Knutson et al., 1998; EC - WSDEN, 2007; Iglesias et al., 2007; FAO & NDMC, 2008; UNISDR, 2009; UNDP, 2011; Wilhite et al., 2014). This shows the need to promote and facilitate drought policy development worldwide towards the Integrated Drought Management Programme (IDMP) (n.d.) (WMO & GWP, 2014), an international initiative that "supports governments and other stakeholders at all levels on implementing Integrated Drought Management by providing policy, technical and management guidance and by sharing scientific knowledge and best practices". This initiative was launched as a result of the High-level Meeting on National Drought Policy (HMNDP) that took place in Geneva, Switzerland, from 11 to 15 March 2013 (Sivakumar et al., 2014). Since then, significant progress has been made worldwide. Very recently, the International Drought Resilience Alliance (IDRA) (n.d.), led by Spain together with Senegal, was launched during COP 27 in 2023 with the aim of catalyzing political momentum and facilitating the mobilization of resources to improve drought resilience in countries, cities and communities. UNCCD has supported the operationalization of the Alliance since its launch.

However, there is an important gap between theory and practice (Fu & Tang, 2013) and the development of drought management plans or other drought-related planning tools are still very limited. In fact, even if this initiative provides a good platform to share knowledge, experiences and information on drought policy development, our knowledge about how we are addressing drought in practice is limited as drought policy analyses are scattered, heterogeneous and insufficient. This is especially the case of South America region, where the gaps of information on drought management are still significant and greater than for other regions like United States of America, Europe or Australia. It should be noted that there has been recent improvements in this regard in the region, especially with the publication of the "Drought Atlas of Latin America and the Caribbean" (Núñez Cobo & Verbist, 2018), which covers most of the countries of the region: Argentina, Bolivia, Brazil, Chile, Colombia, Ecuador, Peru, Uruguay and Venezuela; and Honduras from Central America Region.

Thus, in order to improve drought policy development, we need to better understand how South American countries are dealing with drought management, as a starting point for further improvements. In this study we present a systematic review of drought policy characteristics by country and analyze to what extent they adopt a proactive approach to address drought issues. The results may help understand how drought policy is shaped in different countries and give a picture on the degree of drought policy development in the region and the challenges that they face nowadays. This will result in valuable information for decision-makers in South American countries.

8.2 Methodology

The research aims to understand drought policy characteristics in different countries across the South American region. For this purpose, the following key questions were formulated: (1) Which are the main policy instruments developed for drought management in the country/region? This implies analyzing from which policy area/sector do they emerge, the scale at which they operate, their geographical scope and timeframe for action; (2) Which is the degree of proactive approach adoption? Answering this question involves analyzing drought management characteristics considering key elements on monitoring and early warning systems, impacts and vulnerability assessment as well as preventive, mitigation and recovery measures development.

The geographical scope of the study is the South American region, including all countries except Chile.[5] These are Argentina (ARG); Bolivia (BOL); Brazil (BRA); Colombia (COL); Ecuador (ECU); Guyana (GUY); Peru (PER); Paraguay (PRY); Uruguay (URY); Venezuela (VEN).

This study is based on an exhaustive literature review on drought management and policy for the selected countries included in the analysis and available online. We combine two sources of information: (1) Scientific and academic literature; (2) policy, regulatory and grey literature on drought issues.

As a starting point, two international databases were revised to obtain information on the existing regulatory and normative framework in each country: (1) ECOLEX-The gateway to environmental law[6]; (2) ELAW-Environmental Law Alliance Worldwide.[7] Main laws, strategies and plans on environmental issues related to drought were retrieved. We then followed a snowball approach to update and fill the gap on the missing information country by country. Other sources consulted derived from UN agencies and international agencies specialized in drought, risk, and climate change (e.g. FAO, UNDRR, UNCCD, UNFCCC, WMO, GWP), as well as from national meteorological services, civil protection units and agricultural ministers. Additionally, a systematic revision on academic literature was carried out to complement knowledge acquisition on drought main topics: events characteristics, impacts, vulnerability, mitigation, and response measures.

The analytical framework was adjusted from the WMO and GWP (2014) document "National Drought Management Policy Guidelines: A Template for Action''. This document states which may be the process and the key elements to be

[5] Chile has been excluded from this analysis as it is widely covered by other chapters of the book.

[6] Available at https://www.ecolex.org/

[7] Available at https://www.elaw.org/

considered to develop a National Drought Policy under risk-based approach. Thus, it is used as a framework of reference of what a country is expected to do if it wants to develop its national drought policy adopting a proactive approach. We established as analytical categories the 3 pillars for integrated drought management, which sets the basis for a national drought policy development (WMO & GWP, 2014; Vogt et al., 2018; Crossman, 2018; FAO, 2019) (Table 8.1).

Each pillar has been qualitatively rated on a high (3)-medium (2)-low (1) scale, where 'High' corresponds to a good and complete development of the pillar, 'Medium' value corresponds to a partially incorporated or incomplete development of the pillar and 'Low' value corresponds to limited or insufficient development of the pillar.

Table 8.1 Key aspects of drought policy

Pillars	Description
Pillar 1. Drought monitoring, forecast, and early warning	Drought monitoring allows to track progress and the state of drought. This provides timely and essential information to support drought management. It may include meteorological, hydrological and agronomic indices as well as data on impacts. A forecasting component is also valuable in order to help predict drought evolution. Thus, this information will allow for the development of an Early Warning System (EWS) based on the analysis of drought situations and link it with the activation of measures (via threshold triggers). According to this, we analyzed the existence and characteristics of the following elements at country level: • Type of drought and variables calculated: Meteorological, hydrological, agronomic, socioeconomic • Drought Monitor system characteristics, including drought state surveillance and maps delivering, drought impacts or risk level analysis and forecasting activities • Communication system established, in terms of the publication of a specific drought bulletin or other meteorological, hydrological, agricultural information or alerts
Pillar 2. Vulnerability, risk and impact analysis	Understanding vulnerability and impacts is key in designing an adequate and effective drought management response. They are complex and context-specific issues and a non-universal model for their analysis exists, which adds difficulties to their use for drought policy development. Thus, for this pillar we analyzed the existence and characteristics of impacts and vulnerability assessment both in policy documents and in academic literature
Pillar 3. Mitigation and response	Actions to deal with drought are diverse as they need to cover the different stages of drought events progress (preventive, mitigation, emergency and recovery), different sectors (agriculture, water supply, industry, energy, civil protection, among others) and should consider short-, medium- and long-term strategies and multiple scales (local, regional and national scope). It is also key that they are sufficiently described in terms of its objectives, impacts tackled, resources needed and their activation process and link to drought status and progress. Therefore, we analyzed the existence and degree of description of measures, primarily in drought planning tools available at country level

Source: Own elaboration

Finally, we aggregated the values of the 3 pillars into an equal-weighted index as a proxy to show the degree of integrated and proactive approach development at country level.

8.3 Results

8.3.1 Drought Regulation and Planning Efforts in South America Region

There is an extensive set of regulatory tools and documents approved to deal with drought management in the region, especially linked to water, agriculture, climate change and disaster risk management. The analysis of this documentation provides a good overview of the extent and diversity of regulatory frameworks and tools related to drought management in the region. Even if the legal databases consulted (ECOLEX and ELAW) are not complete nor updated, they provide a good starting point to analyze the characteristics and level of development of regulatory frameworks related to drought. Thus, this information was complemented with more extensive literature review based on grey literature and country-specific data, documents, and websites (Table 8.2).

Table 8.2 Overview of main regulatory and planning tools for drought

Country	Main drought regulation/planning/tools
ARG	Action Plan to reduce the risk of droughts in Argentina (2023) National Plan for Disaster Risk Reduction 2024–2030 (2023) National Plan for Disaster Risk Reduction 2018–2023 (2018) Interinstitutional Information Management Protocol to cope with the threat of droughts in Argentine territory (2022) National Plan for Adaptation and Mitigation to Climate Change (2022) National Action Program to Fight Desertification and Mitigation of the Effects of Drought (PAN) Updated to the 2030 goal (2021)
BOL	Plurinational Climate Change Policy (2023) Plurinational Plan for Immediate Response to Drought (2022) Bolivian Drought Monitor (2020) Metropolitan Master Plans for Drinking Water and Sanitary Sewerage Contingency Plans for disasters and/or local emergencies Water Plan for good living (2020) Risk Management Law 602 (2014) National Strategy for agricultural risk management and adaptation to climate change for resilient agriculture (2017–2020) National Plan to Fight Desertification PAN-LCD (1995–1997)

(continued)

Country	Main drought regulation/planning/tools
Bra	National Plan on Risk Management and Response to Natural Disasters (2012–2014) National Climate Change Policy (PNMC) (2009) National Action Program to Fight Desertification and Mitigate the Effects of Drought (PAN-Brazil 2004)
COL	National Contingency Plan First Dry Season (2022) National Contingency Plan for the first dry season of 2022 with La Niña Phenomenon Condition (2022) Colombia Long-Term Climate Strategy E2050 (2021) National Strategy for the Comprehensive Management of Drought in Colombia (2018) National Disaster Risk Management Plan (PNGR) 2015–2025 (2015) Early Warning System to monitor El Niño Phenomenon 2014–2016. National Disaster Risk Management Policy (2012) National Plan for Adaptation to Climate Change (2012) National Plan to Fight Desertification and Drought in Colombia (2004)
ECU	National Drought Plan 2021–2025 (2021) Institutional Strategic Plan on National Disaster Risk Response Plan (2018) Organic Environmental Code (COA) (2017) National Environmental Authority Standards for Desertification, Land Degradation and Drought (2014) National Climate Change Strategy 2012–2025 (2012) National Action Plan to fight Desertification, Land Degradation and Drought (PAND) (2004)
Guy	Guyana's National Drought Mitigation and Adaptation Plan (2020) National Climate Change Policy and Action Plan (2020–2030) (2020) Drought Early Warning System Protocol (2015) National Integrated Disaster Risk Management Plan and Implementation Strategy for Guyana (2013) Water and Sewerage Act (2002) (Drought Order)
Per	National Disaster Risk Management System (SINAGERD) (Law 29,664) (2011) Gender and Climate Change Plan of Peru (2016) National Strategy to fight Desertification and Drought (ENLCDS), 2016–2030 (2016) National Strategy for Climate Change (ENCC) (2015) National Water Resources Plan (PNRH) (2015) National Plan for Risk Management and Adaptation to Climate Change in the Agriculture Sector, period 2012–2021 (PLANGRACC-A) (2012).
PRY	National Drinking Water and Sanitation Plan (2023) Report: Paraguay, economic and social impacts of drought (2022) Climate Contingency Plan (2021) National Policy for Management and Reduction Disaster Risk (2018) National Action Plan to fight Desertification and Drought of the Republic of Paraguay 2018–2030 (2018) National Climate Change Mitigation Plan and Action Program (2017) Basic handbook for comprehensive Agro Climatic risk management of Paraguay (2017) National Climate Change Plan (2016) National DRM Plan and adaptation to Climate Change of the Agricultural Sector 2016–2022 Drought Characterization in Paraguay using the SPI (2016) Paraguay National Development Plan 2030 (2014)

Table 8.2 (continued)

Country	Main drought regulation/planning/tools
URY	National Emergency System (2020) Uruguay, compendium of risk maps (2020) National Policy for Comprehensive Management of Emergency and Disaster Risks in Uruguay (2019–2030) National Water Plan (2017) National Climate Change Policy (2017) Strategy to Promote the Development of Irrigation agriculture in Uruguay (2015) National Action Plan to fight Desertification and Drought—PAN (2005)
VEN	Organic Law of the Environment (Law No. 5833) (2019) National Land Degradation Neutrality Strategy (NDT), towards 2030, Final Report (2019) Index of Environmentally Sensitive Areas to Desertification (IAASD), San Juan parish, Mérida, Venezuela (2019) National Report on the Implementation of the United Nations Convention to Combat Desertification and Mitigation of Drought (2000)

Source. Own elaboration

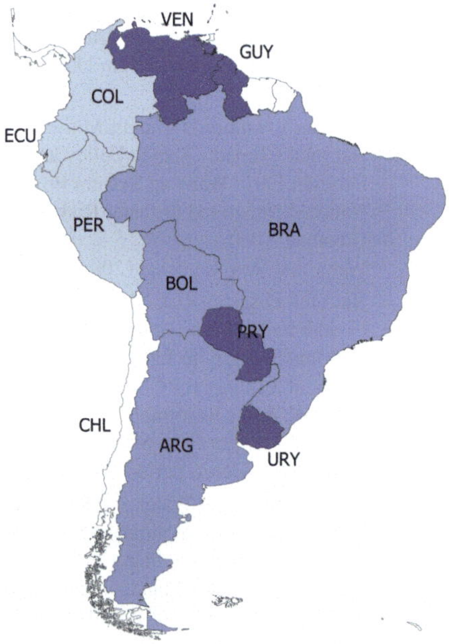

Map 8.1 Drought planning level. (Source: Own elaboration)

Multi-level regulations are available in the South American region, from national level to other sub-national scopes, either regional or local (Map 8.1). Drought planning initiatives mainly encompasses the national scale (GUY, PRY, URY, VEN) while in a limited number of countries, the development of drought-related planning tools have been also developed for specific regional levels (ARG, BOL, BRA) or local levels (COL, ECU, PER).

From a **sectoral perspective**, drought policy has been traditionally linked with agricultural and water resources management policies in all the countries. More

Table 8.3 Main drought planning tools developed in South America countries

Country	Time frame	Specific drought planning tools
ARG	2023	Action Plan for drought risk reduction in Argentina
BOL	2022	Pluriannual plan for rapid drought response
COL	2018	Estrategia Nacional para la Gestión Integral de la Sequía en Colombia
ECU	2021	Drought National Plan 2021–2025
GUY	2020	Guyana's National Drought Mitigation and Adaptation Plan

Source: Own Elaboration

recently, Climate Change (CC) and Disaster Risk Reduction (DRR) or Disaster Risk Management (DRM) have emerged as policy areas to include drought risk management issues as they provide suitable and comprehensive conceptual frameworks that encompass regulation development for drought management. Given the international and national-led initiatives to promote better planning on these topics, most of the countries of the region have developed new regulations and plans within these thematic policies (ARG, BRA, COL, ECU, GUY, PER, PRY, URY), except for VEN. This situation lays good political and conceptual foundations for the future development of specific drought plans or strategies in the countries that currently have nor developed it yet. On the other hand, specific plans to deal with drought have been developed in 5 out of 10 of the countries analyzed, representing 50% of the countries included in the study (ARG, BOL, COL, ECU, GUY) (Table 8.3).

As shown in Table 8.3, the development of these plans is relatively new, where the first drought plan approved was in 2018 and corresponds to COL. The 4 remaining plans have been approved in the last 3 years (2020–2023).

It is worth mentioning the efforts to address drought issues promoted by the United Nations Convention to Combat Desertification (UNCCD) to develop National Action Plans to combat desertification and drought are present in all the countries analysed.

Thus, the extent of regulations related to drought presents a wide development in the region and half of countries have specific drought plans in place, while the rest present good political and conceptual foundations based on disaster risk management and climate change policy developments.

8.3.2 Pillar 1: Drought Monitoring, Forecast, and Early Warning

In examining the state of drought Monitoring and Early Warning Systems (EWS) across South American countries, a diverse range of approaches and levels of development can be observed.

Drought monitors are the first step to develop a complete and pertinent EWS. In the region, different drought monitors have been developed. Only two countries have developed their own drought monitor system: BOL and BRA. In the rest of the countries, we found that they rely on a regional monitor but not fully integrated into national policies (ARG and PER), it may be under development (COL, ECU, GUY) or have developed limited activities to address drought monitoring (PRY, URY, VEN).

The Bolivian Drought Monitor, established in 2020, provides a web-based platform for visualizing drought severity maps through a Drought Combined Index. This monitor is hosted in the Agricultural and Livestock Early Warning System platform, which offers various risk levels and alerts for meteorological, hydrological, and agricultural issues. However, the criteria for transitioning between different drought states (Exceptional, Extreme, Severe, Moderate, Dry, and Normal conditions) are not clearly defined, and no specific measures are associated with each phase.

Brasil has been operating its Brazilian Drought Monitor since 2014, offering a web-based site for visualizing drought severity maps. It should be noted that it was the first drought monitor established in South American Region. Like BOL, the transition mechanisms between different drought states are not explicitly clear, and no specific measures are linked to each condition. BRA also hosts the SimaCaating system, which monitors and alerts based on vegetation coverage for the Caatinga region, utilizing SPI and NDVI calculations.

It should be highlighted that there are some relevant regional drought monitors operating in the region; the Information System on Drought for Southern South American Region (SISSA) covering ARG, BOL, BRA, PRY and URY and the Regional Climate Center for Western South America (CRS-OCA), a regional drought monitor based on SPI for BOL, COL, ECU, PER and VEN. However, even if these tools provide valuable information and data on drought characteristics, they are not integrated into national drought policies, except for ARG and PER.

For instance, ARG does not possess a national drought monitor website but instead depends on a regional monitoring system known as the Information System on Drought for the Southern South American Region (SISSA). Alongside SISSA, ARG has established an "Interinstitutional Protocol for information management on drought hazards - Emergency Phase (2022)," which integrates meteorological and agronomic forecasts for moderate and severe drought phases. However, this system lacks hydrological or socioeconomic analysis and information. Efforts related to hydrological drought are conducted by the National Water Institute (INA) but have not been integrated yet into a comprehensive monitoring system. Additionally, ARG relies on a local EWS known as the "Information and Early Warning System of the Southwest of Buenos Aires (SIAT)".

PER features a drought monitor that integrates meteorological and hydrological information and maps, supported by UNESCO and FAO among others. Although several drought indices are calculated and mapped by SENAMIH (National Meteorological and Hydrological Service of Peru), these two platforms are not integrated into a national monitor, and there is no reference to specific measures.

COL, ECU, GUY, PRY, URY, and VEN currently do not have national drought monitors. In COL and ECU national drought monitors are under development with support from International Center for Research on the El Niño Phenomenon (CIIFEN) and Euroclima+. In COL, relevant data and maps visualization sites exist and also bulletins related to different types of droughts, but they are not integrated into a common drought monitor site, so the information exists but is scattered. ECU National Drought Monitor is currently under development, supported by the UN Climate Technology Centre and Network (CTCN), based on hydroclimatic and meteorological data visors already in place by the INAMHI (Meteorological and Hydrological National Institute). GUY has developed drought monitoring activities and an EWS Protocol but lacks a dedicated drought monitor website. It is a member of the Caribbean Drought and Precipitation Monitoring Network (CDPMN) (n.d.). PRY, URY and VEN present the drought monitoring systems least developed in the region and relies on the regional monitoring systems SISSA and CIIFEN and a specific drought monitoring system website is lacking.

This overview highlights the varied state of drought monitoring and EWS across South American countries, reflecting a mix of regional collaborations, under-development systems, and the absence of national platforms in some cases.

Considering the different types of droughts that can be monitored, BOL, COL, ECU includes monitoring meteorological, hydrological and agricultural drought analysis, while only two types of drought variables are covered by PER (meteorological and hydrological) and ARG and BRA (meteorological and agricultural variables). Only meteorological drought is considered for GUY, PRY, URY and VEN. Efforts to monitor drought impacts and link them to a drought monitor and EWS are available in ARG, BOL, and BRA, being the component of a EWS less developed in the region. In contrast, forecasting activities, particularly based on meteorological variables related to drought, like the Standarized Precipitation Index (SPI), are present in all the countries although to different extents of detail and development and not yet incorporated into a specific drought tool.

Communication about drought situations and its evolution is key for decision making to adopt the most appropriate response to drought. Thus, the existence of a specific drought bulletin is an important tool for a good EWS on drought if it is linked to the drought monitor and with an appropriate delivery frequency, mainly monthly. This is the case of ARG, BOL, BRA, COL, ECU, GUY and PER. The frequency and quality of the bulletins and maps produced varies among countries. Moreover, sometimes they are accompanied by additional monitoring activities. For instance, in ARG, the Office for Monitoring Agricultural Emergencies (OMEGA) informs, through a bulletin, to institutions and the general public on the emergency indicators established. The collected data is analyzed and presents the state of key crops for the development of measures to prevent and mitigate social and productive vulnerability to droughts, fires and floods. BRA uses the Agrometeorological Information System (AGRITEMPO) to complement drought monitor information based on SPI.

It should be noted that only three countries do not have specific drought bulletins: PRY, URY, and VEN. However, other types of agroclimatic or hydrological or agrometeorological bulletins exist, which can be considered a preliminary step in the development of specific drought bulletins in those countries.

Map 8.2 Degree of development of drought Monitoring & EWS. (Source: Own elaboration)

In summary, the level of development of a Drought Monitor and EWS varies among countries in the region where 6 out of 10 present a low score, showing there is room for improvement, particularly for the countries from the northwest of the South American region (Map 8.2).

ARG and BOL present a high level of development even though some challenges should be tackled. Two countries present a moderate level of development: BRA and PER, which need further efforts to enhance drought impacts and risk levels analysis development. ECU, COL and GUY are in the process of developing their own monitoring system and show some progress but more efforts to develop this pillar are needed. Least and limited development is found for PRY, URY, VEN.

The main challenge of the region is to ensure that National Drought Monitors are aligned among countries in the region, so as to provide compatible and shared information systems on drought issues. Currently, CIIFEN provides the basis to do so, and Brazilian Monitor shares the same methodological and calculation approach (CIIFEN Bulletin, 2023).

8.3.3 Pillar 2: Vulnerability, Risk, and Impact Analysis

Vulnerability and impact analysis are key issues to approach when dealing with proactive drought management development as they allow for better understanding of drought effects and their underlying causes. However, their level of inclusion in the drought plans or strategies greatly varies among countries and are in general terms limited.

Some countries have developed specific vulnerability assessments to drought and include them in their drought planning tool. This is the case of ARG, ECU and PER. A group of a few countries present an analysis of vulnerability to drought, but it is approached from a theoretical or conceptual basis and limited details are given for their specific territory (BOL, GUY) or it is done only for specific sectors like the Agriculture (PRY). Vulnerability assessment is lacking for BRA, COL, URY and VEN.

It should be highlighted that drought impacts assessment is more developed than vulnerability research. At a regional level, the "Drought Atlas of Latin America and the Caribbean" (Núñez Cobo & Verbist, 2018) provides a good synthesis of main drought impacts for the countries included (ARG, BRA, COL, ECU, PER, URY, VEN). Moreover, two other relevant initiatives can be highlighted: DESINVENTAR.org (n.d.), a free, open-source Disaster Information Management System which encompasses the major regional database on disaster for South America region, and EM-DAT—The international disasters database. EM-DAT is the major database of disaster events, including hazard inventories and its impacts, with a worldwide coverage and collecting data since 1988. Even if this database is commonly used in some policy plans to present an overview on disaster's impacts, neither of the two are integrated into the national initiatives to address drought.

At national level, some countries clearly link drought impact assessment with a drought emergency declaration process (BOL, BRA). Impact assessments are mainly linked with agricultural impacts (ARG, BOL, PRY, URY), and with water resources impacts to a lesser extent. The analysis done in relation to specific drought events rather than in general terms. It should be noted that drought impacts are addressed in all the countries analyzed except for VEN.

In ARG, both impacts and vulnerability are tackled together. The Action Plan for drought risk reduction (2023) outlines a conceptual framework for addressing drought risks, while more detailed analyses are found in the National Action Plan for Adaptation and Mitigation of Climate Change (2022) and a final report aimed at establishing a national strategy for drought adaptation and mitigation (2020). These documents provide comprehensive insights into drought impacts and vulnerability. Furthermore, academic contributions, such as those by Thomasz et al. (2023), Bert et al. (2021), Solano-Hernández et al. (2020), and Naumann et al. (2019), enrich the understanding of these topics in the Argentine context.

Like ARG, BOL and ECU approaches impacts and vulnerability issues in an integrated manner. In BOL, the National Strategy for Agricultural Risk Management and Adaptation to Climate Change (2017–2020) offers some information on drought impacts and vulnerability, albeit limited. However, the National Report of Drought Effects, published by the Vice Ministry of Civil Defense (VIDECI), provides data on the municipal level impacts of drought, including affected areas and populations. Academic literature, including works by Canedo-Rosso et al. (2021), Heyer et al. (2018), and Walsh-Dilley (2020), further explores these themes. In ECU the National Drought Plan offers a detailed analysis of drought impacts by event and sector, including a thorough vulnerability assessment and risk mapping. However, specific academic literature on drought impacts or vulnerability in Ecuador is scarce.

Some countries present great differences in the extent to which they address these topics, as vulnerability analysis is mainly lacking for BRA and COL. In BRA, the Integrated Disaster Information System (S2ID) and the Brazilian Atlas on Natural Disasters serve as key tools for analyzing drought impacts among other hazards, offering a web-based mapping platform for historical analysis. While it is an innovative and pioneer tool for the region, there is a noted lack of formalized national-level vulnerability and resilience assessments, as highlighted by Gutiérrez et al. (2014). Additional academic sources, such as Dantas et al. (2020), Sena et al. (2014) and de Azevedo Reis et al. (2020), contribute to the discourse on drought impacts and vulnerability in Brazil. COL addresses drought impacts in its National Strategy for Integrated Drought Management but lacks a comprehensive vulnerability assessment. Limited academic works, such as those by Nauditt et al. (2022), Weng et al. (2020), and Zuleta et al. (2017) provide further insights on these topics.

In contrast, in PER vulnerability assessments present an advanced work as it is linked to Prevention and Disaster Risk Reduction Plans, particularly for the Piura region, with vulnerability levels calculated at the district level in the National Plan for Disaster Risk Reduction and Climate Change Adaptation in the agricultural sector (2012–2021). However, only the National Strategy to combat desertification and drought (2016) offers a general analysis of drought impacts. Academic literature, such as Condori-Apaza et al. (2021) and Lasage et al. (2015), provides additional insights.

There is limited development of both impacts and vulnerability assessment in GUY, PRY, URY and VEN. In GUY, vulnerability analysis is anticipated under the State of Drought declaration as per the DEWS protocol. The National Drought Mitigation and Adaptation Plan presents a conceptual framework for vulnerability but lacks concrete calculations. The plan details sector-specific drought impacts, but an impact database is notably absent. For, PRY drought impacts are particularly linked to agriculture and a recent report on drought impacts in 2022, supported by the UNDP, analyzes the economic impacts on agriculture and the food system. The basis for the National Drought Plan outlines drought vulnerability for agricultural and water supply sectors, complemented by the National Plan to combat desertification and drought. In URY vulnerability studies are documented in the Risk Atlas of Uruguay (2020), with academic literature, such as Diaz et al. (2018), focusing on livestock vulnerability case studies. However, official documents on impact assessment are lacking. In VEN, official documents on drought vulnerability and impacts are absent, but academic articles by Paredes-Trejo et al. (2023) and Olivares et al. (2017) provide some insights into the subject.

Research and academic papers on drought impacts and/or vulnerability are very limited for the countries analyzed, except for BRA, the country where more studies on vulnerability and impacts have been found.

To summarize, the analysis of drought impacts is tackled by most of the countries analyzed while vulnerability assessments are less developed, and information contained in policy documents is very limited and mainly theoretical or qualitative at regional level. This is also coherent with the scarcity of specific academic literature found in relation with these topics and for the region. Thus, most of the countries present a moderate level of development of this pillar (Map 8.3), showing a clear need for improvement and research.

Map 8.3 Degree of drought vulnerability and impacts information available. (Source: Own elaboration)

8.3.4 Pillar 3: Mitigation and Response

The analysis of drought measures and response should go beyond the analysis of existing drought management plans or policies, when they exist, as drought measures may be cross-cutting different policy areas such as water resources, agriculture, climate change, disaster risk reduction, and civil protection, among others. Thus, their analysis is even more complex in addition to the lack of guidelincs and practical references on how to assess drought measures.

Among the five existing drought plans in the region, the identification and description of measures is heterogeneous in the type of information provided and its extent. Some do not include information on drought measures, while others do not differentiate among typology of measures - preventive, mitigation, recovery—or do not describe them with enough detail (ARG, COL). In contrast, only three countries provide detailed descriptions of the measures proposed (BOL, ECU, GUY). It should be noted that some of the countries only refer to emergency or recovery measures (BRA, PRY, URY, VEN) linked to the provision of subsidies, financial aid or the establishment of specific emergency funds, especially in agriculture (ARG, BRA, COL, ECU, GUY, PER, URY), or for water supply (e.g. BOL), whether they have a plan or not.

The overall analysis for this pillar reveals that a group of three countries show an appropriate and detailed development of different typology of drought measures (BOL, ECU, GUY) while a group of three countries present a medium stage in terms of establishment of a drought mitigation and response plan (ARG, COL, PER) and the rest, up to four countries (BRA, PRY, URY, VEN) do not have a well-established drought mitigation and response plan. Despite the little information found on Venezuela, a deeper analysis needs to be carried out to better understand its dynamics to manage droughts though, and which are the concrete measures it implements as it is assume that measures can take place even if they are not accessible through strategic and planning documents (Map 8.4).

At the national level, some key issues can be highlighted to better detail the situation in each country.

BOL details drought measures within its drought contingency plan, categorizing them by type (preventive, response, recovery) and by actor, with a clear explanation of the alert and emergency response process across municipal, departmental, and national levels. The plan also outlines sector-specific objectives and actions for water, food security, health, agriculture, and infrastructure. Drought measures are also well described in ECU as its National Drought Plan includes a comprehensive set of measures aimed at prevention, drought risk reduction, and the development of a Drought Monitor system, among other goals. The plan introduces a specific indicator, the Index of Drought Risk Management, calculated for various years to integrate aspects of identification, reduction, management, and governance of drought issues. A National Plan for Disaster Response governs emergency declaration situations. Thus, the policy framework for drought management is very well detailed in the corresponding strategic documents revised. GUY drought plan also encompasses preventive, adaptation, and mitigation measures, which are well described in the policy document. Moreover, response and communication actions are planned, with the Minister of Agriculture authorized to issue drought orders.

Good foundations are observed towards enhancing drought planning tools and procedures for other countries like ARG, COL and PER. In ARG, the National Drought Plan lacks a clear typology of measures and does not provide insights into preparedness, prevention, or mitigation levels. The plan references the Agricultural

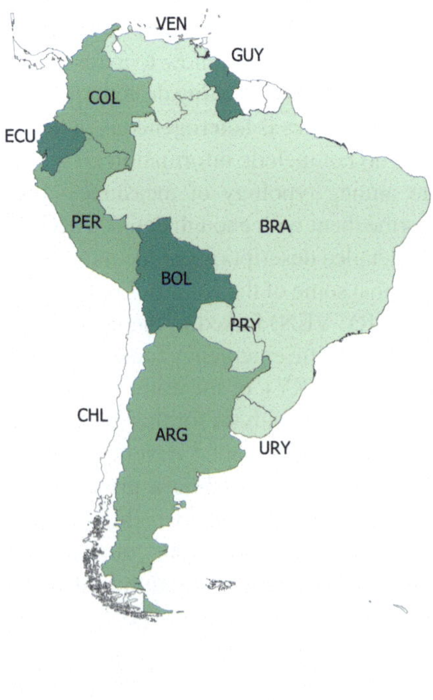

Map 8.4 Level of mitigation and adaptation measures development. (Source: Own elaboration)

Emergency Drought Declaration (Law 26509), outlining the governance and declaration process but not linking measures to specific drought levels. However, the National Strategy for Adaptation and Mitigation of Climate Change, along with its preceding document from 2020, presents a list and description of mitigation and adaptation measures, offering a more comprehensive approach to drought management. Thus, good foundations are observed towards enhancing drought planning tools and procedures in the country. In COL, the National Strategy for Integrated Drought Management (2018) briefly addresses mitigation and adaptation, focusing on specific preventive and mitigation actions for different types of droughts (meteorological and agricultural), particularly within the agricultural sector. Recovery efforts are incorporated into contingency and disaster management planning tools, detailing emergency recovery processes and fund distribution. Additionally, COL utilizes Contingency Plans at both sectoral and territorial levels, with Special Plans developed as needed. Local and regional drought plans, such as the Action Plan to Fight Drought and Forest Fires in the Department of Cordoba (2017), complement national efforts. In PER the National Plan for Disaster Risk Reduction and Climate Change Adaptation in the Agricultural Sector (2012–2021) provides a detailed strategic framework with specific objectives, priorities, and actions for drought. Even if it is not a specific drought plan, the information contained is valuable for drought management. These actions are also supported by various prevention and disaster risk reduction plans at departmental, provincial, and district levels, as they detail actions, responsibilities, budgets, and timeframes for implementation.

There are four countries where drought measures are mainly reactive and focus on the emergency and recovery phases of drought events. In BRA, the approach to drought primarily involves the declaration of emergency situations, enabling the allocation of funds based on the impacts identified through Damages and Losses Reports (FIDE). In URY similar mechanisms exist while drought measures are primarily emergency-oriented with a particular focus on the livestock sector. PRY presents a different situation as drought management approach is outlined in the basis for a National Drought Plan, identifying recommended actions across three pillars, though the plan has not yet been approved. Meanwhile, the existing Climate Contingency Plan focuses on mitigating the consequences and effects of drought, with a reactive approach to emergency measures and a focus on agricultural concerns. In VEN, drought measures pertain to water management and agricultural recovery, but available information is limited.

8.3.5 *Overview of Drought Management Approach*

The joint analysis of the three pillars at country level provides a good overview on how a specific country is developing its own national drought policy and provides valuable insights on to what extent it is incorporating a proactive approach (Map 8.5).

None of the countries analyzed present the highest possible performance, which implies High values for the 3 pillars. However, five countries present a high-medium

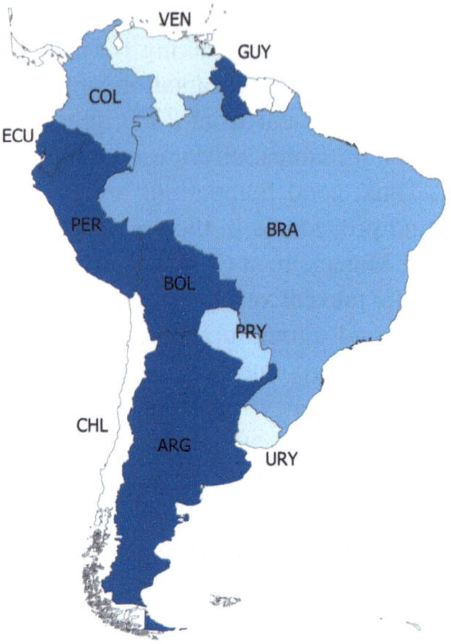

Map 8.5 Analysis of degree of proactive approach development. (Source: Own elaboration)

level score as they show some potential for enhancement in pillar 1 (ECU, GUY, PER), on pillar 2 (BOL) or on pillar 3 (ARG, PER), while their performance is high for the rest of the pillars. On a moderate level of development of a proactive approach for drought management, we find two countries (BRA, COL) as they present relevant advances but still specific shortcomings on a specific pillar or medium rates for all of them. Another country (PRY) presents low and medium scores for the 3 pillars while a group of two countries can be found at the lowest positions due to their limited development on the 3 pillars (URY, VEN), which clearly presents a reactive drought management approach.

Thus, the development of drought policies in South American countries reveals a diverse landscape, with some nations adopting more proactive approaches to drought management, while others remain more reactive. The next overview by country highlights the varied approaches to drought policy development in the region as well as some general issues for improvement detected.

ARG has embraced a proactive drought management framework, grounded in Comprehensive Risk Management laws and plans. The approval of its first National Drought Plan in 2023 marks a significant step forward, offering a thorough overview of drought characteristics, impacts, and associated vulnerabilities. Despite these advancements, the development of specific drought measures remains limited, with coordination mainly focused on monitoring impacts within the agriculture sector. Recent progress in drought planning and the development of management tools over the last 5 years underscore Argentina's proactive stance, though improvements could be made in monitoring systems and linking measures with alert levels.

BOL drought management has historically been tied to water management policies, with a focus on ensuring water access and availability, particularly concerning agricultural impacts. However, there has been a recent shift towards integrating risk management and climate change perspectives. The Pluriannual Plan for Rapid Drought Response (2022) outlines strategies and actions for dealing with drought events, supported by the recent development of a Bolivian Drought Monitor in 2020, linked to the Agricultural Early Warning System. These initiatives indicate BOL moves towards a more proactive approach, though data availability and dissemination remain areas for improvement.

BRA experience in drought management has been characterized by emergency management and relief efforts. Despite having a well-developed monitoring system and impact database, BRA faces challenges in establishing a clear proactive framework for implementing preventive, adaptive, and mitigation measures in an integrated manner.

In COL integrated risk management approach to drought management, established in 2018, encompasses a National Plan for Disaster Risk Management and a National Contingency Plan. These plans cover preparation, attention, recovery, and evaluation stages, but a National Drought Monitor is still under development, with related data and maps being scattered. COL proactive approach to drought management is evident, though enhancements in impact and vulnerability assessments could further strengthen this strategy.

ECU has made strides in adopting a proactive approach to drought management with the development of the National Drought Plan 2021–2025, aligned with the National Plan for Disaster Response (2018). The National Drought Monitor, currently under development, is supported by hydroclimatic and meteorological data, with detailed assessments of drought impacts and vulnerabilities included in the National Plan.

In 2020, GUY approved a National Drought Mitigation and Adaptation Plan, complemented by integrated disaster risk management plans. The proactive management approach is evident in the detailed drought plan and the development of a Drought EWS protocol in 2015, covering the three DRAM framework pillars. The Minister of Agriculture's authority to issue drought orders illustrates the established mechanism for addressing drought emergencies in GUY.

In PER the National Plan for Disaster Risk Reduction and Climate Change Adaptation in the Agricultural Sector (2012–2021) clearly incorporates a proactive approach to drought risk analysis and management, supported by a functioning drought monitor and risk scenarios. Despite the integration of drought management within the broader Disaster Risk Reduction strategy, the lack of a specific drought plan limits its scope, particularly for the agricultural sector.

PRY approach to climate change and disaster risk reduction includes strategic planning but lacks a specific drought management plan. A draft document for a drought management plan has been developed and it presents a proactive approach, but information remains scattered and sectorial, with a focus on agriculture. Enhancements in drought monitoring and a clearer operational framework for drought measures are needed.

URY approach for drought management is primarily reactive, with the recent publication of the Atlas of Risk (2020) offering a potential foundation for a proactive approach. However, drought issues are not yet integrated into related policies like the National Climate Change Policy (2017), and drought monitoring and EWS activities are limited, mainly focusing on the agricultural and livestock sectors.

VEN approach to drought management is characterized by crisis management, with policies scattered across sectors and limited development in drought monitoring, vulnerability and impact assessments, and measures focusing mainly on emergency and recovery.

8.4 Discussion

Drought planning efforts have been promoted in the last decade in Latin America and Caribbean (LAC) region, experiencing a great development in the last 5 years in South American countries. Thus, specific drought plans are relatively new in the region but are accompanied by an extensive set of legislation and regulatory tools that provides an adequate policy framework in which drought issues can be integrated and developed. Notice that only Mexico has previously developed a national drought policy in 2014 and has received notable attention both at academic and policy level (Ortega-Gaucin et al., 2016; Aguilar-Barajas et al., 2016).

For many countries of the region, initiatives, studies, and plans are particularly related to the agricultural sectors, either because they focus on the impacts on agriculture, or because they frame drought management under agriculture policy. Even if there has been a move from including drought issues in agricultural and water resources management towards climate change and disaster risk reduction policies, focusing on the agricultural sector is still a main criticism among vulnerability, impacts and measures and adaptation issues (Savelli et al., 2022). Moreover, there is a clear opportunity for integrating the efforts already being made through these related policies into a specific drought policy which will contribute to increase policy coherence and drought management efficiency, therefore reducing negative consequences of drought.

When focusing on how drought is being managed, we found a different degree of development among the three pillars analyzed, being pillar 1 (drought monitoring, forecasting and early warning) the most developed, followed by pillar 2 (vulnerability, risk and impact assessment) and pillar 3 (mitigation and response). This may be due to several varied reasons which may be different depending on the country. On one hand, some countries like Brazil have an extensive tradition in drought monitoring while others developed their national monitoring systems based on international or regional initiatives launched to support countries to enhance their monitoring capacities. Moreover, understanding drought events and characteristics within a territory is the starting point for building a proactive management system. Thus, some countries that traditionally managed drought through a crisis management approach have taken the opportunity to move quickly towards a proactive approach management and have made great improvements in developing its own national monitoring and early warning systems. Advances made in drought monitoring are seen as a good start to improve early warning systems, but two main challenges emerged in this regard. First, a clear link of drought risk levels and measures should be established to be effective. Second, promoting shared methodologies and data will increase the utility of drought monitoring experiences at regional level as countries may learn from each other, taking into account the extent literature on drought indices (WMO & GWP, 2016). Moreover, further analysis on the specific index and types of drought covered by each monitoring system is needed.

In relation to drought impacts and vulnerability analysis included in pillar 2, these two topics present several issues for improvements in the South America region, especially vulnerability issues, as it is underrepresented in the academic literature (de Azevedo Reis et al., 2020). Impacts can be better integrated into drought monitoring and EWS but some initiatives to collect drought impacts may broaden their scope and go beyond the agricultural sector. There is also room to use available databases like EM-DAT or DESINVENTAR as a source of information to address drought impacts. Moreover, they also provide a good example and starting point to develop country-led impacts databases on drought. This is the example of Brazil, which includes drought impacts in a broader platform for Integrated Disaster Information System (S2ID). However, as Sutanto et al. (2019) highlight, impact-based forecasts are still far from including risk information.

The development of academic research should be promoted to develop national, regional and local vulnerability assessments needed to increase our understanding of drought impacts and causality relationships in the particular context of the South America region, as it is underrepresented in the academic literature in comparison with other regions like Europe (Rossi et al., 2023) and Australia (Wilhite et al., 2005). Moreover, a common framework and guidelines to address drought vulnerability assessments in the region would be highly recommended. Further research is needed to fill gaps on vulnerability and risk assessment, especially its practical implication for drought management (Hagenlocher et al., 2019, 2023). Therefore, additional efforts are needed to include such information into the decision-making process on drought.

Pillar 3 presents the greatest room for improvement, especially within drought plans available. The level of detail and accuracy of the measures could be enhanced and a clear link to drought status and progress or risk level should be included to increase the efficiency of the management system. Moreover, drought measures and response may go beyond agricultural issues and broaden to other relevant sectors impacted by drought like health, industry, tourism, or energy, among others. However, the inclusion of long -term perspective on drought studies and policy tools is needed as they "provide useful insights about the manner in which society and drought risk co-evolve over time and across space" (Savelli et al., 2022, p. 13), which is key to adopt a proactive approach.

Thus, improvements towards the adoption of a proactive approach have been identified in the countries of the South American region. While there is still room for improvement only in a limited number of countries predominates a crisis management approach. Aligning efforts to deal with SDG of the 2030 Agenda, Sendai Framework and Paris Agreement on climate change will contribute to addressing adaptation challenges giving the opportunity to join efforts towards proactive and resilient policy development and to overcome reactive approaches.

8.5 Final Remarks

In recent years, the South American region has advanced in terms of setting drought policy agenda, planning tools and monitoring mechanisms. These positive advances match with an increased political environment engagement and awareness on drought impacts for the agriculture sector, climate change urgency, community resilience and sustainable development needs. Drought is also being included into media coverage as an important topic to discuss, especially within climate change and disaster risk reduction narratives. This increasing attention helps to integrate it into a political regional agenda and has nourished a high-level discussion of the negative consequences of droughts and the need for urgent action. Therefore, it is important to take advantage of this momentum to continue supporting the consolidation of the ongoing experiences of the countries of the region and strengthen the adoption of proactive drought management approaches in the South America region.

There are several valuable initiatives and experiences to promote a proactive approach to deal with drought, both at country and regional level. However, those efforts are scattered, and still insufficient and drought impacts and vulnerability assessments are not efficiently tackled. For instance, information and regulations are embedded in several policies as cross cutting topics. It is cornerstone to integrate all the advances already made into national drought policies to increase its relevance and policy coherence. This may require limited efforts in contrast with the expected potential improvement in drought policy development it may deliver. In addition, it is key to encourage a dialogue between political stakeholders and technicians to enrich the suggested integration process for national drought policy development.

Additionally, best practices compilation and knowledge management must be promoted across the region to enhance coherence and mutual learning among countries. This will respond to the need for practical experiences and knowledge sharing in the region.

Annex. Drought Related Regulations and Plans by Country

Argentina

Ministerio de Agricultura, Ganadería y Pesca (2022). Protocolo Interinstitucional de Gestión de Información ante la amenaza de sequías en el territorio argentino. 17p.

Ministerio de Ambiente y Desarrollo Sostenible (2020). Reporte final sobre las bases para una estrategia nacional de adaptación y mitigación de los efectos de la sequía en la República Argentina. 207p.

Ministerio de Ambiente y Desarrollo Sostenible (2022). Plan Nacional de Adaptación y Mitigación al Cambio Climático.

Sistema de Información sobre Sequías para el Sur de Sudamérica (2023). Plan de Acción para la reducción de riesgo de sequías en Argentina. Documento Final. Documento elaborado por Maria Soledad Ordoqui. 54p.
Sistema Nacional para la Gestión Integral del Riesgo (2017). Plan Nacional para la Reducción del Riesgo de Desastres 2018–2023 (PNRRD). Ministerio de Seguridad de la Nación.
Sistema Nacional para la Gestión Integral del Riesgo (2023).Plan Nacional para la Reducción del Riesgo de Desastres 2024–2030. Ministerio de Seguridad de la Nación, Noviembre 2023, 133p.
Secretaría de Gobierno de Ambiente y Desarrollo Sustentable (2021). Programa de Acción Nacional de Lucha contra la Desertificación y Mitigación de los Efectos de la Sequía (PAN) Actualizado a la meta 2030. 126p.

Bolivia

Ministerio de Desarrollo Rural y Tierras (2017). Estrategia Nacional de gestión de riesgo agropecuario y adaptación al cambio climático para una agricultura resiliente (2017–2020). La Paz, Bolivia 2018. 102p.
MMA, MDRyT, MD (2022). Plan Plurinacional de Respuesta Inmediata a la Sequía. La Paz, Bolivia, noviembre 2022. 96 págs.
Viceministerio de Defensa Civil (2023). Reporte Nacional de Afectación por la Sequía 2023.
Servicio Nacional de Meteorología e Hidrología (SENAMHI) (2020) Monitor de la Sequía de Bolivia. Available at http://monitorsequias.senamhi.gob.bo/#/
Gobierno de Bolivia (2023). Planes Maestros Metropolitanos de Agua Potable y Alcantarillado Sanitario.
Gobierno de Bolivia (2023). Planes de contingencia de desastres y/o emergencias locales.
Gobierno de Bolivia (2020–2023). Plan Agua para Vivir Bien.

Brazil

Integrated Disaster Information System (S2ID). Available at https://s2id.mi.gov.br/
Brazilian Atlas on Natural Disasters. Available at https://atlasdigital.mdr.gov.br/.
National Plan on Risk Management and Response to Natural Disasters (2012–2014).
National Climate Change Policy (PNMC) (2008).
National Action Program to Combat Desertification and Mitigate the Effects of Drought (PAN-Brazil 2004)

Colombia

Unidad Nacional para la Gestión del Riesgo de Desastres (2022). Plan Nacional de Contingencia para la primera temporada seca de 2022 con Condición Fenómeno La Niña. Periodo del plan 01/02/2022–14/03/2022). Bogotá D.C., Colombia 2022. 123p.

Unidad Nacional para la Gestión del Riesgo de Desastres (2014). Plan Nacional de Gestión del Riesgo de Desastres-PNGR 2015–2025.

Gobierno de Colombia (2021). Estrategia Climática de Largo Plazo de Colombia E2050. Estrategia climática de largo plazo de Colombia E2050 para cumplir con el Acuerdo de París. 137p.

Gobierno de Colombia (2018). Estrategia Nacional para la Gestión Integral de la Sequía en Colombia. 79p.

Unidad Nacional para la Gestión del Riesgo de Desastres (2022). Plan Nacional de Contingencia Primera Temporada Seca.

Gobierno de Colombia (2012). Plan Nacional de Adaptación al Cambio Climático. 100p.

Ministerio de Ambiente, Vivienda y Desarrollo Territorial (2004). Plan Nacional de lucha contra la Desertificación y la Sequía en Colombia (PAN). Bogotá D.C., Colombia 2022. 176p.

Unidad Nacional para la Gestión del Riesgo de Desastres (2016). Sistema de alerta temprana seguida en el Fenómeno El Niño 2014–2016.

Unidad Nacional para la Gestión del Riesgo de Desastres (2012). Política Nacional de Gestión de Riesgos de Desastres.

Ecuador

República del Ecuador. Ministerio del Ambiente (2012). Estrategia Nacional de Cambio Climático 2012–2025. 153p.

Ministerio del Ambiente, Agua y Transición Ecológica (2021). Plan Nacional de Sequía 2021–2025. Juntos construyendo resiliencia. 172p.

Secretaría de Gestión de Riesgos (2018). Plan Nacional de Respuesta ante Desastres. 446p.

Ministerio del Medio Ambiente (2004). Plan de Acción Nacional para la Lucha contra la Desertificación, Degradación de Tierras y Sequía (PAND).

Guyana

Epstein, M. (2015). Guyana Drought Early Warning System Protocol. Guyana Land and Surveys Commission (GLSC) and United Nations Development Programme (UNEP). 37p.

Guyana Land and Surveys Commission and Co-operative Republic of Guyana (2020). Guyana's National Drought Mitigation and Adaptation Plan. July 2020, 133p.
Office for climate Change (2019). National Climate Change Policy and Action Plan (2020–2030). Draft 2.0. May 14, 2019.

Peru

Ministerio de Agricultura (2012). Plan Nacional de Gestión de Riesgo y Adaptación al Cambio Climático en el Sector Agrario, periodo 2012–2021 (PLANGRACC-A). Lima, Peru. 98p.
Ministerio de Ambiente (2015). Estrategia Nacional ante el Cambio Climático (ENCC). Peru. 88p.
Ministerio de Ambiente (2016) Estrategia Nacional de Lucha contra la Desertificación y la Sequía (ENLCDS) 2016–2030.

Paraguay

Benitez, V.C.D (2016). Caracterización de la sequía en el Paraguay mediante la utilización del SPI. CONACYT. 90p.
Ministerio de Agricultura y Ganadería & FAO (2016). Plan Nacional de GRD y adaptación al Cambio Climático del Sector Agrícola 2016–2022. Asunción, 2016. 80p.
SEAM, PNUD, FMAM. (2017). Plan Nacional de Adaptación al Cambio Climático. Proyecto TCN e IBA. Asunción, Paraguay. 160p.
SEAM, PNUD (2017). Plan Nacional de Mitigación al Cambio Climático y Programa de Acción. Paraguay. 166p.
Ministerio del Ambiente y Desarrollo Sostenible (2017). Plan de Acción Nacional de Lucha contra la Desertificación y Sequía de la República del Paraguay 2018–2030.
Gauto Ortigoza, J (2019). Bases para la Estrategia Nacional de Sequía de Paraguay. Ministerio del Ambiente y Desarrollo Sostenible, Paraguay. 87p.
PNUD (2022). Informe: Paraguay, impactos económicos y sociales de la sequía. Asunción, Paraguay. 33p.
Dirección de Agua Potable y Saneamiento (2023). Plan Nacional de Agua Potable y Saneamiento.
Presidencia de la República del Paraguay (2014). Plan Nacional de Desarrollo Paraguay 2030 (Decreto N° 2.794).
Ministerio de Agricultura y Ganadería, IICA (2017). Manual Básico de gestión integral de riesgos Agroclimáticos de Paraguay.

Uruguay

República Oriental del Uruguay (2020). Sistema Nacional de Emergencia. Reglamento de la ley 18.621.

Cardona, O.D., Bernal, G., Escovar, M.A., Gonzáles, D., Grajales, S., Marulanda, P., Molina, J., Rincón, D., y Villegas, C. (2020). Uso de datos masivos para la eficiencia del Estado y la integración regional. Compendio de mapas de riesgo de Uruguay. Preparado para Fundación Ricaldoni. INGENIAR Risk Intelligence Ltda. Bogotá Plan Nacional de Aguas 2017 (2017).

Sistema Nacional Ambiental (2017). Política Nacional de Cambio Climático. 32p.

Ministerio de Agricultura, Ganadería y Pesca (2015). Estrategia de Fomento del Desarrollo de la Agricultura Regada en Uruguay. 38p.

Gobierno de Uruguay (2005). Plan de Acción Nacional de Lucha contra la Desertificación y la Sequía—PAN.

Venezuela

Asamblea Nacional de la República Bolivariana de Venezuela (2009). Ley Orgánica del Ambiente (Ley No. 5.833).

Ministerio del Poder Popular para el Ecosocialismo (2019). Estrategia Nacional Neutralidad en la Degradación de las Tierras (NDT), hacia el 2030, Informe final.

Universidad de los Andes (2019). Índice de Áreas Ambientalmente Sensibles a la Desertificación (IAASD), parroquia San Juan, Mérida, Venezuela.

Franklin Paredes, Universidad de Carabobo (2012). Sistema para la Alerta Temprana de Sequías Meteorológicas en Venezuela. Tesis Doctoral.

Ministerio del Ambiente y de los Recursos Naturales (2000). Informe Nacional sobre la Implementación de la Convención de Naciones Unidas de Lucha contra la Desertificación y Mitigación de la Sequía.

References

Aguilar-Barajas, I., Sisto, N. P., Magaña-Rueda, V., Ramírez, A. I., & Mahlknecht, J. (2016). Drought policy in Mexico: A long, slow march toward an integrated and preventive management model. *Water Policy, 18*(S2), 107–121.

Bert, F., de Estrada, M., Naumann, G., Negri, R., Podestá, G., de los Milagros Skansi, M., et al. (2021). *The 2017–18 drought in the Argentine Pampas—Impacts on agriculture*. United Nations Office for Disaster Risk Reduction. GAR Special Report on Drought.

Canedo-Rosso, C., Hochrainer-Stigler, S., Pflug, G., Condori, B., & Berndtsson, R. (2021). Drought impact in the Bolivian Altiplano agriculture associated with the El Niño–Southern Oscillation using satellite imagery data. *Natural Hazards and Earth System Sciences, 21*(3), 995–1010.

Caribbean Drought and Precipitation Monitoring Network (CDPMN). http://www.cimh.edu.bb/cami/files/cdpmn.pdf

Centro Internacional para la Investigación del Fenómeno de El Niño (CIIFEN). (2023). https://ciifen.org/

Condori-Apaza, V., Mamani-Luque, O. R., Alfaro-Alejo, R., Laqui, W., & Condori, W. F. (2021). Analysis and impact of meteorological droughts in the agriculture of Puno region, Peru. In *E3S web of conferences* (Vol. 304, p. 03002). EDP Sciences.

CRED. (2021). *Disasters in numbers* (p. 2022). CRED.

Crossman, N. D. (2018). *Drought resilience, adaptation and management policy (DRAMP) framework* (p. 20).

Dantas, J. C., da Silva, R. M., & Santos, C. A. G. (2020). Drought impacts, social organization, and public policies in northeastern Brazil: A case study of the upper Paraíba River basin. *Environmental Monitoring and Assessment, 192*, 1–21.

de Azevedo Reis, G., de Souza Filho, F. A., Nelson, D. R., Rocha, R. V., & da Silva, S. M. O. (2020). Development of a drought vulnerability index using MCDM and GIS: Study case in São Paulo and Ceará, Brazil. *Natural Hazards, 104*, 1781–1799.

DESINVENTAR. https://db.desinventar.org/

Diaz, I., Achkar, M., & Mazzeo, N. (2018). Drought vulnerability assessment of cattle producers in the sierras del Este-Uruguay: Interactions between actors and agents. *Outlook on Agriculture, 47*(4), 315–325.

EC - WSDEN. (2007). *Drought management plan report including agricultural, drought indicators and climate change aspects* (p. 2007). Water Scarcity and Droughts Expert Network.

EM-DAT. The international disasters database. https://www.emdat.be/

FAO. (2019). *Proactive approaches to drought preparedness—Where are we now and where do we go from here?* FAO.

FAO & NDMC. (2008). *The near East drought planning manual: Guidelines for drought mitigation and preparedness planning* (p. 59). Food and agriculture Organization of the United Nations Regional Office for the near East, Cairo, Egypt and National Drought Mitigation Center University of Nebraska-Lincoln.

Fu, X., & Tang, Z. (2013). Planning for drought-resilient communities: An evaluation of local comprehensive plans in the fastest growing counties in the US. *Cities, 32*, 60–69.

Gutiérrez, A. P. A., Engle, N. L., De Nys, E., Molejón, C., & Martins, E. S. (2014). Drought preparedness in Brazil. *Weather and Climate Extremes, 3*, 95–106.

Hagenlocher, M., Meza, I., Anderson, C. C., Min, A., Renaud, F. G., Walz, Y., Siebert, S., & Sebesvari, Z. (2019). Drought vulnerability and risk assessments: State of the art, persistent gaps, and research agenda. *Environmental Research Letters, 14*(8), 083002. https://doi.org/10.1088/1748-9326/ab225d

Hagenlocher, M., Naumann, G., Meza, I., Blauhut, V., Cotti, D., Döll, P., et al. (2023). Tackling growing drought risks—The need for a systemic perspective. *Earth's Future, 11*(9), e2023EF003857.

Hayes, M. J., Wilhelmi, O. V., & Knutson, C. L. (2004). Reducing drought risk: Bridging theory and practice. *Natural Hazards Review, 5*(2), 106–113.

Heyer, J. P., Power, M. J., Field, R. D., & van Marle, M. J. E. (2018). The impacts of recent drought on fire, forest loss, and regional smoke emissions in lowland Bolivia. *Biogeosciences, 15*(14), 4317–4331. https://doi.org/10.5194/bg-15-4317-2018

Iglesias, A., Moneo, M., & Quiroga, S. (2007). *Methods for evaluating social vulnerability to drought. OPTIONS Méditerranéennes. Série B: Etudes et Recherches*. CIHEAM.

Information System on Drought for Southern South American Region - SISSA (Sistema de Información sobre Sequías para el sur de Sudamérica). (2022). https://sissa.crc-sas.org/

Integrated Drought Management Programme (IDMP). https://www.droughtmanagement.info/

International Drought Resilience Alliance (IDRA). https://idralliance.global/

IPCC. (2022). Climate change 2022: Impacts, adaptation and vulnerability. In H.-O. Pörtner, D. C. Roberts, M. Tignor, E. S. Poloczanska, K. Mintenbeck, A. Alegría, et al. (Eds.),

Contribution of working group II to the sixth assessment report of the intergovernmental panel on climate change (p. 3056). Cambridge University Press.
Kallis, G. (2008). Droughts. *Annual Review of Environment and Resources, 33*, 85–118.
Kampragou, E., Apostolaki, S., Manoli, E., Froebrich, J., & Assimacopoulos, D. (2011). Towards the harmonization of water-related policies for managing drought risks across the EU. *Environmental Science & Policy, 14*(7), 815–824.
Knutson, C., Hayes, M., & Phillips, T. (1998). *How to reduce drought risk*. National Drought Mitigation Center.
Lasage, R., Muis, S., Sardella, C. S. E., Van Drunen, M. A., Verburg, P. H., & Aerts, J. C. J. H. (2015). A stepwise, participatory approach to design and implement community based adaptation to drought in the Peruvian Andes. *Sustainability, 7*, 1742–1773.
Nauditt, A., Stahl, K., Rodríguez, E., Birkel, C., Formiga-Johnsson, R. M., Kallio, M., et al. (2022). Evaluating tropical drought risk by combining open access gridded vulnerability and hazard data products. *Science of the Total Environment, 822*, 153493.
Naumann, G., Vargas, W. M., Barbosa, P., Blauhut, V., Spinoni, J., & Vogt, J. V. (2019). Dynamics of socioeconomic exposure, vulnerability and impacts of recent droughts in Argentina. *Geosciences, 9*(1), 39.
Núñez Cobo, J., & Verbist, K. (Eds.). (2018). *Atlas de Sequía de América Latina y el Caribe* (p. 204). UNESCO y CAZALAC.
Olivares, B., Cortez, A., Lobo, D., Parra, R., Rey, J., & Rodriguez, M. (2017). Evaluation of agricultural vulnerability to drought weather in different locations of Venezuela. *Revista de la Facultad de Agronomía, Universidad del Zulia, 34*(1), 103–129.
Ortega-Gaucin, D., Pérez, M. L., & Cortés, F. A. (2016). Drought risk management in Mexico: Progress and challenges. *International Journal of Safety and Security Engineering, 6*(2), 161–170.
Paredes-Trejo, F., Olivares, B. O., Movil-Fuentes, Y., Arevalo-Groening, J., & Gil, A. (2023). Assessing the spatiotemporal patterns and impacts of droughts in the Orinoco River basin using earth observations data and surface observations. *Hydrology, 10*(10), 195.
Rossi, L., Wens, M., De Moel, H., Cotti, D., Sabino Siemons, A., Toreti, A., Maetens, W., Masante, D., Van Loon, A., Hagenlocher, M., Rudari, R., Naumann, G., Meroni, M., Avanzi, F., Isabellon, M., & Barbosa, P. (2023). *European drought risk atlas*. Publications Office of the European Union. https://doi.org/10.2760/608737
Savelli, E., Rusca, M., Cloke, H., & Di Baldassarre, G. (2022). Drought and society: Scientific progress, blind spots, and future prospects. *Wiley Interdisciplinary Reviews: Climate Change, 13*(3), e761.
Sena, A., Barcellos, C., Freitas, C., & Corvalan, C. (2014). Managing the health impacts of drought in Brazil. *International Journal of Environmental Research and Public Health, 11*(10), 10737–10751.
Sivakumar, M. V. K., Stefanski, R., Bazza, M., Zelaya, S., Wilhite, D., & Magalhaes, A. R. (2014). High Level Meeting on National Drought Policy: Summary and Major Outcomes. Weather and Climate Extremes, 3, 126–132. https://doi.org/10.1016/j.wace.2014.03.007.
Solano-Hernández, A., Bruzzone, O., Groot, J., Laborda, L., Martínez, A., Tittonell, P., & Easdale, M. H. (2020). Convergence between satellite information and farmers' perception of drought in rangelands of north-West Patagonia, Argentina. *Land Use Policy, 97*, 104726.
Stahl, K., Kohn, I., Blauhut, V., Urquijo, J., De Stefano, L., Acácio, V., et al. (2016). Impacts of European drought events: Insights from an international database of text-based reports. *Natural Hazards and Earth System Sciences, 16*(3), 801–819.
Sutanto, S. J., van der Weert, M., Wanders, N., Blauhut, V., & Van Lannen, H. A. J. (2019). Moving from drought hazard to impact forecasts. *Nature Communications, 10*(1), 4945. https://doi.org/10.1038/s41467-019-12840-z
Thomasz, E. O., Vilker, A. S., Pérez-Franco, I., & García-García, A. (2023). Impact valuation of droughts in soybean and maize production: The case of Argentina. *International Journal of Climate Change Strategies and Management, 16*, 63–90.

UNCCD. (2023). *Global drought snapshot 2023. The need for proactive action*. UNCCD.

UNISDR. (2009). *Drought risk reduction framework and practices contributing to the implementation of the Hyogo framework for action* (p. 213). United Nations Secretariat of the International Strategy for Disaster Reduction (UNISDR).

UNDDR. (2021). *GAR special report on drought 2021*. UNISDR.

UNDP. (2011). Mainstreaming drought risk management – a primer. United Nations Development Programme, 73pp.

Van Loon, A. F., Stahl, K., Di Baldassarre, G., Clark, J., Rangecroft, S., Wanders, N., Gleeson, T., Van Dijk, A. I., Tallaksen, L. M., Hannaford, J., & Uijlenhoet, R. (2016). Drought in a human-modified world: Reframing drought definitions, understanding, and analysis approaches. *Hydrology and Earth System Sciences, 20*(9), 3631–3650.

Vogt, J. V., Naumann, G., Masante, D., Spinoni, J., Cammalleri, C., Erian, W., Pischke, F., Pulwarty, R., & Barbosa, P. (2018). *Drought risk assessment. A conceptual framework*. EUR 29464 EN, Publications Office of the European Union. https://doi.org/10.2760/057223

Walsh-Dilley, M. (2020). Resilience compromised: Producing vulnerability to climate and market among quinoa producers in southwestern Bolivia. *Global Environmental Change, 65*, 102165.

Weng, W., Becker, S. L., Lüdeke, M. K., & Lakes, T. (2020). Landscape matters: Insights from the impact of mega-droughts on Colombia's energy transition. *Environmental Innovation and Societal Transitions, 36*, 1–16.

Wilhite, D. A., Hayes, M. J., Knutson, C., & Smith, K. H. (2000). Planning for drought: Moving from crisis to risk management. *JAWRA Journal of the American Water Resources Association, 36*(4), 697–710.

Wilhite, D. A., Botterill, L., & Monnik, K. (2005). National drought policy: Lessons learned from Australia, South Africa, and the United States. In *Drought and water crises: Science, technology and management issues* (pp. 137–172). CRC.

Wilhite, D. A., Sivakumar, M. V., & Pulwarty, R. (2014). Managing drought risk in a changing climate: The role of national drought policy. *Weather and Climate Extremes, 3*, 4–13.

World Meteorological Organization (WMO) and Global Water Partnership (GWP). (2014). National Drought Management Policy Guidelines: A template for action (D.A. Wilhite). In *Integrated Drought Management Programme (IDMP) tools and guidelines series 1*. WMO and GWP.

World Meteorological Organization (WMO) and Global Water Partnership (GWP). (2016). Handbook of drought indicators and indices. In M. Svoboda & B. A. Fuchs (Eds.), *Integrated drought management Programme (IDMP), integrated drought management tools and guidelines series 2*. WMO and GWP.

Zuleta, D., Duque, A., Cardenas, D., Muller-Landau, H. C., & Davies, S. J. (2017). Drought-induced mortality patterns and rapid biomass recovery in a terra firme forest in the Colombian Amazon. *Ecology, 98*(10), 2538–2546.

Chapter 9
Drought Policy and Governance in Spain: New Tools and Remaining Challenges

Jesús Vargas Molina and Pilar Paneque Salgado

Abstract Spain experiences drought as a natural and recurrent phenomenon. Climate change models foresee an escalation in the frequency and intensity of this risk in the coming years. There is an imperative need to shift the drought management strategies from a crisis-management approach to a proactive risk management paradigm, incorporating anticipation, prevention, and mitigation measures. Significant progress has been accomplished in this domain. Primarily by implementing compulsory drought plans at both the river basin district and urban supply system levels. Despite these efforts, the current water management model tends to exert additional pressure on water resources, thereby augmenting the vulnerability of these systems to withstand prolonged drought periods. This chapter critically examines drought management tools and asses the progress made thus far. Nonetheless, it also addresses the persisting challenges that require attention to establish an effective and preventive drought management approach.

Keywords Drought · Risk management · Crisis management · Drought plans · Drought management tools · Spain

9.1 Introduction

Droughts and floods are the most frequent and impactful risks in Spain, leading to substantial socioeconomic consequences. Projected regionalized climate change models for Spain indicate a decline in available water resources, exacerbating as the twenty-first-century advances, and an escalation in the frequency and intensity of droughts (Centro de Estudios y Experimentación de Obras Públicas, CEDEX,

J. V. Molina (✉)
Universidad de Málaga, Málaga, Spain
e-mail: jvarmol@uma.es

P. P. Salgado
Universidad Pablo de Olavide, Sevilla, Spain

D. Rivera, G. Donoso (eds.), *Droughts in Chile*, Global Issues in Water Policy 31,
https://doi.org/10.1007/978-3-031-85040-0_9

2017). Thus, effectively managing and governing drought risk has become one of the primary challenges facing Spain in the current context of climate change. To comprehensively analyze the challenges posed by drought management, it is essential to distinguish between drought as a natural phenomenon and drought as a risk. Drought, as a natural phenomenon, is defined as a temporary period of decreased precipitation in a specific area compared to its usual values (Wilhite, 2000). This natural phenomenon is recurrent and inherent to Spain's climatic conditions (Pita, 2007). When this temporary decrease in rainfall threatens to impede meeting water demands, it transforms into a risk that requires management. Nonetheless, this transition from a natural phenomenon (meteorological drought) to a risk is a complex process, evident at different temporal (agronomic, hydrological, and/or socioeconomic drought) and spatial scales. The transition from a natural phenomenon (meteorological drought) to a risk is a complex process, manifesting itself at various temporal (edaphic, hydrological, and/or socioeconomic drought) and spatial scales. This transition is not solely dictated by the frequency, duration, and intensity of precipitation decrease but is also influenced by the management of water resources and the formulation of strategies and tools to cope with such events. These factors determine the vulnerability of different systems before the onset of a drought period and their capacity to respond effectively when the phenomenon occurs.

The significance of studying drought as a natural hazard lies in the consequences that variations in precipitation can have on society and the environment. When a decrease in precipitation extends beyond mere rainfall reduction and affects the volumes of water required for crop maintenance, drinking water supply, and ecosystem conservation, the impacts can be substantial (Morales Gil et al., 2000). Indeed, it is the convergence of abnormal conditions that bestows drought with the potential to inflict damages, as it disrupts the regular weather patterns (Pita, 1989).

Spain has experienced severe drought episodes that significantly impacted the entire country during specific periods, such as 1941–45, 1979–83, 1990–95, 2004–08, 2015–2018, and 2020–23. Given this regular occurrence, it becomes essential to recognize and acknowledge the normality of drought periods and integrate them into the routine management of water resources. This approach aims to reduce the vulnerability of systems to the occurrence of such phenomena (Vargas & Paneque, 2019).

Human activities that alter the hydrological cycle, such as water extraction, deforestation, and land-use changes, can exacerbate drought conditions. These anthropogenic influences make it challenging to attribute drought solely to meteorological factors that impact precipitation frequency and intensity.

The distinctive characteristics of drought as a risk phenomenon, including its prolonged effects, uncertain boundaries, and difficulty in quantifying impacts, require management strategies to go beyond traditional approaches (Wilhite & Glantz, 1985; Wilhite, 2000; Wilhite et al., 2014). Instead, effective drought management must consider the social perception and interpretation of the phenomenon. Understanding how communities perceive and respond to drought can play a crucial role in formulating appropriate and sustainable strategies for mitigating its impacts and building resilience in the face of this natural hazard. This integration of social factors is vital for developing comprehensive drought management plans that address the complexities of the drought risk and its implications for both society and the environment.

To point out the challenges that Spain faces concerning drought risk management and governance this chapter will follow a structured approach. First, we will provide a climatic characterization of the country, focusing on precipitation patterns and the drought regime. This analysis will also explore the potential effects of climate change on the variables influencing drought. Then, we will delve into an examination of the model employed for water resources management and exploitation. This will involve considering the implications of this model on the vulnerability of the systems to endure periods of reduced precipitation. Moving forward, we will present and analyze the different drought plans that have been implemented in Spain since 2001. These plans serve as the primary tools employed to prevent and mitigate the impacts of drought. Lastly, we will engage in a comprehensive discussion about the progress made so far and the remaining challenges that Spain must confront in order to achieve effective drought risk management and governance. By addressing these challenges head-on, Spain can enhance its resilience and response capacity to mitigate the impacts of drought effectively.

9.2 A Natural and Recurrent Phenomenon in a Climate Variability Context

Precipitation distribution in Spain exhibits a range of distinct characteristics, encompassing interannual variability, irregularity, and diverse seasonal rainfall regimes, alongside a complex spatial pattern (Martín-Vide, 1996). On average, Spain receives an annual precipitation of approximately 650 mm (reference series 1981–2010). However, this value displays a significant interannual variability, with sequences of both wet and dry years being a natural occurrence. Moreover, Spain's precipitation varies considerably across its geographical expanse. The northwestern regions receive more than 2000 mm of precipitation per year, whereas some southeastern regions witness a meager accumulation of about 250 mm per year (Fig. 9.1). Such marked contrast in precipitation distribution emphasizes the country's climatic heterogeneity, thereby presenting unique challenges for effective drought risk management and water resource planning.

Precipitation patterns in Spain also exhibit a pronounced seasonality, with the majority occurring during the autumn and spring months, and a notable absence during the summer months. This irregularity is attributed, on one hand, to the latitudinal position of the Iberian Peninsula, situated in the transition zone between two distinct climatic zones, the Atlantic and the Mediterranean. On the other hand, Spain's extensive territorial expanse and diverse topography also contribute to this irregular precipitation pattern (Gil Olcina & Olcina, 2001).

Within this irregular precipitation pattern, meteorological droughts are a natural and recurring phenomenon in Spain's Mediterranean climate. These droughts

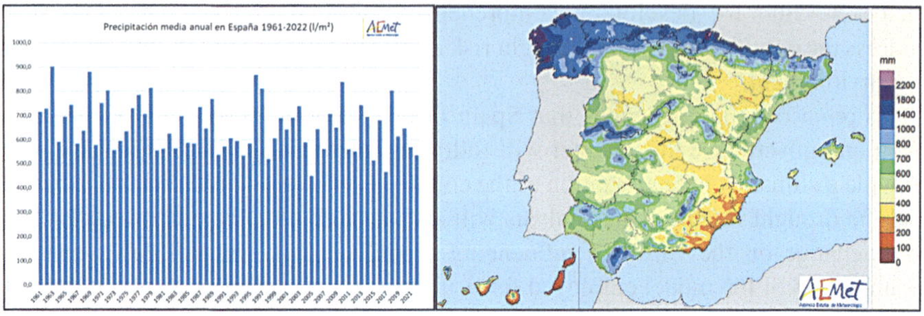

Fig. 9.1 Annual (1961–2022) and spatial (1981–2010) distribution of mean annual precipitation in Spain. (AEMET, 2018, 2022)

display diverse temporal and spatial variations, leading to uneven impacts across different regions. The factors that contribute to the occurrence of meteorological droughts in Spain are complex and multifaceted, encompassing hydro-atmospheric processes as well as geographical elements like orography and altitude. As a result, all regions in Spain experience episodes of meteorological droughts with varying intensity, frequency, and territorial significance (Olcina, 2008).

Significant spatial variations exist in the frequency, mean duration, and magnitude of droughts, displaying a general north-south gradient. The northern regions experience more frequent drought episodes compared to central and southern Spain. However, in the central and southern regions drought episodes tend to be of longer duration and greater magnitude (Domínguez-Castro et al., 2019; Vicente-Serrano et al., 2022).

The potential impacts of different drought episodes are contingent on the time scale over which the drought is assessed. Initially, a decrease in precipitation leads to the lack of soil moisture availability, known as agronomic drought, which promptly affects vegetation, rainfed crops, and livestock farms. If the precipitation deficit persists over an extended period, natural and artificial water reserves, both on the surface and underground, start to deplete, giving rise to a hydrological drought. In both cases, if the water scarcity results in economic, environmental, or social consequences, it is categorized as a socioeconomic drought. Meteorological drought serves as the origin of the other types of drought, with the introduction of the human component that demands and manages resources. While meteorological droughts stem from natural factors, hydrological, agronomic, and socioeconomic droughts arise from a combination of environmental and socioeconomic factors (UNISDR, 2009). At this point, the temporary decrease in precipitation poses a risk as it limits societies' capacity to meet their demands. The response time to such meteorological conditions varies significantly between regions and basins, primarily depending on the physiographic characteristics of the basin (geology, topography, soils, and vegetation), preceding meteorological conditions (such as evapotranspiration rates, snow cover, and rainfall intensity), and water management strategies (Lorenzo-Lacruz et al., 2013; Vicente-Serrano & López-Moreno, 2005). Two main types of hydrological drought responses to meteorological droughts have been identified in

Spain. In the northern region, a pattern is observed indicating higher hydrological sensitivity at short time scales. Conversely, in the central and southern regions, a sustained response is observed, implying greater sensitivity to meteorological droughts at longer time scales (Lorenzo-Lacruz et al., 2013).

Concerning the foreseeable changes in the drought regime in the future, the sixth IPCC report (2023) reinforces, with a higher level of confidence than its predecessor, the projection that droughts will increase in frequency, severity, duration, and spatial extent. The Mediterranean area has experienced more frequent and severe meteorological and hydrological droughts in recent decades, aligning with the anticipated trend toward a higher frequency of drought periods in a warmer climate in the future.

Estimating the impacts of climate change on water resources and droughts entails significant uncertainties. On one hand, these uncertainties arise from different projections of greenhouse gas (GHG) emissions, which present various scenarios for variables like precipitation and temperature. On the other hand, the water cycle's complexity, involving a multitude of hydrometeorological and anthropogenic factors, makes it challenging to establish direct relationships between climate variability and hydrological variables, such as flow, lake levels, and reservoir storage rates (Räisänen, 2007; Raje & Mujundmar, 2010; Alkama et al., 2011).

In the case of Spain, the State Agency for Metrology (AEMET) has undertaken a regionalization of the climate scenarios Representative Concentration Pathways (RCP) 4.5, 6.0, and 8.5 derived from the fifth IPCC report (2014). Building upon these scenarios, the Public Works Research and Experimentation Center (CEDEX) published the report "Evaluation of the Effects of Climate Change on Water Resources and Droughts in Spain"in 2017. This comprehensive report evaluates the potential future evolution of various hydrological variables (precipitation, runoff, aquifer recharge, and potential evapotranspiration) for the RCP4.5 and RCP8.5 scenarios, spanning the time scales of 2010–2040, 2040–2070, and 2070–2100.

Despite the inherent uncertainties, a consistent trend emerges showing a reduction in precipitation (ranging between −2% and −4% for 2010–2040, −6% and −8% for 2040–2070, and −7% and −14% for 2070–2100). Additionally, there is a decrease in runoff (between −3% and −7% for 2010–2040, −11% and −14% for 2040–2070, and −13% and −24% for 2070–2100) and aquifer recharge (between −3% and −7% for 2010–2040, −11% and −14% for 2040–2070, and −13% and −24% for 2070–2100). Potential evapotranspiration is projected to increase due to the effect of temperature rise (between 3% and 4% for 2010–2040, 7% and 10% for 2040–2070, and 9% and 12% for 2070–2100). Table 9.1 displays the regionalized results, organized by river basin district, showcasing the projected changes in surface runoff for the RCP 4.5 and RCP 8.5 scenarios concerning the historical reference series from 1940 to 2006.

Under the RCP 4.5 scenario, a relatively more optimistic outlook is portrayed, with notable reductions in surface runoff observed in the northern and central peninsular regions of the atlantic watershed. Particularly significant reductions are highlighted in the Balearic and Canary archipelago. However, under the more severe RCP 8.5 scenario, a much more aggravated reduction in water resources is foreseen,

Table 9.1 Changes in available resources for the RCP 4.5 and RCP 8.5 scenarios in the Spanish river basin districts (period 2010–2040) in relation to the historical reference series 1940–2006

River Basin District	RCP 4.5	RCP 8.5
Miño-Sil	−3%	−7%
Cantábrico Occidental	−2%	−6%
Cantábrico Oriental	−4%	−8%
Duero	−5%	−10%
Tajo	−5%	−10%
Guadiana	1%	−5%
Guadalquivir	0%	−7%
Segura	3%	−4%
Júcar	−8%	−15%
Ebro	−1%	−15%
Cuencas Mediterráneas Andaluzas	1%	−7%
Tinto-Odiel-Piedras	3%	−7%
Guadalete-Barbate	3%	−4%
Cuencas Internas de Catalunya	7%	−3%
Galicia Costa	−4%	−7%
Cuencas Internas del País Vasco	−4%	−8%
Islas Baleares	−29%	−36%
Canarias	−12%	−20%

Source: (CEDEX, 2017)

with pronounced decreases once again in the two archipelagos and the Ebro and Júcar Mediterranean basin regions standing out.

The report also evaluates the foreseeable changes in droughts of return periods of 2 and 5 years in relation to the drought regime. The majority of climate projections indicate a future in which droughts will become more frequent and worsen as the twenty-first century progresses. Under the RCP 4.5 scenario, 2-year droughts are projected to be more frequent (shorter return period for the same deficit) in 2070–2100 and 2040–2070 (based on three projections), with greater uncertainty for 2070–2100. The same trend is observed for 5-year droughts. For the RCP 8.5 scenario, there are few differences compared to RCP 4.5, although droughts would tend to be more frequent, especially during the last periods of the twenty-first century. Regarding the spatial distribution of droughts in the future, all scenarios indicate an increase in frequency as the twenty-first century progresses in all Spanish river basin districts. However, this increase in drought frequency will be less pronounced in the northern regions and more significant in the central, southern, and southeastern peninsular river basin districts.

9.3 Droughts Within the Framework of Water Management in Spain

The spatial, interannual, and seasonal irregularity of rainfall in Spain, coupled with the frequent occurrence of meteorological drought periods, has provided a rationale for promoting a hydraulic intervention policy on water resources. This policy entails the construction of large hydraulic infrastructures designed to enable water storage, regulation, and transfer between different regions (Del Moral et al., 2003). Throughout the twentieth century, Spain has actively pursued this hydraulic policy, resulting in the country ranking as the third European nation with the highest number of river barriers, totaling 171,203, as reported by the Adaptive Management of Barriers in European Rivers project (Amber, 2020). Among these barriers, there are more than 1200 dams and regulation reservoirs. As a whole, these infrastructures provide a total storage capacity of 56,000 hm^3, which accounts for approximately 50% of the country's river flow (MITERD, 2023), though with notable territorial variations. Figure 9.2 illustrates the spatial distribution of reservoirs, their regulation capacities in each river basin district, and the geographical layout of canals and water transfer systems.

The center and southern river basin districts have the highest regulation capacity due to the greater irregularity of rainfall in these regions. Furthermore, the rivers on the Atlantic slope, with their longer length and favorable topography for constructing large dams and reservoirs, possess greater regulation capacity compared to those on the Mediterranean and Cantabrian watersheds, except for the Ebro River basin.

This network of infrastructures has historically been a driving force of the Spanish economy, contributing significant benefits to various sectors such as irrigation, population supply, hydroelectric generation, and the tourism industry, among others, through the regulation of water resources. Additionally, it has profoundly altered the natural hydrological response to meteorological droughts by delaying the impacts of precipitation deficits over extended time scales, particularly in hydrographic demarcations with higher regulation capacity and flood attenuation. Furthermore, since the mid-twentieth century, groundwater abstraction increased, particularly in the southern and southeastern regions of Spain. This expansion has facilitated the extension of irrigation and the development of the tourism sector in these areas (Estela & Vargas, 2016). While this usage has contributed to the social and economic advancement of these regions, it has also given rise to significant challenges, including contamination and overexploitation of aquifers, leading to adverse effects on associated ecosystems.

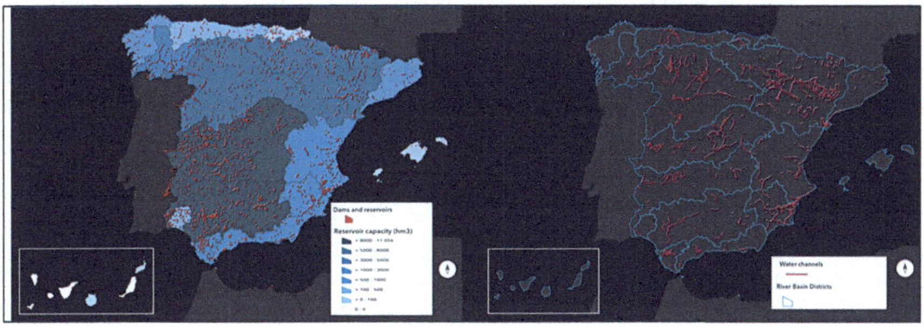

Fig. 9.2 Reservoir infrastructure location and capacity (left) and water channels (right). (Own elaboration based on MITERD, 2023)

The water infrastructure model led to an increase in water demands and, consequently, a rise in pressure on these resources. However, at the onset of the twenty-first century, this water management model is being questioned due to its environmental, economic, and social externalities. Additionally, the experience of the 1991–1995 drought, which severely affected a significant portion of the country, brought about restrictions on thousands of hectares of irrigated land and supply cuts for over ten million people, along with serious environmental repercussions that were not adequately addressed. More recently, the debate on water policies in Spain has started to incorporate the effects of climate change on water resources and droughts, although some reluctance remains. As a result, the current discourse is characterized by a shift towards a hydrological transition from the traditional paradigm, which relied heavily on intervention in hydrological systems through the construction and management of hydraulic infrastructure, mainly consisting of water transfers and regulation reservoirs. This approach is gradually being abandoned, although it persists to some extent. Instead, there is a growing focus on paradigms aimed at the sustainable management of water demand, the preservation and conservation of aquatic ecosystems, and the integration of risk management and preventive strategies linked to resource management. This transition seeks to address the challenges posed by climate change and to promote more efficient and ecologically sustainable water management practices in the face of evolving hydrological conditions.

An important step toward this paradigm shift was the approval of the Water Framework Directive (WFD) in 2000, which is mandatory for all European Union member states. The main objective of the WFD is to promote an integrated community water policy aimed at preserving and improving the aquatic environment within the European Union. As a result, the WFD encourages a new approach to water resources management among the Member States, shifting away from traditional policies centered on increasing water supply towards measures focused on halting deterioration and safeguarding water resources and their associated ecosystems. To achieve these goals, the WFD introduces the concept of the river basin district as the new territorial framework for water management and planning. A river basin district is defined as "the marine and terrestrial area comprised of one or more adjacent river basins and the associated groundwater and coastal waters" (Article 2.15 of the WFD). The river basin district management plans (HP) serve as essential management tools for each river basin district. These plans undergo a public information and consultation process, and they are subject to review and update every 6 years in a cyclical planning process. By adopting this approach, the WFD seeks to foster a coordinated and comprehensive approach to water resource management, enabling sustainable use and protection of water resources across Europe.

The DMA does not substantially address the management of drought risk. However, it does include among the objectives of ordinary hydrological planning the mitigation of the impacts of both droughts and floods. Additionally, in Article 4.6, concerning the environmental objectives of bodies of water, the WFD states that "temporary deterioration in the status of bodies of water shall not constitute a breach of the provisions of this Directive if it is due to natural causes or force majeure, which are exceptional or could not reasonably have been foreseen, in particular severe floods and prolonged droughts, or the result of circumstances arising from accidents which could not reasonably have been foreseen." This provision is applicable only if all necessary measures have been taken to avoid deterioration, if the water plan specifically considers the conditions under which such circumstances may be declared, if the measures to be taken are included in the program of measures of the water plan, if the situations are exceptional and could not reasonably have been foreseen, and if a summary of the effects produced by such circumstances is included in the next plan. At the European level, there is no specific Directive that establishes the basic obligations and guidelines for drought management in the member states, as is the case with flood risk management under Directive 2007/60 on the assessment and management of flood risks.

The transposition of the WFD into the Spanish legal framework was carried out in 2003. Prior to this, Spain already had some advantage over other European countries in certain aspects of the Directive, particularly with respect to the traditional water management in hydrographic basins. In Spain, the monitoring of hydrographic basins dates back to the year 1926, and the use of hydrological plans as management tools had been in practice since 1998. However, the transposition of the WFD necessitates a territorial reorganization of the traditional river basins. As shown in Fig. 9.3 in Spain, a distinction is made between inter-community and intra-community river basin districts. Inter-community river basin districts are those whose surface area exceeds the limits of an Autonomous Community and are managed by the Spanish Government. There are nine inter-community basin organizations known as Hydrographic Confederations, which are attached to the Ministry for Ecological Transition and Demographic Challenge (MITERD). On the other hand, intra-community river basin districts are those whose surface area is entirely within the territorial scope of an Autonomous Community, and the competence for management and planning lies with the water administrations of the respective Autonomous Communities. In total, there are twelve management areas in intra-community basins under the jurisdiction of the Autonomous Communities. Additionally, Spain has four transboundary river basin districts, shared with Portugal. These river basin districts are governed by the Albufeira Convention, and on the Spanish side, their management responsibility also falls under the State through the Hydrographic Confederations (Guadiana, Tajo, Duero and Miño-Sil).

The incorporation of the Water Framework Directive (WFD) into the Spanish legal framework not only seeks to achieve the good status of water bodies and protect the public water domain but also includes the objective of harmonizing regional and sectoral development, which in turn allows for an increase in the availability of water resources. Consequently, the objective of hydrological planning in Spain is characterized by a dual and somewhat contradictory nature. While aiming to attain the good status of all water bodies, it also acknowledges the need to accommodate new demands that may exert pressure on the resource, thereby conditioning and limiting the achievement of the primary objective. This transposition of the WFD into Spanish law has been instrumental in legitimizing the continuation of a traditional model that emphasizes increasing water supply, as evidenced by the

Fig. 9.3 Territorial and jurisdictional framework for water management and planning in Spain. (Own elaboration) 1: El Hierro; 2: La Palma; 3: La Gomera; 4: Tenerife; 5: Gran Canaria; 6: Lanzarote; 7 Fuerteventura

persistence of construction policies. For instance, the hydrological plans of the first (2016–2021) and second planning cycles (2016–2021) prioritize options in their programs of measures such as the commissioning and projection of new reservoirs and the expansion of irrigable areas. This reflects the continuity of inertia associated with the traditional interventionist paradigm, hindering the full realization of the potential offered by the WFD to reorient water planning and management models (Sampedro & Del Moral, 2014). While hydrological plans of the third planning cycle (2022–2027) exhibit a significant change by limiting the construction of new hydraulic infrastructures, the expansion of irrigation continues in various Spanish hydrographic demarcations. This expansion of irrigation is rationalized based on the anticipated water savings resulting from the modernization of irrigation techniques, transitioning from localized irrigation systems to more efficient localized systems that consume less water. Nevertheless, empirical evidence demonstrates that in districts where modernization has been implemented, no actual water savings have been realized. This is due to irrigation modernization has not been accompanied by a review of water concessions, so that the possible water savings that could be obtained in the water used are often used to produce more, intensifying the crop and often also expanding the irrigated area. This is why the increased water use efficiency of irrigation modernization often leads in the end to higher consumption, rather than savings (Aliod & Martínez, 2022). As a result of the interventionist

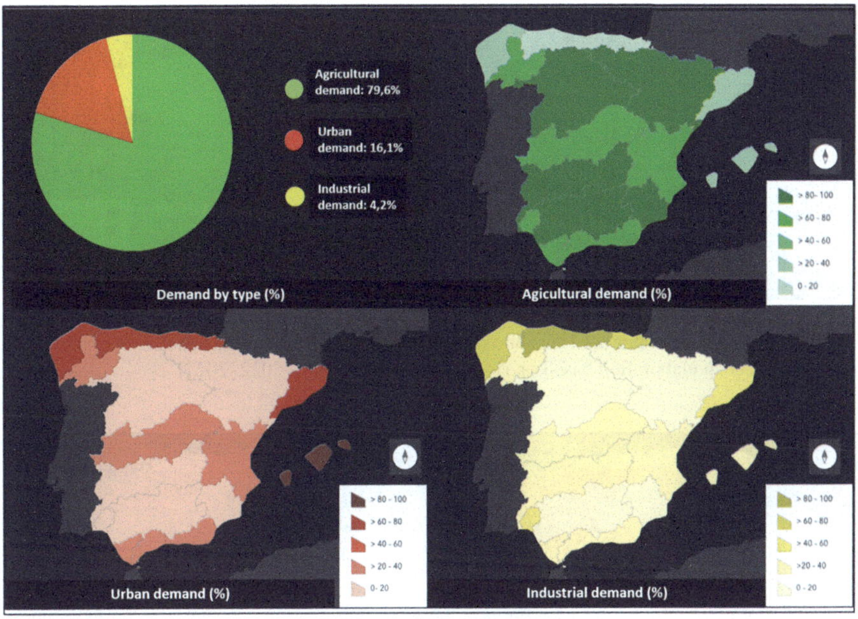

Fig. 9.4 Water demands by river basin districts in the third hydrological planning cycle (2022–2027). (Own elaboration based on river basin districts plans 2022–2027)

model of water policy in Spain, water demand has been increasing in the first two planning cycles and has been consolidated in the third. This increase in demands has generally been in favor of agriculture, which currently accounts for 79.6% of total water demands, compared to urban (16.1%) and industrial (4.2%) with some territorial differences that can be seen in Fig. 9.4.

The increase in water demand has mainly occurred at the expense of renewable natural resources. As depicted in Fig. 9.5, the utilization of alternative resources in Spain, such as desalination and wastewater reclamation, remains insignificant.

The decision to expand irrigation through the promotion of modernization has been subject to scrutiny and debate due to the lack of observable water conservation benefits. This raises questions about the effectiveness and validity of such policies and prompts a reevaluation of water management strategies to ensure their alignment with sustainable practices and the principles of the WFD. Increased water demands have led to elevated pressure on water resources, which in turn restricts the capacity of systems to withstand periods of drought (EEA, 2021). Drought exacerbates the pressure on water resources, as higher pressure makes it more challenging to meet demands when the availability of resources diminishes due to drought. The extent of pressure on water resources within a river basin district can be characterized by two key variables: the level of water stress and the status of water bodies, which serve as the sources for fulfilling various demands.

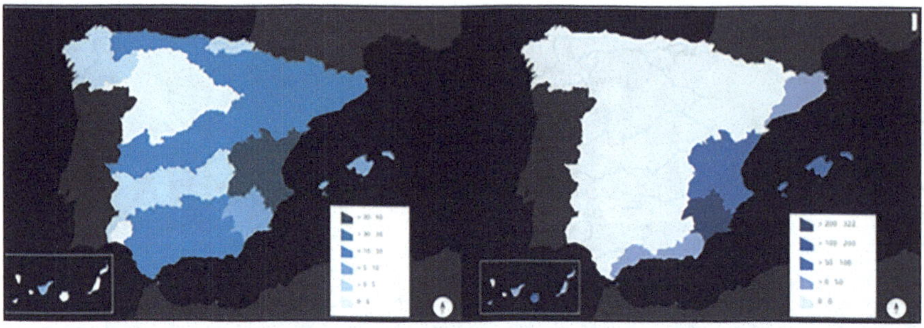

Fig. 9.5 Regenerated water use (left) and desalination capacity (right) of Spanish river basin districts (hm^3). (Own elaboration based on river basin districts plans 2022–2027)

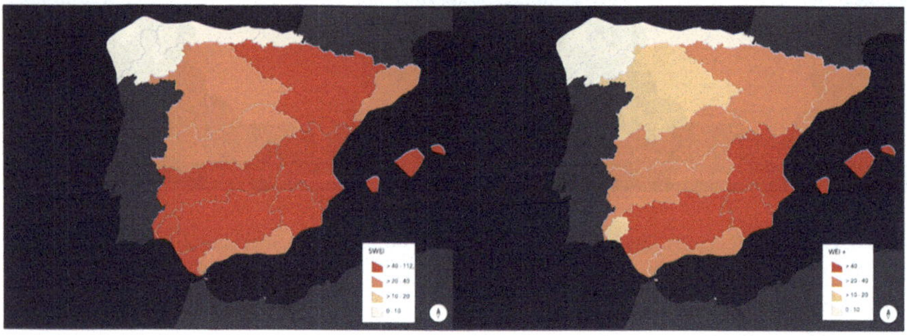

Fig. 9.6 S-WEI (left) and WEI+ (right) values for Spanish river basin districts. (Own elaboration based on MITECO, 2018)

Water stress affects 20% of the European territory and 30% of the European population each year, a situation that is aggravated in southern Europe, with marked problems of structural water stress (EEA, 2021). In Spain, many of the river basin districts are in this situation, with the water consumed by agriculture and the tourism sector being the keys to the pressure on water resources. Figure 9.6 presents the results of the water stress level of all Spanish river basin districts as a result of the application of two indicators. On the one hand, the results are presented using the S-Water Exploitation Index (S-WEI) indicator, which relates the percentage that water allocations represent with respect to the available resource, i.e., the foreseeable extraction versus the resource. The second indicator used is the Water Exploitation Index+ (WEI+) proposed by the European Environment Agency in 2012 as the reference indicator to calculate the level of water stress. This indicator relates the total amount of resources consumed (consumption without return) to the total amount of renewable resources (EEA, 2012). For both indicators, WEI values below 10% imply that the system under analysis is not under any type of stress, if the value is between 10–20% it implies a low level of water stress, if it exceeds 20% it is considered to be alarmed by water stress levels, and if it exceeds 40% severe stress.

S-WEI indicator shows severe water stress values for 15 river basin districts and 4 more in alarm due to water stress. Only the river basin districts of the northwestern peninsular show low water stress levels. The values obtained for WEI+ generally show lower stress values in almost all the river basin districts. The differences are due to particularities in the calculations that are explained in each of the river basin districts plans. In particular, the methodology used to calculate the WEI+ in some of the river basin districts may underestimate the level of water stress when comparing hyperannual resources, given the large regulation capacity of some districts with annual demands.

On the other hand, the assessment of water bodies status serves as an indicator that reveals the level of pressure and the impact of various demands on water resources. The global status of each surface water body (river, lake, transition, and coastal) is evaluated based on its ecological status or potential (for artificial or highly modified water bodies that cannot reach good status by their nature). This ecological status can be classified as very good, good, moderate, deficient, or poor, while the chemical status is characterized as either good or poor. Similarly, for each groundwater body, the quantitative status (good or bad) and the chemical status (good or bad) are analyzed. In both cases, whether for surface or groundwater bodies, the final definition of the global status for each water body is conditioned by the most unfavorable value obtained from the two indices used. Figure 9.7 shows the global status of all surface and groundwater bodies for the third planning cycle (2022–2027).

For Spain as a whole, out of the 5162 surface water bodies, only 52% of them reach an overall good status. In terms of ecological status or potential, 12.7% are classified as very good, 42% as good, 29% as moderate, 9.3% as poor, 5% as bad, and 2% have not yet been evaluated. Regarding chemical status, 88% of the water bodies attain a good chemical status, whereas 6% are in poor status, and 6% are yet to be evaluated. In relation to groundwater, out of the 768 bodies in Spain, only 55% reach a good overall status. For quantitative status, 75.7% are classified as good, while the remaining 24.3% have a poor quantitative status. As for the chemical condition, 64% are considered to have good chemical status, and the remaining 36% are

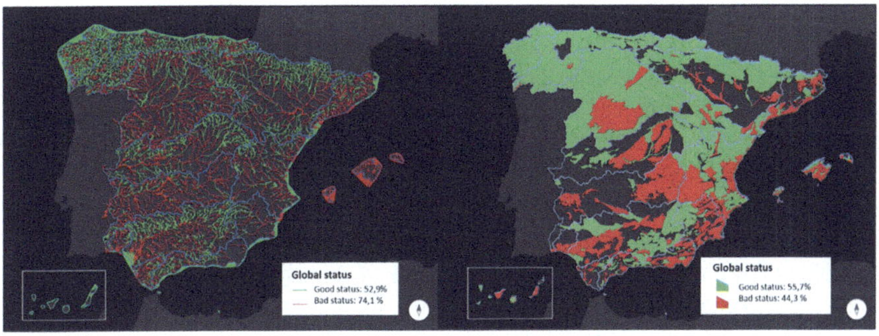

Fig. 9.7 Global status of surface (left) and groundwater bodies (right). (Own elaboration based on river basin districts plans 2022–2027)

characterized as having a poor chemical condition. As evident from the provided data, more than 20 years after the implementation of the WFD, which aimed to achieve an overall good status in all water bodies by 2015, Spain is still far from accomplishing this objective.

9.4 Paradigm Shift. From Crisis Management to Risk Management

Traditional drought management strategies have been framed within an interventionist water management model. Within this approach, drought has been viewed as a disaster, leaving little room for action other than the application of reactive and emergency measures once the impacts have already manifested. This crisis management approach has been extensively discussed in the literature (Wilhite, 2000; UNISDR & NDMC, 2009; WMO & GWP, 2014). In this perspective, drought is considered an exceptional situation, and its impacts are seen as a crisis resulting from the interventionist hydraulic system. Consequently, such an approach has contributed to justifying the continuation of the supply-oriented policy (Hernández-Mora & Del Moral, 2015). The primary political and management instruments employed within this framework are exceptional emergency measures, the construction of new water infrastructures (such as reservoirs, transfers, and wells) to augment water supply, and economic compensations to alleviate damages and losses (Paneque, 2015; Urquijo et al., 2015; Estela & Vargas, 2016).

These measures have been articulated through the approval of Royal Decrees that take advantage of the alarm situation and utilize the declaration of "general interest" to justify, legitimize, and reinforce the consolidation of water supply policies and the construction of infrastructures as the sole viable solution to address water scarcity (Paneque, 2015). The sense of urgency serves to divert attention from the fundamental causes leading to a decrease in precipitation, thus generating situations of scarcity. This approach attributes the causality of drought exclusively to the natural phenomenon, without questioning the management and exploitation of the resource. In Spain, the sense of urgency and exceptionality has also been used to justify decision-making that circumvents the mechanisms established in planning, such as the economic justification of measures, determination of environmental consequences, evaluation of alternatives, and public participation processes (Urquijo et al., 2015).

Due to the increase in losses associated with drought risk, and despite the efforts made to mitigate its impacts, in the last decades of the last century, some authors began to question the effectiveness of this type of reactive approach to dealing with droughts (Wilhite, 2000). The main criticisms of this approach are as follows: (1) the ineffectiveness of the reactive approach to reducing losses associated with droughts; (2) the limitation of solutions to technical aspects, lacking evaluation of alternatives or participation of stakeholders in their design; (3) the limited capacity to react and adapt by applying emergency measures once the event has occurred; and (4) the emphasis on the impacts instead of addressing the underlying causes that determine them. As an alternative to this reactive approach, the risk management approach arises, based on a better understanding of the underlying causes that determine drought risk (both natural and anthropogenic). This approach forms the basis for risk prevention, mitigation, and adaptation strategies (Wilhite, 2000). The focus is on identifying the causes and consequences of vulnerability in various sectors, regions, communities, or population groups, allowing for the implementation of context-specific mitigation and adaptation measures.

9.5 Advances and New Tools for Drought Management in Spain

Following Poljansek et al. (2017), three aspects are fundamental for adequate preventive risk management: (1) improving scientific knowledge on risk determinants; (2) ensuring proper risk communication; and (3) expanding tools oriented towards prevention and adaptation.

Significant progress has been made in understanding the meteorological component of drought risk (Domínguez-Castro et al., 2019; Donaire, 2006; Jimenez-Donaire et al., 2020; Lorenzo-Lacruz et al., 2013; Noguera et al., 2020; Vicente-Serrano et al., 2017, 2020). Particularly relevant to this issue is the meteorological drought monitor carried out by the Spanish National Research Council (CSIC) and made available to the public in open access through https://monitordesequia.csic.es/monitor/. This monitor provides data on the duration and intensity of meteorological droughts at different time scales (1, 3, 6, 9, 12, and 24 months) for the SPI and SPEI drought indices from January 1961 to the present. The drought monitor is regularly updated with real-time information collected from automatic stations, and historical information is also utilized to identify and track drought events (Vicente-Serrano et al., 2022).

Drought risk communication is an indispensable element for the social legitimization of public water policies and the application of measures in risk situations, as it allows for a more educated and informed population. However, there are still many pending challenges in this area. For instance, the survey conducted by the Drought Citizan Observatory on the perception of water and drought in Spain (Lafuente et al., 2023) reveals a lack of knowledge among the general population regarding water use and drought management in the country. A particularly illustrative example is that only 31% of the population considers agriculture is the sector with the highest water consumption when in reality. This sector consumes almost 80% of the resources. Achieving a water transition requires progress in enhancing the social understanding of the problem of widespread overexploitation of water resources (Del Moral, 2023).

Regarding the operational management tools, the paradigm shift is consolidated at the institutional level after the approval of Law 10/2001, of July 5, 2001, of the National Hydrological Plan. Article 27 of this law is dedicated to drought management and establishes the following provisions: (1) The obligation to establish a system of indicators to forecast drought situations; (2) the obligation of the basin organizations to develop special plans of action in situations of alert and possible drought (hereinafter, PES) within the scope of the hydrological plans. These plans include the rules of operation of the systems and the measures to be applied in relation to the use of the public water domain; (3) the requirement for public administrations responsible for urban supply systems serving a population equal to or greater than 20,000 inhabitants to have an emergency plan for drought situations (hereinafter, PEM). This PEM must take into account the rules and measures in the PES and must be operational by 2005.

Drought plans are based on the establishment of a system of indicators that enable the prediction and monitoring of the evolution of a drought period. This allows for the implementation of a series of progressive measures as the phenomenon progresse with the objective of minimizing the environmental, economic, and social impacts. Both PES and PEM must be coordinated with each other and with the ordinary hydrological planning, in a way that the PES are dependent on the river basin district hydrological plan, and the PEM must be adapted and coherent with these.

9.5.1 Special Warning and Eventual Drought Plans (PES)

PES at the river basin district level is the responsibility of the basin organizations and should encompass the rules for the operation of the systems and the measures related to the use of the public water domain (as stated in Article 27 of the NHP Law). The first PES for the intercommunity river basin districts were approved in 2007, and they underwent revisions and updates in 2017 and recently in 2023. As for the PES of the intra-community demarcations, they have been approved and updated mostly at present, although some have experienced slight delays compared to their intercommunity counterparts. However, an exceptional situation arises with the 7 river basin districts of the Canary Islands archipelago, as they still do not have an active PES.

In 2007, the PES were approved. These PES relied on the concept of conducting straightforward hydrological balances. These balances compared available resources and their projected evolution with established demands to estimate the likelihood of imbalances between resources and demands, thereby determining their potential severity. Different thresholds were established as a reference for applying various types of measures when the demand cannot be met with the established contribution probabilities. The PES introduced two key aspects. Firstly, it identified, designed, and proposed preventive measures to be applied during a drought period, reducing or eliminating the sense of urgency and the unilateral application of measures typical of other reactive procedures like drought decrees. Secondly, the PES set forth distinct measures for each phase or scenario, including the normality phase. This flexibility allows for adapting measures to each context and specific situation.

The first PES included the approval of drought Royal Decrees as one of the possible measures to alleviate the effects of drought in the alert and emergency phases. Although certain measures may require the activation of drought decrees (e.g., application of economic and fiscal measures beyond the water administration's competence), considering them as ordinary measures and their contents when approved (e.g., extraordinary works, indiscriminate fiscal aid, legal reforms surpassing the scope of the drought event, etc.) perpetuate the reactive and emergency approach to drought management. This approach reflects a level of urgency and unilateralism in decision-making, allowing the bypassing of legal guarantees, which are characteristic of the traditional model of managing such events.

In response to the limitations of the PES approved in 2007, the Ministry of Agriculture and Fisheries, Food and Environment (now Ministry for Ecological Transition and Demographic Challenge) published the Technical Instruction for the elaboration of the special drought plans and definition of the global system of indicators of prolonged drought and scarcity (IT) in 2017. The TI proposes a fundamental change in the new PES by introducing differentiation between situations of prolonged drought and situations of conjunctural scarcity. This differentiation leads to the establishment of new territorial management units for each situation (Fig. 9.8), new indicators capable of identifying, evaluating, and monitoring the states of prolonged drought and conjunctural shortages, and the implementation of different types of measures for each of these two situations.

PES in Spain are based on two indicators: the prolonged drought indicator based on the precipitation variable, and the conjunctural shortage indicator based on hydrological balances between water resources and demands, which could be adapted or not to the case of Chile. In addition, the aforementioned PEMs are a component of the PES and are under the supervision of the river basin authorities.

Prolonged drought refers to situations where there is a temporary decrease in precipitation that significantly reduces water inputs, regardless of demands. The

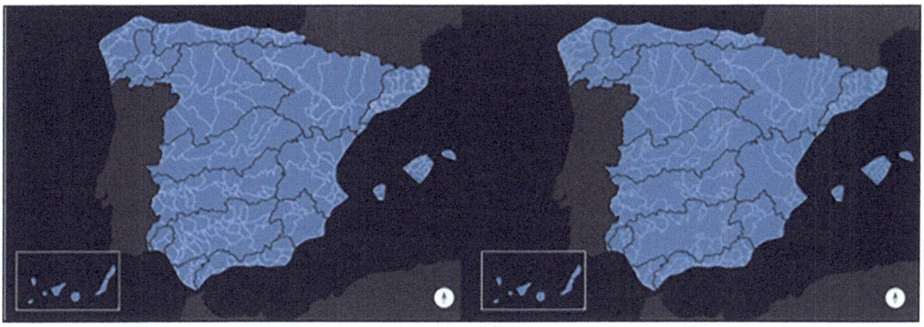

Fig. 9.8 Prolonged Drought Territorial Units (UTS) (left) and Shortage Territorial Units (UTE) (right). (Own elaboration based on MITERD, 2023)

assessment of prolonged drought is conducted on a monthly basis for each of the Prolonged Drought Territorial Units (UTS) within each River Basin District (RBD). The indicator used for evaluating prolonged drought is the Standardized Precipitation Index (SPI), which is adapted by each RBD to the time scale that best suits its characteristics (3, 6, 9, or 12 months). The SPI values are rescaled to normalize the index, ranging between a minimum value of 0 and a maximum value of 1 for each Prolonged Drought Territorial Units (UTS) category. Based on the normalized SPI values, each UTS is categorized into one of two possible scenarios: prolonged drought, when the normalized SPI value (I_e) is less than 0.3, and absence of drought when the value of I_e is equal to or greater than 0.3.

Conjunctural shortages refer to situations where there is a temporary decrease in available water resources that may pose a risk to meeting existing demands for socioeconomic purposes. Short-term shortages are assessed on a monthly basis for each of the Shortage Territorial Units (UTE) within each River Basin Discticts (RBD). The indicators used to assess the guarantee of demands include the reservoir level, piezometric levels, or a combination of both. Based on the obtained indicator value, which is representative of each UTE, the state index (I_e) is calculated. The state index is a dimensionless numerical value ranging from 0 to 1, capable of quantifying the current situation with respect to the proximity or severity of a shortage. Depending on the values acquired by I_e, each UTE is categorized into one of the possible scenarios: normality, pre-alert, alert, or emergency. Each scenario has specific pre-determined measures associated with it (Table 9.2).

The measures associated with prolonged drought situations include two main strategies: (1) Allowing for the temporary deterioration of water bodies: during exceptional natural events like severe floods or prolonged droughts, there might be a temporary decline in the condition of water bodies. In such cases, this temporary

Table 9.2 Summary of scenarios, thresholds and types of measures included in PES for situations of prolonged drought and conjunctural shortages

Situación	Territorial Unit	Index (I_e)	Scenarios	Mesures tipology
Prolonged drought	Prolonged Drought Territorial Units (UTS)	$I_e < 0.30$	Absence	Monitoring
		$I_e \geq 0.30$	Prolonged drought	Temporary deterioration of the status of water bodies Less demanding ecological flow regime
Cojuntural shortage	Shortage Territorial Units (UTE)	$I_e > 0.50$	Normality	Monitoring
		$0.50 < I_e \geq 0.30$	Pre-alert	Awareness, savings and monitoring
		$0.30 < I_e \geq 0.15$	Alert	Management measures (demand and supply), and control and monitoring
		$I_e < 0.15$	Emergency	Intensification of measures considered to be on alert and possible adoption of exceptional measures

Source: Author's own elaboration

deterioration is considered acceptable and does not lead to a breach of environmental standards. (2) Reducing the ecological flow regime in areas outside protected regions (usually Natura 2000 Network). Regarding conjunctural shortage measures, the PES includes a diverse set of specific actions tailored to each Shortage Territorial Unit (UTE). These measures encompass monitoring and surveillance, demand management, and restrictions on allocations for socio-economic uses of water in favor of urban supply demands. It also includes avoiding the construction of new hydraulic infrastructures that may not provide immediate benefits and whose potential environmental, economic, and social impacts should be thoroughly assessed during the ordinary hydrological planning process. It's important to note that these measures primarily aim to mitigate the impacts of hydrological droughts and may not directly address soil drought effects. Consequently, they may not encompass measures to minimize the impacts on rainfed agriculture or extensive livestock farming, despite these sectors being particularly vulnerable when faced with meteorological droughts.

Moreover, PES introduced a third scenario known as "exceptional extraordinary drought." This scenario arises when one or several Conjunctural shortage Territorial Units (UTEs) experience scarcity situations during alert or emergency scenarios that overlap both in time and geographical area with a territorial region facing prolonged drought. The situation of extraordinary drought can be declared even if there is no prolonged drought, which allows for the possibility of implementing measures such as temporary deterioration of water bodies or the reduction of ecological flows, as outlined in Article 4.6 of the Water Framework Directive (WFD). In this case, the threshold of 0.3 for the I_e of prolonged drought is not necessarily relevant, as the application of these measures is triggered by a resource-demand imbalance caused by inadequate planning rather than by a meteorological drought event (FNCA, 2023).

The latest update of the PES (2023) has not brought any significant new developments, maintaining the concepts, applied indicators, territorial scopes and measures to be implemented. Progress has been based on the improvement of indicators to characterize prolonged drought and conjunctural shortage through the incorporation of new meteorological stations and the adaptation of more updated reference time series, in the case of prolonged drought, and the incorporation of complementary indicators in some of the UTE for the characterization of conjunctural scarcity (FNCA, 2023).

9.5.2 Drought Emergency Plans for Urban Water Supply (PEM)

The objective of the PEM is not to deal with structural shortages but to mitigate the effects of specific shortages resulting from a period of drought. The functioning of these EMPs is similar to that of the PEM but only in relation to shortage situations and they must be capable of preventing operational drought situations, with the aim of taking gradual measures in the different scenarios of normality, pre-alert, alert and emergency in order to avoid or delay as much as possible the restrictions to human consumption.

The responsibility for water supply and sanitation lies with the municipalities, which are considered basic entities of the territorial organization of the Spanish State. Thus, the competence for the preparation of the PEM rest with the municipalities that individually provide water supply services or with consortiums (groups of municipalities) that jointly provide services to more than one municipality, provided that the total population served exceeds the threshold of 20,000 inhabitants. It is worth noting that the division of administrative powers between inter-community and intra-community river basin districts, as well as the presence of specific water laws in those Autonomous Communities with water management and supply competences, have contributed to the establishment of a new water management and supply system.

In the case of the Autonomous Community of Andalusia,[1] reduces the population threshold that determines the obligation to draw up an PEM to 10,000 inhabitants (the rest of the Autonomous Communities with their own water legislation maintain the criterion of 20,000 inhabitants). This generates a situation in which within the Autonomous Community of Andalusia there are different legal regimes, as it forms part of several intercommunity river basin districts (Guadalquivir, Segura and Guadiana) for which the threshold of 20,000 inhabitants is applied and three intra-community districts (Andalusian Mediterranean Basins and Guadalete-Barbarte and Tinto-Odiel-Piedras) for which it is reduced to 10,000 inhabitants.

In addition to this, the framework of competence in the management of the integral urban water cycle (CIAU) in Spain presents a peculiarity. The CIAU comprises four complementary services whose management may be concessioned to a Mancomunidad or consortium, or to a public or private company. These services are as follows: Upstream supply, which involves collecting water from various sources (rivers, reservoirs, aquifers, seawater) and transporting it to the headwater reservoirs of the specific supply system (municipal and/or Mancomunidad reservoirs, if applicable). Additionally, the potabilization of the water takes place in drinking water treatment plants (DWTP); Downstream water supply or distribution, which is responsible for distributing water from the headwater reservoirs to the end-users, such as domestic, commercial, industrial, and institutional consumers; Downstream sewerage and rainwater collection and channeling, which includes the collection of wastewater from users (homes, businesses, companies, etc.) and its transport to the sewers and/or wastewater treatment plant. Additionally, it involves the collection, storage, and possible reuse of rainwater, as well as its discharge.; Sewerage and wastewater treatment, which encompasses the transportation of wastewater to the collectors and its subsequent treatment in wastewater treatment plants (WWTP). The objective is to return the treated water to the environment in suitable conditions. If applicable, reclamation and reuse measures may also be considered.

[1] Law 9/2010 on Water for the Autonomous Community of Andalusia.

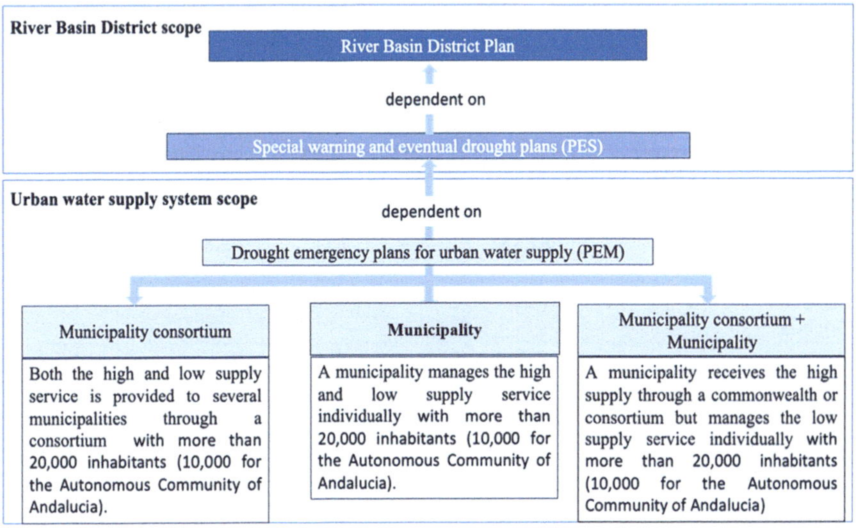

Fig. 9.9 Linkage between hydrological planning and drought plans. (Own elaboration)

Each of these four services can be provided jointly or individually. The PEM must only include the part of these that includes the high and low supply, and there are three different situations that result in different obligations when preparing an EMP: (1) when the high and low supply service is provided to several municipalities through a commonwealth or consortium; (2) when a municipality manages the high and low supply service individually; and (3) when a municipality receives the high supply through a commonwealth or consortium but manages the low supply service individually. Figure 9.9 summarizes the linkage between ordinary hydrological planning and the different drought management plans (PES and PEM) and presents the different casuistry and obligations derived from the different management models when preparing a PEM.

Table 9.3 presents a comprehensive overview of compliance with the PEM in the entire Spanish RBD. It includes the number of municipalities or consortiums mandated to prepare a PEM and the corresponding population they should cover. Additionally, the table showcases the number and percentage of approved PEM, along with the total population covered by these approved plans.

PEM were supposed to come into force in 2005. Nowadays, only 19.7% of the 305 supply systems in Spain that were required to develop an PEM have done so. The current population covered by PEM amounts to 63.1% of the total. This contrast between the percentage of approved PEM and the population covered by them is mainly due to the approved PEM belonging to large supply systems that serve a significant portion of the population. There are significant differences between RBD, and some of them do not even provide data on the obligations and administrative status of the PEM.

Given the non-compliance and the challenges faced by many municipalities and supply systems in preparing these PEM, two methodological guides have been

Table 9.3 Supply systems with obligation to prepare PEM and their administrative situation by river basin district

River Basin District	Number of systems required to prepare PEM	Total population	Number of approved PEM	% Approved PEM	Total population covered by PEM	% Population covered by PEM
Miño-Sil	5	370,177	0	0.0	0	0.0
Cantábrico Occidental	14	1,087,837	1	7.1	42,877	3.9
Cantábrico Oriental	5	1,814,206	2	40.0	1,108,511	61.1
Duero	15	1,311,288	6	40.0	814,544	62.1
Tajo	15	7,660,728	4	26.7	6,977,276	91.1
Guadiana	25	1,024,632	9	36.0	418,394	40.8
Guadalquivir	21	3,926,982	10	47.6	3,227,065	82.2
Segura	4	1,881,455	2	50.0	1,816,999	96.6
Júcar	40	3,444,663	9	22.5	306,341	88.9
Ebro	18	3,516,946	12	66.7	302,684	86.0
Cuencas Mediterráneas Andaluzas	62	2,764,221	0	0.0	0	0.0
Tinto-Odiel-Piedras	11	342,216	0	0.0	0	0.0
Guadalete-Barbate	22	972,965	2	9.1	126,850	13.0
Cuencas Internas de Catalunya	No data	No data	No data	No data	No data	No data
Galicia Costa	19	1,198,156	3	15.8	219,369	18.3
Cuencas Internas del País Vasco	5	273,114	0	0.0	0	0.0
Baleares	24	1,454,519	0	0.0	0	0.0
Canarias[a]	No data	No data	No data	No data	No data	No data
TOTAL	305	33,044,105	60	19.7	20,841,410	63.1

Source: Author's own elaboration from Special warning and eventual drought plans (PES)
[a]Data for the 7 Canarian RBD have been aggregated

published to facilitate their development. The "Guide for the preparation of emergency plans in drought situations in urban supply systems" (AEAS, 2019) and the "Methodological guide for the participatory preparation of drought management plans in small and medium-sized municipalities" (Hernández-Mora & Vargas, 2018) aim to assist in the preparation of PEM. The latter guide introduces innovations and tools to enhance participatory processes and empower local communities, focusing on smaller supply systems with limited technical, economic, and human resources to address the development of such plans.

9.6 Pending Challenges for Adequate Drought Risk Management in Spain: Discussion and Conclusions

Spain is predicted to experience reduced precipitation and water resource availability due to climate change. This reduction is not solely attributed to decreased precipitation and runoff, but also to a general increase in temperatures, resulting in a negative balance between precipitation and evapotranspiration. Consequently, there will be lower water resource availability under normal conditions so the imbalances between water demands and available resources will exacerbate over time. Within this context, meteorological droughts, which are already a characteristic feature of the Spanish climate, are anticipated to become more frequent and severe as we progress into the twenty-first century. Coping with droughts and climate change effects on water resources availability is one of the main challenges facing Spain in the future. While there have been notable efforts to improve drought risk management in Spain, several challenges persist.

Drought risk vulnerability in Spain is significantly influenced by the high level of pressure on water resources, as evidenced in this chapter through water stress indices and water bodies status. This pressure, stemming from the interventionist and guaranteeing model of water management, directly impacts the resilience of these resources in times of drought.

In recent years, significant strides have been made in understanding meteorological phenomena, and the development of open access tools has facilitated the use of meteorological data by the general population. Of particular importance is the meteorological drought monitor (Vicente-Serrano et al., 2020). Additionally, substantial efforts have been made to regionalize the effects of climate change on water resources and the drought regime in Spain, providing insights into future scenarios (CEDEX, 2017). Since the beginning of the twenty-first century, there have also been important advances in the way drought risk is perceived and managed, with drought plans being the most important exponent of these changes. However, the application of the risk management paradigm through drought plans still presents important limitations. On the one hand, there are some RBD (the 7 of the Canary Islands archipelago) that have not yet drafted their PES. A worse scenario is observed if we look at the PEM, where only 19.7% of the supply systems have their EMPs approved and in force. On the other hand, the PES should be a tool that makes it possible to identify and anticipate situations of shortage of resources to meet the demands as an effect derived from the reduction of resources due to meteorological drought. The main weakness of PES lies in their differentiation between prolonged drought and conjunctural shortages. Particularly, in the temporal and spatial inconsistencies where conjunctural scarcity often occurs without prolonged drought situations. This suggests that conjunctural shortages is often more related to deficient management rather than the direct effect of decreased precipitation. The logic behind the establishment of thresholds to determine prolonged drought is to adhere to Article 4.6 of the WFD, but their incorrect adaptation to the regulations on the PES allows for the deterioration of water bodies or the reduction of ecological flows even in situations of prolonged drought when no restrictions on water use have been imposed.

Significant advancements have been also made in public participation processes, particularly concerning access to information and public consultation of the RBD plans and the PES. All RBD publish updated documents and provide spatial data infrastructures accessible to the public for consulting and downloading geographic information. However, both the hydrological plans and the PES remain highly technical documents, challenging for non-specialized individuals to comprehend, which hinders broader public involvement and participation in the consultation processes. The situation is different concerning PEM where the processes of information, consultation, and access to official documents are not effective in most cases.

Continuing to strengthen these governance and public participation processes is crucial for adequate risk management for several reasons: (1) Drought risk, unlike others, stems from the decrease of a vital resource essential for maintaining living conditions, economic activities, and ecosystem health, making it a risk that concerns the entire society; (2) when managing a risk, one is essentially managing future situations fraught with uncertainty, revolving around a resource, water, that holds diverse social, economic, and environmental values and interests. Therefore, the pooling of knowledge from experts and non-experts and considering different legitimate assessments of the resource and the risk contribute to improving diagnostic processes and seeking consensual solutions; (3) information and participation processes promote social empowerment, fostering the generation of informed opinions and citizen involvement—vital aspects in dealing with risks; and (4) these processes also serve to legitimize decision-making, potentially averting conflicts when implementing measures that might be less accepted by the population. Legitimizing the decision-making processes can help avoid conflicts in the implementation of less popular measures among the population.

Improvements in scientific knowledge regarding meteorological phenomena, the development of regionalized climate change scenarios in Spain, the realization that traditional drought management mechanisms have been neither effective nor efficient, and the shift in mentality at the regulatory and institutional levels on how to perceive and manage droughts, all open a window of opportunity for anticipating and adapting to the effects of climate change on water resources availability in Spain in the future. However, the ability to seize this window of opportunity cannot solely rely on the effectiveness of drought plans; it must be accompanied by a real and effective paradigm shift in water resources management. This shift should entail adjusting to climate change scenarios through controlling and reducing water demands.

Drought risk vulnerability is significantly influenced by the high level of pressure on water resources, as demonstrated in this chapter through water stress indices and water bodies status. The prevailing interventionist and guaranteeing model of water management has a significant impact on the capacity of these resources to withstand drought periods. The transition towards a hydrological paradigm of this nature requires deep reflection and debate on the management model of water and soil. It must be capable of acknowledging resource scarcity and drought, in line with the recommendations made by the European Union years ago (EU, 2007). The improper allocation of water among economic sectors creates an imbalance between needs and available water resources, particularly affecting the agricultural sector in the future. If the current model persists, its sustainability is under serious threat. In general terms, there is a lack of coherence between sectoral policies, and in some cases, certain policies may have counterproductive effects on water resource protection.

This debate must involve all social actors, as only through consensual solutions will it be possible to effectively advance towards a just hydrological transition. Opening up this reflection and analyzing management alternatives to all citizens is still a pending issue, as traditionally only economic agents have had a voice in this debate.

The debate should also consider the possibility of increasing the use of alternative resources, such as desalination and the reuse of reclaimed water from urban supply, which, despite their potential in Spain, remain insignificant. Addressing a social challenge of this magnitude requires information, facilitation, and deliberation, so that new discourses on water and drought can be developed.

References

AEAS (Asociación Española de Abastecimientos de Agua y Saneamiento). (2019). *Guía para la elaboración de planes de emergencia ante situaciones de sequía en sistemas de abastecimiento urbano*. AEAS. https://www.aeas.es/images/Doc_Manu_Guia/GUIA_PARA_LA_ELABORACIN_DE_PLANES_DE_EMERGENCIA-small.pdf

AEMET (Agencia Estatal de Meteorología) (2018). *Mapas climáticos de España (1981-2010) y ETo (1996-2016)*. Ministerio para la Transición Ecológica. Gobierno de España.

AEMET (Agencia Estatal de Meteorología) (2022). *Resumen anual climatológico 2022. Ministerio para la Transición Ecológica y Reto Demográfico*. Gobierno de España.

Aliod, R., & Martínez, J. (2022). Los proyectos de modernización de regadíos y el concepto de interés general. In F. La Roca & J. Martínez (Eds.), *Retos de la planificación y gestión del agua en España. Informe 2022*. Observatorio de Políticas Públicas. Fundación Nueva Cultura del Agua. https://fnca.eu/biblioteca-del-agua/documentos/documentos/Informe-OPPA-2022.pdf

Alkama, R., Decharnme, B., Douville, H., & Ribes, A. (2011). Trends in global and basin-scale runoff over the late twentieth century: Methodological issues and sources of uncertainty. *Journal of Climate, 24*, 3000–3014. https://doi.org/10.1175/2010JCLI3921.1

Amber (Adaptive Management of Barriers in European Rivers). (2020). https://amber.international/european-barrier-atlas/

CEDEX (Centro de estudios hidrográficos y experimentación de obras públicas). (2017). *Evaluación del cambio climático en los recursos hídricos y las sequías en España*. Ministerio de Agricultura, Alimentación y Medio Ambiente. https://ceh.cedex.es/web_ceh_2018/documentos/CAMREC/2017_07_424150001_Evaluaci%C3%B3n_cambio_clim%C3%A1tico_recu.pdf

Del Moral, L. (2023). Retos a la planificación y gestión del agua en España ante situaciones de crisis hídrica. *Temas para el debate, 343*, 29–31.

Del Moral, L., Van der Werff, P., Bakker, K., & Handmer, J. (2003). Global trends and water policy in Spain. *Water International, 28*(3), 358–366. https://doi.org/10.1080/02508060308691710

Domínguez-Castro, F., Vicente-Serrano, S. M., Miquel Tomás-Burguera, M. T., Peña-Gallardo, M., Beguería, S., El-Kenawy, A., Luna, Y., & Morata, A. (2019). High spatial resolution climatology of drought events for Spain: 1961-2014. *International Journal of Climatology, 39*(13), 5046–5064. https://doi.org/10.1002/joc.6126

Donaire, J. (2006). Pluviometric indicators of drought in Spain (1865-2004). *Observatorio Medioambiental, 9*, 187–213. ISSN: 1139-1987.

EEA (European Environmental Agency). (2012). *EEA Report N° 11/2012. Water resources in Europe in the context of vulnerability EEA 2012 state of water assessment*. Office for Official Publications of the European Union. https://doi.org/10.2800/65298

EEA (European Environmental Agency). (2021). *EEA Report No 12/2021 Water resources across Europe. confronting water stress: an updated assessment*. Publications Office of the European Union. https://doi.org/10.2800/320975

Estela, T., & Vargas, E. (2016). Drought management plans in the European Union. The case of Spain. *Water Resources Management, 26*(6), 1537–1553. https://doi.org/10.1007/s11269-011-9971-2

EU (European Union). (2007). Communication from the Commission to the European Parliament and the Council—Addressing the challenge of water scarcity and droughts in the European Union {SEC(2007) 993} {SEC(2007) 996}.

FNCA (Fundación Nueva Cultura del Agua). (2023). *Observaciones a la propuesta de Plan Especial de Sequías de las demarcaciones intercomunitarias españolas*. Fundación Nueva Cultura del Agua. https://fnca.eu/biblioteca-del-agua/documentos/documentos/20230628Obs-PES.pdf

Gil Olcina, A., & Olcina, J. (2001). Circulación atmosférica general y diversidad climática. In A. Gil Olcina & J. Gómez Mendoza (Eds.), *Geografía de España* (pp. 87–122). Ariak. ISBN 9788434434684.

Hernández- Mora, N., & Vargas, J. (2018). *Guía metodológica para la elaboración participada de planes de gestión de riesgo por sequía en pequeñas y medianas poblaciones*. Fundación Nueva Cultura del Agua. https://fnca.eu/biblioteca-del-agua/directorio/file/2840?search=1

Hernández-Mora, N., & Del Moral, L. (2015). La experiencia de sequías en España: inercias del pasado y nuevas tendencias en la gestión de riesgos. In *Conference: 5° Water Governance International Meeting, Water Governance Practices under Water Scarcity. November 10th to 13th 2015*. Universidade de São Paulo.

IPCC. (2023). Global warming of 1.5 °C. An IPCC special report on the impacts of global warming of 1.5 °C above pre-industrial levels and related global greenhouse gas emission pathways. In P. Masson-Delmotte, H. O. Zhai, D. Pörtner, D. Roberts, J. Skea, P. R. Shukla, A. Pirani, W. Moufouma-Okia, C. Péan, R. Pidcock, et al. (Eds.), *The context of strengthening the global response to the threat of climate change, sustainable development, and efforts to eradicate poverty*. IPCC.

Jimenez- Donaire, M., Giraldez, J., & Vanwalleghem, T. (2020). Impact of climate change on agricultural droughts in Spain. *Water, 12*(11), 3214. https://doi.org/10.3390/w12113214

Lafuente, R., Paneque, P., Ganuza, E., & Vargas, J. (2023). *Informe de resultados de la encuesta sobre el agua y la sequía*. Observatorio Ciudadano de la Sequía. https://doi.org/10.46661/rio.20230720_1

Lorenzo-Lacruz, J., Vicente-Serrano, S. M., González-Hidalgo, J. C., López-Moreno, J. I., & Cortesi, N. (2013). Hydrological drought response to meteorological drought in the Iberian Peninsula. *Climate Research, 58*, 117–131. https://doi.org/10.3354/cr01177

Martín-Vide, J. (1996). Decálogo de la pluviometría española, clima y agua. In J. Mazol (Ed.), *La Gestión de un Recurso Climático* (pp. 15–24). Tabapress.

MITECO (Ministerio para la transición ecológica). (2018). *Síntesis de los planes hidrológicos españoles Segundo ciclo de la DMA (2015–2021)*. NIPO: 013–18–124-7 DL: M-28707-2018. https://www.miteco.gob.es/es/agua/temas/planificacion-hidrologica/libro_sintesis_pphh_web_tcm30-482083.pdf

MITERD (Ministerio para la Transición Ecológico y el Reto Demográfico). (2023). *Boletín Hidrológico Nacional*. https://www.arcgis.com/apps/dashboards/912dfee767264e3884f7aea8eb1e0673

Morales Gil, A., Olcina, J., & Rico Amorós, A. M. (2000). Diferentes percepciones de la sequía en España: adaptación, catastrofismo e intentos de corrección. *Investigaciones Geográficas, 23*, 5–46.

Noguera, I., Dominguez-Castro, F., & Vicente-Serrano, S. M. (2020). Characteristics and trends of flash droughts in Spain, 1961–2018. *Annals of the New York Academy of Sciences, 1472*(1), 155–172. https://doi.org/10.1111/nyas.14365

Olcina, J. (2008). Prevención de riesgos: cambio climático, sequías e inundaciones. *Panel científico-técnico de seguimiento de la política de aguas en España*.

Paneque, P. (2015). Drought management strategies in Spain. *Water, 7*(12), 6689–6701. https://doi.org/10.3390/w7126655

PES (Planes Especiales de Sequía) (2023). *Ministerio para la Transición Ecológica y Reto Demográfico*. Gobierno de España. https://www.miteco.gob.es/es/agua/temas/observatorio-nacional-de-la-sequia/planificacion-gestion-sequias.html

Pita, M. F. (1989). Reflexiones en torno a la sequía. *Boletín de la Asociación de Geógrafos Españoles, 10*, 21–40.

Pita, M. F. (2007). Recomendaciones para el establecimiento de un sistema de Indicadores para la previsión, el seguimiento y la gestión de la sequía. In *La Sequía en España. Directrices Para Minimizar su Impacto* (pp. 69–85). Comité de Expertos en Sequía del Ministerio de Medio Ambiente; Ministerio de Medio Ambiente.

Poljansek, K., Marin Ferrer, M., De Groeve, T., & Clark, I. (2017). *Science for disaster risk management 2017: Knowing better and losing less*. Publications Office of the European Union.

Räisänen, J. (2007). How reliable are climate models? *Dynamic Meteorology and Oceanography, 59*(1), 2–29. https://doi.org/10.1111/j.1600-0870.2006.00211.x

Raje, D., & Mujundmar, P. (2010). Reservoir performance under uncertainty in hydrologic impacts of climate change. *Advances in Water Resources, 33*(3), 312–326. https://doi.org/10.1016/j.advwatres.2009.12.008

Sampedro, D., & Del Moral, L. (2014). Tres décadas de política de aguas en Andalucía. Análisis de procesos y perspectiva territorial. *Cuadernos Geográficos, 53*(1), 36–67. e-ISSN 2340–0129.

UNISDR. (2009). *Drought risk reduction framework and practices: contributing to the implementation of the hyogo framework for action* (p. 213). United Nations secretariat of the International Strategy for Disaster Reduction (UNISDR).

UNISDR & NDMC (United Nations secretariat of the International Strategy for Disaster Reduction & National Drought Mitigation Center). (2009). *Drought risk reduction framework and practices contributing to the implementation of the hyogo framework for action*. United Nations. https://www.unisdr.org/files/11541_DroughtRiskReduction2009library.pdf

Urquijo, J., De Stefano, L., & La Calle, A. (2015). Drought and exceptional laws in Spain: the official water discourse. *International Environmental Agreements: Politics, Law and Economics, 15*(3), 273–292. https://doi.org/10.1007/s10784-015-9275-8

Vargas, J., & Paneque, P. (2019). Challenges for the integration of water resource and drought-risk management in Spain. *Sustainability, 11*(2), 308. https://doi.org/10.3390/su11020308

Vicente-Serrano, S., & López-Moreno, J. (2005). Hydrological response to different time scales of climatological drought: an evaluation of the Standardized Precipitation Index in a mountainous Mediterranean basin. *Hydrology and Earth System Sciences, 9*, 523–533. https://doi.org/10.5194/hess-9-523-2005

Vicente-Serrano, S., Tomas-Burguera, M., Beguería, S., Reig, F., Latorre, B., Peña-Gallardo, M., Luna, M., Morata, A., & González-Hidalgo, J. (2017). A high resolution dataset of drought indices for Spain. *Data, 2*(3), 22. https://doi.org/10.3390/data2030022

Vicente-Serrano, S. M., Quiring, S. M., Peña-Gallardo,M., Yuanb, S. & Domínguez-Castro, F. (2020). A review of environmental droughts: Increased risk under global warming? Earth-Science Reviews, 201(102953). https://doi.org/10.1016/j.earscirev.2019.102953

Vicente-Serrano, S., Domínguez-Castro, F., Reig, F., Beguería, S., Tomas-Burguera, M., Latorre, B., Peña-Angulo, D., Noguera, I., Rabanaque, I., Luna, M., Morata, A., & El Kenawy, A. (2022).

A near real-time drought monitoring system for Spain using automatic weather station network. *Atmospheric Research, 271*(106), 095. https://doi.org/10.1016/j.atmosres.2022.106095

Wilhite, D. (2000). *Drought: A global assessment*. Routledge. ISBN1 0415168333, ISBN29780415168335.11.

Wilhite, D. A., Sivakumarb, M., & Pulwartyc, R. (2014). Managing drought risk in a changing climate: The role of national drought policy. *Weather and Climate Extremes, 3*, 4–13.

Wilhite, D. A. & Glantz, M. H. (1985). Understanding the Drought Phenomenon: The Role of Definitions. *Drought Mitigation Center Faculty Publications. 20*. http://digitalcommons.unl.edu/droughtfacpub/20

WMO & GWP (World Metheorological Organization and Global Water Partnership). (2014). National drought management policy guidelines: A template for action. In D. A. Wilhite (Ed.), *Integrated Drought Management Programme (IDMP) Tools and Guidelines Series 1*. WMO and GWP.

Chapter 10
Understanding Drought in California

Jeanine Jones

Abstract This chapter begins by providing background on the setting for water management in California, then reviews the state's most significant historical droughts. Outcomes and lessons learned from these droughts are then discussed, along with expectations about future drought conditions. Drought management in California has historically focused on the concept of drought being a period of consecutive dry years of statewide spatial scope. The adaptive capacity provided by the state's extensive water infrastructure and widespread access to groundwater makes a single dry year not a significant water management concern.

Keywords Drought · California · Climate change · Aridification

10.1 Introduction and Background

California's population growth and agricultural economy have been facilitated by an extensive system of water infrastructure developed in response to the state's highly variable climate and the imbalance between locations of water supply availability and water use. Drought is a recurring climate feature of California's climate and has resulted in substantial operational experience in drought response and management; the state's most significant droughts have led to institutional adaptations intended to improve response capability in future droughts. Impacts experienced in twenty-first century droughts have been significantly more severe than those in twentieth century droughts, driven by climate warming.

Throughout this chapter droughts and hydrologic data are expressed in terms of water years. The term water year is used by U.S. water agencies for hydrologic data reporting. It extends from October 1 through September 30. Water Year 2024, for example, is the period from October 1, 2023, through September 30, 2024.

J. Jones (✉)
California Department of Water Resources, Sacramento, CA, USA
e-mail: Jeanine.Jones@water.ca.gov

D. Rivera, G. Donoso (eds.), *Droughts in Chile*, Global Issues in Water Policy 31,
https://doi.org/10.1007/978-3-031-85040-0_10

10.2 Physical and Hydroclimate Setting

Variability of topography and climate is a characteristic feature of California. The state contains the point of lowest elevation in the United States (Death Valley at more than 60 m below sea level) and highest in the contiguous U.S. (Mount Whitney at an elevation of almost 4400 m), 11 geomorphic provinces, and the nation's greatest annual precipitation variability (Dettinger et al. 2011) (Fig. 10.1). Spatially, average annual precipitation varies from more than 2500 mm in some mountain ranges to about 50 mm in the most arid deserts.

California's proximity to the Pacific Ocean and its major mountain ranges (Fig. 10.2) shapes its hydroclimate setting. Most of California's precipitation originates from atmospheric water vapor traveling across the Pacific Ocean; water vapor crossing mountain ranges like the Coast Range or Sierra Nevada condenses and falls as rain or snow. Historically mountain snowpack in the Cascade Range and Sierra Nevada has provided winter storage for water that arrives in the spring as snowmelt runoff feeding the state's largest rivers and groundwater basins in the Central Valley.

Most of California experiences a Mediterranean climate with dry summers that are warm or hot and wet winters that are cool or cold. Summers are characterized by a blocking high pressure zone that diverts atmospheric water vapor away from the state, and the proximate cause of California's droughts is similarly a persistent

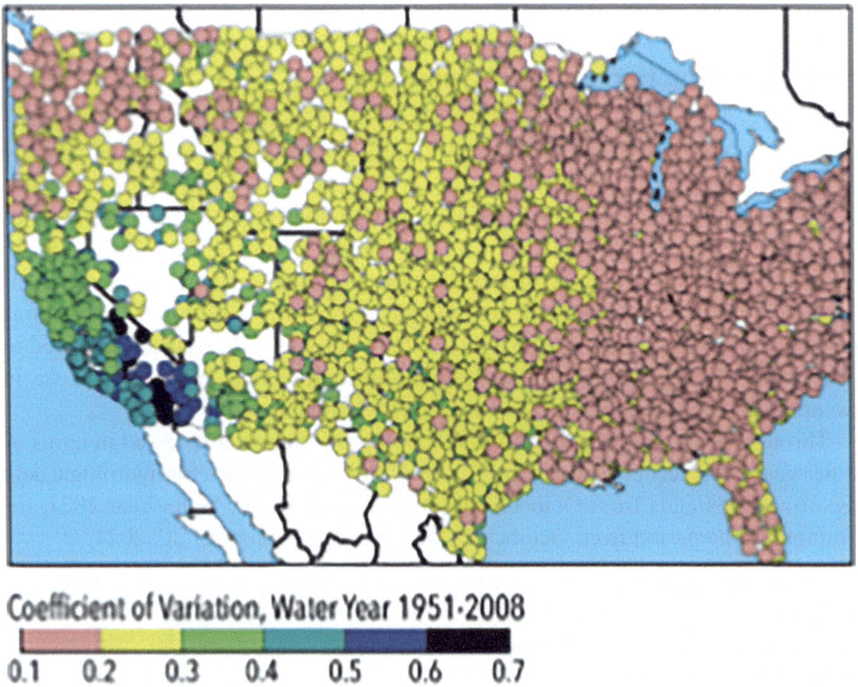

Fig. 10.1 Annual variability of California precipitation. (Mike Dettinger, U.S. Geological Survey)

Fig. 10.2 A map of California's location

Pacific high-pressure zone during the winter's normally wettest months. On average, about 70% of the state's average annual precipitation of nearly 560 mm falls between November and March, with about half of that amount occurring between December and February (Jones, 2020, p. 13). The state's annual water budget is determined by the presence or absence of a relatively small number of larger storms. A shortfall of a few major winter storms results in a dry year while a few very wet storms tilt the scale to a wet year. El Niño-Southern Oscillation (ENSO) conditions are not predictive of California water supply conditions, as ENSO accounts for only about 25% of annual precipitation variability (Jiang et al., 2022).

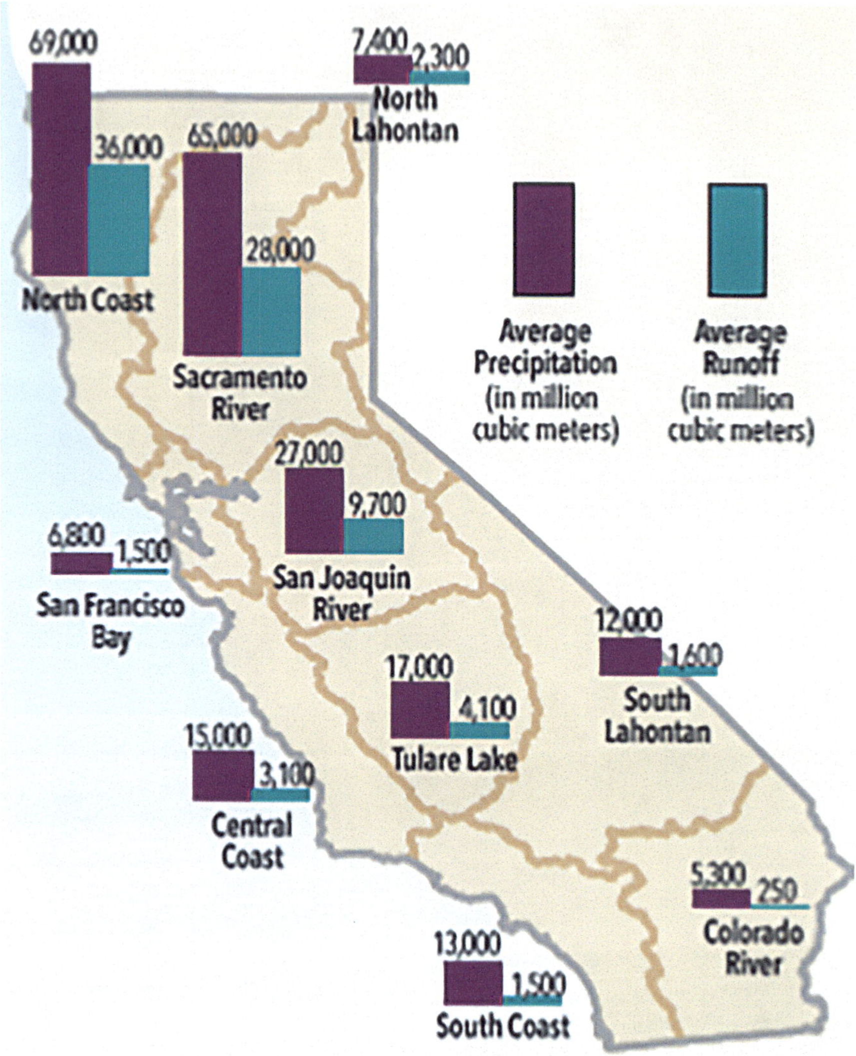

Fig. 10.3 Average annual precipitation and runoff by Hydrologic Region

The northern part of the state receives substantially more precipitation than the southern part (Fig. 10.3), and much of the state's developed water supplies originate as rain and snow that fall on the Cascade Range and Sierra Nevada mountains.

However, most of the demand for water for urban and agricultural use occurs farther south, which has resulted in the development of the large-scale water projects shown on Fig. 10.4. Most of the state's water supply comes from intrastate

Fig. 10.4 California water infrastructure

sources; there is one large interstate source (the Colorado River) that supplies parts of Southern California and is shared among seven U.S. states and Mexico.

California has more than 1500 dams and reservoirs (federal, state, local, or private ownership) (USACE, n.d.) but most are relatively small; much of the state's total reservoir storage of more than 53,000 million cubic meters (MCM) (PPIC, 2018) is contained in a few hundred reservoirs. Almost all the largest reservoirs (capacity of 1200 MCM or more) are located on the western slopes of the Cascade

Range and Sierra Nevada and are operated for both water supply and flood risk reduction. The large reservoirs feeding into the northern part of the Central Valley (the Sacramento Valley) operate in a mixed rain-snow regime depending on the characteristics of individual storms, as 95% of the area of major Sacramento Valley watersheds is below 2300 m in elevation. The large reservoirs feeding the southern part of the Central Valley (the San Joaquin Valley) operate in a snowmelt dominated regime, with 50% of the area of the major watersheds being above 2100 m in elevation (Jones, 2021, p. 23). Storage in the larger mountain reservoirs is restricted during the fall and winter months to provide capacity for flood management, and they are then allowed to fill in the spring with snowmelt runoff for water supply.

Far more water is stored in the state's more than 500 groundwater basins than in surface reservoirs. Ninety-four of these basins have been identified as having special priority for state groundwater management requirements and must prepare local management plans as discussed later in this chapter; these 94 basins plus an additional 30 areas with court-administered groundwater extraction adjudications account for 98% of groundwater extraction in California (Fig. 10.5). On average, groundwater provides about 40% of California's agricultural and urban water supplies, an amount that increases to 60 per cent in dry years when surface supplies are short (CDWR, 2020, pp. H14–H18). Groundwater is the dominant supply source for most small and medium drinking water suppliers, and a partial source for many larger ones. California has numerous managed groundwater recharge and storage projects, the earliest of which began in the 1910s; most use constructed basins or river channel segments for recharge rather than injection wells. The largest have managed storage capacities approximately of 1200 MCM, comparable in size to a large surface water reservoir.

The California Department of Water Resources (CDWR) has estimated statewide annual water use for many years for general planning purposes; average annual water use from 1998–2020 has been estimated at 99,000 million m^3 for urban, agricultural, and environmental purposes (CDWR, 2023, pp. 2–10). CDWR defines environmental water use as water that has been set aside by statute or regulation for managed wildlife refuges, instream flows, flows in rivers defined as wild and scenic, and required freshwater outflow to the San Francisco Bay. Actual water use varies widely from year to year based on hydrologic conditions, as demonstrated by the example in Fig. 10.6 for recent wet and dry years. Agriculture and the environment are the largest categories of water use but environmental water use declines significantly in dry years because it is dominated by the quantity of flow in rivers set aside as wild and scenic. The water classified as environmental in these rivers is largely not supported by storage, so the quantity drops in dry years.

California has for decades been the number one U.S. agricultural production state (based on cash revenues) because of its favorable climate and large amount of arable land. CDWR estimates that the state's irrigated acreage peaked at almost four million hectares around 1980 (Jones, 2020, p. 46). Subsequent urbanization of

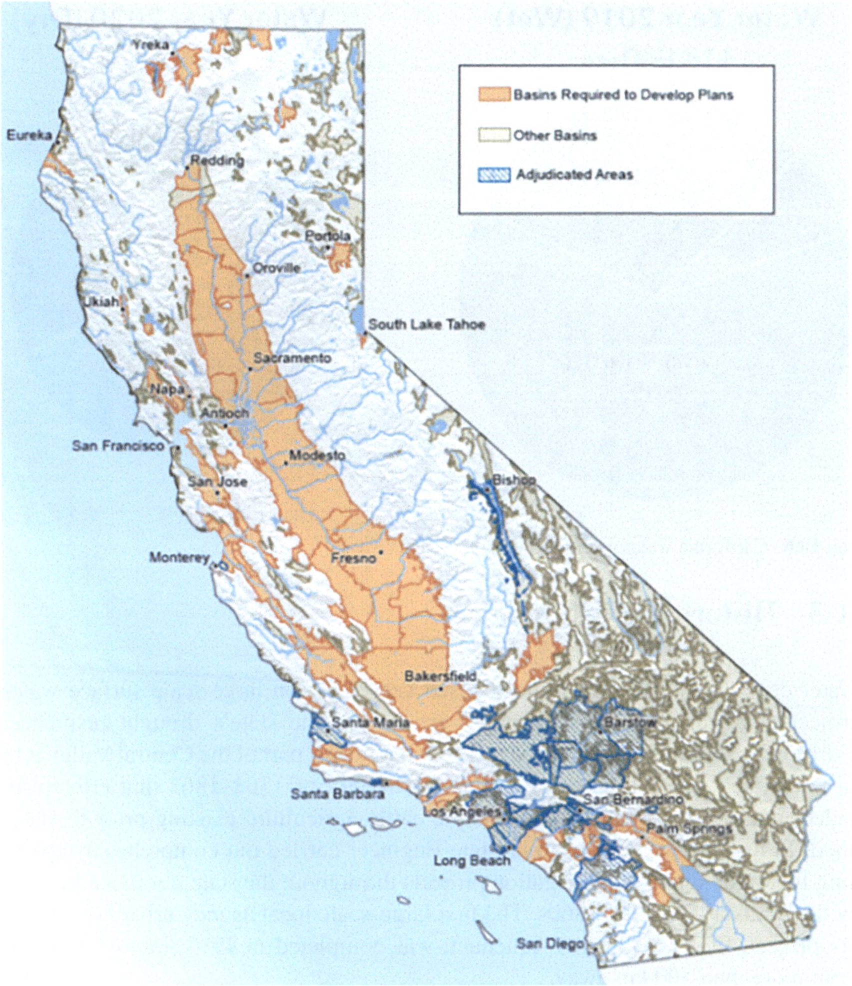

Fig. 10.5 California groundwater basins and adjudicated areas

agricultural land and other factors have reduced irrigated acreage to closer to 3.6 million hectares in average water years, although there can be significant year-to-year variability because of crop prices and water supply availability. Most crop production relies on irrigation because there is no significant precipitation during about half of the year. Crops are grown year-round in some areas where the climate is favorable. In recent past droughts it has been estimated that agricultural land fallowing in response to surface water shortages has been about 200,000 hectares annually (Jones, 2020).

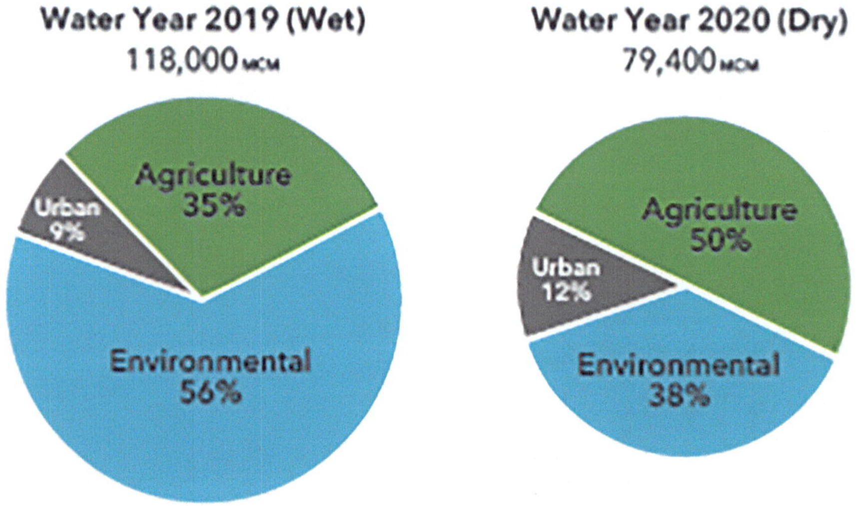

Fig. 10.6 California water use in wet and dry years

10.3 Historical Background

Water development in California focused very early on large-scale surface water projects, projects that subsequently were key to the state's drought resilience. California's so-called Great Flood of 1862 that turned part of the Central Valley into an inland sea was followed by the Great Drought of 1863–1864 that effectively ended much of the Spanish rancho-style cattle agriculture existing prior to statehood. In the 1880s the California State Engineer carried out comprehensive planning for water supply and irrigation projects throughout the state, motivated in part by the devastation of the 1860s. The first large-scale local agency urban water supply project, the Los Angeles Aqueduct, was completed in 1913, importing water from more than 300 km away.

State government began detailed planning for a state-owned water project in the 1920s, with publication of a State Water Plan in 1931 (CDPW, 1931) that detailed facilities and costs for a massive infrastructure project to move water from the water-rich Sacramento Valley to areas farther south. The state was unable to finance the project during the U.S. Great Depression of the 1930s and sought assistance from the federal government. USBR began construction later that decade on initial facilities of the federal Central Valley Project (CVP), a mostly agricultural water supply project and California's largest intrastate water project. California experienced major population growth as World War II was ending, prompting state government to begin further planning and to construct the State Water Project (SWP), a mostly urban water supply project and the state's second-largest intrastate project, in the 1960s. The SWP also develops water in the Sacramento Valley and transports it through an aqueduct more than 650 kilometers south to urban Southern California.

Local water agencies were also active in building water projects during this time, with cities on the eastern side of San Francisco Bay and the City of San Francisco completing aqueducts to convey water from Sierra Nevada reservoirs to their areas in 1929 and 1934. A consortium of Southern California cities in the Los Angeles completed an aqueduct about 400 kilometers to the Colorado River in 1941. By the 1970s most of California's backbone of major state, local, and federal water storage and conveyance infrastructure had been constructed. Far less visible, but important for expansion of irrigated agriculture, was the development and widespread installation by the 1920s–1930s of deep-well turbine pumps that enabled extraction of substantial volumes of groundwater from the state's previously only lighted tapped aquifers.

From the 1970s onward there was substantially less construction of major large-scale water infrastructure and more focus on institutional/regulatory actions and smaller-scale local urban water agency projects. By this time the negative impacts of groundwater overdraft had been recognized and reacted to by local agencies in a few groundwater basins. State goals for water recycling and ocean desalination were being formulated, an increasing number of urban water recycling projects were being constructed, and mandatory urban water conservation measures were being imposed. Additional regulatory measures were put in place to manage or control water use in response to the dry conditions of the twenty-first century.

California's present system of administering surface water rights was established by legislation in 1914, but a state system for regulating groundwater rights was not established then; it was not until 2014 that state legislation was adopted requiring mandatory local management of groundwater. Negative impacts of long-term groundwater overdraft had been recognized for many decades, but achieving legislative consensus on a solution was elusive. Groundwater overdraft in the intensively farmed San Joaquin Valley averaged 1200 to 2400 MCM annually for decades; some localized areas experienced as much as 9 meters of land subsidence between the 1920s and the present, damaging large-scale water conveyance infrastructure and destroying well casings. The 2014 legislation required formation of new local agencies that were to adopt management plans by 2020–2022 and would have a 20-year period to reduce excessive extractions and achieve compliance with their plans. It has been estimated that 200,000 hectares of agricultural land in the San Joaquin Valley may need to be fallowed to achieve compliance (Hanak et al. 2023). Plan implementation will influence local agencies' response to future droughts; in the past unrestricted groundwater pumping was a common solution to surface water shortages during droughts.

Statutory and regulatory requirements for environmental protection relative to water management have evolved over time. Basic legal protections for fish passage in waterways were enacted shortly after statehood and were further codified in 1937. Fish hatcheries were built as mitigation for construction of major dams, especially those along the Cascade and Sierra Nevada foothills that blocked migration of anadromous fish. State and federal Endangered Species Acts were adopted in the 1970s, authorizing stringent regulatory actions to protect listed species. State and federal Wild and Scenic Rivers Acts were adopted in the late 1960s and early 1970s to prohibit construction of new water infrastructure on designated river segments. Instream flow requirements imposed through water rights and court action based on the public trust doctrine have been additional tools used to reduce existing water supplies of urban and agricultural water agencies for environmental purposes as, for example, use of the public trust doctrine to reduce the City of Los Angeles' diversions from Mono Lake tributaries (SWRCB, 1994). Federal legislation in 1992 dedicated almost one million MCM of CVP yield for fisheries purposes and placed those purposes on an equal footing with urban and agricultural deliveries and additionally mandated baseline deliveries for specified wildlife refuges. This legislation and other authorizations have allowed federal and state agencies to expend hundreds of millions of dollars for habitat improvement and ecosystem restoration measures.

10.4 Institutional Setting

State governments in the U.S. have the primary responsibility for managing water supply availability through issuance of water rights that authorize agencies, private companies, and individuals to use specified amounts of water for specified purposes. California (like most states in the relatively arid western U.S.) follows the prior appropriation doctrine, which grants the earliest diverters of water from a particular surface water source the highest priority to use water during times of shortages. California also recognizes riparian water rights, which allow landowners adjacent to a waterway to divert water onto the adjacent lands; riparian water users share shortages equally. State laws authorize the formation of various types of special districts (local agencies) with specified powers related to water management. Districts can be formed under generic legislative acts (for example, irrigation districts, water districts, or county water authorities) or under an individual legislative act that is designed for a particular geographic area with powers specific to the circumstances of that area. Overall, there are several thousand special districts in California, representing about 50 main types of districts. There are more than 1000 special districts in California having some type of water-related authority. Special districts can vary greatly in size and scope. The Metropolitan Water District of Southern California, for example, provides supplemental water supplies in a service area having more than 19 million people, while a county service area may serve water to only 50 homes. Special districts, together with cities and counties, are considered political subdivisions of the state. There are more than 7000 water systems (both governmental and privately owned) regulated under federal and state laws to provide drinking water (SWRCB, 2022, p. 18).

California is unusual among U.S. states in owning a state water project operated by CDWR; most water projects in the U.S. are owned and operated by local agencies or by the federal government. Two federal agencies, the U.S. Bureau of Reclamation (USBR) (which covers only the western U.S.) and U.S. Army Corps of Engineers (USACE), are the primary federal water management agencies. Other federal agencies can significantly affect California water management through regulatory programs, especially programs related to environmental protection. Both federal and state governments regulate drinking water quality.

Drought is not defined through an explicit statutory approach, rather it is the impacts of drought (water shortage for a particular purpose or application, or wildfire risk) that may be addressed in statute or regulation. The major health and safety impact of drought in California has been related to wildfire which is addressed through federal and state statutes covering forestry and emergency management. Other drought impacts may also be handled through federal or state emergency management authorities, although the primary federal authority for major disasters (the Stafford Act) is generally intended for rapid-onset events and is almost never used for drought specifically because drought impacts can be mitigated at local or state levels. The main federal drought-specific response is financial assistance provided to agricultural producers under a program covering a broad variety of conditions affecting agriculture (e.g. flood, frost, heat, pests). State law requires that drinking water suppliers prepare and update plans detailing how they will respond to shortages from any condition (e.g. drought, earthquake, wildfire), with larger suppliers being required to prepare more complex plans than smaller ones.

10.5 Past California Droughts

CDWR has published a series of reports describing past California droughts (Santos and Godwin 1978; Jones, 2000, 2010, 2020, 2021); this section summarizes some of that information. Readers are encouraged to consult those reports for more detail.

High-resolution paleoclimate streamflow and precipitation data from tree ring reconstructions have provided a good understanding of drought severity and duration for roughly the past millennium. CDWR has funded such reconstructions to help water agencies understand drought risk in the context of natural climate variability (Meko et al. 2017). Generally, the reconstructions show that long-duration droughts are relatively uncommon, although these and other reconstructions show that there were periods of multi-decadal drought across large parts of western North America that would present major water management challenges were they to reoccur in modern times. In an analysis prepared for CDWR of more than 1000 years of reconstructed streamflow in the state's largest rivers (the Sacramento and San Joaquin) an observed dry period in the 1920s–1930s was shown to rank among the top ten driest decades in that extended period (Meko et al. 2014).

In the observed hydroclimate record, the 1920s–1930s were a very dry time in California (Fig. 10.7) but they were not then managed as a drought in the modern sense. Reported impacts were relatively few and bore little resemblance to impacts experienced in the state's modern droughts. The difference was water use. In 1930 California's population was only 5.7 million and estimated irrigated acreage was less than half of modern levels. This period was a time of active construction of new water projects. Subsequently the six-year period of Water Years 1924–1929 was

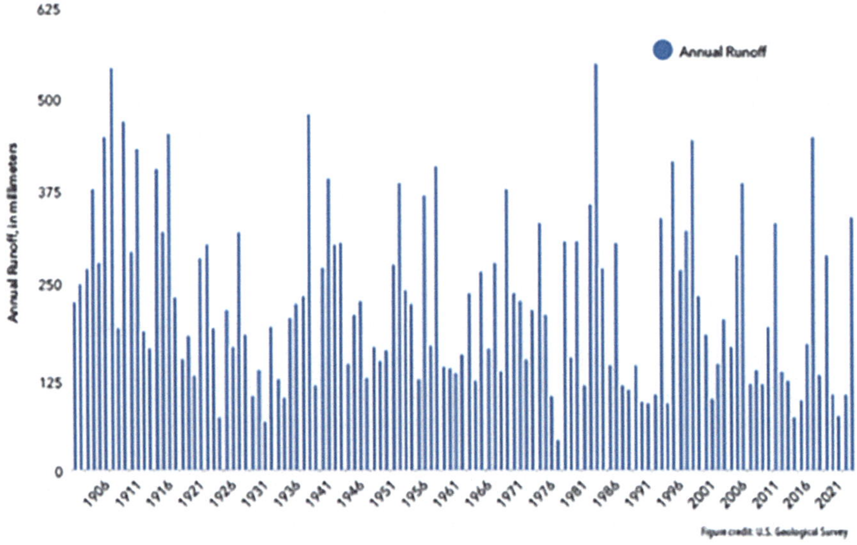

Fig. 10.7 Annual California Runoff

classified as one of the state's most severe droughts (in terms of hydrology, not impacts), and the hydrology of this drought was widely used as design or reservoir operations criteria for many of the large water projects constructed from the 1940s onward.

Drought was not a significant issue in the decades immediately following the 1920s–1930s; although there were multi-year statewide dry periods their hydrology was not severe, and their impacts were unremarkable. California's next significant statewide was the two-year event of Water Years 1976–1977. Water agencies were caught by surprise by the severe hydrology after a long period of relative abundance and were unprepared for managing shortages. California's population was 22 million then and requirements for dedication of water supplies for environmental purposes were modest by today's standards, although irrigated acreage was close to present amounts. Widespread urban conservation and mandatory rationing were hallmarks of the drought, and measures such as emergency pipelines and temporary interconnections were needed for some urban areas. State response action included aggressive enforcement of surface water rights and CDWR's construction of temporary emergency rock barriers in estuarine channels draining to San Francisco Bay to control inland saltwater intrusion. Urban water agencies which had historically relied only on a single local surface water source learned an important lesson about the need to diversify their water supply portfolios. A state commission established to review California water rights law in response to the drought identified many statutory changes that could be made to improve water management under shortage conditions, including managing groundwater extractions and establishing a legal framework for water conservation and water transfers (Governor's Commission, 1978).

The 6-year event of Water Years 1987–1992 was California's first extended dry period since the 1920s–1930s, and it remains California's longest drought (measured as consecutive dry years) at a near-modern level of development. State population was roughly 30 million and irrigated acreage was essentially at present levels. Dedication of water supplies for environmental purposes was greater than in 1976–1977, but still modest by present standards; new requirements had been enacted near the drought's end but were not yet in effect. Water users served by most of the state's larger suppliers did not begin to experience shortages until the third or fourth year of the drought. The drought's long duration spurred progress on establishing an institutional framework for water transfers (short-term leases of surface water rights), and CDWR operated a first-ever state water bank to purchase water from willing sellers and then sell the water to entities experiencing shortages. Making large-scale transfers is facilitated by the state's extensive water conveyance infrastructure, and the drought fostered many new interconnections among the largest urban and agricultural water suppliers. The drought was also an impetus for adoption of a package of state legislation that codified many of the recommendations about water conservation and water transfers made following 1976–1977.

The 2007–2009 drought was California's first drought for which state emergency management provisions were used for drought response, and a statewide emergency proclamation was issued to support drought response. A proclamation allows waiver of state (but not federal) laws and regulations to enable faster response. At the time, this drought was only the seventh driest three-year period in the observed record based on precipitation, but an impetus for the emergency proclamation was then-unprecedented restrictions in supplies for the state's two largest water projects because of new federal requirements for protection of certain fish species, a regulatory circumstance that significantly exacerbated the impacts of hydrologic drought. Agricultural water users in the San Joaquin Valley were especially affected and the drought coincided with a major economic recession in the U.S. which was hitting the San Joaquin Valley hard. Agricultural jobs that might have otherwise cushioned the recession's impacts were being reduced in response to water shortage, and this drought was California's first for which a separate state emergency proclamation was issued to order emergency social services assistance (such as food banks) for the affected area. The drought provided an impetus for state legislation that required owners of specified groundwater wells to periodically report water level data to the state. Previously state agencies and the public had no access to water level data in private and local agency wells, significantly hampering the ability to plan emergency response. State legislation enacted in 2009 also called for a 20% reduction in urban water use by 2020. California's population was approximately 37 million at the end of the drought.

The 5-year 2012–2016 drought was notable for its hydrologic severity, setting a record for the driest four consecutive years of statewide precipitation and requiring state response actions not needed since 1976–1977. It was also notable for record warm statewide temperatures, which contributed to new lows for statewide snow water content, routinely measured by CDWR since 1950. The previous record low 25% of average set in 1977 was tied in 2014 and a new record of only 5% of average was set in 2015. New and more severe impacts occurred in this drought as compared to prior droughts, driven by a warming climate as discussed in more detail below. The federal CVP for the first time made zero deliveries to its agricultural contractors in 2014 and the SWP made only 5% deliveries. Ensuing surface water shortage in the intensively farmed San Joaquin Valley caused increased well construction and groundwater pumping, resulting in expansion of areas experiencing land subsidence and subsidence rates that tied historical maxima. Thousands of shallow domestic wells in rural San Joaquin Valley areas went dry. CDWR developed a pioneering effort to use airborne interferometric synthetic aperture radar throughout the Central Valley to monitor the land subsidence, and in another first was directed to provide emergency assistance to dry domestic well owners in hard hit areas by constructing distribution systems to connect them to larger drinking water providers. Following the drought, state legislation was enacted requiring urban water suppliers to achieve stringent reductions in their urban per capita water use, requirements in addition to those enacted in 2009. State legislation was also enacted authorizing an appropriation of $130 million dollars per year from the General Fund for ten years for addressing problems at small water systems, and the State Water Resources Control Board (SWRCB) (the agency regulating drinking water providers) was given additional authority to require mandatory consolidation of failing small systems with larger systems.

The 2012–2016 drought was soon followed by the three-year 2020–2022 drought, and by a sense of déjà vu as extreme hydrology and impacts were again experienced. These three years set another record for the driest three consecutive years for statewide precipitation. The CVP had two consecutive years of zero deliveries to its agricultural contractors, and a first-ever minimum health and safety-only delivery to its municipal contractors. The SWP had two consecutive years of 5% deliveries. The local agency serving much of Southern California made health and safety-only deliveries to part of its service area. Numerous small water systems (drinking water suppliers) experienced water shortage problems and state financial assistance was provided to them; CDWR continued with projects to connect dry domestic well owners to municipal water system. For the first time notable groundwater impacts (land subsidence and widespread areas of dry domestic wells) were experienced in the water-rich Sacramento Valley. This drought was ended by one of California's snowiest winters in Water Year 2023, when water agencies struggled to maximize groundwater recharge opportunities with the massive volumes of unexpected snowmelt runoff floodwater. State legislation enacted during the drought-imposed shortage planning requirements on small water systems that were a scaled-down version of the shortage planning required of larger systems, as well as shortage planning requirements for counties on behalf of the smallest of small systems and private domestic well owners (CDWR & SWRCB, 2022).

10.6 Climate Change and the Xericene

The term Anthropocene has been suggested as an alternate name for the Holocene epoch, or as a term for the more recent part of the Holocene in which the impacts of anthropogenic warming of the climate system are becoming apparent. Following this line of thought, California's two most recent droughts could be described as examples of the Xericene from a water management perspective. Evolution of the climate system to aridification in a warmer and drier future is being imposed on top of the state's normally high precipitation variability. California's past approach of treating drought as an emergency management problem is not designed to address this problem.

The 2012–2016 and 2020–2022 droughts differed fundamentally from earlier droughts with new and more severe impacts attributable to climate warming to date (Fig. 10.8). Historically drought in California has been viewed as a normal, recurring feature of the state's climate, and one which could be handled through the typical emergency management institutional structure used for shorter duration events such as floods, wildfire, or earthquakes. All three of California's twenty-first century droughts were managed through state emergency proclamations that by their nature are intended to be a short-term response to relatively brief episodic events. Eleven of the 23 years in this century through the end of 2020–2022 drought were years of multiyear statewide drought. If standalone statewide dry years such as Water Years 2001 or 2018 were added to this total, more than half the years in this century would be dry. Tools such as the U.S. Geological Survey's Basin Characterization Model (Flint et al. 2021) allow quantitative examination of how warming temperatures are influencing hydrologic processes. The model uses gridded precipitation and temperature data in calculation of water and energy balances that can show changes over time in conditions such as runoff efficiency or climatic water deficit (potential minus actual evapotranspiration) at a fine spatial scale. Using this model has allowed CDWR to see that climatic water deficit has worsened over

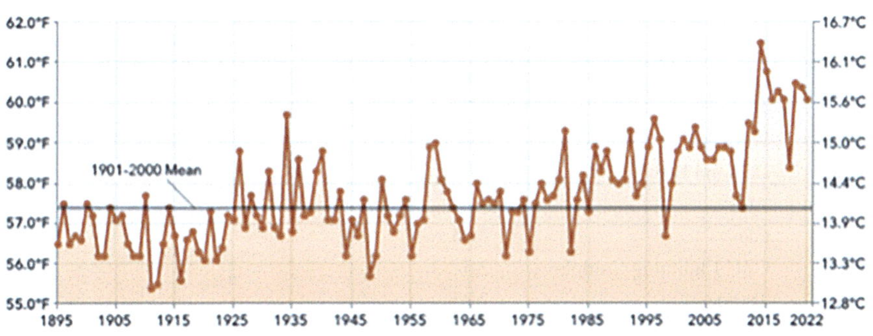

Fig. 10.8 California Average Temperature. (NOAA National Centers for Environmental Information)

time in severe drought years such as 1977, 2014, and 2021, confirming the reason for the worsening drought impacts.

Climate change affects more than just precipitation-runoff relationships that determine water supplies. It also affects the natural environment and unmanaged systems, with drought intensifying these impacts. Two examples are intensified drought impacts on wildfire and fisheries, both of which were notable in recent droughts. Historically, California's largest quantifiable economic impacts of drought have been associated with loss of timber resources and wildfires, but wildfires have now risen to a new threat level. The top ten costliest wildland fires in the U.S. (Table 10.1) have all occurred in California during or following droughts that provided ideal conditions for megafires (III, 2022). Elevated wildfire risk persists after hydrologic drought has ended, because of the large fuel mass (dead and dying woody vegetation) remaining on the landscape. A wet year following drought promotes ample growth of grasses that, when they die down in the summer, provide additional fuel for starting fires. The 2012–2016 drought resulted in widespread tree mortality in the Sierra Nevada mountains, with trees weakened by lack of water becoming vulnerable to bark beetle infestation. A U.S. Forest Service aerial survey following the 2012–2016 drought estimated that 129 million trees had died in California's forests since 2010.

All but two of California's most damaging wildfires occurred from 2000 onward as have all but two of the largest fires based on acreage burned (CAL FIRE, n.d.). Wildfire damage to water systems, previously mostly confined to the smallest of systems in remote areas, has increased and affected larger systems in more urbanized areas as fire sizes have increased. The 2017 Tubbs Fire and the 2018 Camp Fire shown in the table caused impacts to water infrastructure that would have been unimaginable previously. The Tubbs Fire occurred in a highly urbanized community and destroyed the water distribution system in affected residential neighborhoods. It also caused contamination from volatile organic compounds (VOCs) in the distribution system (e.g. benzene from melted plastic piping) resulting in portions of the system being under do not drink or boil water mandates for nearly a year. The

Table 10.1 Top Ten Costliest Wildfires in the U.S.

Year	Fire name	Estimated insured losses (nominal U.S. $ billions)
2018	Camp	10
2017	Tubbs	8.7
2018	Woolsey	4.2
1991	Oakland Hills	1.7
2017	Atlas	3
2020	Glass	2.95
2020	CZU Lightning Complex	2.5
2017	Thomas	2.25
2020	LNU Lightning Complex	2.25
2007	Witch	1.6

Source: (III, 2022)

2018 Camp Fire destroyed much of the small town of Paradise with similar damage to the water distribution system and VOC contamination.

Droughts affect ecosystems, especially aquatic ones, and are an additional stressor for fish populations that may already be in decline because of human development, competition from non-native introduced fish species, and other factors. Salmonid species, which in the 1800s were abundant in coastal waterways and major rivers draining into the Central Valley, have been particularly affected. During droughts state and federal fishery agencies have taken response actions such as fish rescues, closures of recreational and commercial fisheries (ocean and inland), and trucking of juvenile hatchery-raised fish to San Francisco Bay for release; regulatory authorities have also been used to reduce water agencies' diversions. Inland fishery closures (either recreational or commercial) have become common in drought years. In 2023 the California commercial and recreational ocean fishery for Chinook salmon was closed because of low ocean fish abundance.

Salmonids are particularly problematic in a warming climate because of the temperature sensitivity of their eggs and fry. The exceptionally warm settings of the 2012–2016 and 2020–2022 droughts resulted in periods when river temperatures were lethal for eggs and fry; it was estimated, for example, that 75 to 86% of winter-run Chinook salmon eggs in the Sacramento River died in 2014, 2015, and 2021 (the driest and hottest years of these droughts) when there was not enough cold reservoir water left to keep the river temperature cold enough for egg survival (NOAA Fisheries, n.d.). As anadromous fish, salmonids are also at risk when ocean temperatures become too warm, disrupting marine food webs and causing massive harmful algal blooms. The largest documented marine heatwave in the eastern North Pacific Ocean occurred in 2014–2016 and large marine heatwaves have occurred annually in 2019–2022 (IEA, n.d.).

Climate change is increasingly a motivation for future water supply planning. Although exact metrics for change vary as climate models are updated and different representative greenhouse gas concentration pathways are examined, the general themes remain clear. Temperatures are expected to continue warming, precipitation may remain roughly the same (although the number of large storms may decrease with the remaining large storms becoming more extreme), and warming temperatures and evapotranspiration will increase. Mountain snowpack will decrease by more than a third by mid-century and by more than half by the end of the century (CCAS, n.d.).

10.7 Discussion and Lessons Learned

The unexpectedly severe hydrology of 1976–1977 was both a surprise and a challenge to water suppliers who lacked shortage response tools such as system interconnections and an institutional framework for conservation and water transfers. A review of drought response actions at the time did not ascribe impacts experienced to lack infrastructure investment because much new infrastructure had been completed in the relatively recent past (Santos and Godwin, 1978). One basic lesson learned in the1976–1977 drought was simply that severe multi-year droughts were possible. Table 10.2 compares statewide reservoir storage at the end of California's significant droughts for the roughly 150 large reservoirs whose storage is routinely tracked by CDWR. Water agencies managed their reservoirs in 1976–1977 to maximize yield, not to ensure carry-over storage. Many agencies subsequently changed their reservoir operations criteria to be more conservative during dry years to provide greater carry-over storage in case the following year might be dry. Additionally, regulatory protections for environmental purposes have required some reservoir operators to maintain greater carry-over storage for fishery flows.

To some extent the 1987–1982 drought was also a surprise for water users because of its extended duration; the extended dry period and severe hydrology of the 1920s–1930s were a distant memory. By the time of the 2007–2009 drought, however, the changed institutional setting had significantly increased water management challenges. The state's population had grown by 15 million since 1976–1977, irrigated acreage was close to the same, and increased restrictions were placed on operations of the CVP and SWP for fishery or environmental purposes, reducing deliveries from the projects to urban and agricultural water users. (Following 1977 both projects provided their contractors with their full requested delivery amounts through 1989, but subsequently meeting full delivery requests became increasingly rare.) As previously noted, few large-scale or regional water development projects were constructed in the twenty-first century as compared to the twentieth century. Water conservation and water recycling were increasingly being advocated as urban agency drought solutions, such as the 20% by 2020 conservation mandated in 2009. California has been successful in using statutory and regulatory mandates to reduce urban per capita water use over time, although at the risk of demand hardening in future droughts as many communities have used their saved water to support additional population growth. The most recent California water use reductions enacted in 2018 required urban water suppliers to achieve an indoor per capita water use standard of 208 liters per capita daily (LPCD) in their service areas by 2025, but 2022 legislation reduced this value to 178 LPCD in 2025 and 159 LPCD in 2031. Although water recycling projects are increasingly common, their contribution to overall state water supplies remains relatively small. CDWR estimated that in 2021 there were 287 local facilities producing recycled water statewide, generating 900 MCM of supply (CDWR, 2023).

Table 10.2 End of water year statewide reservoir storage at end of multi-year drought

Year	Storage as % of average
1977	36
1992	58
2009	79
2016	82
2022	67

California Department of Water Resources 2023

California's experience has been that large urban water agencies are able to successfully manage multi-year droughts because they have large enough ratepayer bases to be able to afford water supply reliability. Small water systems, systems that may serve only tens to hundreds to a few thousand connections, often do not, and many rural systems rely on groundwater from fractured bedrock sources which can be unreliable during drought. Drinking water shortage problems during droughts have historically been associated with small systems and private domestic wells. The 2012–2016 and 2020–2022 droughts were catalysts for focusing attention on this problem and for enactment of legislation that should improve outcomes in future droughts.

Although significant progress has been made, one problem that remains is the lack of improvement in subseasonal to seasonal precipitation forecasting. If skillful, precipitation forecasts beyond conventional seven- to ten-day weather forecasts would enable more efficient operation of water infrastructure and improved drought preparedness and response by providing longer lead times for taking mitigation measures. Negotiating annual water transfers, obtaining permits for temporary groundwater recharge operations, or spinning up emergency assistance well drilling programs for small water systems are examples of drought response actions that require lead time to put in place. Adequate forecasts would additionally support emerging technologies such as forecast-informed reservoir operations (Reauthorizing the Weather Act, 2023). NOAA, the federal agency responsible for weather and climate forecasting, has issued operational outlooks for seasonal precipitation since the mid-1990s but their skill is minimal (Fig. 10.9). Improving precipitation forecasting at longer lead times is scientifically challenging and requires a commitment to supporting the needed research, but such research is far less costly than emergency drought response actions or federal agricultural disaster assistance programs.

Because California experiences such high annual precipitation variability, average water supply loss over all water year types at some future time is not a key metric from a water management perspective. The more relevant question is water availability during multiyear dry periods. Climate model studies performed to assess impacts of warming on water availability during drought periods on major rivers draining the Central Valley have suggested that future deficits can be much more severe than those observed in the recent (already severe) droughts (Stewart et al. 2020; Lynam and Piechota 2021).

Loss of mountain snowpack increases water management risk, affecting California's largest rivers that provide much of the state's developed water supplies. Those dams and associated conveyance infrastructure were designed based on mid-twentieth century hydrology which does not reflect current conditions. California's recent groundwater management legislation and the need to compensate for lost snowpack storage point to substantial new infrastructure needs to maximize groundwater storage opportunities from short-term flood events that are expected to occur less frequently but in greater volumes. Most of California's large reservoirs store less than the average annual flow of their watersheds; operations of the large CVP

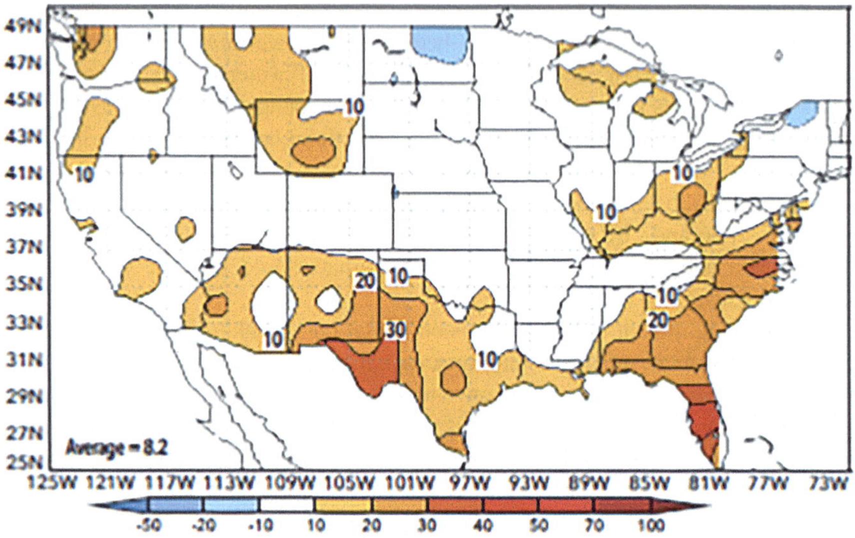

Fig. 10.9 Historical Skill of NOAA Seasonal Precipitation Outlooks. (NOAA Climate Prediction Center) Seasonal (Lead 0.5 Months) Precipitation Heidke Skill Score (DJF Manual Forecasts from 1995 to 2022)

and SWP reservoirs in the recent droughts have increasingly prioritized use of carry-over storage for fishery and environmental flows, emphasizing the importance of maximizing groundwater recharge to support urban and agricultural water users.

Expected reductions of dry year water availability in future droughts illustrate why the historical paradigm of treating droughts as an occasional emergency is losing its effectiveness. In its most recent droughts California has already experienced impacts to water users that would have been unthinkable in the past. Addressing the expected future shortfalls in a warming climate will require substantial investment in new infrastructure that will realistically take years to put in place. Adapting to a warming climate is a different planning paradigm than short-term emergency management, although emergency response tools can still be a tool for drought response.

The historical approach of conceptualizing droughts as isolated emergency response incidents is not equipped to address long-term aridification of the climate system and its impacts on water infrastructure and water management.

Disclaimer The views expressed in this chapter are those of the author and not of the State of California.

References

CAL FIRE. (n.d.). *Statistics*. Retrieved November 29, 2023, from https://www.fire.ca.gov/our-impact/statistics

California Climate Adaptation Strategy. (n.d.). *Summary of projected climate change impacts on California*. Retrieved November 29, 2023, from https://climateresilience.ca.gov/overview/impacts.html

California Department of Public Works. (1931). *Bulletin no. 25 report to legislature of 1931 on state water plan 1930*. California Department of Public Works. https://ia804701.us.archive.org/26/items/state1931egislat25calirich/state1931egislat25calirich.pdf
California Department of Water Resources. (2020). *California's groundwater: Update 2020*. California Department of Water Resources. https://data.cnra.ca.gov/dataset/calgw_update2020/resource/d2b45d3c-52c0-45ba-b92a-fb3c90c1d4be
California Department of Water Resources. (2023). *California water plan 2023 update, Public Review Draft*. California Department of Water Resources. https://water.ca.gov/-/media/DWR-Website/Web-Pages/Programs/California-Water-Plan/Docs/Update2023/PRD/California-Water-Plan-Update-2023-Public-Review-Draft.pdf
California Department of Water Resources & State Water Resources Control Board. (2022). *Primer of senate bill 552: Drought planning for small water suppliers and rural communities*. California Department of Water Resources and State Water Resources Control Board. https://water.ca.gov/-/media/DWR-Website/Web-Pages/Programs/Water-Use-And-Efficiency/Urban-Water-Use-Efficiency/SB-552/Primer-of-SB-552-052522_final.pdf
Dettinger, M. D., Ralph, F. M., Das, T., Neiman, P. J., & Cayan, D. R. (2011). Atmospheric Rivers, floods, and the water resources of California. *Water, 3*(2), 455–478. https://doi.org/10.3390/w3020445
Flint, L. E., Flint, A. L., & Stern, M. A. (2021). *The basin characterization model—a monthly regional water balance software package (BCMv8) data release and model archive for hydrologic California (ver. 3.0, June 2023)*. U.S. Geological Survey Data Release. https://doi.org/10.5066/P9PT36UI
Governor's Commission to Review California Water Rights Law. (1978). *Governor's commission to review California water rights law: Final report: California agencies*. https://digitalcommons.law.ggu.edu/caldocs_agencies/426/
Hanak, E., Ayres, A., Peterson, C., Escriva-Bou, A., Cole, S., & Joaquin Morales, Z. (2023). *Managing water and farmland transitions in the San Joaquin Valley*. Public Policy Institute of California. https://www.ppic.org/?showpdf=true&docraptor=true&url=https%3A%2F%2Fwww.ppic.org%2Fpublication%2Fmanaging-water-and-farmland-transitions-in-the-san-joaquin-valley%2F
Insurance Information Institute. (2022). *Facts + Statistics: Wildfires*. Retrieved November 29, 2023, from https://www.iii.org/fact-statistic/facts-statistics-wildfires#Top%2010%20Costliest%20Wildland%20Fires%20In%20The%20United%20States%20(1)
Integrated Ecosystem Assessment. (n.d.). *California current integrated ecosystem assessment*. Retrieved November 29, 2023, from https://www.integratedecosystemassessment.noaa.gov/regions/california-current/california-current-marine-heatwave-tracker-blobtracker
Jiang, X., Waliser, D. E., Gibson, P. B., Chen, G., & Guan, W. (2022). Why seasonal prediction of California winter precipitation is challenging. *Bulletin of the American Meteorological Society, 103*(12), E2688–E2700. https://doi.org/10.1175/BAMS-D-21-0252.1
Jones, J. (2000). *Preparing for California's next drought: Changes since 1987–92*. California Department of Water Resources. https://water.ca.gov/-/media/DWR-Website/Web-Pages/Water-Basics/Drought/Files/Resources/Preparing-for-california-Next-Drought-1987-92.pdf
Jones, J. (2010). *California's drought of 2007–2009: An overview*. California Department of Water Resources. https://water.ca.gov/-/media/DWR-Website/Web-Pages/Water-Basics/Drought/Files/Resources/California-Drought-of-200709.pdf
Jones, J. (2020). *California's Most significant droughts: Comparing historical and recent conditions*. California Department of Water Resources. https://water.ca.gov/-/media/DWR-Website/Web-Pages/What-We-Do/Drought-Mitigation/Files/Publications-And-Reports/CalSigDroughts19_v9_ay11.pdf
Jones, J. (2021). *California's drought of 2012–2016: An overview*. California Department of Water Resources. https://water.ca.gov/-/media/DWR-Website/Web-Pages/Water-Basics/Drought/Files/Publications-And-Reports/033021_2012-16-Drought-Report_v4_ay11.pdf
Lynam, L., & Piechota, T. (2021). California drought outlooks based on climate change models' effects on water availability. *Water, 13*(22), 3211. https://doi.org/10.3390/w13223211

Meko, D. M., Woodhouse, C. A., & Touchan, R. (2014). *Klamath/San Joaquin/Sacramento hydroclimate reconstructions from tree-rings*. Prepared for California Department of Water Resources under Agreement 4600008850. https://cawaterlibrary.net/wp-content/uploads/2017/05/DWR-Tree-Ring-Report.pdf

Meko, D. M., Woodhouse, C. A., & Bigio, E. R. (2017). *University of Arizona Southern California Tree-Ring Study, Final Report*. Prepared for California Department of Water Resources under Agreement 4600011071. https://cwoodhouse.faculty.arizona.edu/sites/cwoodhouse.faculty.arizona.edu/files/FinalReport2018NoAppendices.pdf

National Oceanic and Atmospheric Administration Fisheries. (n.d.). *River temperatures and survival of endangered California Winter-Run Chinook Salmon in the 2021 Drought*. Retrieved November 29, 2023, from https://www.fisheries.noaa.gov/west-coast/climate/river-temperatures-and-survival-endangered-california-winter-run-chinook-salmon

Public Policy Institute of California. (2018). *Storing water*. Retrieved November 29, 2023, from https://www.ppic.org/wp-content/uploads/californias-water-storing-water-november-2018.pdf

Reauthorizing the Weather Act. (2023). Users of Weather Data and Areas for Improvement by Sector: Hearing Before the Subcommittee on Environment Committee on Science, Space, and Technology United States House of Representatives, 118th Cong. (testimony of Jeanine Jones). https://science.house.gov/2023/6/environment-subcommittee-hearing

Santos, J., & Godwin, A. F. (1978). *The 1976–1977 California drought: A review*. California Department of Water Resources. https://water.ca.gov/-/media/DWR-Website/Web-Pages/Water-Basics/Drought/Files/Resources/The-19761977-California-Drought-A-Review.pdf

State Water Resources Control Board. (1994). *Mono Lake Basin water rights decision 1631*. State Water Resources Control Board. https://www.waterboards.ca.gov/publications_forms/publications/general/docs/monolake_wr_dec1631_a.pdf

State Water Resources Control Board. (2022). *Drinking water program annual compliance report*. State Water Resources Control Board. https://www.waterboards.ca.gov/waterrights/board_decisions/adopted_orders/decisions/d1600_d1649/wrd1631.pdf

Stewart, I. T., Rogers, J., & Graham, A. (2020). Water security under severe drought and climate change: Disparate impacts of the recent severe drought on environmental flows and water supplies in Central California. *Journal of Hydrology X, 7*, 100054. https://doi.org/10.1016/j.hydroa.2020.100054

U.S. Army Corps of Engineers. (n.d.). *Dams of California*. National Inventory of Dams. Retrieved November 29, 2023, from https://nid.sec.usace.army.mil/#/

Chapter 11
Drought Management Policies in Australia

Alec Zuo and Sarah Ann Wheeler

Abstract Australia faces recurring challenges due to its highly variable climate, characterized by periods of drought and heavy rainfall events. Climate change exacerbates these events, impacting agriculture, rural communities, and ecosystems. Drought policy, historically crisis-driven, focuses on farm welfare but commonly neglects broader socio-economic impacts and preventative measures. Recent policies emphasize ad-hoc responses rather than long-term risk management. The Millennium Drought in the 2000s, plus other serious droughts, are a stark example of the country's vulnerability, prompting temporary relief measures and highlighting the need for sustainable policy frameworks. Although current drought policies recognize the importance of risk management and adaptation, they lack cohesion and fail to address the underlying causes of financial distress among farmers. Alternative solutions (such as Revenue Contingent Loans) have been proposed and aim to balance fiscal responsibility with adequate support for farmers. Policy revisions should focus on evidence-based approaches to manage farm poverty and promote resilience across the agricultural sector. Moving forward, a comprehensive approach to drought policy should prioritize risk-sharing mechanisms, promote sustainable agricultural practices, understand behavioural economic issues, reprioritise supporting farm exit for those that need it and address broader socio-economic issues.

Keywords Drought policy · Drought risk management · Preventative measures · Australia

A. Zuo (✉) · S. A. Wheeler
Centre for Social Impact, College of Business, Government and Law, Flinders University, Adelaide, South Australia

National Centre for Groundwater Research & Training, Flinders University, Adelaide, South Australia
e-mail: alec.zuo@flinders.edu.au

D. Rivera, G. Donoso (eds.), *Droughts in Chile*, Global Issues in Water Policy 31,
https://doi.org/10.1007/978-3-031-85040-0_11

11.1 Introduction

Australia has an exceptionally variable climate, with periods of drought punctuated by heavy rainfall events that can cause widespread flooding (King et al., 2020; Wheeler et al., 2020b). The country continues to regularly experience low rainfall, high temperatures, and protracted droughts—with climate change only set to exacerbate these events' frequency and severity (Ault, 2020; King et al., 2020; IPCC, 2021a). Drought policy is often only of periodic interest to the Australian public—rather than a matter of ongoing concern like, for example, interest rate policy (Peterson, 2016). As a crisis, it is complicated by aspects of the national ethos that react to images of agrarianism and which, when evoked, can hinder the rationality of debate over appropriate policy responses (Botterill, 2003).

As described by Fleming-Munoz et al. (2023), there are four 'types' of drought often distinguished: (i) meteorological: a measure of dryness, often defined when rainfall in a region is at the lower extreme of experienced variation for a period of time; (ii) agricultural: results from meteorological drought and happens when available moisture affects plant growth and development; (iii) hydrological: a reduction in surface flow, storage or groundwater levels relative to the historical levels of the region—particularly affecting irrigated crops; and (iv) socio-economic: when societies experience economic impacts from meteorological, agricultural or hydrological drought.

Arguably, Australia's most impacted group is the significant number of small and medium-sized family farms, as they may not have the same financial resources as their corporate counterparts. For this reason, Australia's National Drought Policy in place for over three decades centred on farms and the welfare needs of farming communities (Botterill, 2021). While the stated principles supporting the policy have remained relatively stable—focusing on risk management and farmer self-reliance—the policy largely remains crisis-driven, with changes to the policy, particularly big-ticket spending, often announced during severe drought due to the political imperative to respond (Botterill et al., 2017; Botterill, 2021).

In addition to the economic impact, the negative social impacts of drought on rural communities are gaining increasing attention—particularly on the wellbeing of farmers and others who directly depend on farm income (Austin et al., 2018; Edwards et al., 2019; Yazd et al., 2020). State-based Emergency Drought Relief Package has provided drought relief mental health support programs. However, relief packages often focus on mitigation rather than prevention, leaving vulnerable rural communities susceptible to mental health risks from increasingly frequent and severe droughts (Wheeler et al., 2019).

Understanding the complexities of drought policy in Australia unveils its intricate connections with several other critical policy domains such as environmental preservation, agricultural competitiveness, and rural structural adjustment (e.g. McColl & Young, 2005; Grafton & Wheeler, 2018; Wheeler et al., 2020b). Environmental policy, with its overarching goal of safeguarding fragile ecosystems and biodiversity, has implications for drought policy by influencing land use patterns, water resource management strategies, and conservation initiatives. Notably, the Australian Government's Murray-Darling Basin Plan aims to reallocate water from consumptive agricultural use to environmental conservation through strategic measures like water entitlement buybacks and investments in water-efficient infrastructure. While the buyback approach facilitates rural structural adjustment by offering support for less viable farms to scale down operations or exit the industry, the infrastructure-focused approach incentivizes primary producers to expand, particularly towards high-value, water-intensive crops like perennial crops. However, this strategy presents a potential contradiction with drought policy objectives, as it may expose perennial crop producers to heightened risks during future drought events (Nauges et al., 2016; Wheeler, 2022; Wheeler et al., 2020a). Achieving coherence among these policy realms demands a comprehensive approach that harmonizes economic, environmental, and social imperatives, thereby ensuring the sustainability and resilience of Australia's agricultural sector amidst the evolving challenges posed by drought.

11.2 Background History on Climate/Drought Occurrence in Australia

Drought is a widely acknowledged climatic feature of rain-fed Australian agriculture (Kingwell & Xayavong, 2017). Australia's primary grain crop, wheat, is largely grown in regions where summers are typically hot and dry, and winters are mild and wet. Prolonged periods of low rainfall, high temperatures, high evapotranspiration, and soils that inadequately store moisture, often lead to problematic low yields in crops and pastures (Ockwell, 1990). Such poor yields are the general outcome of drought; however, drought also sets the scene for bushfires, dust storms, and land degradation (BoM, 2023a). Furthermore, Australia's climate is highly variable across the continent—as well as during seasons, years, and decades. The characteristics of individual droughts in Australia also differ considerably. As outlined previously by the four different types of droughts there are, droughts vary significantly in duration and intensity, the time between occurrences, temperature, rainfall, and other climatic factors between events. They also contrast in their geographic spread; and it is not uncommon for some regions to experience good rainfall while others are in drought (Peterson, 2016).

Australia's worst drought on record, the Millennium Drought, resulted in reduced irrigation water allocations and contributed to diminished dry-land productivity, particularly during 2002–03 (Fig. 11.1) and from 2006 to 2009 (Fig. 11.2) (Van Dijk et al., 2013; Wittwer & Young, 2020). During the Millennium Drought, southeast Australia suffered the driest period since 1900 based on several measures—with consequences for ecosystems, the economy, and society as a whole (van Dijk et al., 2013). Commonwealth Scientific and Industrial Research Organisation (CSIRO) (2010) affirmed that the period 1997–2009 had indeed witnessed the lowest average

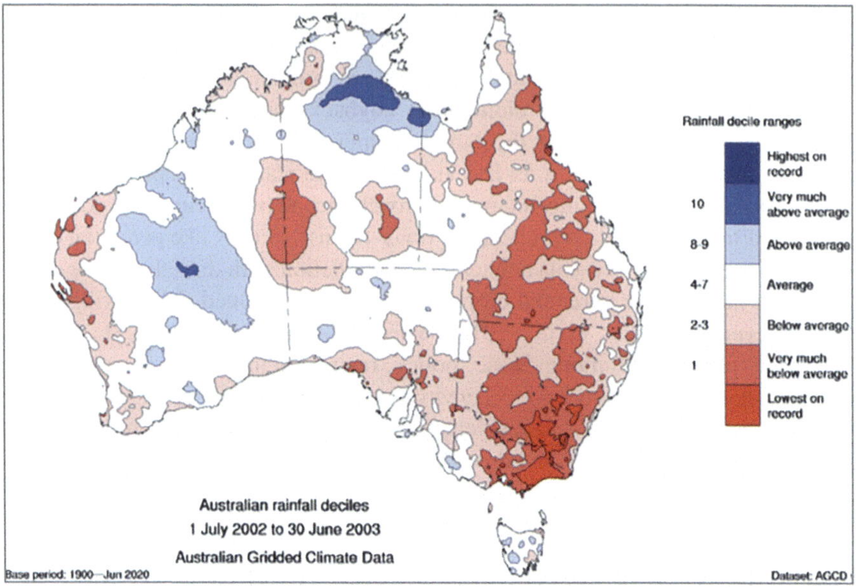

Fig. 11.1 Rainfall deciles for Australia during 01/07/2002–30/06/2003. (Bom, 2023b)

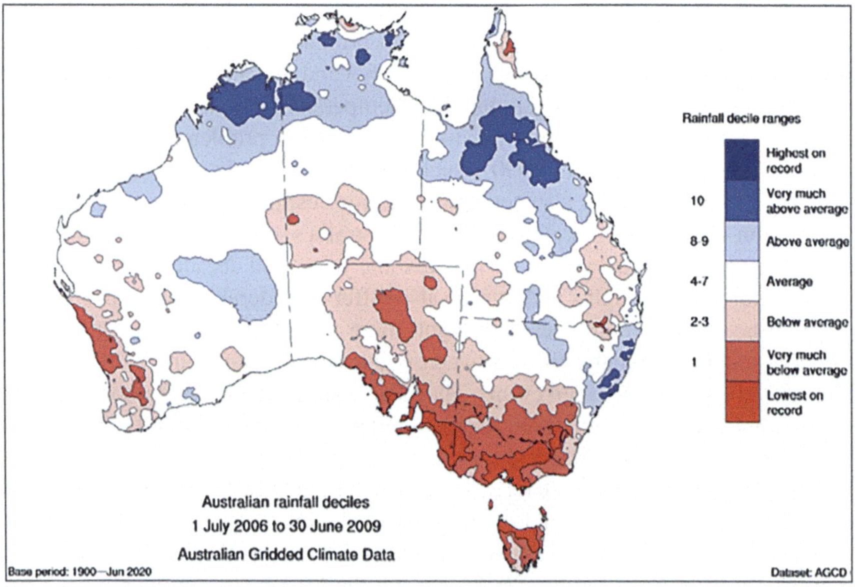

Fig. 11.2 Rainfall deciles for Australia during 01/07/2006–30/06/2009. (Bom, 2023b)

national rainfall on record (albeit with some above-average precipitation years). In contrast, Van Dijk and Renzullo (2009) found that a scarcity of surface water resources had already developed from around 1994.

The Millennium Drought is, therefore, typically defined as the period 2001–2009, specifically the longest uninterrupted sequence of years with below-median rainfall in southeast Australia since at least 1900 (BoM, 2023a). The end of the drought is less ambiguous: an intense La Niña event in early 2010 brought very high precipitation and large-scale flooding to many parts of southeast and east Australia (Beard et al., 2011). However, vast swathes of the country—in particular the Murray–Darling Basin (MDB) region of southeastern Australia—were in drought once more during 2017–2019 (King et al., 2020; Bom, 2023a). See Fig. 11.3.

Drought occurrence in Australia, especially in regional areas, is a recurrent and pressing issue that profoundly affects agricultural productivity, water resources, and rural livelihoods. Fleming-Munoz et al. (2023) provide a recent review of the various impacts and costs of drought. Regional communities, often reliant on agriculture as a primary economic activity, bear the brunt of prolonged dry spells and erratic rainfall patterns, exacerbating vulnerabilities in water security and food production. As seen in Fleming-Munoz et al. (2023), the drought of 1982–1983 was estimated to cost approximately 200,000 jobs, the 2006–2007 drought some 6000 jobs, and the 2018–2020 drought around 34,000 jobs in NSW. In response to these challenges, drought policy in Australia has undergone evolution and adaptation over the years, aiming to strike a balance between supporting affected communities and promoting resilience and self-reliance among farmers. However, the effectiveness

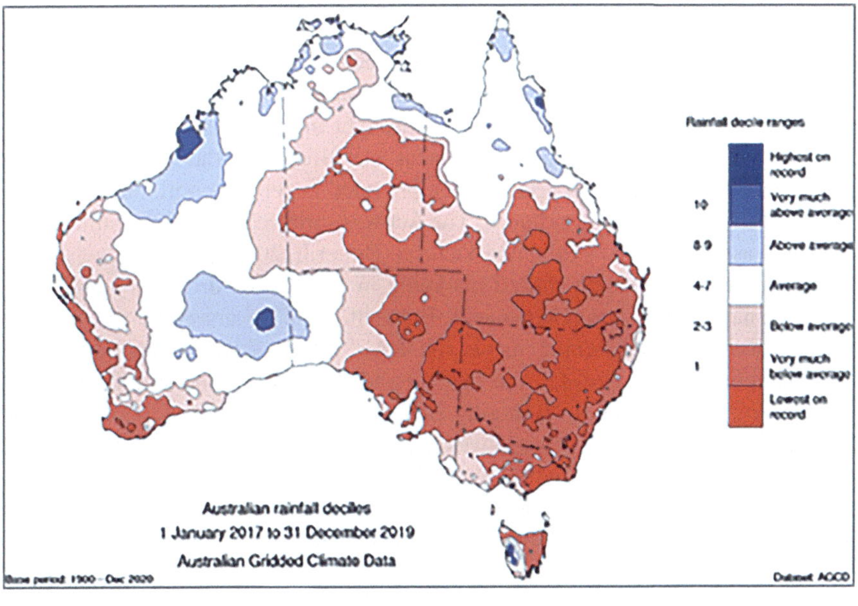

Fig. 11.3 Rainfall deciles for Australia during 01/01/2017–31/12/2019. (Bom, 2023b)

of these policy responses often comes under scrutiny, particularly in addressing the complex and multifaceted impacts of drought on regional economies and communities. As climate change intensifies, with projections of more frequent and severe droughts in the future, there is a growing imperative for policymakers to re-evaluate and refine drought policies, ensuring they are adaptive, holistic, and inclusive of the diverse needs of regional areas and their communities.

11.3 Drought Policy in Australia

11.3.1 Historical Drought Policy

Drought policy reform in Australia has a long history, dating back to the mid-1800s. The early focus was on attempts to drought-proof agriculture by expanding irrigation (Peterson, 2016). Constitutional responsibility for natural disasters has always lied with the states; however, in 1939, the Federal Government became involved, initially via a decision to grant £1000 to Tasmania for bushfire relief (Botterill, 2003). The drought was one of several reasons for Federal support for struggling wheat growers in 1934, but it was not the primary reason for the subsequent legislation. Over the following three decades, Federal involvement in disaster relief was *ad hoc* until a policy emerged that saw the Federal Government matching state spending on personal hardship relief and the restoration of public assets (Butler, 1979).

The broad policy of dollar-for-dollar matching for disaster relief was extended by special Federal Government arrangements for several disasters—including the drought of 1965–66 (Botterill, 2003). In late 1965, Prime Minister Robert Menzies told the House of Representatives that he had advised the Premiers of New South Wales and Queensland that "assistance would cover whatever deficit they ultimately might have in their budgets as a result of drought measures that might be taken" (Menzies, 1965, p. 3647). This commitment was affected in the two *State Grants (Drought Assistance) Acts* of 1966, which provided Federal support under five categories of drought-related assistance. The Act was followed in 1968 by similar legislation to assist South Australia and Victoria (Botterill, 2003).

Government policy then shifted markedly in 1971 when droughts came to be treated as natural disasters (Chapman & Botterill, 2002), overseen by the standing Natural Disaster Relief Arrangements (NDRA).[1] These arrangements defined the shared responsibilities between the Federal and State Governments for managing and funding disaster relief (Botterill, 2021). However, it was not until almost 20 years later, in 1989, that drought was acknowledged different from other natural disasters—slow moving and without an easily discernible start or finish (Peterson, 2016). Around the same time, there were revelations that the Queensland Government had been using drought declarations to misappropriate funds (Botterill,

[1] Now known as the Natural Disaster Relief and Recovery Arrangements (NDRRA, 2023).

2021). Subsequently, the Federal Government announced that drought would be removed from the listed events covered by the NDRA. A task force was established to identify policy options to encourage the agricultural sector to adopt self-reliant approaches in the management of drought—and to consider the integration of drought policy with other relevant policies, such as land management, social welfare, structural adjustment, conservation of breeding stock, and animal welfare (Kerin, 1989).

A series of recommendations followed based on the underlying principle that drought forms part of the Australian farmer's operating environment. Indeed, the Task Force argued that managing drought was about managing risks involved in carrying out an agricultural business, given climate variability (Drought Policy Review Task Force, 1990). This led to the country's first National Drought Policy in 1992. The policy was also a world first: a comprehensive suite of measures based on risk management principles. This stands in contrast with the USA, for example, where advocates have been campaigning for a national drought policy based on similar principles since the 1980s but have yet to succeed in their endeavours (Wilhite & Easterling, 1987; Wilhite et al., 2006; Botterill, 2013). Fundamental to the new policy approach was the understanding that drought is a normal part of the Australian climate, and one of a variety of risks facing farm and pastoral businesses (Botterill, 2021). Table 11.1, provides a brief history of Australian drought policy, while a more detailed list of federal and state legislation can be found in Table 11.2 under Appendix.

Botterill et al. (2017) characterise drought policy as taking two broad forms. First, *contingency-based measures*, including interest rate subsidies designed to reduce the burden around the repayment of commercial loans. A discontinued grant system of this form was the drought relief provisions under the Exceptional Circumstances (ECs) legislation, which was replaced in 2015 by the Drought Concessional Loans Program. Second, *income smoothing devices*, such as drought

Table 11.1 Australian drought policy—a brief history

1971	Federal Natural Disaster Relief Arrangements (NDRA)—Drought is recognised as a natural disaster.
1989	Major review of drought policy—Drought is removed from listed events covered by the NDRA.
1992	National Drought Policy—Emphasis shifts to drought preparedness, maintaining the resource base and drought recovery in agriculture.
1997	Exceptional circumstances (ECs) interest rate Subsidy and relief payment—Supports to farmers in drought-declared areas.
2000	ECs mechanisms are replaced—Support to farmers replaced by the farm management deposit Scheme and rural financial counselling service.
2013	Intergovernmental agreement on National Drought Program Reform—New approach declared to focus on preparing farm businesses for drought.
2018	National Drought Agreement—Strengthens the focus on development of measures that support farm business risk management.

Source: adapted from NACC (2023)

bonds and other instruments have evolved into the current Farm Management Deposits Scheme (FMD). These broad policy areas are discussed in the next two sub-sections, followed by drought research and development and an overview of drought policy governance arrangements in Australia.

11.3.2 Drought Contingency Policies

The key objectives of the landmark 1992 policy were to encourage primary producers, and rural Australia as a whole, to adopt self-reliant approaches in managing climate variability, maintain and protect the agricultural and environmental resource base during periods of climatic stress, and enable early recovery of agricultural and rural industries—to be consistent with long-term sustainable levels (DAFF, 2021). The new national policy was centred around the Federal Government's then recently reviewed Rural Adjustment Scheme—notably its new Exceptional Circumstances (ECs) provisions—which enabled assistance in 'exceptional downturns' for otherwise viable farms facing temporary financial difficulty (Productivity Commission, 2009). Like previous Rural Adjustment Schemes, this version was established under Federal legislation. However, State Governments were primarily responsible for implementation and contributing towards the cost of interest rate subsidies—the main form of financial support offered by the scheme. This contribution was greater for the subsidies paid under ECs provisions (*Rural Adjustment Act 1992, s21*), ensuring that states did not request ECs declarations without carefully considering their merits, given the financial implications for state budgets (Botterill, 2021).

The policy also included arrangements for short-term household support, programs to aid income smoothing and the accumulation of cash reserves, training and education, and natural resource management measures (Botterill, 2021). However, the definition of an 'exceptional circumstances drought' was especially contentious and became the central focus of policy debate, together with the geographic boundaries for areas declared to be in ECs, which became known as the 'lines on maps' problem. The debate intensified after introducing the *Drought Relief Payment Act 1994* (DRP)—a welfare payment accessible to farm households in areas declared to be in ECs (Botterill, 2003, 2021).

The DRP can be viewed as a concession to agrarian values and is a prime example of the 'cycling' between values that occurs within the policy process, as governments seek to deal with conflicting policy goals (Botterill, 2003; Thacher and Rein, 2004). Eventually, following a change of Federal Government from the Labour Party to the Liberal-National Party Coalition in 1996 and a review conducted in 1997 (McColl et al., 1997), the Rural Adjustment Scheme was terminated. The ECs provisions were retained as interest rate subsidies, while the DRP was extended and renamed the Exceptional Circumstances Relief Payment (Botterill, 2021).

In 2008, a wide-ranging review of the National Drought Policy was undertaken.[2] The Productivity Commission (2009) examined the effectiveness of government programs; the Commonwealth Scientific and Industrial Research Organisation (CSIRO) and Bureau of Meteorology (BoM) reported on the likely impact of climate change on future Australian droughts (Hennessy et al., 2008); while an expert panel inquired into the various social impacts of drought (Expert Social Panel, 2008). Following the review, the ECs interest rate subsidies ceased, with the Commission asserting that 'the rationale underlying most of the National Drought Policy's objectives are sound.' It cautioned against 'the use of reactive business assistance measures such as interest rate subsidies and other transactions-based subsidies, including waivers for legitimate business expenses' (Productivity Commission, 2009, p.44).

In early 2013, the Federal and State Governments signed an Intergovernmental Agreement on National Drought Program Reform. Botterill (2021) asserts that the agreement was very light on detail, with 'outputs' including continued access to the Farm Management Deposit (FMD) scheme, a farm household support payment, a national approach to farm business training, better tools and technologies to assist farmer decision making, and a coordinated approach to the provision of social support services (DAFF, 2013). Later that year, the Federal government announced a suite of farm assistance measures, including concessional loans, an enhanced FMD scheme, continued funding for the Rural Financial Counselling Service, together with a nationally consistent approach to debt mediation (DAFF, 2021). Subsequently, in December 2018, Federal, State and Territory Governments signed a new National Drought Agreement that described how governments would collaborate on drought-related issues (NDA, 2018). The new agreement outlined responsibilities when supporting farming businesses, families, and communities—the aim is to strengthen the focus on developing measures that support farm business risk management. (DAFF, 2021).

11.3.3 Income Smoothing Policies

Australian governments have intervened in the agricultural credit market for decades, the overarching aim being income stabilisation (Botterill & Chapman, 2009a). The first of these initiatives was the Drought Bond, introduced by the Federal Government following the major drought of the 1960s. The Drought Bond scheme was explicitly targeted at livestock producers to encourage the accumulation of financial reserves during high-income years. The Bonds paid interest and matured after 10 years. However, they were redeemable before maturity in the case of drought (Glau, 1970).

[2] Smaller drought policy reviews were conducted in 1997 and 2004 however these were not as comprehensive as either the 1990 Task Force review (leading to the 1992 National Drought Policy) or subsequent 2008 review.

The Drought Bond program was replaced with the Income Equalisation Deposit (IED) scheme from July 1975, given it was found that 'comparatively few primary producers'took advantage of the Drought Bond scheme (Harris, 1974, p.78). While the sole focus of Drought Bonds had been livestock producers, the IED scheme had a far greater reach, applying to all primary producers (Douglas & Davenport, 1993). The IED also incorporated the principle of government intervention to stabilise farm incomes—the Green Paper on Rural Policy, which was a precursor to the scheme, concluding that 'instability of farm incomes involves farmers and those in rural areas in substantial welfare costs and leads to a less than fully effective use of farm resources' (Harris, 1974, p.81). The paper further suggested that 'income equalisation arrangements are worthy of detailed and expert examination' (Harris, 1974, p.81).

Despite the deregulatory policy trend during the 1970s, such interventions have continued to the present day, in the form of interest rate subsidies offered within industry adjustment programs and—as noted above—under drought policy (Botterill & Chapman, 2009a). However, several reviews have highlighted that such subsidies are an inequitable and ineffective means of providing support (e.g., Freebairn, 1983; McColl et al., 1997). Botterill and Chapman (2009a) argue that such interventions can result in considerable transfers of public money to individual farmers and, unsurprisingly, only provide relief to those with debt or prepared to take on debt. Furthermore, interest rate subsidies have the potential to be regressive as larger farms with greater incomes and higher levels of debt attract greater subsidies (Botterill & Chapman, 2009a).

Towards the end of the 1980s, the tone of drought policy debate shifted, including the rationale for income smoothing mechanisms—which moved from an emphasis on income stabilisation to risk management (Botterill & Chapman, 2009a). The 1992 National Drought Policy was negotiated through a joint Federal-State ministerial council, the Agriculture Council of Australia, and New Zealand (ACANZ). The policy would be centred on 'principles of sustainable development, risk management, productivity growth and structural adjustment in the farm sector' (ACANZ, 1992, p.13). Notably, the policy entailed a proviso relating to 'severe downturns' during which support would be delivered to 'those with sound prospects who are temporarily in difficulty' (ACANZ, 1992, p.13). As outlined in Table 11.1, this response to severe events, including drought, was given effect in the Exceptional Circumstances (ECs) provisions of the *Rural Adjustment Act 1992*.

A review, however, of the *Rural Adjustment Scheme* in 1997 highlighted the issue of linking welfare and business support (Botterill & Chapman, 2009b), namely that 'welfare assistance should not be delivered through instruments that assist businesses. Such an approach confuses the objective of the intervention, does not effectively target the welfare problem, and distorts market signals to farm businesses receiving assistance' (McColl et al., 1997, p.38). Furthermore, the IED scheme was to be supplemented by a new Farm Management Bond initiative, which was more generous in its tax treatment but with much tighter withdrawal conditions—being accessible only during 'periods of financial stress caused by factors such as drought, commodity price collapse, severe disease outbreak, etc.' (ACANZ, 1992, p.15).

In early 1998, the Federal Government announced the details of the new Farm Management Deposit (FMD) program, which would eventually replace both the IED and FMB schemes. An essential feature of FMDs is that they are held by commercial financial institutions—in contrast with previous deposits in income smoothing schemes, which had been held by the government (Botterill & Chapman, 2009b). Furthermore, interest rates applying to the deposits under FMDs were set commercially rather than paid at the three-year government bond rate (Anderson, 1998). In introducing the FMD scheme, the Federal Government—once again—emphasised its use as a financial risk management tool, stating that the bill had been designed to 'encourage increased financial self-reliance among farmers, while also taking account of the high variability of farm income streams and the vulnerability of farming businesses to natural events' (Scott, 1998, p.4066).

The Productivity Commission concluded that the FMD scheme had 'encouraged farmers to save and to be more self-reliant' (Productivity Commission, 2009). In the most recent review of the FMD scheme (earlier reviews were undertaken in 2002, 2006 and 2012) stakeholder views on the scheme were comparable to those expressed in previous years—in that FMDs were considered an effective tool to assist primary producers in planning and managing their operations. The scheme was considered by some, however, as inflexible, which may limit the extent to which the scheme is used by eligible participants (DAWE, 2021).

Figure 11.4 displays the quarterly value of FMD holdings, together with the number of holders, from June 1999 to December 2023. It was only until June 2002 for the value of FMD holdings and number of holders to increase dramatically from the initial level in 1999. After 2002, while the number of holders remain relatively stable between 40,000 and 50,000, the total value had a sustained increase from Dec

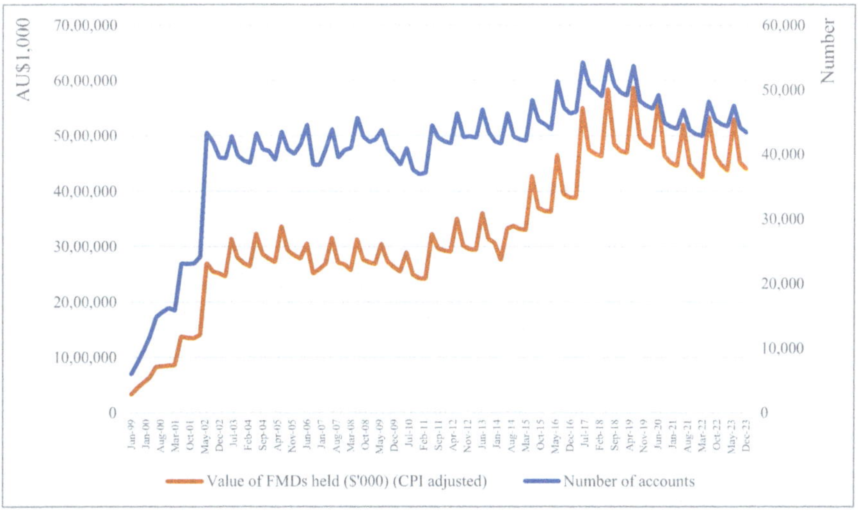

Fig. 11.4 Value of FMD holdings and number of holders, 1999–2023. (Authors' own graph. Data sourced from DAFF (2024), CPI 2011–12 = 100)

2010 to September 2019. Both value and account number exhibit an obvious seasonal pattern: highest in the June quarter and lowest in the December quarter, suggesting that farmers deposit fund before the financial year end in June and withdraw fund during mid-financial year. The decline between 2019 and 2022 coincided with a fall in average farm cash incomes of broadacre farms, attributed to drought across much of Australia (Litchfield et al., 2021) and may also reflect the uncertainty arising from the COVID-19 pandemic.

In terms of the primary use for FMDs, the Department of Agriculture, Water and the Environment (DAWE) surveyed primary producers (n = 183) and showed that FMDs were frequently used for planning for and managing through drought, followed by income smoothing and tax management (DAWE, 2021).

11.3.4 Drought Research and Development

Drought research was advanced significantly in Australia during the 1970s and 1980s, particularly around providing leading scientific publications in climate science. These advances led to an improved understanding of climate variability and climate forecasting research, instigating quasi-operational climate forecasting outputs by the BoM and state-based organisations primarily focussed on the need to manage primary industries (Stone, 2014). Meanwhile, the Australian Federal Government provided incentives to improve self-reliance and farm management, enabling preparedness for the seasons and years of drought when they occurred. As we have seen, government policy development over this period was able to provide fundamental changes in drought assistance through new policy initiatives that, at the time, were unequalled in the world (O'Meagher et al., 2000; Wilhite, 2005).

The establishment of the Australian National Drought Policy in 1992 saw further advancement of seasonal climate predictions and investments in agricultural systems research aimed at increasing the drought resilience of farm operations. As noted above, these developments marked a significant shift away from earlier crisis-driven subsidies, towards a more proactive risk management approach—aimed at enabling self-reliance in managing climate variability (Laughlin & Clark, 2000). There remains an urgent need to address climate-related risk and vulnerability, particularly regarding drought and water security across much of urban and rural Australia (e.g., Nelson et al., 2008; Fragar et al., 2010; Rickards, 2012; Kiem & Austin, 2013).

Lellyett et al. (2022) recently conducted a comprehensive review of drought early warning systems in Australia. They identified several advancements in climate information, forecasting, and current and future research and development areas. One such advancement was improved retrospective analyses, enabled through high-resolution grid-based investigations (Lewis & Karoly, 2013; Frost et al., 2016; Evans et al., 2020) along with better definitions of drought using both meteorological and non-meteorological variables (UNIDSR, 2009; WMO, 2016). Recent advancements have also been achieved in predictive outputs and projections, namely improvements in seasonal and inter-annual forecasting (Wang et al., 2011; Hudson et al., 2017) and more accurate climate projections (IPCC, 2021b).

Current and future areas of research highlighted by Lellyett et al. (2022) include the forecasting of soil moisture (Zhao et al., 2017; Griffiths et al., 2017; Frost et al., 2018) and evapotranspiration (Nguyen et al., 2017; Frost et al., 2018); better distinguishing between 'flash drought' and the onset of longer-term widespread drought (Frost et al., 2018; Nguyen et al., 2019); and improved predictions of drought duration across multi-year forecasts (Sheen et al., 2017; Luo et al., 2018). However, the latter remains a significant challenge and considerable uncertainty still exists regarding the quantitative prospects of cessation of meteorological drought—and even more uncertainty regarding agricultural drought.

Other Australian research on farmers' drought related behaviour has focussed on behavioural economics aspects and in particular, studying the link between farmers' climate change attitudes and their adaptation behaviour on-farm (e.g. Seidl et al., 2021). For example, Wheeler et al. (2021) showed that there has been a significant change in farmers attitudes to climate change over time, increasing from a low base (Wheeler et al., 2013) to more accepting the science. However, this study also discovered a feedback loop between climate change attitudes and adaptation behaviour on farm. It was found that farmers who agreed climate change was a risk in 2010–11, were more likely to undertake farm production decisions to reduce that risk (e.g. changing crop mix, reducing irrigated area and consequently selling water entitlements)—which negatively fed back and reducing their stated climate change risk perceptions in 2015–16. Conversely, farmers who were originally deniers were more likely to undertake somewhat riskier farm production decisions (e.g. increasing water utilisation rates and irrigation areas)—which consequently had the impact of positively increasing their climate change risk perceptions in 2015–16. This study, along with Wheeler and Marning (2019), suggested that there may be a difference in the type of adaptation behaviour undertaken by farmers, and that more natural resource management or soft skills changes did not have the same level of feedback effect. Wheeler et al. (2015) also found some evidence that organic irrigation farms use less absolute water than conventional farms; use a smaller percentage of water received; and are more water-use productive (i.e. water use divided by net farm income). Both Wheeler et al. (2015) and Wheeler and Marning (2019) advocated a need for governments to reorientate irrigation policy towards more multi-layered and inclusive practices that promote better soil conditions and water and natural resource management, rather than focussing on providing subsidies for technology adoption as a water security measure.

11.3.5 Drought Policy Governance Arrangements

The Australian government's current drought policy and governance arrangements are outlined in Fig. 11.5 by Samnakay (2023, p.152). The lead drought policy implementing agencies are identified, with the 2018 National Drought Agreement (NDA) positioned as the strategic sustainable policy, and subordinate policies and

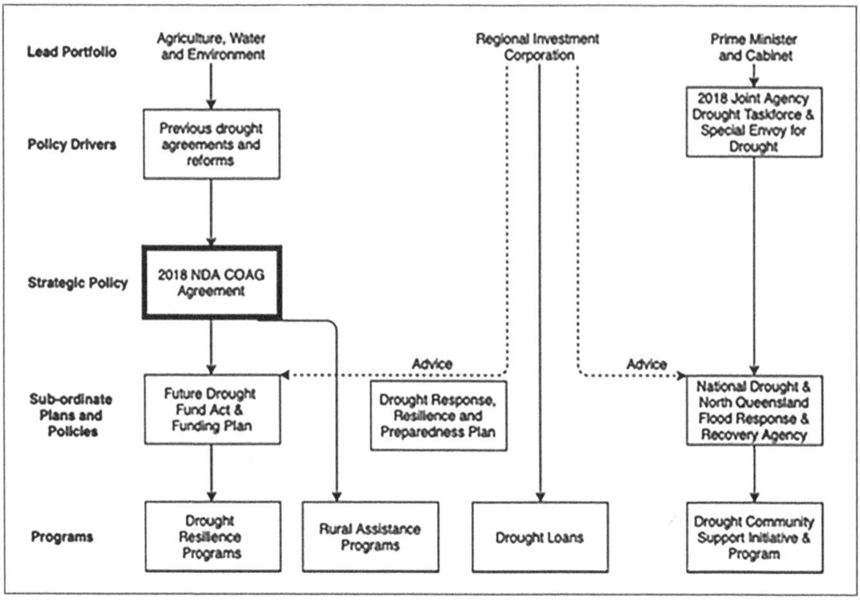

Fig. 11.5 Australian drought policy governance arrangements. (Samnakay, 2023, p.152)

programmes delivered by Federal Government agencies (NDA, 2018; Samnakay, 2023). While state initiatives are not represented here, it is recognised that states are signatories to the NDA with their own implementing responsibilities, and that some nationally coordinated programmes are run in partnership with State Governments, such as the Rural Financial Counselling Service (DAFF, 2019).

Following the 2018 agreement, Botterill (2021) highlights that the most recent drought of 2017–2019 has spurred announcements of a range of measures by both State and Federal Governments—involving substantial amounts of money. For example, the New South Wales (NSW) Government announced AUD$190 million for transport subsidies (Grain Central, 2018)—a form of support the states agreed to phase out in 1992. Transport subsidies had attracted criticism from the Productivity Commission, which found that similar to ECs interest subsidies, 'state-based transactions subsidies are ineffective, can perversely encourage poor management practices and should be terminated' (Productivity Commission, 2009, p.20).

More recently, the NSW Government announced an additional AUD$310 million to support farmers through the drought recovery phase, including concessional loans, further transport subsidies (this time for restocking) and a doubling of its interest-free drought loans to AUD$100,000 (ABC News, 2020). Meanwhile, the Federal Government's most recent response has included the extension of concessional low-interest loans, additional Household Allowance payments, the establishment of the Future Drought Fund, and the appointment of a Coordinator General for Drought (Botterill, 2021). The NDA has also been reviewed before its expiry on 30 June 2024, and recommended changes to 'ensure farmers and farming communities are well supported before, during and after drought' (NDA, 2022, p.5). Figure 11.6

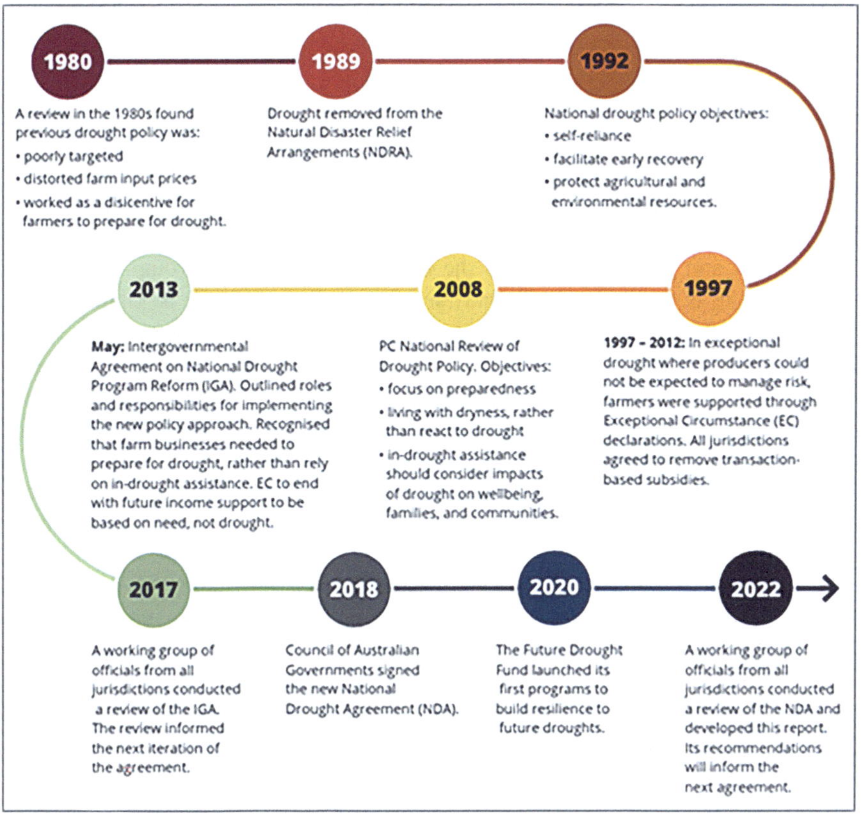

Fig. 11.6 Progression of Australian Drought Policy. (NDA, 2022, p.11)

below provides an overview of drought policy progression up to and beyond the 2018 National Drought Agreement.

Botterill (2021) compares the language of the original 1992 National Drought Policy with the 2018 NDA. While the objectives appear quite similar, the implementation of the two policies is noticeably different. Unlike the original policy, the current agreement provides a framework that opens the door for knee-jerk policy responses. Any measures introduced during drought are intended to comply with the principles set out in the agreement. However, the principles are sufficiently vague that various possible approaches could be undertaken.

While the exceptional circumstances provisions under the 1992 policy became contentious and problematic over time due to the 'lines on maps' issue, any drought-related measures were established well in advance—as was the detail of the support that would become available during periods of drought. A crucial element of any national drought policy focused on preparedness and self-reliance would be certainty regarding the type of government support likely to be forthcoming and when.

Botterill (2021) asserts that the three-phase approach within the 2018 NDA still leaves open the possibility of *ad hoc* government responses.

11.4 Discussion

There are a range of challenges that confront governments in developing drought policy. First and foremost, there is the difficulty in defining drought, including distinguishing between "normal" and "severe" drought (Wilhite & Glantz, 1985; Chapman & Botterill, 2002). Another issue is the high level of integration between the farm household and farm business, which poses the question of whether drought support should be directed at the entire farm household or be limited to the farm business (McColl et al., 1997; Botterill & Chapman, 2009a, b). A further problem confronting policymakers is around structural adjustment, which has significant implications regarding eligibility for drought support programs—essentially, whether support should only be made available to farmers who are viable in the long term (Botterill, 2000; Chapman & Botterill, 2002). Given drought (or water scarcity) is found to be a significant factor for farm exit (Wheeler et al., 2020b; Zuo et al., 2015), particularly for the poorly performing farming enterprises (Wheeler & Zuo, 2017), drought support should focus more on the farm household than the business, which assists rural structural adjustment and increase the resilience and competitiveness of the whole sector. Issues with farm exit need far greater discussion—especially in areas where certain farming systems (e.g. irrigation) may no longer be viable in a climate change and water reallocation future (Wheeler, 2023).

While the original risk management principles continue to underpin Australian drought policy, Botterill and Fenna (2019) contend that much of the intent and cohesion of the seminal 1992 policy has been missing in recent years. Indeed, the conflict between an approach centred on-farm risk management and political pressure to act during severe drought—due to strong, sympathetic public sentiment towards farmers—has led to the demise of arguably the policy's most important goal: the shift away from ad hoc policy responses during periods of drought (Botterill & Fenna, 2019). The Productivity Commission (2009, p.24) identified a clear mismatch between the National Drought Policy's objectives and programs. From its inception, the policy has:

> ...ostensibly centred on helping farmers build their self-reliance to manage climate variability and preparedness for droughts. Program expenditures, however, have not been directed to this end but have mainly flowed as emergency payments to a minority of farmers in hardship and to stressed farm businesses.

There are two primary concerns regarding the Federal Government's approach to drought policy (Botterill et al., 2017). First, direct grants and interest rate subsidies provided to farm properties are considered budgetary expenses lacking the necessary redistributive characteristics. Taxpayers, who often face financial constraints themselves, fund these costs to support farmers, which may not effectively address the root causes of financial distress. Secondly, although the current concessional loans scheme is less burdensome than traditional commercial loans, it offers limited relief from borrowing risks. In fact, it may exacerbate solvency concerns as the loans eventually require full repayment, potentially leading to their conversion into commercial loans, thereby adding to financial pressures.

Building upon earlier work undertaken by Chapman et al. (2004), Botterill et al. (2017) propose a drought policy approach aimed at addressing the aforementioned concerns: a Revenue Contingent Loan (RCL). The RCL is a financial instrument crafted to minimise taxpayer subsidies while simultaneously offering farmers protection against default. Following the principles underpinning the Higher Education Contribution Scheme (HECS), an RCL provides financial aid that is repayable based on the farm's financial capacity, thereby reducing the likelihood of loan defaults (Chapman, 2006). Through modelling contemporary RCL revenue streams, the authors illustrate strong stability concerning repayment timelines and minimal government subsidies required. Additionally, the modelling highlights the income-smoothing benefits inherent in an RCL.

As previously discussed, the Fund Management Deposit (FMD) program has proven largely effective, enabling farms to save during prosperous years and utilize these savings as a financial buffer during challenging periods (DAFF, 2022). In contrast, an RCL operates in reverse; it allows farmers to borrow from anticipated future good years to mitigate current drought impacts (Chapman, 2006). Thus, a potential future drought policy could integrate RCLs with FMDs into a unified financial risk management initiative. Under this approach, farmers could first draw from their FMDs and then access an RCL once their reserves are depleted, potentially retaining a portion of their FMD funds for working capital (Botterill et al., 2017). Such a policy framework could also address a significant limitation of FMDs, particularly concerning new entrants who may not have sufficient time to accumulate reserves before encountering financial downturns (DAWE, 2021). Additionally, RCLs could provide assistance to those with inadequate FMDs during drought periods (Botterill, 2021).

Finally, a pressing and overdue necessity is the consideration of the welfare aspect within drought policy (McColl et al., 1997; Botterill, 2007; Botterill & Chapman, 2009b; Botterill, 2021). Until farm poverty is managed as an independent policy issue, concerns about the impact of drought on the welfare of farm families will continue to hamper attempts to implement the policy separation of the needs of the farm family from those of the farm business (Botterill, 2007). Botterill (2021) asserts the inadequate research into the nature and scope of farm poverty in Australia, leading to policy formulations based on assumptions rather than empirical evidence, thereby complicating the delivery of drought policy and potentially exacerbating disparities with other community groups.

11.5 Conclusion

Australia's National Drought Policy is internationally recognized for its acknowledgement of the reality of climate uncertainty and variability and emphasis on farm adaptation to climatic uncertainty. The policy does not subsidise agriculture in low rainfall areas. But, while the underlying principles of the policy appear sound, after three decades of implementation, incremental changes to the schemes beneath it have led to the dilution of its risk management message, continued reliance on exceptional circumstances provisions, and the emergence of disparities among farmers and between farmers and non-farmers.

Due to widespread agrarian sentiment among Australian voters, governments will inevitably continue to provide necessary financial assistance to farmers affected by drought. Therefore, the critical question for drought policy should focus on identifying the farmers genuinely in need, without compromising equity and risk management capacity. The impact and true costs of droughts should rely on rigorous research to guide the relief support (Fleming-Munoz et al., 2023). One avenue of research to improve risk management could involve the development of more accurate and reliable early warning systems, integrating climate data, remote sensing and machine learning algorithms to predict drought onset and severity with greater precision. More emphasis on interdisciplinary research is needed to examine the socio-economic implications of droughts, including food security, education performance and public health (Hanigan et al., 2012; Xu et al., 2023). Behavioural economic research (e.g. Wheeler et al., 2021, 2020a) also illustrates the importance of understanding the link between farmers attitudes towards climate change and drought, and their adaptation behaviour on-farm. Subsidising hard infrastructure (such as irrigation infrastructure, dams etc) can leave farmers in a worse-off position than focussing on building natural capital on farm and developing more 'soft' risk management adoption skills (Wheeler & Marning, 2019). In addition, understanding, and facilitating farm exit (Zuo et al., 2015), is also a critical factor that will need far greater attention in coming years.

The ongoing efforts of drought policy should still prioritize the development of mechanisms for risk sharing across seasons (e.g. FMD) and among farmers (e.g. index insurance), ensuring a collective approach to managing the impacts of drought. This entails fostering a culture of preparedness and resilience within the agricultural community. Long-term programs of drought policy should focus on the provision of accurate drought conditions at finer geographic scales and the promotion of farm practices that are inherently resilient to drought. Policy-makers should also consider other issues facing the agricultural and regional communities such as environment degradation, rural adjustment, and international trade environment. By focusing on these aspects, drought policy can contribute to building a more resilient and sustainable agricultural sector capable of withstanding future climate challenges.

Acknowledgements The authors gratefully acknowledge Adam Wheeler for his valuable research assistance and the Australian Research Council for funding this research (FT220100695).

Appendix

Table 11.2 Federal Government drought-related legislation in Australia from 1940 onwards

Legislation	In Force	Objective(s)
States Grants (Drought Relief) Act 1940 (six states (NSW, QLD, SA, TAS, VIC, WA)	No	For alleviating the hardship suffered by wheat-growers in consequence of drought.
Loan (Drought Relief) Act 1940	No	Authorisation to raise money to be loaned for drought relief in NSW, VIC, QLD, SA and WA.
States Grants (Drought Relief) Act 1944	No	States having effects of drought on cereal crops.
Drought Relief Act 1945 (amend the State Grants (Drought Relief) Act 1944)	No	Increase the financial assistance from 1.5 M pounds to 1.86 M pounds, add ACT into the ACT.
States Grants (Drought Relief) Act 1946	No	States having effects of drought on cereal crops.
New South Wales Grant (Drought Relief) Act 1947	No	Grant financial assistance to NSW for drought relief.
States Grants (Drought Assistance) Act 1966	No	Make provision for the grant of financial assistance to NSW and QLD about the alleviation of the effects of drought.
States Grants (Drought Assistance) Act 1966 (No. 2)	No	Grant financial assistance to NSW and QLD.
States Grants (Drought Reimbursement) Act 1968	No	Grant financial assistance to VIC and SA to meet the cost of measures for alleviating the effects of drought.
States Grants (Drought Assistance) Act 1968	No	Grant financial assistance to NSW, VIC, QLD, and SA.
Loan (Drought Bonds) Act 1969	No	Create and issue stock called "Drought bounds".
Income Tax Assessment Act (No. 3) 1969	No	Amend the law relating to income tax on Drought Bonds.
Income Tax (Drought Bonds) Act 1969	No	Impose income tax about certain redemptions of drought bonds by reference to the income tax saved because of the purchase thereof.
Loan (Drought Bonds) Amendment Act 1976	Yes	Amend the Loan (Drought Bonds) Act 1969.
Drought Assistance (Primary Producers) Act 1982	No	States (including NT) to receive drought assistance for primary producers.
Rural Adjustment Act 1992	Yes	Foster the development of a more profitable farm sector that is able to operate competitively in a deregulated financial and market environment; and to improve the competitiveness of the farm sector in a sustainable manner.
States Grants (Rural Adjustment) Amendment Act 1992	Yes	Contains three amendments to the above *Rural Adjustment Act 1992* in relation to drought.

(continued)

Table 11.2 (continued)

Legislation	In Force	Objective(s)
Drought Relief Payment Act 1994	Yes	Amend *the Farm Household Support Act 1992* about the qualification for drought relief payment, the drought exceptional circumstances certificate, rate of farm household support, and rate of drought relief payment. Amend *Health Insurance Act 1973* and *Income Tax Assessment Act 1936*.
Taxation Laws Amendment (Drought Relief Measures) Act 1995	No	Involves amendments to the *Income Tax Assessment Act 1936* related to the Drought Investment Allowance and encompasses amendments to *the Loan (Income Equalization Deposits) Act 1976*, specifically relating to Farm Management Bonds.
Farm Household Support Amendment (Additional Drought Assistance Measures) Act 2008	Yes	Amend the *Farm Household Support Act 1992* to improve access to exceptional circumstances relief payment, and increase portability of exceptional circumstances relief payments
Appropriation (Drought and Equine Influenza Assistance) Act (No. 1 & No. 2) 2007–2008	No	Appropriates money out of the consolidated revenue fund for the ordinary annual services of the government (and for the certain expenditure) regarding drought, equine influenza assistance, and for related purposes.
Future Drought Fund Act 2019	Yes	Enhances the Federal Government's ability to make arrangements with persons or bodies in relation to drought resilience; and make grants to persons or bodies in relation to drought resilience. The future drought fund consists of the future drought fund special account; and the investments of the future drought fund.
Future Drought Fund Investment Mandate Direction 2019	Yes	Gives guidance to the board about its investment strategy for the fund.
Future Drought Fund (Consequential Amendments) Act 2019	Yes	Deals with consequential matters arising from the enactment of the Future Drought Fund Act 2019 and for other purposes. Amendments include general amendments, abolition of the Building Australia fund and technical corrections.
Regional Investment Corporation (Agribusiness Natural Disaster Loans—2019 North Queensland Flood) Rule 2019	No	Prescribes the terms and conditions that the regional investment corporation may include in loans provided to farm businesses that have suffered direct damage as a result of the North Queensland floods of January and February 2019.
Regional Investment Corporation (Small Business Drought Loans) Rules 2020	Yes	Allows for loans to be made available to drought-affected small businesses that can demonstrate that they have a direct dependency on farmers through providing primary production-related goods and services to farm businesses currently in drought.
Regional Investment Corporation (Drought Loans Expansion) Rule 2020	Yes	Expands the eligibility criteria for drought loans to include farm businesses located outside the affected area defined by the United Nations Convention to Combat Desertification.

(continued)

Table 11.2 (continued)

Legislation	In Force	Objective(s)
Regional Investment Corporation (Drought and Small Business Drought Loans) (Cessation of Interest-Free Period) Amendment Instrument 2020	No	Farmers and farm-related small businesses had until 30 September 2020 to take advantage of 2 years interest-free from the RIC before the drought-related loan terms change.)
Future Drought Fund (Drought Resilience Funding Plan 2020 to 2024) Determination 2020	Yes	The Drought Resilience Funding plan (the plan) sets out an approach for making arrangements or grants about drought resilience or entering into agreements about such grants.

Source: Adapted from Federal Register of Legislation (Australian Government, 2023)

References

ABC News. (2020). *NSW Government announce new $310 million drought support package as farmers struggle 'for at least another year'*. Australian Broadcasting Corporation. Accessed at: https://www.abc.net.au/news/rural/2020-07-01/restockers-farmers-rejoice-new-nsw-drought-support-package/12406672

ACANZ. (1992). *Record and resolutions: 138th meeting, Mackay 24 July 1992*. Australian Government.

Anderson, J., the Hon MP. (1998). Farm management deposits scheme even better for farmers. Media Release by Minister for Primary Industries and Energy DPIE98/44A, 17 April 1998.

Ault, T. (2020). On the essentials of drought in a changing climate. *Science, 368*(4688), 256–260.

Austin, E. K., Handley, T., Kiem, A. S., Rich, J. L., Lewin, T. J., Askland, H. H., et al. (2018). Drought-related stress among farmers: Findings from the Australian rural mental health study. *Medical Journal of Australia, 209*(4), 159–165.

Australian Government. (2023). *Federal register of legislation*. Australian Government. Accessed at: https://www.legislation.gov.au (Advanced Search).

Beard, G., Chandler, E., Watkins, A. & Jones, D. (2011). How does the 2010–11 La Niña compare with past La Niña events. Bulletin of the Australian Meteorological and Oceanographic Society, 24, 17–20.

BoM. (2023a). *Previous droughts*. Bureau of Meteorology. Accessed at: http://www.bom.gov.au/climate/drought/knowledge-centre/previous-droughts.shtml#:~:text=The%202019%20cool%20season%20was,Basin%20and%20New%20South%20Wales.

BoM. (2023b). *123 years of Australian rainfall*. Bureau of Meteorology. Accessed at: http://www.bom.gov.au/climate/history/rainfall/.

Botterill, L. C. (2000). Government responses to farm poverty: The policy development process. *Rural Society, 10*(1), 15–27.

Botterill, L. C. (2003). Uncertain climate: The recent history of drought policy in Australia. *Australian Journal of Politics & History, 49*(1), 61–74.

Botterill, L. C. (2007). Responding to farm poverty in Australia. *Australian Journal of Political Science, 42*(1), 33–46.

Botterill, L. C. (2013). Are policy entrepreneurs really decisive in achieving policy change? Drought policy in the USA and Australia. *Australian Journal of Politics & History, 59*(1), 97–112.

Botterill, L. C. (2021). From national drought policy to national drought agreement: The evolution of Australia's drought response. *The Rangeland Journal, 43*(3), 77–86.

Botterill, L. C., & Chapman, B. (2009a). A revenue contingent loan instrument for agricultural credit with particular reference to drought relief. *Australian Journal of Labour Economics, 12*(2), 181–196.

Botterill, L. C., & Chapman, B. (2009b). *Income contingent loan for drought relief: Delivering better outcomes for farmers and taxpayers*. Centre for Economic Policy Research, Australian National University.

Botterill, L. C., & Fenna, A. (2019). *Interrogating public policy theory: A political values perspective* (pp. 1–11). Edward Elgar Publishing.

Botterill, L. C., Chapman, B., & Kelly, S. (2017). Revisiting revenue contingent loans for drought relief: Government as risk manager. *Australian Journal of Agricultural and Resource Economics, 61*(3), 367–384.

Butler, J. R. G. (1979). *The economics of natural disaster relief in Australia*. Centre for Research on Federal Financial Relations, ANU Press.

Chapman, B. (2006). *Government managing risk: Income contingent loans for social and economic progress* (Vol. 40, p. 246). Routledge.

Chapman, B., & Botterill, L. C. (2002). *Developing equitable and affordable government responses to drought in Australia* (Discussion Paper No. 455). Centre for Economic Policy Research, Australian National University.

Chapman, B., Botterill, L., & Egan, M. (2004). Income related loans for drought relief Centre for Economic Policy Research discussion paper, *Discussion Paper No. 472*. Accessed at: https://ideas.repec.org/p/auu/dpaper/472.html

DAFF. (2013). *Intergovernmental agreement on National Drought Program Reform*. Department of Agriculture, Fisheries and Forestry, Australian Government. Accessed at: https://www.agriculture.gov.au/sites/default/files/sitecollectiondocuments/agriculture-food/drought/drought-program-reform/iga.pdf

DAFF. (2019). *Internal review into the 2016 to 2020 Rural Financial Counselling Service Program*. Department of Agriculture, Fisheries and Forestry. Accessed at: https://www.agriculture.gov.au/sites/default/files/sitecollectiondocuments/ag-food/drought/rfcs/internal-review-2016-2020-rfcs-program.pdf

DAFF. (2021). *History of drought policy*. Department of Agriculture, Fisheries and Forestry, Australian Government. Accessed at: https://www.agriculture.gov.au/agriculture-land/farm-food-drought/drought/drought-policy/history#daff-page-main

DAFF. (2022). *Farm management deposits*. Department of Agriculture, Fisheries and Forestry, Australian Government. Accessed at: https://www.agriculture.gov.au/agriculture-land/farm-food-drought/drought/assistance/fmd

DAFF. (2024). *Historical farm management deposits statistics*. Department of Agriculture, Fisheries and Forestry, Australian Government. Accessed at: https://www.agriculture.gov.au/agriculture-land/farm-food-drought/drought/fmd/historical-fmd-statistics

DAWE. (2021). *Farm management deposits scheme: 2021 evaluation*. Department of Agriculture, Water and the Environment, Australian Government. Accessed at: https://www.agriculture.gov.au/sites/default/files/documents/fmds-evaluation-2021-report_0.pdf

Douglas, R., & Davenport, S. (1993). *A review of the income equalisation deposit scheme*. New South Wales Agriculture. Accessed at: https://catalogue.nla.gov.au/catalog/1852897

Drought Policy Review Task Force. (1990). *National Drought Policy*. National Library of Australia. Accessed at: https://catalogue.nla.gov.au/Record/1924829

Edwards, B., Gray, M., & Hunter, B. (2019). The social and economic impacts of drought. *Australian Journal of Social Issues, 54*(1), 22–31.

Evans, A., Jones, D., Smalley, R., & Lellyett, S. (2020). *An enhanced gridded rainfall analysis scheme for Australia*. Bureau of Meteorology.

Expert Social Panel. (2008). *It's about people: Changing perspectives on dryness, an assessment of the social impacts associated with drought*. Australian Government. Accessed at: https://www.agriculture.gov.au/sites/default/files/sitecollectiondocuments/ag-food/drought/publications/dryness_report.pdf

Fleming-Muñoz, D., Whitten, S., & Bonnett, G. (2023). The economics of drought: A review of impacts and costs. *Australian Journal of Agricultural and Resource Economics, 67*(4), 501–523.

Fragar, L., Stain, H. J., Perkins, D., Kelly, B., Fuller, J., Coleman, C., et al. (2010). Distress among rural residents: Does employment and occupation make a difference? *Australian Journal of Rural Health, 18*(1), 25–31.

Freebairn, J. W. (1983). Drought assistance policy. *Australian Journal of Agricultural Economics, 27*(3), 185–199.

Frost, A. J., Ramchurn, A., & Smith, A. (2016). *The bureau's operational AWRA landscape (AWRA-L) model. Bureau of Meteorology Technical Report.* Bureau of Meteorology.

Frost, A. J., Hendon, H., Lellyett, S. C., & Griffiths, M. (2018). *Sydney water ACCESS-S downscaled seasonal forecast/AWRA-L project.* Bureau of Meteorology.

Glau, T. E. (1970). The commonwealth drought bond. *Australian Journal of Agricultural Economics, 14*(2), 121–130.

Grafton, R. Q., & Wheeler, S. A. (2018). Economics of water recovery in the Murray-Darling Basin, Australia. *Annual Review of Resource Economics, 10*, 487–510.

Grain Central. (2018). *NSW offers freight subsidies as drought's grip tightens.* Grain Central. Accessed at: https://www.graincentral.com/news/nsw-offers-freight-subsidies-as-droughts-grip-tightens/

Griffiths, M., Shi, L., & Hendon, H. (2017). *Preliminary evaluation of ACCESS-S calibrated predictions for the Sydney water ACCESS-S downscaling/AWRA-L project.* Bureau of Meteorology.

Hanigan, I., Butler, C., Kokic, P., & Hutchinson, M. (2012). Suicide and drought in new South Wales, Australia, 1970–2007. *Proceedings of the National Academy of Sciences, 109*(35), 13950–13955.

Harris, S. (1974). *The principles of rural policy in Australia: A discussion paper (No. 92).* Australian Government Publishing Service.

Hennessy, K. J., Fawcett, R., Kirono, D., Mpelasoka, F., Jones, D., Bathols, J., et al. (2008). *An assessment of the impact of climate change on the nature and frequency of exceptional climatic events.* CSIRO and Bureau of Meteorology.

Hudson, D., Alves, O., Hendon, H., Lim, E., Liu, G., Luo, J. J., et al. (2017). ACCESS-S1 the new Bureau of Meteorology multi-week to seasonal prediction system. *Journal of Southern Hemisphere Earth Systems Science, 67*(3), 132–159.

IPCC. (2021a). Climate change 2021: The physical science basis. In: Masson-Delmotte, V., P. Zhai, A. Pirani, S.L. Connors, C. Péan, S. Berger, … & B. Zhou (eds.) *Contribution of Working Group I to the sixth assessment report of the Intergovernmental Panel on Climate Change.* Cambridge University Press. In Press.

IPCC. (2021b). Summary for policy makers. In climate change 2021: The physical science basis. Contribution of working group I to the sixth assessment report of the IPCC. In V. Masson-Delmotte, P. Zhai, A. Pirani, S. L. Connors, C. Péan, S. Berger, N. Caud, Y. Chen, L. Goldfarb, M. Gomis, et al. (Eds.), *Intergovernmental panel on climate.*

Kerin, J. (the Hon MP) (1989). Task force to review drought policies. Media Release by Minister for Primary Industries and Energy PIE 89(102K), 12.

Kiem, A. S., & Austin, E. K. (2013). Drought and the future of rural communities: Opportunities and challenges for climate change adaptation in regional Victoria, Australia. *Global Environmental Change, 23*(5), 1307–1316.

King, A. D., Pitman, A. J., Henley, B. J., Ukkola, A. M., & Brown, J. R. (2020). The role of climate variability in Australian drought. *Nature Climate Change, 10*(3), 177–179.

Kingwell, R. S., & Xayavong, V. (2017). How drought affects the financial characteristics of Australian farm businesses. *Australian Journal of Agricultural and Resource Economics, 61*(3), 344–366.

Laughlin, G., & Clark, A. (2000). *Drought science and drought policy in Australia: A risk management perspective.* In proceedings of the world meteorological organisation expert group meeting on early warning Systems for Drought Preparedness and Drought Management, Lisbon, Portugal, 5–7 September 2000; pp. 32–44.

Lellyett, S. C., Truelove, R. N., & Huda, A. K. (2022). Improving early warning of drought in Australia. *Climate, 10*(7), 91.

Lewis, S. C., & Karoly, D. J. (2013). Anthropogenic contributions to Australia's record summer temperatures of 2013. *Geophysical Research Letters, 40*(14), 3705–3709.

Litchfield, F., Weragoda, A., Coelli, R., & Ashton, D. (2021). *Farm management deposits: Analysis of survey results*. ABARES research report 21.06. Australian Bureau of Agricultural and Resource Economics.

Luo, J., Nguyen, H., Hendon, H., Alves, O., Liu, G., Dunstone, N., & Mac Lachlan, C. (2018). *Assessment of multi-year climate predictions for potential use by Sydney water*. Bureau of Meteorology.

McColl, J., & Young, M. (2005). *Managing Change: Australian structural adjustment lessons for water*. Natural Resource Management Economics, Policy and Economic Research Unit, CSIRO Land and Water.

McColl, J., Donald, R., & Shearer, C. (1997). Rural Adjustment: Managing Change. In *Mid-term review of the rural adjustment scheme. Australian government*.

Menzies, R (the Hon MP) (1965). Drought Assistance: Ministerial Statement to the Commonwealth Parliamentary Debates, by the Rt Hon. Sir Robert Menzies, MP, House of Representatives, 7 December 1965, p.3647.

NACC. (2023). Resources for drought resilience plan in the mid-west. Northern Agricultural Catchments Council. Accessed at: https://www.nacc.com.au/resources-for-drought-resilience-plan-in-the-mid-west/.

Nauges, C., Wheeler, S., & Zuo, A. (2016). Elicitation of irrigators' risk preferences from observed behaviour. *Australian Journal of Agricultural and Resource Economics, 60*(3), 442–458.

NDA. (2018). National Drought Agreement. Department of Agriculture. Accessed at: https://www.agriculture.gov.au/agriculture-land/farm-food-drought/drought/drought-policy/national-drought-agreement.

NDA. (2022). Review of the National Drought Agreement. Department of Agriculture. Accessed at: https://www.agriculture.gov.au/sites/default/files/documents/nda-review-report.pdf.

NDRRA. (2023). Natural disaster relief and recovery arrangements. Australian Government. Accessed at: https://www.disasterassist.gov.au/disaster-arrangements/natural-disaster-relief-and-recovery-arrangements.

Nelson, R., Howden, M., & Smith, M. S. (2008). Using adaptive governance to rethink the way science supports Australian drought policy. *Environmental Science & Policy, 11*(7), 588–601.

Nguyen, H., Zhang, H.-Q., Hendon, H., & Lellyett, S. (2017). *ACCESS-S1 seasonal forecast skill assessment for evapotranspiration and other hydro-climate variables over Sydney region. Bureau Research Report*. Bureau of Meteorology.

Nguyen, H., Wheeler, M. C., Otkin, J. A., Cowan, T., Frost, A., & Stone, R. (2019). Using the evaporative stress index to monitor flash drought in Australia. *Environmental Research Letters, 14*(6), 064016.

O'Meagher, B., Stafford-Smith, M., & White, D. H. (2000). Approaches to integrated drought risk management: Australia's national drought policy. In D. A. Wilhite (Ed.), *Drought: A global assessment* (Vol. 2, pp. 115–128). Routledge.

Ockwell, A. (1990). The economic structure of Australian agriculture. In D. B. Williams (Ed.), *Agriculture in the Australian economy* (3rd ed., pp. 27–49). Sydney University Press.

Peterson, D. C. (2016). Drought policy: Lessons and strategies to change the transformational game. In M. Young & C. Esau (Eds.), *Transformational change in environmental and natural resource management* (pp. 134–149). Routledge.

Productivity Commission. (2006). Industry commission submission to the mid-term review of the rural adjustment scheme. Productivity Commission. Accessed at: https://www.pc.gov.au/research/supporting/rural-adjustment-scheme/rural-adjustment-scheme.pdf.

Productivity Commission. (2009). Government drought support. *Report No. 46, Final Inquiry Report*. Productivity Commission. Accessed at: https://www.pc.gov.au/inquiries/completed/drought/report/drought-support.pdf

Rickards, L. (2012). Critical breaking point: The effects of climate variability, climate change and other pressures on farm households. *Report for the Birchip Cropping Group and the Sustainable Agriculture Initiative Platform Australia*.

Samnakay, N. (2023). Evaluating Australian drought policy from the perspective of good-practice strategic policymaking. *Australasian Journal of Water Resources, 27*(1), 149–159.

Scott, B., the Hon MP (1998). Taxation Laws amendment (farm management deposits), Bill 1998. House of Representatives Hansard, p.4066, 28 May 1998.

Seidl, C., Wheeler, S. A., & Zuo, A. (2021). The drivers associated with Murray-Darling basin irrigators' future farm adaptation strategies. *Journal of Rural Studies, 83*, 187–200.

Sheen, K. L., Smith, D. M., Dunstone, N. J., Eade, R., Rowell, D. P., & Vellinga, M. (2017). Skilful prediction of Sahel summer rainfall on inter-annual and multi-year timescales. *Nature Communications, 8*(1), 14966.

Stone, R. C. (2014). Constructing a framework for national drought policy: The way forward–the way Australia developed and implemented the national drought policy. *Weather and Climate Extremes, 3*, 117–125.

Thacher, D., & Rein, M. (2004). Managing value conflict in public policy. Governance, 17(4), 457–486.

UNISDR. (2009). Drought risk reduction framework and practices: Contributing to the implementation of the Hyogo framework for action. In *United Nations secretariat of the international strategy for disaster reduction (UNISDR)*.

Van Dijk, A., Beck, H., Crosbie, R., De Jeu, R., Liu, Y., Podger, G., et al. (2013). The millennium drought in Southeast Australia (2001–2009): Natural and human causes and implications for water resources, ecosystems, economy, and society. *Water Resources Research, 49*(2), 1040–1057.

Van Dijk, A., & Renzullo, L. J. (2009). Evaluation of alternative model-data fusion approaches for retrospective water balance estimation. 18th World IMACS Congress and MODSIM 2009 Proceedings.

Wang, G., Hudson, D., Yin, Y., Alves, O., Hendon, H., Langford, S., et al. (2011). POAMA-2 SST skill assessment and beyond. *CAWCR Research Letters, 6*, 40–46.

Wheeler, S. (2022). Debunking Murray-Darling basin water trade myths. *Australian Journal of Agricultural and Resource Economics, 66*(4), 797–821.

Wheeler, S. (2023). Informal and formal markets in meeting water needs. *Nature Sustainability, 6*(11), 1291–1292.

Wheeler, S., & Marning, A. (2019). Turning water into wine: Exploring water security perceptions and adaptation behaviour amongst conventional, organic and biodynamic grape growers. *Land Use Policy, 82*, 528–537.

Wheeler, S., Zuo, A., & Bjornlund, H. (2013). Farmers' climate change beliefs and adaptation strategies for a water scarce future in Australia. *Global Environmental Change*, 23(2), 537–547.

Wheeler, S., & Zuo, A. (2017). The impact of drought and water scarcity on irrigator farm exit intentions in the Murray–Darling Basin. *Australian Journal of Agricultural and Resource Economics, 61*(3), 404–421.

Wheeler, S., Zuo, A., & Loch, A. (2015). Watering the farm: Comparing organic and conventional irrigation water use in the Murray–Darling Basin, Australia. *Ecological Economics, 112*, 78–85.

Wheeler, S., Zuo, A., Xu, Y., & Grafton, Q. (2019). Emergency drought relief package—Health and resilience services: An evidence check rapid review brokered by the sax institute (www.saxinstitute.org.au) for the NSW Ministry of Health.

Wheeler, S., Carmody, E., Grafton, Q., Kingsford, R., & Zuo, A. (2020a). The rebound effect on water extraction from subsidising irrigation infrastructure in Australia. *Resources, Conservation and Recycling, 159*, 104755.

Wheeler, S., Xu, Y., & Zuo, A. (2020b). Modelling the climate, water and socio-economic drivers of farmer exit in the Murray-Darling basin. *Climatic Change, 158*(3), 551–574.

Wheeler, S., Nauges, C., & Zuo, A. (2021). How stable are Australian farmers' climate change risk perceptions? New evidence of the feedback loop between risk perceptions and behaviour. *Global Environmental Change, 68*, 1–14.

Wilhite, D. A. (2005). Drought policy and preparedness: The Australian experience in an international context. In L. C. Botterill & D. A. Wilhite (Eds.), *From disaster response to risk management: Australia's national drought policy* (pp. 157–176). Springer.

Wilhite, D. A., & Easterling, W. E. (1987). Drought policy: Toward a plan of action. In D. A. Wilhite, W. E. Easterling, & D. A. Wood (Eds.), *Planning for drought: Toward a reduction of societal vulnerability* (pp. 573–583). Westview Press.

Wilhite, D. A., & Glantz, M. H. (1985). Understanding: The drought phenomenon: The role of definitions. *Water International, 10*(3), 111–120.

Wilhite, D. A., Diodato, D., Jacobs, K., Palmer, R., Raucher, B., et al. (2006). Managing drought: a roadmap for change in the United States. In *A conference report from managing drought and water scarcity in vulnerable environments* (pp. 18–20).

Wittwer, G., & Young, M. D. (2020). *Distinguishing between policy, drought, and international events in the context of the Murray-Darling basin plan (No. g-295).* Victoria University, Centre of Policy Studies/IMPACT Centre.

WMO. (2016). Handbook of drought indicators and indices. In M. Svoboda & B. A. Fuchs (Eds.), *Integrated Drought Management Programme (IDMP), integrated drought management tools and guidelines series 2.* World Meteorological Organisation Global Water Partnership (GWP).

Xu, Y., Wheeler, S., & Zuo, A. (2023). *Drought and hotter temperature impacts on suicide: Evidence from the Murray-Darling Basin, Australia.* Climate Change Economics, 2350024.

Yazd, S. D., Wheeler, S. A., & Zuo, A. (2020). Understanding the impacts of water scarcity and socio-economic demographics on farmer mental health in the Murray-Darling Basin. *Ecological Economics, 169*, 106564.

Zhao, M., Zhang, H. Q., & Dharssi, I. (2017). *Impact of land-surface initialization on ACCESS-S1 and comparison with POAMA, Research Report No. 023.* Bureau of Meteorology.

Zuo, A., Wheeler, S. A., Boxall, P., Adamowicz, W. L., & Hatton MacDonald, D. (2015). Identifying water prices at which Australian farmers will exit irrigation: Results of a stated preference survey. *The Economic Record, 91*, 109–123.

Part III
Improving Drought Management in Chile: Policy Proposals and Adaptation Options

Chapter 12
General Assessment of Drought Management and Proposals

Guillermo Donoso, Daniela Rivera, and Cristian Jordán

Abstract Current drought management in Chile is fragmented, with various public and private institutions addressing drought issues at multiple levels, often without coordination or a clear strategy. Chile's response to drought is primarily reactive, focusing on damage mitigation. An effective strategy is essential to decentralize drought management, engage affected communities, enhance local capacities, and transition from crisis response to preventative risk management. This chapter systematizes the main gaps, challenges and problems that currently exist in drought management in Chile, from a legal, institutional, economic, and environmental point of view. Subsequently, we evaluate the performance of drought management in Chile by examining six policy dimensions using a qualitative methodology that incorporates various information sources, including empirical data and stakeholder insights. This assessment reviews drought management through the lens of risk management, pinpointing crucial policy deficiencies. Finally, a set of public policy recommendations to improve drought management in Chile is proposed.

Keywords Drought · Drought management performance · Policy proposals · Chile

---

G. Donoso (✉)
Faculty of Agronomy and Natural Systems, Pontificia Universidad Católica de Chile, Santiago, Chile

Center for Water Law and Management, Pontificia Universidad Católica de Chile, Santiago, Chile
e-mail: gdonosoh@uc.cl

D. Rivera
Center for Water Law and Management, Pontificia Universidad Católica de Chile, Santiago, Chile

Faculty of Law, Pontificia Universidad Católica de Chile, Santiago, Chile

C. Jordán
Faculty of Agronomy and Natural Systems, Pontificia Universidad Católica de Chile, Santiago, Chile

D. Rivera, G. Donoso (eds.), *Droughts in Chile*, Global Issues in Water Policy 31,
https://doi.org/10.1007/978-3-031-85040-0_12

12.1 Introduction

Despite the frequent occurrence of drought and its significant impact on societies and development, it has traditionally been regarded globally as an abnormal event requiring emergency responses typical of disaster or catastrophe management. Thus, drought has been treated with reactive policies and instruments (Wilhite & Pulwarty, 2005). This reactive approach addresses the impacts of drought once these have occurred. They do not tackle the underlying causes of vulnerabilities associated with these impacts (Wilhite et al., 2014). The increasing frequency of droughts and their significant impacts underscore the inadequacies of the current crisis-based approach to drought management. A proactive approach is thus needed to prevent, mitigate, and recover from droughts, and to better prepare for future events (Hayes et al., 2004; Wilhite et al., 2000, 2014).

Some countries have initiated efforts to shift this paradigm from crisis or emergency management to risk management (Wilhite et al., 2000, 2005; Neuman, 2003; Wilhite, 2009; Brufao, 2012; Courtenay & Hayes, 2012; Fu et al., 2013; Paneque, 2015; Stein et al., 2016; Embid, 2018; Setuaín, 2018; Berbel & Esteban, 2019; Hervás-Gámez & Delgado-Ramos, 2019; Iglesias et al., 2019; Martín, 2019–2020, 2022, 2024; Arana & Miranzo, 2021). Various authors have noted that the crisis management paradigm has been characterized by a lack of effectiveness, efficiency, coordination, and timeliness. This has resulted in limited progress in mitigating the risks associated with droughts (Wilhite et al., 2000, 2005; Neuman, 2003; Paneque, 2015; Martín, 2022).

For example, Spain's progress with mandatory drought plans for river basins and urban supplies but highlight challenges in establishing effective preventive measures (Vargas & Paneque, 2025—Chap. 9). Jones (2025) in Chap. 10 points out that California's emergency response approach is inadequate for long-term climate aridification. Similarly, Australia's current drought policies need to prioritize risk management to lower vulnerability (Zuo & Wheeler, 2025—Chap. 11).

To evaluate the development and challenges of drought policy in South America, an analytical framework adapted from WMO and GWP (2014) is employed in Urquijo et al. (2025) in Chap. 8. This framework consists of three pillars for proactive drought management: (1) monitoring, forecasting, and early warning systems; (2) vulnerability, risk, and impact analysis; and (3) mitigation and response strategies (Fig. 12.1). The authors identified multiple initiatives and experiences that aim to promote a proactive approach to drought management in the South American region. Despite these advancements in establishing proactive drought management policies, the efforts remain fragmented and insufficient, leading to inefficient management of drought impacts and vulnerabilities.

This fragmented approach is also the case in Chile. Rivera and Donoso (2025) in Chap. 7 identified numerous public and private entities that have a role in drought management across different levels. The main features observed are fragmentation, dispersion, disorder, and a lack of coordination and unified plan. To enhance drought policy in Chile, Sect. 12.2 of this chapter presents an overview of the performance

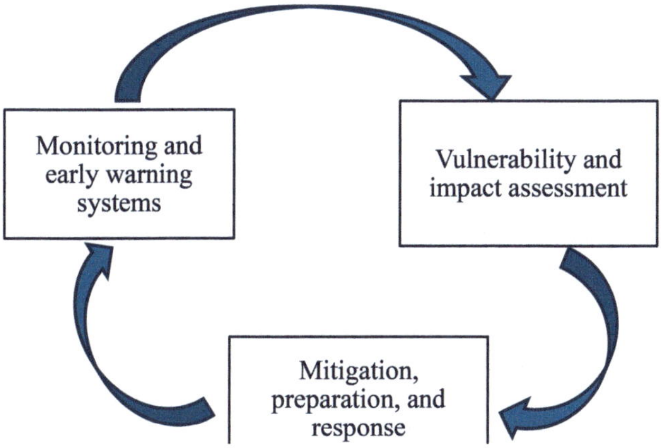

Fig. 12.1 Pillars for proactive drought management. (WMO and GWP, 2014)

assessment methodology employed. Subsequently, Sect. 12.3 presents the performance assessment of Chile's drought policy. Based on these results, Sect. 12.4 sets forth a set of public policy recommendations to improve drought management in Chile. Finally, Sect. 12.5 concludes the chapter with some key messages.

12.2 Methodology

The methodology contemplated two main activities:

1. Identification and evaluation of gaps and challenges of the current state of public policies on drought.
2. Drought management assessment.

12.2.1 Identification and Evaluation of Gaps and Challenges of the Current State of Public Policies on Drought

We utilize the analytical framework presented in Chap. 8 (Urquijo et al., 2025) to examine Chile's progress on the three pillars of proactive drought management (Table 12.1). This analysis is based on data collected from government agencies through semi-structured interviews.

A set of questions was developed, corresponding to the three pillars. These were compiled into a questionnaire that guided the interview. The questionnaire was reviewed and approved by the Scientific Ethics Committee of Social Sciences, Arts and Humanities of the Pontificia Universidad Católica de Chile.

Table 12.1 Questionnaire

Pillar	Questions
Pillar 1	Does your institution generate information that is used in drought monitoring/warning (hydrological and agricultural)?
	Does the information system implemented, or the information generated by your organization interact or link with another information system?
	Does the information generated comply with any rules, regulations or protocols?
	Does your institution carry out any action to disseminate information generated by other stakeholders?
	Does your institution elaborate or issue any type of warning/recommendation to authorities or the public in relation to droughts?
	Does your institution integrate information generated by other institutions into its analysis?
Pillar 2	Does your institution have instruments/programs to deal with future droughts?
	Does your institution generate studies, research and/or scientific evidence that contribute to the generation of information regarding the impact, management and long-term adaptation to droughts or scarcity of water resources?
Pillar 3	Does your institution have a mandate to implement drought management measures?
	Does your institution develop or is it part of the development of drought response plans?
	Does your institution have a strategy or plan that includes drought or climate change in its formulation?
	Does your institution implement measures to mitigate the effects of droughts?
	Does your institution implement support measures (economic, technical, etc.) to mitigate the effects of droughts?
	Does your institution coordinate with other public or private entities for the implementation of drought support measures?
	Is there a formally regulated procedure for the application of the aforementioned tasks?

Authors own elaboration

We conducted 23 interviews with institutions involved in drought management, 17 public and 6 private, some of which were identified in Chap. 7. Based on the results of the analysis of the collected data, each pillar was qualitatively rated on a high (3)-medium (2)-low (1) scale, where 'High' corresponds to a good and complete development of the pillar, 'Medium' value corresponds to a partially incorporated or incomplete development of the pillar, and 'Low' value corresponds to limited or insufficient development of the pillar.

In order to compare our results for Chile with those reported in Chap. 8 (Urquijo et al., 2025) for other South American countries, we followed their methodology and aggregated the values of the 3 pillars into an equal-weighted index as a proxy to show the degree of integrated and proactive approach development at the country level.

Table 12.2 Description of the six dimensions of the analysis framework

Pillars	Dimension	Description	
Pillar 1	1	Drought definition	How drought is defined in official documents.
		Drought monitoring	Drought Monitor system characteristics, including drought state surveillance and maps and forecasting activities.
Pillar 2	2	Evaluation and learning	Review of the practices and actions carried out, identifying gaps and failures that should be corrected to improve drought management.
Pillar 3	3	Institutional context	Scenario for policy implementation, which is found in strategic plans, programs, normal or exceptional laws, etc.
	4	Drought measures	Actions implemented to reduce the effects of drought, which should be planned in advance, including strategic, management and recovery measures.
	5	Resources	Availability of financial and human resources to deal with the drought.
	6	Coordination and participation	Articulation between institutions and participation of stakeholders and civil society in drought decision-making.

Author's own elaboration based on Urquijo et al. (2017) and Chap. 8

12.2.2 Drought management Assessment

We assess drought management through the analysis of the six key policy dimensions proposed by Urquijo et al. (2017) using a qualitative approach that combines both factual data and stakeholders' perceptions. The conceptual model comprises six dimensions of analysis that align with policy cycle stages. The model's dimensions are presented in Table 12.2.

This analysis is based on survey data from government and private agencies. Following Urquijo et al. (2017), the six dimensions were defined by 20 criteria (see Table 12.3), forming the basis of the survey questions. The Scientific Ethics Committee of Pontificia Universidad Católica de Chile reviewed and approved the survey, which was completed by 22 institutions.

Based on the results of the analysis of the collected data, each criterion was qualitatively rated on a high (3)-medium (2)-low (1) scale, where 'High' corresponds to a good and complete development of the criteria, 'Medium' value corresponds to a partially incorporated or incomplete development of the criteria, and 'Low' value corresponds to limited or insufficient development of the criteria. We then determined the fuzzy set values for each dimension.

Table 12.3 Assessment criteria for each dimension

Dimension	Criteria
Drought definition	Existence of an official definition of drought
	Presence of operational definition of drought
	Degree of practical implications of drought definitions in management practices
	Extent to which stakeholders (stakeholders) understand the concept of drought
Evaluation and learning	Degree of improvement in drought management compared to previous droughts
	Scope of assessment practices
Institutional context	Suitability of the regulatory framework
	Level of development of the drought policy
	Existence of an assessment of vulnerability to drought
Drought measures	Presence of different types of instruments (strategic, management and recovery)
	Perception of the implementation of the measures (stakeholders)
	Degree of contribution of the measures to mitigate impacts
Resources	Availability of financial resources for drought management
	Adequate level of financial resources
	Degree of balance in the distribution of resources between sectors and/or needs
	Degree of adequacy of human resources responsible for drought management
Coordination and participation	Existence of an institution responsible for drought management
	Adequacy of stakeholder participation (during) droughts
	Degree of formalization of mechanism/processes to allow participation
	Degree of inclusion and balance of representation

Author's own elaboration based on Urquijo et al. (2017)

12.3 Results

12.3.1 Identification and Evaluation of Gaps and Challenges of the Current State of Public Policies on Drought

Chile's progress on the three pillars of proactive drought management is presented in Fig. 12.2 This analysis is based on data collected from government agencies through semi-structured interviews. The three pillars show varying development levels: Pillar 3 (mitigation and response) is the most advanced, followed by Pillar 1 (drought monitoring and early warning). Pillar 2 (vulnerability, risk, and impact assessment) remains undeveloped.

As can be seen from Fig. 12.2, Pillar 1 is assessed as "Low" since drought monitoring, forecast, and early warning systems for drought monitoring are at an embryonic stage of development. The interview results show that only 22% of the

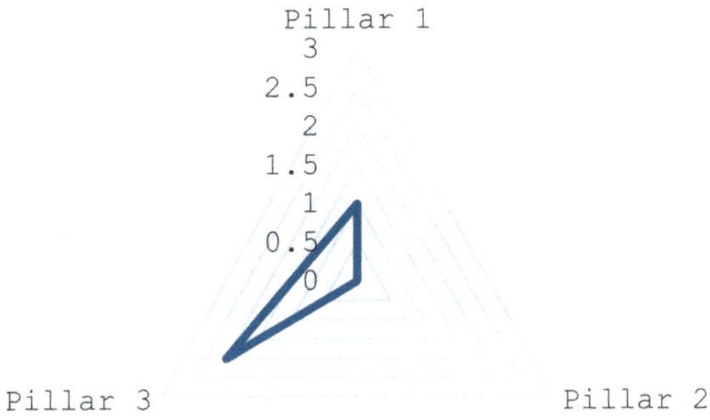

Fig. 12.2 Development among the three pillars in Chile

institutions generate and use information. It is important to point out that, with few exceptions, drought information is not shared between institutions and, thus, there is no integrated drought information system. There is insufficient dissemination of drought information to other stakeholders, partially due to certain distrust among stakeholders and lack of fluidity in the exchange of information among them.

There is no clear definition of drought and, thus, actions are undertaken with multiple drought indices with different levels of severity for the classification of drought. Of the surveyed institutions, 4% elaborate a working definition of drought.

There is no official monitoring system or one managed by the DGA, the central water authority. Additionally, there is no early warning system.

Assessing the social and economic impacts of drought is crucial for informed decision-making and the development and implementation of proactive plans. However, as illustrated in Fig. 12.2, Pillar 2 (vulnerability and impact assessment) is absent. This absence is attributed to the fact that no institution has the mandate to conduct assessments on drought vulnerability and impacts. Only a few studies have been conducted by academic and research institutions.

Pillar 3 presents a partial development and thus is rated as "medium". Drought management in Chile involves many fragmented and dispersed actors, both public and private, who operate at different scales with little coordination (see Chap. 7). Additionally, there are numerous public policies and actions with varying approaches and objectives, but without a unified roadmap.

As Fig. 12.3 shows, the primary drought policy instruments are concentrated in two institutions: MOP (Ministry of Public Works), and MINAGRI (Ministry of Agriculture). It is important to point out that these instruments are mostly reactive, as their objective is to mitigate the effects of different types of droughts. The exception is Agricultural insurance, a meteorological based insurance, which is the only preventive policy instrument.

The joint analysis of the three pillars at country level indicates that Chile presents an overall low score. This highlights Chile's reliance on reactive measures typical of disaster drought management rather than a proactive approach. Thus, Chile aligns with Uruguay and Venezuela's reactive drought management approach, as they show low scores on all pillars (Urquijo et al., 2025—Chap. 8). As Chap. 8 shows, only five South American countries (Argentina, Bolivia, Ecuador, Guyana, and

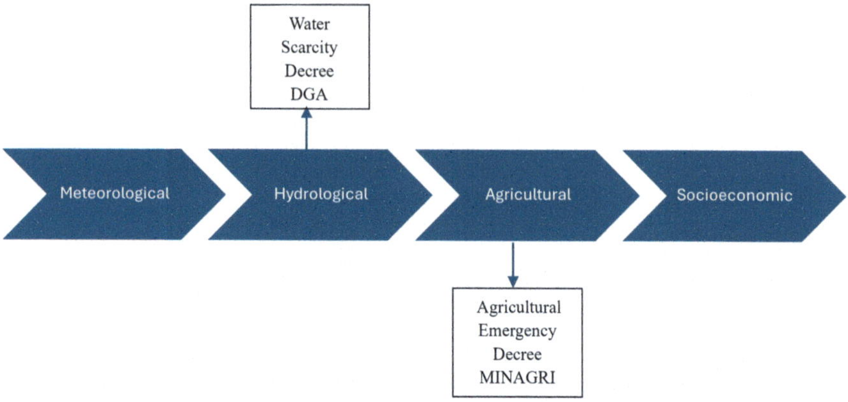

Fig. 12.3 Main drought measures implemented in Chile. (Author's own elaboration based on Chap. 7)

Fig. 12.4 Drought management assessment

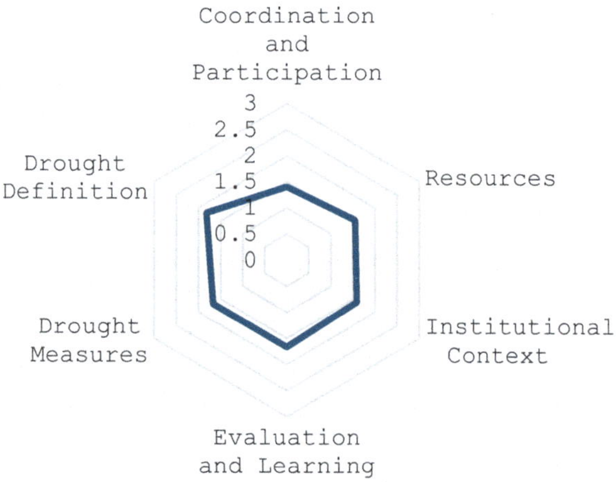

Peru) have a high-medium drought management level. Brazil, Colombia, and Paraguay show moderate progress towards proactive drought management. Consequently, Chile is among the two countries with the lowest rankings due to their limited progress on the three pillars and a predominantly reactive approach to drought management.

12.3.2 Drought Management Assessment

The assessment depicted in Fig. 12.4 highlights Chile's strengths and weaknesses in drought governance and management, with higher scores indicating a more robust risk-management approach. The shape and size of the profile indicate that Chile has adopted a crisis-management strategy, rather than a risk management approach. The dimensions reveal an overall low to medium level of development, with minimal variation. The development scores for the dimensions range from 1.41 to 1.82. Dimension 1 (Drought Definition) is the most advanced, followed by Dimensions 2 (Evaluation and Learning) and 4 (Drought Measures). Dimensions 3 (Institutional Context) and 5 (Resources) exhibit intermediate levels of progress with respect to the other dimensions. Dimension 6 (Coordination and Participation) shows the lowest level of development.

Dimension 1, which pertains to the definition of drought, received the highest rating, scoring 1.82 on a scale of 1–3. This low to medium rating is consistent with the previous assessment of the three pillars for an effective proactive drought management. Thus, despite our assessment indicating no clear definition of drought, the surveyed government and private agencies acknowledge its presence. This could partially be explained by the fact that "drought" is expressly mentioned four times in the Water Code and, and a DGA (General Directorate of Water) resolution establishes the criteria to qualify a drought as severe, which makes it possible to declare a water shortage zone (Rivera & Donoso, 2025—Chap. 7), which could be considered sufficiently developed. The score indicates that the absence of a precise operational definition of drought has impeded effective drought management in Chile. Specifically, there is a lack of clarity in distinguishing between drought and long-term water scarcity. In addition, the Water Code only contemplates one univocal conception of drought: severe drought. There is no gradation of different states of drought, which would allow distinguishing and applying different measures, according to the intensity or severity of the phenomenon.

Drought measures, Dimension 4, is the second highest, scoring 1.67. Thus, its performance has been low to medium. This is due to the multiplicity of public policies and actions, with different approaches and objectives, without a clear and common roadmap. Additionally, most of these instruments have been implemented in response to drought episodes that have already impacted society. Many measures involve multiple administrative authorities or agencies, emphasizing the need for better coordination to prevent duplication and inconsistencies. Thus, the score of this dimension reflects the dispersion of efforts and resources, resulting in the effects of actions taken against droughts being either diluted or not easily perceptible.

The valuation of Dimension 2—Evaluation and Learning—is noteworthy, as it is rated from low to medium (1.66). This contrasts with our assessment of the pillars for effective proactive drought management, which indicated that this dimension was non-existent in Chile. The lack of assessments on drought vulnerability and impacts is due to the absence of an institution with this specific mandate. Only a limited number of studies have been conducted by academic and research institutions. For example, Fernández and Vásquez-Lavín (2025) in Chap. 5 presents a comprehensive framework that incorporates Infrastructure Capacity (IC), the Amount of Water in Storage (AWS), and adaptation strategies, to conduct a detailed analysis of resilience mechanisms and the economic impacts of drought conditions. The low to medium score can indicate a positive evaluation of academic studies. However, it is important to note that this result is worldwide. González Tánago et al.

(2016) reviewed literature on vulnerability assessments, identifying 46 drought vulnerability assessments, indicating such studies are not a common practice.

Institutional context (Dimension 3) received a score of 1.59, indicating significant fragmentation within Chile's water institutions and organizations. This reflects the existing institutional dispersion in water issues in Chile, which is characterized by institutions with overlapping responsibilities and unclear allocation of duties. The fragmented approach has led to the lack of a strong agency that identifies, formulates and implements drought policy and gives coherence to the actions of the various other agencies. This low to medium score is also a reflection of the lack of effective horizontal and vertical coordination.

Dimension 5 (Resources) received the second to lowest score (1.55). The perception of the public and private actors is that the implementation of drought measures has involved high public expenditure, although it would be insufficient to face the serious impacts of the drought. This reveals a dissemination of efforts and resources, so that the effects of all the actions deployed concerning droughts are diluted or are not cost-effective. In addition, the resources and human capacities in the public agencies responsible for drought management would be insufficient, which affects the effectiveness and timeliness of the measures adopted.

Coordination and participation (Dimension 6) is the lowest rated dimension. This indicates that the 17 entities of the State Administration exercising the main competences in drought management in Chile, identified in Chap. 7, do not act in an articulated manner. In fact, a series of unconnected and dispersed actions can be observed, reflecting weak institutional cooperation and fragmentation of governmental roles. However, this lack of coordination is not specific to drought management but is a common feature of water management in Chile, that has been repeatedly recognized as a weakness in water governance in Chile (Banco Mundial, 2011, 2013; OECD, 2011). Additionally, there have been challenges in adopting a participatory framework for water resources management in Chile. Several advances and reforms of Chile's institutional and legal framework for water management have not fully addressed participatory water resource management framework (Donoso, 2014; Peña, 2018). Consequently, Chile's institutional and legal framework for water management has not progressed significantly in incorporating stakeholder and civil society participation in drought management decisions.

This assessment provides a comprehensive overview of drought management in Chile. It indicates that the current strategy primarily employs a crisis management approach. Specifically, the assessment effectively identifies challenges that need to be addressed to transition towards a risk-management approach. Furthermore, it highlights a range of policy instruments that must be considered to address previously inadequately managed challenges. This policy proposal is presented in the next section.

Table 12.4 Policy proposals for dimension 1: Drought definition and monitoring

Policy proposal	Type of measure		
	Preparation	Response	Recovery
Incorporate a definition of drought at the legal or regulatory level, which allows the specifications to be made at the level of national, basin, and sectoral plans.	X		
Create a system for drought monitoring, tracking, and management based on quantitative drought indices and their thresholds to trigger the different instruments in the event of different drought categories, allowing for the particularities to be considered in the national, basin, and sectoral plans.	X	X	

12.4 Public Policy Proposals for Proactive Drought Management

Public policy proposals for proactive drought management are formulated based on previous assessments. These proposals align with a risk management approach for drought policy, which necessitates preparations for society before droughts occur, responsive actions to mitigate impacts during droughts, and an evaluation of the policy to suggest reforms based on lessons learned. These proposals were validated through collaboration with public and private stakeholders, as well as researchers, during a workshop.

12.4.1 Dimension 1: Drought Definition and Monitoring

Table 12.4 outlines policy proposals to enhance drought definition and monitoring, primarily focusing on preparation measures.

Firstly, due to the unclear definition of drought in Chile's policy, we propose defining it at the legal or regulatory level. Defining the phenomenon of drought is essential due to its significant impact on the development and utilization of water resources in drought affected regions. Effective management of this resource necessitates thorough characterization of drought conditions. Gironas and Nuñez (2025) in Chap. 2 define drought and examine the conditions for its study in Chile.

Secondly, based on the results presented in Chap. 13 (Gironas et al., 2025), a system for drought monitoring, tracking, and management based on quantitative drought indices must be implemented. The platform should consider multiple drought indices with different levels of severity and triggersfor their classification. These indices should be automatically calculated using as much information as possible, both from national and international sources, from point sources as well as

Table 12.5 Policy proposals for dimension 2: Evaluation and Monitoring

	Type of measure		
Policy proposal	Preparation	Response	Recovery
Ex-post evaluation of integrated drought management (led by the DGA).	X		X
Estimate of social, economic and environmental vulnerability by basin or territorial unit.	X		
Develop drought vulnerability assessment	X		X

from satellite and remote sensing products, sentinel hydrological process,[1] and even obtained from modeling. Additionally, these indices should allow for a simple analysis and comparison of drought conditions over time and in different spatial locations.

In addition, this platform should be open to the public and designed to address the needs of different users. Thus, the visualization should be simple, straightforward, and flexible. The tool should also focus on identifying a drought event from quantitative information and provide information on its manifestation in the territory and its short-term forecast. All these attributes are essential to lay the foundations for public policies focused on preventive and proactive drought management.

12.4.2 Dimension 2: Evaluation and Learning

As noted by Urquijo et al. (2017), evaluation is a crucial phase of the policy cycle as it enhances learning capacity, enabling the identification of gaps and shortcomings that need to be addressed to improve drought management. Our assessment of the pillars for effective proactive drought management, indicated that this dimension was non-existent in Chilean drought management policies. Thus, we propose that the DGA be the lead agency mandated to conduct ex-post evaluations of integrated drought management (Table 12.5).

12.4.3 Dimension 3: Institutional Context

Our assessment identified significant fragmentation within Chile's water institutions and organizations which is characterized by overlapping responsibilities and unclear allocation of duties. Table 12.6 presents the proposals to face these challenges.

Chile lacks a national water plan that serves as a guiding and binding instrument, developed with standards of participation, information dissemination, and

[1] See Chap. 15 (Gaxiola et al., 2025) for studies on sentinel variables of the hydrological cycle.

Table 12.6 Policy proposals for dimension 3: Institutional Context

Policy proposal	Type of measure		
	Preparation	Response	Recovery
National water plan, with a chapter or section on drought.	X	X	X
Update of the basin water resources plan, incorporating a drought chapter.	X	X	X
Review and adjust the Water Code on drought issues, to define the basic elements or criteria for drought management.	X	X	X
Elaborate and dictate a drought regulation, with criteria, procedures, thresholds, measures, instruments (preparedness, response and recovery approach), with effective articulation between institutions and their respective instruments.	X	X	X
Elaborate a regulatory, institutional and financial framework that facilitates development of non-conventional sources such as desalinization.	X	X	

inclusion. Thus, we propose that a national water planning framework be developed with a clear reference to drought management, to ensure long-term water security. Additionally, water policy and national plans are not regulated in Chile, so their inclusion should be encouraged. Water planning must also consider supplementary water sources such as desalination to ensure secure a long-term water supply considering climate change projections (Molinos-Senante et al., 2025—Chap. 18).[2] The creation of national or territorial-scale programs and policies for the planning and execution of complementary water sources are necessary for their coordinated development (Voutchkov, 2022) Additionally, this comprehensive water planning framework must include mechanisms for monitoring the plan and provisions for its periodic updates.

Additionally, the strategic water management plans at the basin level that have been established and are continuing to be formulated across basins within the country should be properly aligned with the national plan and consider local drought management plans.

As pointed out, the fragmented institutionality has led to the lack of a strong agency that identifies, formulates and implements drought policy and provides coherence to the actions of the other agencies. Additionally, there is a multiplicity of actions, with different approaches and objectives, without a clear and common roadmap. Many of these measures involve multiple administrative authorities or agencies, emphasizing the need for better coordination to prevent duplication and

[2] For example, hydrological water management plans in Spain give greater prominence to desalination, thus ensuring supply and balancing demands and environmental requirements. In turn, the National Urban Water and Desalination Plan in Australia helped major cities secure their water supplies and reduce their dependence on natural continental sources by supporting infrastructure and research projects in desalination, water recycling, and rainwater collection and reuse.

inconsistencies. Thus, we propose a comprehensive review and adjustment of the Water Code to include drought issues and to define the essential elements or criteria for drought management.

It is also necessary to elaborate and dictate specific drought regulation, with criteria, procedures, thresholds, measures, instruments covering the three pillars for a proactive management, with effective instances of articulation between institutions and their respective instruments. We propose specific measures in Dimension 4.

Along the same lines, it is important to create a regulatory, institutional and economic framework that makes the implementation of complementary sources, such as desalination (Molinos-Senante et al., 2025—Chap. 18) and the reuse of treated wastewater, viable. So far, both issues remain largely unregulated, and it is critical to make progress in this area.

12.4.4 Dimension 4: Drought Measures

From a normative and regulatory point of view, drought has traditionally been addressed as an emergency, with multiple instruments and measures that operate under a reactive framework. Of all the existing instruments, the two most used in drought scenarios are the declaration of a water scarcity zone due to severe drought and the declaration of an agricultural emergency due to water deficit.

To offer proactive instruments that avoid drought damages, we propose a mix of policy instruments (Table 12.7). In the first place, a sectoral meteorologic and hydrologic indexed insurance must be implemented. Chile has non-indexed agricultural meteorological insurance. However, it does not cover all crops. Further, there is no insurance for hydrological or agricultural drought. An indexed hydrological insurance is proposed; the basis for its design is a clear definition of drought (Chap. 2) and indexes (Chap. 13).

Table 12.7 Policy proposals for dimension 4: Drought Measures

	Type of measure		
Policy proposal	Preparation	Response	Recovery
Develop and implement sectoral meteorological and hydrological insurance (indexed).	X		
Tariffs for water use (holders of water rights).	X		
Implement accurate water accounting systems.	X	X	
Market price revealing mechanism to facilitate the market of water volumes and water rights as a drought adaptation strategy.	X	X	
Develop financial instruments such as contingent agreements or option contracts between users and sectors.		X	
Economic instruments to support vulnerable groups			X

Water consumption has increased in the past decades. Thus, decoupling economic growth from water demands in Chile has not been an automatic by-product of growth in national incomes and requires dedicated policies to improve water allocation between competing uses so as to not limit future economic growth. Furthermore, per capita water consumption in major urban areas increased during the recent drought period. This suggests that users did not respond to water scarcity. Water pricing has become a key issue in both the developed and developing world, largely due to increasing water scarcity. Dinar et al. (2015) indicates that experiences in both developing and developed countries suggest a shift in water pricing from focusing primarily on cost recovery or efficiency to also considering social parameters and environmental factors. There has been an increase in the use of incentive tariffs across all sectors compared to previous reports. Chile has not used water pricing to indicate scarcity. Thus, we suggest creating and applying water tariffs as a general measure, not only for drought periods.

In the presence of an increased probability of droughts due to climate change, water systems need to be more resilient. One potential response to drought is the implementation of a flexible reallocation system for water resources among competing demands (Koopman et al., 2017). During periods of drought, water markets have emerged as a critical tool for the effective and efficient redistribution of water resources among various uses and users (Loch et al., 2013). These markets have increasingly served as a flexible reallocation mechanism during drought conditions (Bjornlund, 2003; Grafton et al., 2010). According to Holley and Sinclair (2016), to maximize the benefits of water markets as a flexible reallocation mechanism, robust regulation is essential. This includes accurate water accounting, effective pricing mechanisms, and the consideration of environmental and social impacts. Thus, we propose implementing a robust, transparent, and reliable price revealing mechanism to reduce price variability and an accurate water accounting system, so as to strengthen water markets as an adaptation strategy to drought.

An additional proactive measure involves financial instruments, such as contingent agreements or option contracts for the temporary transfer of water volumes between water supply providers and other holders of water rights. The activation of the agreement is based on clear definition of drought indicators and trigger thresholds. Vicuña et al. (2025) in Chap. 16 presents a case study demonstrating its feasibility for implementation in Chile.

Goodwin et al. (2022) found that specific support interventions maintained household incomes during droughts, while others encouraged adaptive behaviors. Since drought impacts cannot be entirely avoided, a mix of financial instruments is proposed to support impoverished, vulnerable groups and reduce drought vulnerability. The design of these schemes should consider consultations with beneficiaries, perform *ex ante* assessment of the potential impacts and implementation costs of different support instruments, and conduct need-based assessments (Bevan and Cook, 2014; Clark et al., 2017; Goodwin et al., 2022).

Table 12.8 Policy proposals for dimensions 5 and 6

Dimension	Proposal	Type of measure		
		Preparation	Response	Recovery
5: Resources	Create a drought fund to finance preparedness, response and recovery measures, with effective transparency and control mechanisms.	X	X	X
6: Coordination and participation	Include at the legal level, as a function of the DGA, the articulation of all agencies and functions related to drought management.	X	X	X
	Inclusion of participation mechanisms in water plans at different levels (national, basin, sectoral).	X		

12.4.5 Dimension 5 (Resources) and 6 (Coordination and Participation)

The proposals to strengthen Dimensions 5 and 6 are summarized in Table 12.8.

The assessment found that Chile lacks a robust funding for droughts. Public agencies face limited resources, leading to ineffective drought management. Thus, we propose the creation of a drought fund to finance preparedness, response and recovery measures, with effective transparency and control mechanisms. The design of the drought fund must consider the three primary sources of government revenue: taxes, user fees and transfers (typically development assistance in the form of grants). Therefore, this proposal is closely aligned with the drought mitigation measure of implementing water tariffs.

Finally, as our assessment showed that Dimension 6 (Coordination and Participation) presents the lowest level of development, we propose two measures:

1. Include at the legal level, as an explicit function of the DGA, the articulation of all agencies and functions related to drought management.
2. Inclusion of participation mechanisms in water plans at different levels (national, basin, sectoral). There have been advancements in participation at the basin level through the Strategic Water Resources Committees (Mesas Estratégicas de Recursos Hídricos). The Strategic Water Resources Committees are led by the DGA and include representatives from public agencies and basin stakeholders. These committees will collaborate on developing the Strategic Water Resources Basin Plans. This initial form of participation requires evaluation, and probably reinforcement.

12.5 Conclusions

The findings of this chapter reveal a fragmented and reactive approach to drought management in Chile that is ill-equipped to address the intensifying impacts of recurrent droughts. Despite some progress in specific areas, such as the establishment of agricultural insurance and strategic water resources plans at the basin level, significant gaps persist across the three pillars of proactive drought management: monitoring and early warning, vulnerability and impact assessment, and mitigation and response.

The analysis highlights that while Chile formally recognizes drought in policy documents, there is no unified operational definition, leading to inconsistencies in its application across institutions. Coordination among the various public and private actors remains weak, further exacerbated by institutional overlaps and a lack of coherent frameworks. This situation has resulted in inefficiencies, with resources being dispersed and measures often being reactive rather than preventive.

To address these challenges, this chapter proposes a comprehensive revision of Chile's drought management approach. The recommendations advocate for a clear legal and institutional framework that integrates drought as a core component of national water policy, not as an emergency, but as a regular component of the water reality. A critical starting point is the establishment of a legally binding definition of drought, that considers different degrees of drought, coupled with the development of a national water plan that includes a dedicated section on drought. Such a plan should incorporate robust monitoring systems, quantitative indices, and thresholds to trigger early interventions.

Furthermore, institutional reforms are essential to address fragmentation. Strengthening the role of existing entities like the DGA is crucial for ensuring coordinated action and clear accountability. Complementary measures, such as financial instruments to support vulnerable populations, indexed insurance schemes, and flexible water reallocation mechanisms, are also vital to enhance resilience.

The transition to a proactive risk-management framework requires a cultural shift towards preventive action. This includes fostering greater stakeholder participation at all levels, from national to basin and sectoral, and enhancing transparency in decision-making processes. Additionally, investments in research and capacity-building are necessary to improve the understanding of drought vulnerabilities and adaptive strategies.

In conclusion, while Chile's current drought management approach underscores the challenges of reactive systems, it also presents an opportunity to redefine its water governance landscape. By adopting the proposed measures, Chile can transition towards a sustainable and integrated drought management strategy that not only mitigates current vulnerabilities and also builds resilience for future challenges. Such a transformation will align Chile with international best practices and contribute significantly to its climate adaptation and water security goals.

References

Arana, E., & Miranzo, J. (2021). La gestión de la escasez de agua y de las sequías por parte de las entidades locales. *Guía para una gestión eficiente*. (Navarra, Editorial Aranzadi), 156.

Banco Mundial. (2011). *Diagnóstico de la gestión de los recursos hídricos*. Santiago. Available at: https://www.academia.edu/download/62341620/Diagnostico_gestion_de_recursos_hidricos_en_Chile_Banco_Mundial20200311-25794-1vn2e5i.pdf

Banco Mundial. (2013). *Estudio para el mejoramiento del marco institucional para la gestión del agua*. Santiago. Available at: https://bibliotecadigital.ciren.cl/bitstream/20.500.13082/33281/1/BANCO_MUNDIAL_2013-Estudio-para-el-Mejoramiento-del-Marco-Institucional-para-la-Gestion-del-Agua-DGA-BM-2013.pdf

Berbel, J., & Esteban, E. (2019). Droughts as a catalyst for water policy change. Analysis of Spain, Australia (MDB), and California. *Global Environmental Change, 58*, 101969.

Bevan, D., & Cook, S. (2014). *Public expenditure following disasters* (Policy Research Working Paper 7355). World Bank SDRFI Impact Appraisal Project.

Bjornlund, H. (2003). Efficient water market mechanisms to cope with water scarcity. *Water Resources Development, 19*(4), 553–567.

Brufao, P. (2012). El régimen jurídico de las sequías: crítica a la regulación extraordinaria y urgente de un fenómeno natural y cíclico propio del clima. *Revista de Administración Pública, N°187*, 199–239.

Clark, D. J., Mahul, O., Poulter, R., & Teh, T. L. (2017). Evaluating sovereign disaster risk finance strategies: A framework. *The Geneva Papers on Risk and Insurance-Issues and Practice, 42*, 565–584.

Courtenay, L., & Hayes, M. (2012). Drought triggers and declarations: Science and policy considerations for drought risk management. *Natural Hazards, 64*, 139–151. https://doi.org/10.1007/s11069-012-0231-4

Dinar, A., Pochat, V., & Albiac-Murillo, J. (Eds.). (2015). *Water pricing experiences and innovations*. Springer.

Donoso, G. (2014). Integrated water management in Chile. In R. Llamas & A. Garrido (Eds.), *Integrated water resources management in the 21st century: Revisiting the paradigm*. RC Press/Balkema—Taylor & Francis Group.

Embid, A. (2018). Elementos para la construcción de una teoría general de las sequías e inundaciones como fenómenos hidrológicos extremos. In Embid, A. (Dir.), *Sequía e inundación como fenómenos hidrológicos extremos* (pp. 25–84). Editorial Aranzadi.

Fernández, F. J., & Vásquez-Lavín, F. (2025). Economic impacts of droughts on agriculture and urban water supply sector. In D. Rivera & G. Donoso (Eds.), *Droughts in Chile: Impacts, monitoring, and adaptation (management) policies*. Springer.

Fu, X., Svoboda, M., Tang, Z., Dai, Z., & Wu, J. (2013). An overview of US state drought plans: Crisis or risk management? *Natural Hazards, 69*, 1607–1627. https://doi.org/10.10007/s11069-013-0766-z

Gaxiola, A., del Río, C., Miranda, M., & Suárez, F. (2025). Sentinel hydrological processes: Informing drought management strategies in Chile. In D. Rivera & G. Donoso (Eds.), *Droughts in Chile: Impacts, monitoring, and adaptation (management) policies*. Springer.

Gironás, J., & Nuñez, M. (2025). Definition and characterization of drought. In D. Rivera & G. Donoso (Eds.). Droughts in Chile: Impacts, monitoring, and adaptation (management) policies, Springer.

Gironás, J., Núñez, M., Vicuña, S., Aedo, S., Morales, D., Vargas, J., Guzmán, J., & Suárez, F. (2025). Towards a drought monitoring, tracking, and management platform based on quantitative indices. In D. Rivera & G. Donoso (Eds.), *Droughts in Chile: Impacts, monitoring, and adaptation (management) policies*. Springer.

Goodwin, D., Holman, I., Pardthaisong, L., Visessri, S., Ekkawatpanit, C., & Rey Vicario, D. (2022). What is the evidence linking financial assistance for drought-affected agriculture and resilience in tropical Asia? A systematic review. *Regional Environmental Change, 22*(1), 12.

González Tánago, I., Urquijo, J., Blauhut, V., Villarroya, F., & De Stefano, L. (2016). Learning from experience: A systematic review of assessments of vulnerability to drought. *Natural Hazards, 80*, 951–973.

Grafton, R. Q., Landry, C., Libecap, G. D., McGlennon, S., & Brien, R. O. (2010). An integrated assessment of water markets. *Review of Environmental Economics and Policy, 5*, 219–239.

Hayes, M. J., Wilhelmi, O. V., & Knutson, C. L. (2004). Reducing drought risk: bridging theory and practice. *Natural Hazards Review, 5*(2), 106–113.

Hervás-Gámez, C., & Delgado-Ramos, F. (2019). Drought management planning policy: From Europe to Spain. *Sustainability, 11*, 1862. https://doi.org/10.3390/su11071862

Holley, C., & Sinclair, D. (2016). Governing water markets: Achievements, limitations and the need for regulatory reform. *Environmental and Planning Law Journal, 33*(4), 301–324. Available at: https://ssrn.com/abstract=3159081

Iglesias, A., Garrote, L., & Granados, A. (2019). On the institutional framework for drought planning and early action. In A. Iglesias, D. Assimacopoulos, & H. Van Lanen (Eds.), *Drought. science and policy*. Wiley Blackwell.

Jones, J. (2025). Understanding drought in California. In D. Rivera & G. Donoso (Eds.), *Droughts in Chile: Impacts, monitoring, and adaptation (management) policies*. Springer.

Koopman, J. F., Kuik, O., Tol, R. S., & Brouwer, R. (2017). The potential of water markets to allocate water between industry, agriculture, and public water utilities as an adaptation mechanism to climate change. *Mitigation and Adaptation Strategies for Global Change, 22*, 325–347.

Loch, A., Wheeler, S., Bjornlund, H., Beecham, S., Edwards, J., Zuo, A., & Shanahan, M. (2013). *The role of water markets in climate change adaptation* (p. 142). National Climate Change Adaptation Research Facility, Gold Coast.

Martín, L. (2019–2020). Emergencia hídrica, ambiental y climática. Elementos para su configuración y relación con el Estado de Derecho. *Revista Derechos en Acción*, Año 5, N°5: 224–253.

Martín, L. (2022). From scarcity to emergency. A comparative study on water and drought management in the Americas. In Nicolas-Artero, C., Velut, S., Schneier-Madanes, G., Poupeau, F., & Chavarochette, C. (Dir), *Luttes pour l'Eau dans les Amériques* (pp. 43–57). Éditions de l'IHEAL.

Martín, L. (2024). Emergencia climática. Desafíos jurídicos y políticos. In Setuaín, B. (Dir.), *Perspectivas jurídicas sobre clima, agua y energía. Estudios en reconocimiento al magisterio del profesor Antonio Embid Irujo* (pp. 29–55). Editorial Aranzadi.

Molinos-Senante, M., Rivera, D., Donoso, G., & Melo, O. (2025). Desalination in times of drought. In D. Rivera & G. Donoso (Eds.), *Droughts in Chile: Impacts, monitoring, and adaptation (management) policies*. Springer.

Neuman, J. (2003). Drought proofing water law. *University of Denver Water Law Review, 7*(1), 92–110.

OECD. (2011). *Water governance in OECD countries: A multi-level approach, OECD studies on water*. OECD Publishing. https://doi.org/10.1787/9789264119284-en

Paneque, P. (2015). Drought management strategies in Spain. *Water, 7*, 6689–6701. https://doi.org/10.3390/w7126655

Peña, H. (2018). Integrated water resources management in Chile: Advances and challenge. In G. Donoso (Ed.), *Water policy in Chile* (pp. 197–208). Springer. https://doi.org/10.1007/2F978-3-319-76702-4

Rafey, W. (2023). Droughts, deluges, and (river) diversions: Valuing market-based water reallocation. *American Economic Review, 113*(2), 430–471.

Rivera, D., & Donoso, G. (2025). Institutionality, governance, and drought management policies. In D. Rivera & G. Donoso (Eds.), *Droughts in Chile: Impacts, monitoring, and adaptation (management) policies*. Springer.

Setuaín, B. (2018). Respuestas jurídicas a la reciente situación de sequía en España. In Embid, A. (Dir.), *Sequía e inundación como fenómenos hidrológicos extremos* (pp. 149–193). Editorial Aranzadi.

Stein, U., Özerol, G., Tröltzsch, J., Landgrebe, R., Szendrenyi, A., & Vidaurre, R. (2016). European drought and water scarcity policies. In H. Bressers, N. Bresssers, & C. Larrue (Eds.), *Governance for drought resilience* (pp. 17–43). Springer.

Urquijo, J., Pereira, D., Dias, S., & De Stefano, L. (2017). A methodology to assess drought management as applied to six European case studies. *International Journal of Water Resources Development, 33*(2), 246–269.

Urquijo, J., Paizano-Potoy, J., López-Santiago, J., & Pereira, D. (2025). Drought policy overview: Insights from South America region. In D. Rivera & G. Donoso (Eds.), *Droughts in Chile: Impacts, monitoring, and adaptation (management) policies*. Springer.

Vargas, M. J., & Paneque, P. (2025). Drought policy and governance in Spain. New tools and remaining challenges. In D. Rivera & G. Donoso (Eds.), *Droughts in Chile: Impacts, monitoring, and adaptation (management) policies*. Springer.

Vicuña, S., Ávila, A. Gironás, J., Melo, O., Rivera, D., & Molinos, M. (2025) Contingent agreements to reduce the impacts of droughts on human water consumption, In Rivera, D., and Donoso, G. (Eds.), *Droughts in Chile: Impacts, Monitoring, and Adaptation (Management) Policies* (Springer).

Voutchkov, N. (2022). Desalinated water. In M. Qadir, V. Smakhtin, S. Koo-Oshima, & E. Guenter (Eds.), *Unconventional water resources* (pp. 233–254). Springer.

Wilhite, D. (2009). Drought monitoring as a component of drought preparedness plan. In A. Iglesias, L. Garrote, A. Cancelliere, F. Cubillo, & D. Wihite (Eds.), *Coping with drought risk in agriculture and water supply systems* (pp. 3–19). Springer.

Wilhite, D. A., & Pulwarty, R. S. (2005). *Drought and water crises: Lessons learned and the road ahead. Drought and water crises: Science, technology, and management issues* (pp. 389–398).

Wilhite, D., Hayes, M. J., & Knutson, C. (2000). Planning for drought: Moving from crisis to risk management. *Journal of the American Water Resources Association, 36*(4), 697–710.

Wilhite, D. A., Hayes, M. J., & Knutson, C. L. (2005). Drought preparedness planning: Building institutional capacity. Drought and water crises: Science, technology, and management issues, 107, 93–136.

Wilhite, D. A., Sivakumar, M. V., & Pulwarty, R. (2014). Managing drought risk in a changing climate: The role of national drought policy. *Weather and Climate Extremes, 3*, 4–13.

WMO (World Meteorological Organization) and GWP (Global Water Partnership). (2014). *National drought management policy guidelines: A template for action (D.A. Wilhite). Integrated drought management Programme (IDMP) tools and guidelines series 1*. WMO/GWP.

Zuo, A., & Wheeler, S. (2025). Drought policy in Australia. In D. Rivera & G. Donoso (Eds.), *Droughts in Chile: Impacts, monitoring, and adaptation (management) policies*. Springer.

Chapter 13
Towards a Drought Monitoring, Tracking, and Management Platform Based on Quantitative Indices

Jorge Gironás, Marialina Núñez, Sebastián Vicuña, Sebastián Aedo, David Morales, Javier Vargas, Juan de Dios Guzmán, and Francisco Suárez

Abstract Drought is a complex phenomenon that affects the development and use of the country's water resources. Characterization through indices is critical in planning, management, and decision-making to mitigate the effects of drought, mainly when these are implemented in online platforms for monitoring, warning, and tracking the phenomenon. These tools serve as a basis for implementing a set of measures focused on the short, medium, and long term, oriented to different audiences, and with differentiation of procedures and entities involved depending on the case. This work proposes the technical basis for elaborating preventive drought monitor-

J. Gironás (✉)
Departamento de Ingeniería Hidráulica y Ambiental, Pontificia Universidad Católica de Chile, Santiago, Chile

Centro de Desarrollo Urbano Sustentable, Santiago, Chile

Centro de Investigación para la Gestión Integrada del Riesgo de Desastres, Santiago, Chile

Centro de Cambio Global, Pontificia Universidad Católica de Chile, Santiago, Chile
e-mail: jgironas@uc.cl

M. Núñez
Departamento de Ingeniería Hidráulica y Ambiental, Pontificia Universidad Católica de Chile, Santiago, Chile

S. Vicuña
Departamento de Ingeniería Hidráulica y Ambiental, Pontificia Universidad Católica de Chile, Santiago, Chile

Centro de Investigación para la Gestión Integrada del Riesgo de Desastres, Santiago, Chile

Centro de Cambio Global, Pontificia Universidad Católica de Chile, Santiago, Chile

S. Aedo
Latin America Centre, Stockholm Environment Institute, Bogotá, Colombia

D. Rivera, G. Donoso (eds.), *Droughts in Chile*, Global Issues in Water Policy 31,
https://doi.org/10.1007/978-3-031-85040-0_13

D. Morales
Centro de Cambio Global, Pontificia Universidad Católica de Chile, Santiago, Chile

Facultad de Agronomía y Sistemas Naturales, Pontificia Universidad Católica de Chile, Santiago, Chile

J. Vargas
Centro de Cambio Global, Pontificia Universidad Católica de Chile, Santiago, Chile

J. de Dios Guzmán
Departamento de Ingeniería Hidráulica y Ambiental, Pontificia Universidad Católica de Chile, Santiago, Chile

Centro de Cambio Global, Pontificia Universidad Católica de Chile, Santiago, Chile

F. Suárez
Departamento de Ingeniería Hidráulica y Ambiental, Pontificia Universidad Católica de Chile, Santiago, Chile

Centro de Desarrollo Urbano Sustentable, Santiago, Chile

ing in Chile through a visualization platform. Based on a comparative analysis of international tools and the results of a consultation with experts, the relevant characteristics of a system for drought monitoring, follow-up, and management based on standardized drought indexes are analyzed. It is concluded that such a platform should consider multiple drought indices with different levels of severity for their classification. These indices should be automatically calculated using as much information as possible, both from national and international sources, from in-situ gauges as well as from satellite and remote sensing products, and even obtained from modeling. On the other hand, the visualization should be simple, straightforward, and flexible, and the information delivered should provide information on the manifestation of drought in the territory and its short-term forecast.

Keywords Drought indices · Platform · Monitoring · DGA resolution 1331 · Chile

13.1 Introduction

The World Meteorological Organization (WMO, 2017) has recently defined the new standard climatological normal, or reference period for evaluating climatic conditions, the time between 1991 and 2020. This period was also adopted by Chile (MOP, 2021). During these 30 years, the effects of climate change have begun to be experienced in the country; on average, at the national scale, annual temperature has increased while precipitation has decreased. In parallel, there have been reductions in aquifer levels, to which increased demands on groundwater have also contributed. Additionally, since 2009, Chile has faced a megadrought (e.g., Boisier et al., 2018; Garreaud et al., 2019), the longest drought affecting the region since records have been kept, aggravating a trend of increasing dryness and putting Chile at the forefront of the region's water crisis (WMO, 2022). This phenomenon has generated that, as of March 2022, 188 out of 346 municipalities (i.e., approximately 50% of the population) have been under water scarcity zone resolutions (DGA, 2022a).

Drought is a complex phenomenon that significantly affects the development and use of water resources in a region (Fernandez, 1997). It can occur in any climatic regime (e.g., Bordi and Sutera, 2007; Svoboda & Fuchs, 2016), and affects on average, 55 million people each year. Moreover, drought is considered to be the most significant hazard to livestock and crops almost worldwide (WHO, 2022). This phenomenon can be classified according to the perspective or focus of interest, typically as meteorological, hydrological, agricultural, and socioeconomic (Wilhite & Glantz, 1985). However, a general definition (Fernandez, 1997) corresponds to prolonged periods of scarcity of water resources, which are capable of causing adverse effects on society or productive systems.

To mitigate drought impacts, its characterization becomes crucial in supporting planning, management, and decision-making by the corresponding authorities (Fernández & Gironás, 2017). In this line, drought indicators and indices emerge to facilitate the monitoring of droughts. According to the WMO (2016), the former corresponds to variables or parameters used to describe drought conditions, such as precipitation, temperature, flow discharges, groundwater levels, etc., while the latter are numerical representations that report on the severity of droughts, allowing the phenomenon to be analyzed over a given period. In general, drought indices have certain advantages over drought indicators, as they provide a quantitative approach to the severity, geographical location, chronological development, and duration of drought episodes, simplifying complex relationships so that the phenomenon can be monitored. The respective information can be communicated to different audiences and users (WMO, 2016) and inform decision-making. This is why various drought indices have been proposed to provide a synthetic and objective description of the phenomenon (Hao et al., 2018).

Drought indicators and indices can be calculated both historically and in real-time. Drought monitoring and early warning systems are standard in countries seeking to address the phenomenon preventively. These systems serve as a basis for implementing measures focused on the short, medium, and long term, oriented to different audiences or groups, and with differentiation of procedures and entities involved depending on the case. This chapter aims to propose the technical bases to generate preventive monitoring of drought in Chile through a visualization platform, taking advantage of the growing availability of data in the country, and the new technological tools available worldwide. It seeks to address an increasingly imperative need to face drought not as an extraordinary phenomenon to be reacted to, but as a permanent phenomenon that requires a continuum of actions of an adaptive and preventive approach.

13.2 Drought Monitoring

13.2.1 Drought Indices

The standardized indices stand out within the drought indices, notable for their simplicity, clarity, and ease of spatial and temporal comparisons. These indices were developed from the Standardized Precipitation Index (SPI) proposed by McKee et al. (1993) and recommended by the WMO for widespread use in countries to monitor meteorological drought (Hayes et al., 2011). This index can take various accumulation periods, whose scale can range from 1 to 72 months (Edwards & McKee, 1997), although for practical purposes, it is recommended to be 1–12 months. This duration will depend on the purposes for which the index is calculated since certain users, such as farmers, may be interested in shorter periods (3 or 6 months). Conversely, water resource managers may be more interested in durations from 6 to 24 months (Wu et al., 2001). Subsequently, other similar standardized indices were proposed for different hydrological variables, such as flow discharges (i.e., standardized streamflow index, SSI) (Modarres, 2007; Telesca et al., 2012), local precipitation versus evapotranspiration balance (i.e., standardized precipitation evapotranspiration index, SPEI) (Vicente-Serrano et al., 2010), and soil moisture (i.e., standardized soil moisture index, SSMI) (Hao & AghaKouchak, 2013; Carrão et al., 2016).

To calculate this index, it is necessary to estimate the probability function of the variable in the study area of interest with the stipulated accumulation period (for example, cumulative precipitation of 6 months until March). This is done based on a series of, hopefully, at least 30 years, with no missing data (McKee et al., 1993). This distribution makes it possible to establish how high or low is an observed magnitude. The fitted distribution function can vary, although it typically has a certain asymmetry; it may even be necessary to use mixed probability distributions to describe variables that, in many cases, have a value of zero, such as the 3-month precipitation during the summer in arid and semi-arid climates (DGA, 2020b). The cumulative distribution is determined once the probability function of the data has been fitted. Then, an equiprobability transformation of this distribution is performed to a standard normal distribution with a mean of zero and a variance of one, with the quantile for this probability in the domain of the standard normal variable being the corresponding value of the standardized indicator (Wu et al., 2001; DGA, 2022b). This transformation can be seen graphically in Fig. 13.1. This figure shows the value of the SPI associated with a value of 10 mm for two different accumulations. Thus, for a short accumulation of, for example, 1 month, 10 mm of precipitation is a magnitude slightly higher than what is observed 50% of the time, which implies a value of the SPI close to 0.5. On the other hand, if 3 months of accumulation are considered, 10 mm of precipitation is a low amount and is therefore associated with an SPI value lower than zero.

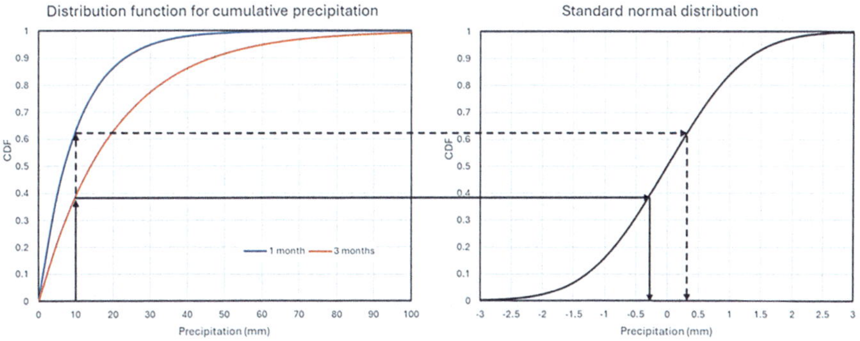

Fig. 13.1 Examples of transformation of precipitation magnitudes using the probability distribution for the rainfall variable to a standard normal distribution of the SPI. The example illustrates the value of the SPI corresponding to 10 mm of rainfall at a specific location and for two accumulation periods

Table 13.1 Return periods associated with different SPI values

SPI Value	Return Period (years)
0	2
−0.5	3.2
−1	6.3
−1.5	15
−2	44
−2.5	161
−3	740

Source: Own elaboration

As the values of a standardized index correspond to quantiles of the standard normal distribution, the occurrence of an indicator value can be associated with return periods, as shown in Table 13.1.

The value of the standardized index, therefore, quantifies the deviation of the value of a variable concerning an average condition for a cumulative period defined in the calculation. Due to the normalization and standardization of the calculation, its values can be compared among different regions and used throughout the year (Bordi & Sutera, 2007), quantifying the occurrence and severity of droughts. A classification of this severity was initially proposed by McKee et al. (1993) and shown in Table 13.2. The calculation and the classification initially proposed for the SPI index are analogous to other standardized indicators, particularly the SSI, which has also been used in Chile.

Table 13.2 Classification of severity or severity of RLS according to its value

SPI Value	Classification
$2.0 \leq SPI$	Extremely wet
$1.5 \leq SPI < 2.0$	Very wet
$1.0 \leq SPI < 1.5$	Moderately wet
$-1.0 < SPI < 1.0$	Normal
$-1.5 < SPI \leq -1.0$	Moderately dry
$-2.0 < SPI \leq -1.5$	Severely dry
$IPE \leq -2.0$	Extremely dry

Adapted by WMO (2012) from McKee et al. (1993)

13.2.2 National Context

To identify and administratively declare droughts in the country, the recently enacted resolution 1331 DGA (2022c) defines four macrozones with similar hydro-climatic characteristics (Fig. 13.2). These macrozones allow consideration of the precipitation gradient from north to south, ranging from <10 mm in the hyper-arid north to >3000 mm in the wettest part of the country's south (Aceituno et al., 2021). The macrozones are as follows:

- Macrozone 1 (18° S–26° S, from the Arica and Parinacota region to the Antofagasta region), which has an arid to semi-arid climate, with scarce precipitation that increases slightly towards the south.
- Macrozone 2 (26° S–36.3° S, from the Atacama to the Maule region), which has a semi-arid and Mediterranean climate, with moderate precipitation concentrated in winter and a dry season of 7–8 months.
- Macrozone 3 (39.3° S–39.5° S, from the Ñuble to the Araucanía region), which has temperate rainy and oceanic rainy climates, with abundant rainfall that increases towards the south.
- Macrozone 4 (39.5° S–55.6° S, from the Los Ríos region to Chilean Antarctica), which has a year-round rainy climate, resulting in the most significant runoff volumes and discharges in Chile (~70% of the country's runoff), as well as the existence of ~80% of the country's water bodies.

Resolution 1331 establishes the SPI and SSI as the two indices to declare droughts according to accumulation periods defined for the four macro-zones of the country (Table 13.3). Regardless of the index, the resolution defines a threshold of −1.04 to identify and declare a drought as severe. To these drought index values, conditions in the country's reservoirs and aquifers can be added, which can also motivate the declaration. This declaration triggers an administrative procedure defined in the water code and directed by the DGA to deal with the drought event.

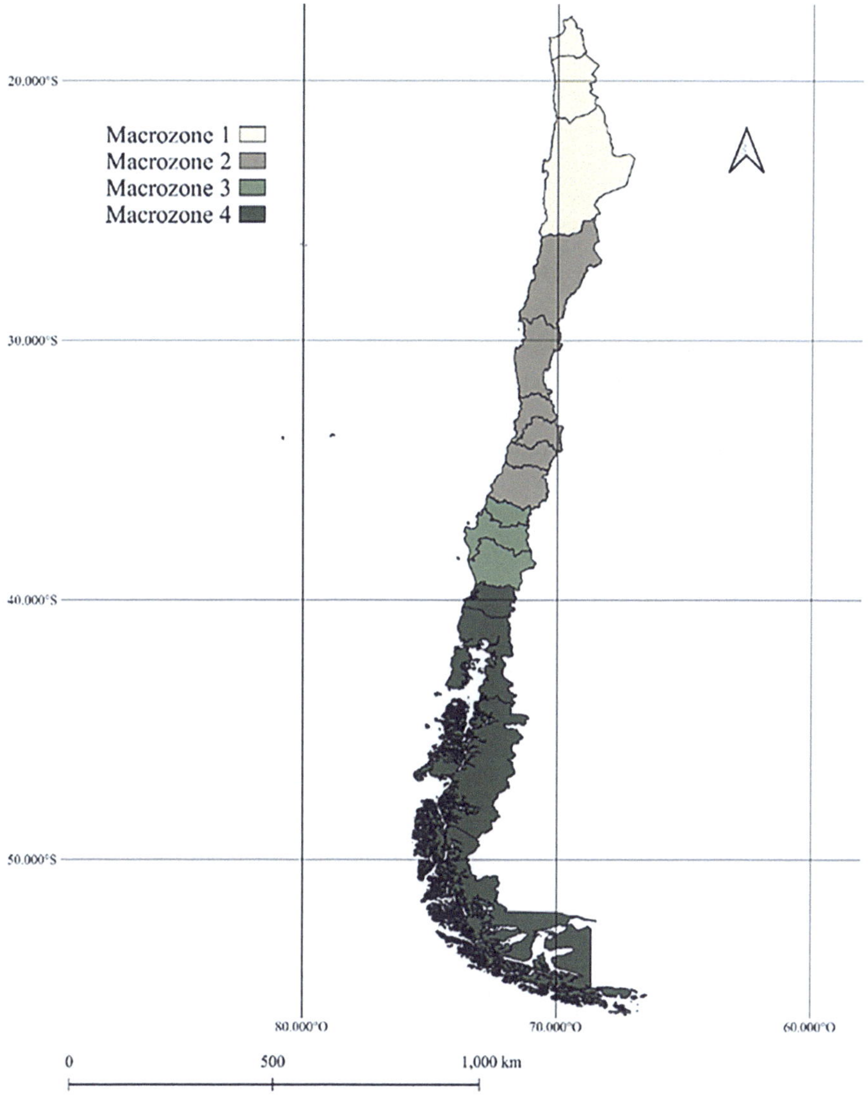

Fig. 13.2 Division of macrozones in Chile for drought management

Table 13.3 Threshold values of drought indexes to declare severe drought for the different Macrozones of Chile according to Resolution 1331 (DGA, 2022c)

Macrozone	Condition
1	SSI 1 for three consecutive months ≤ -1.04
2	SPI 12 or SSI 6 ≤ -1.04
3	SPI 6 or SSI 6 ≤ -1.04
4	SPI 3 or SSI 3 ≤ -1.04

13.3 Background for Drought Monitoring Platform

To propose a monitoring platform to present relevant information on the state of drought in a region, a recent study (DGA, 2022b) proposed two lines of action: (1) a review of the state of the art of international monitoring platforms, and (2) a public consultation or focus group to understand the needs of the intended users of the platform.

13.3.1 Review of the International State of the Art

There are several drought observatories or monitoring platforms with different characteristics and purposes. Some of these tools are the following (DGA, 2022b):

- National Integrated Drought Information System (NIDIS): U.S. platform that collects data from other agencies on current conditions, forecasts, impacts, soil moisture, vegetation, fires, temperature, precipitation, agricultural situation, and snow drought, among others.
- United States Drought Monitor (USDM): The United States platform presents a map of drought conditions in the country according to various indices. On the platform, drought status is classified according to five degrees of intensity. In addition, drought impact tables are presented for each state.
- European Drought Observatory (EDO): European platform that presents maps of indicators, where tools allow displaying and analyzing information through graphs and comparative layers. This platform works with six drought classifications.
- Global Drought Monitor (SPEI) is a Spanish platform that provides spatial information at different time scales (between 1 and 48 months) of the SPEI Index. The platform covers the entire world and uses nine drought classifications.
- North American Drought Monitor (NADAM): Platform for North America (Canada, United States, and Mexico) that provides a continuous and comprehensive drought assessment. It presents monthly drought maps and reports on what has happened in the various regions.

- Climate and Society Map Room (IRI): U.S. platform presenting many climate observations. Concerning droughts, it allows monitoring of the SPI for 1, 3, 6, 9 and 12 months worldwide.
- NSW Department of Primary Industries (DPI): Australian platform that presents maps of the Combined Drought Indicator (CDI), which weights a rainfall index, a soil moisture index, and grass/crop growth indices. This considers different variables that affect drought conditions beyond rainfall deficiencies, including temperatures and their effect on evapotranspiration, runoff, and soil infiltration. The platform considers five phases of drought: no drought, recovery, drought-affected, drought, and severe drought.
- Regional Climate Center for Western South America (CRC-OSA): Platform containing several daily observations at meteorological stations throughout Latin America. Concerning drought, it has a Drought Monitor with a map showing the drought status of the participating countries (Bolivia, Chile, Colombia, Ecuador, Peru, and Venezuela).
- Regional Climate Center for South America (CRC-SAS): A platform that uses meteorological stations in various South American countries (Argentina, Brazil, Bolivia, Chile, Paraguay, and Uruguay), where a series of indexes are calculated monthly to monitor and quantify these events. These indices are updated monthly and are associated with different time scales. The drought indices considered are SPI, SPEI, Percentage of Normal Precipitation, and Precipitation Deciles.
- Agroclimatic Observatory of the Ministry of Agriculture of Chile: National platform that presents a set of maps and figures of current drought conditions, which provide information on the frequency of past droughts and projections of future climatic conditions. The variables considered are precipitation, SPI, observed flows, normalized vegetation difference index, and soil water balance. Finally, a combined helpful index for the general public is also reported.
- Platform of the Chilean Ministry of National Assets: This platform presents the precipitation indicator, reporting precipitation deficits and surpluses in the national territory. The tool allows the definition of spatial filters for visualization.

The objective of the platforms varies from generating and disseminating information to supporting the application of public policy or research. In addition, they consider various services, such as monitoring, forecasting, historical record keeping, and general and specific reports. The spatial coverage of the platforms ranges from local-administrative to national and even planetary scales. These monitors presented various indicators, such as precipitation, flow rates, soil moisture, evapotranspiration, and vegetation indices, whose source of information varies, considering public and private sources and satellite information.

A comparison of the attributes of the different platforms reveals some common characteristics and differences, summarized below:

- All platforms generate and disseminate information on drought behavior. However, most do so for research purposes, without a commitment to enforcing user policies or standards.
- Only three (those of the U.S. and Australia) are explicitly applied to public policy or promote how users could use the information to achieve special conditions.
- Many platforms are updated monthly; some are updated every 1 or 2 weeks.
- Six of the platforms reviewed have a forecasting component.
- The reference period is not always precise. For five platforms, this period starts in 1980 ± 1 year, while for three of these, the end of the period is 2010. The other two platforms extend their reference period to the latest most recent year.
- Most platforms maintain a historical record available for consultation.
- Only the North American platforms produce general reports weekly to quarterly. The NIDIS, EDO, CRC-OSA, and CRC-SAS platforms issue occasional reports at times of particular interest.
- Five platforms can report answers to specific queries, e.g., indicate the particular conditions of a specific location in the face of droughts.
- All platforms, except SPEI and IRI, use information from public stations, although the NSW and CNR platforms also use private information. In addition, all platforms use satellite information from public sources.
- The most commonly used drought indices are meteorological, with the Australian platform being the only one that does not consider them. While no single index is common to all platforms, the SPI is repeated in all but the SPEI platform. These meteorological indices are evaluated and displayed for several cumulative past periods ranging from 1 to 24 months.
- Consequently, precipitation is the most used raw variable in the index calculations in the different platforms. The other two relevant variables, which are much less used, are soil water content and flow discharges.
- Six platforms report some indicators obtained with remote sensing tools that characterize the response of vegetation to periods of drought.

13.3.2 Focus Group

The focus group DGA (2022b) reported included representatives of academic institutions, the private sector, the public sector, international organizations in Chile, civil society, and users. A series of questions were answered, and the visions and experiences of the participants concerning the drought issue were compiled, with a particular focus on the needs and scope of active monitoring of the phenomenon. Here, we summarize the results of the focus group

- For experts and users, the key indicators to consider in drought monitoring and identification are rainfall, flow rates, and well levels, with soil moisture and evaporation as secondary variables. These results reflect a majority interest in detecting droughts early based on observing variables whose measurement is more straightforward and available in the territory. Additionally, there is an understanding of the process of drought propagation from its meteorological dimension to the surface and subway hydrological dimensions.

- The accumulations of 3 and 6 months are considered the most desirable, reflecting the need for calculation windows wide enough to identify a stable drought condition but not so long as to identify the phenomenon late. It is worth mentioning that these accumulations are currently considered in the Resolution 1331 resolution to identify severe drought (DGA, 2022c).
- Most participants (81%) identify the need to update the reference period to incorporate structural changes in the climate. Of this percentage, 38% consider a moving window of 30-year period updated every year. The remaining 43% prefer a 30-year a moving window that can be updated at the beginning of each decade, as recommended by the World Meteorological Organization and adopted by resolution 1331 (DGA, 2022c).
- Seventy-five percent of people believe that drought resolutions should not have a fixed administrative duration. This view then implies the need to have continuous and real-time monitoring of drought conditions to quantitatively and, according to previously defined criteria, be able to declare the end of a drought. This task is facilitated by using drought indicators that are calculated periodically and visualized in a monitoring platform.
- All users expect any drought platform to be public, recognizing that there may be a differentiation in the information accessible by different user groups and their needs.
- Most potential users prefer the finest possible spatial scale (i.e., a grid pixel), allowing the indices' value to be aggregated to other scales or spatial domains of interest. Although this spatial visualization capability is desirable due to its possibilities, its implementation is complex. It requires information not produced and periodically updated by state agencies or algorithms to disaggregate hydroclimatic variables and indices with associated uncertainty. It is noteworthy to verify that the basin far exceeds other administrative spatial units such as regions and communes, reflecting a need to fully consider the hydrological cycle in managing water resources.
- Half of the respondents expect a monthly update of the platform information, with an additional 26% expecting updates every 15 days. These high update frequencies imply a significant effort in terms of resources for a tool and human capital that must be operating continuously. Finally, there is a real need to have at least two levels of drought severity, with 87% of the respondent considering defining three or more of these levels. This is not by the current situation considered by the current regulations in Chile, which only defines a severe drought state. More severity levels would allow the implementation of a portfolio of measures differentiated according to the severity level, which would facilitate a preventive, and not reactive, management of the phenomenon.

13.4 Platform Prototype

Based on the state of the art and the visions compiled in the Focus group described in DGA (2022b), a general diagnosis is proposed regarding the need for a drought monitoring platform. Among the main attributes that this platform should have, the following stand out: (1) it should be open, globally accessible, and usable across different sectors; (2) it should incorporate different sources of national and international information, and (3) it should have the capacity to produce a series of indicators and valuable quantitative and qualitative findings beneficial to all users impacted by and concerned with the issue of water scarcity. Regarding this final aspect, there is a recognized necessity extending beyond technical requirements (such as the quantitative data needed to identify and declare a drought), driven by the demand for accessible information and background for the general public.

The background reviewed allows us to identify the following core characteristics for a drought monitoring, tracking, and management platform. These characteristics are discussed in more detail in DGA (2022b):

- The platform must be articulated with other initiatives of interest and relevant to the theme. In particular, the platform must be articulated with various data sources, both local (stations) and regional (remote sensing), for the automatic calculation of indicators and with initiatives that require the information generated by the platform. Critical non-automatic actions that may restrict the capabilities and uses of the platform should be minimized.
- The platform should be user-friendly, with an attractive, simple, and straightforward visualization. In this context, it should be responsive to the needs of users and stakeholders. The visualization should consider interactive tables, graphs, and maps. It is proposed to follow the FAIR principles of findability, accessibility, interoperability, and data reuse, typical of tools of this type.
- The platform should be helpful to a broad community of users interested in or affected by the drought problem in its different dimensions, especially to guide adaptation actions in the face of scarcity conditions. From this point of view, the platform should report different unidimensional drought indicators (i.e., meteorological, hydrological, etc.), a composite indicator, or a qualitative heuristic of global drought characterization.
- As a reference time window, it is proposed to consider a moving window of 30, which can be updated at the beginning of each decade. The selection of this reference period aligns with the recommendations of the DMC and the WMO, as well as those implemented by the DGA. Such a moving window aligns with the typical actions of the actors involved, who cannot continuously change their protocols, designs, and procedures.
- The platform should consider different drought severities, with at least three conditions different from the normal or non-scarcity condition. This differentiation allows the development of measures with a different focus, all within a context of preventive management with monitoring and early detection of potentially critical conditions.

- The platform should consider a spatialized description of droughts and their attributes at different spatial scales. On the other hand, identifying and classifying hydrological and agricultural droughts eventually requires information on groundwater, snowpack, and soil moisture. This information is not entirely measurable and, in many cases, it is obtained from hydrological models. Finally, the platform should have a forecasting capability based on climate forecasts and general forecasting models.

13.5 Conclusions

Based on a comparative analysis of international tools and the results of expert consultation, this chapter analyzes the expected characteristics of a system for drought monitoring, tracking, and management based on quantitative drought indices. In particular, standardized indices are considered, allowing a simple analysis and comparison of drought conditions over time and in different spatial locations. Both conditions are essential to facilitate management within a national context.

An adequate platform should consider multiple drought indices with different levels of severity for their classification. These indices should be automatically calculated using as much information as possible, both from national and international sources, from point sources as well as from satellite and remote sensing products, and even obtained from modeling. In addition, a good platform should be open to the public and designed to address the needs of different users. The visualization should be simple, straightforward, and flexible. In addition, the tool should focus on identifying a drought event from quantitative information and provide information on its manifestation in the territory and its short-term forecast. All these attributes are essential to lay the foundations for public policies focused on preventive and proactive drought management.

Acknowledgments The authors gratefully acknowledge funding from ANID project FSEQ210018.

References

Aceituno, P., Boisier, J. P., Garreaud, R., Rondanelli, R., & Rutllant, J. A. (2021). Climate and weather in Chile. In B. Fernández & J. Gironás (Eds.), *Water resources of Chile. World water resources, 8* (pp. 7–29). Springer. https://doi.org/10.1007/978-3-030-56901-3_2

Boisier, J. P., Alvarez-Garretón, C., Cordero, R., Damiani, A., Gallardo, L., Garreaud, R. D., Lambert, F., Ramallo, C., Rojas, M., & Rondanelli, R. (2018). Anthropogenic drying in Central-Southern Chile evidenced by long-term observations and climate model simulations. *Elementa Science of the Antropocene, 6*, 74. https://doi.org/10.1525/elementa.328

Bordi, I., & Sutera, A. (2007). Drought monitoring and forecasting at large scale. In G. Rossi, T. Vega, & B. Bonaccorso (Eds.), *Methods and tools for drought analysis and management* (Water science and technology library) (Vol. 62, pp. 3–27). Springer. https://doi.org/10.1007/978-1-4020-5924-7_1

Carrão, H., Russo, S., Sepulcre-Canto, G., & Barbosa, P. (2016). An empirical standardized soil moisture index for agricultural drought assessment from remotely sensed data. *International Journal of Applied Earth Observation and Geoinformation, 48*, 74–84. https://doi.org/10.1016/j.jag.2015.06.011

Dirección General de Aguas (DGA). (2022a). Escasez Hídrica para el 47.5% de la población. https://dga.mop.gob.cl/noticias/Paginas/DetalledeNoticias.aspx?item=835#:~:text=Estas%20188%20comunas%2C%20en%20las,nacional%20(231.056%20km2)%E2%80%9D%2C. Acceso el 15/03/2023.

Dirección General de Aguas (DGA). (2022b). Estudio de perfeccionamiento de las capacidades de la DGA en gestión de ciclos de sequía y escasez" S.I.T. N° 483. Ministerio de Obras Públicas, Dirección General de Aguas, : DICTUC S.A.

Dirección General de Aguas (DGA). (2022c). *Resolución 1331 exenta deja sin efecto la resolución D.G.A. n° 1.674 (exenta), de 12 de junio de* 2012, y establece criterios que determinan el carácter de severa sequía, de conformidad a lo dispuesto en el artículo 314 del código de aguas. Ministerio de Obras Públicas, Santiago, Chile.

Edwards D. C., & McKee, T. B. (1997). Characteristics of 20th century drought in the United States at multiple scales. Climatology Report 97-2, Department of Atmospheric Science, Colorado State University, Fort Collins. https://apps.dtic.mil/sti/pdfs/ADA325595.pdf. Accessed on 23/01/13.

Fernández, B., & Gironás, J. (2017). Drought characteristics. Chapter 79. In V. P. Singh (Ed.), *Handbook of applied hydrology* (pp. 79-1–79-7). McGraw-Hill Education.

Fernández Larrañaga, B. (1997). Identificación y caracterización de sequías hidrológicas en Chile central. *Ingeniería del agua, 4*(4), 37–46.

Garreaud, R., Boisier, J. P., Rondanelli, R., Montecinos, A., Sepúlveda, H. H., & Veloso-Aguila, D. (2019). The Central Chile mega drought (2010–2018): A climate dynamics perspective. *International Journal of Climatology, 1–19*, 421–439. https://doi.org/10.1002/joc.6219

Hao, Z., & Agha Kouchak, A. (2013). Multivariate standardized drought index: A parametric multi-index model. *Advances in Water Resources, 57*, 12–18. https://doi.org/10.1016/j.advwatres.2013.03.009

Hao, Z., Singh, V. P., & Xia, Y. (2018). Seasonal drought prediction: Advances, challenges, and future prospects. *Reviews of Geophysics, 56*(1), 108–141. https://doi.org/10.1002/2016RG000549

Hayes, M., Svoboda, M., Wall, N., & Widhalm, M. (2011). The Lincoln declaration on drought indices: Universal meteorological drought index recommended. *Bulletin of the American Meteorological Society, 92*(4), 485–488. https://doi.org/10.1175/2010BAMS3103.1

McKee, T. B., Doesken, N. J., & Kleist, J. (1993). The relationship of drought frequency and duration to time scales. In *Proceedings of the 8th Conference on Applied Climatology, 17*(22), 179–183.

Ministerio de Obras Públicas (MOP). (2021). *Nuevas normales climáticas 1991–2020*. https://dga.mop.gob.cl/dgadocumentos/Nuevas_Normales_Climaticas.pdf Acceso el 17/03/2023.

Modarres, R. (2007). Streamflow drought time series forecasting. *Stochastic Environmental Research and Risk Assessment, 21*(3), 223–233. https://doi.org/10.1007/s00477-006-0058-1

World Meteorological Organization (WMO). (2016). *Handbook of Drought Indicators and Indices*. Integrated Drought Management Programme (IDMP), Integrated Drought Management Tools and Guidelines Series 2. Geneva.

Organización meteorológica mundial (OMM). (2017). Directrices de la Organización Meteorológica Mundial sobre el cálculo de las normales climáticas. https://library.wmo.int/doc_num.php?explnum_id=4167%20Organizaci%C3%B3n%20Meteorol%C3%B3gica%20Mundial. Acceso el 16/03/2023.

World Meteorological Organization (OMM). (2022). La Organización Meteorológica Mundial presenta el informe "El estado del clima en América Latina y el Caribe." https://www.cepal.org/es/noticias/la-organizacion-meteorologica-mundial-presentainforme-estado-clima-america-latina-caribe. Acceso el 16/03/2023.

World Health Organization (WHO). (2022). Sequía. https://www.who.int/health-topics/drought?gclid=Cj0KCQiAx6ugBhCcARIsAGNmMbhLxG613NA77XlRb-3p2T3ttAdrMGmpnRdU2tu3ZgGqEx1n3_R-4fsaAkTXEALw_wcB#tab=tab_1 Acceso el 15/03/2023.

Svoboda, M., & Fuchs, B. (2016). Handbook of drought indicators and indices. In D. Wilhite & R. S. Pulwarty (Eds.), *drought and water crises: Integrating science, management, and policy, second edition* (1st ed., pp. 1–44). https://doi.org/10.1201/b22009-11

Telesca, L., Lovallo, M., Lopez-Moreno, I., & Vicente-Serrano, S. (2012). Investigation of scaling properties in monthly streamflow and standardized streamflow index (SSI) time series in the Ebro basin (Spain). *Physica A: Statistical Mechanics and its Applications, 391*(4), 1662–1678. https://doi.org/10.1016/j.physa.2011.10.023

Vicente-Serrano, S. M., Beguería, S., & López-Moreno, J. I. (2010). A multiscalar drought index sensitive to global warming: The standardized precipitation evapotranspiration index. *Journal of Climate, 23*(7), 1696–1718. https://doi.org/10.1175/2009jcli2909.1

Wilhite, D. A., & Glantz, M. H. (1985). Understanding: The drought phenomenon: The role of definitions. *Water International, 10*(3), 111–120. https://doi.org/10.1080/02508068508686328

World Meteorological Organization. (2012). *Standardized precipitation index user guide* (p. 1090). World Meteorological Organization.

Wu, H., Hayes, M. J., Weiss, A., & Hu, Q. I. (2001). An evaluation of the standardized precipitation index, the China-Z index and the statistical Z-score. *International Journal of Climatology: A Journal of the Royal Meteorological Society, 21*(6), 745–758.

Chapter 14
Remote Sensing Phenology as a Tool for Monitoring the Impact of Drought on Forest Ecosystem Productivity in Temperate Forests of South America

Marcelo D. Miranda, Aurora Gaxiola, and Valentina Bravo

Abstract Drought's impact on temperate forests' primary productivity is a well-established phenomenon. Monitoring changes in ecosystem phenology is paramount for understanding drought's effects on these forests. To this end, various remote sensing metrics have been developed to track vegetation changes and phenology. In this chapter, we present a selection of these metrics and their applications in monitoring changes in two temperate forests of Chile. Overall, our findings highlight the utility of remote sensing phenology as a tool for accurately assessing the impact of drought on forest ecosystem productivity.

Keywords Forest productivity phenology · Remote sensing · Temperate forest · Drought · South America

14.1 Introduction

Vegetation-related dynamics play a fundamental role in understanding the changes in ecosystems'structure and functioning in response to natural and anthropogenic disturbances (Berdugo et al., 2022; Miranda et al., 2023). In particular, the Temperate

M. D. Miranda (✉)
Facultad de Agronomía y Sistemas Naturales, Pontificia Universidad Católica de Chile, Santiago, Chile

Center of Applied Ecology and Sustainability (CAPES), Pontificia Universidad Católica de Chile, Santiago, Chile
e-mail: mmirands@uc.cl

A. Gaxiola
Facultad de Ciencias Biológicas, Pontificia Universidad Católica de Chile, Santiago, Chile

V. Bravo
Facultad de Agronomía y Sistemas Naturales, Pontificia Universidad Católica de Chile, Santiago, Chile

D. Rivera, G. Donoso (eds.), *Droughts in Chile*, Global Issues in Water Policy 31,
https://doi.org/10.1007/978-3-031-85040-0_14

region of South America (32°–43° S Lat., 71°–74° W Long.) is affected by disturbances derived from climate change, including increased aridity and extreme events such as droughts and heatwaves in recent years (Garreaud et al., 2020). Additionally, it faces human-induced disturbances such as changes in land use, intensive agriculture, pollution, and overexploitation of resources (Hernández et al., 2016; Root-Bernstein et al., 2016). These disturbances result in a decrease in the availability of ecosystem services, compromising their integrity (Smith-Ramírez et al., 2023).

Monitoring vegetation dynamics has become an essential strategy for detecting the impacts of global change on terrestrial ecosystems. Vegetation plays a fundamental role in the hydrological cycle by influencing evapotranspiration and interception and in biogeochemical cycles by participating in photosynthesis and respiration where nutrients are inputted, transported, and fixed. These characteristics make it an active mass and energy exchange regulator between the atmosphere and the land surface (Kariyeva et al., 2012). Therefore, if the relationship between climate and vegetation determines its distribution and dynamics, any change in climatic conditions can lead to modifications in structure and functioning, affecting the ecosystem balance (Malhi et al., 2020).

Remote sensing plays a central role in studying the spectral response of leaves and vegetation canopies to electromagnetic radiation. This response, related to primary productivity, allows continuous monitoring of ecosystems at different spatial scales (Ustin & Middleton, 2021). Satellite image analysis facilitates obtaining information on large-scale temporal variations in forests, thereby aiding the study of their spatiotemporal trajectories (Kwork, 2018). Primary productivity can be estimated from time series based on vegetation indices generated from satellites, such as the Normalized Difference Vegetation Index (NDVI), Enhanced Vegetation Index (EVI), among others (Huang et al., 2019).

Using time series based on vegetation indices allows for identifying of areas where significant disturbances have occurred, such as vegetation browning due to drought and heatwaves and changes in land use associated with human intervention (Miranda et al., 2023). NDVI and EVI indices have been widely used to study vegetation responses to climatic variation in semiarid and temperate areas, proving to be efficient indicators due to their easy obtention and interpretation (van Leeuwen et al., 2013); Glade et al., 2016; Huang et al., 2019).

Satellite sensors such as AVHRR, MODIS, TM, and OLI associated with NOAA satellite series, Terra and Aqua platforms, and Landsat satellites have been crucial for studying the evolution of primary productivity and phenology. Phenology has received significant attention in the last decade as many organisms adapt their life cycles to ongoing climate change (Gordo & Sanz, 2010).

Satellite phenology, which involves using spectral satellite data to track seasonal variations in land surfaces, provides consistent estimates of primary productivity (PP), as well as temporal events associated with the onset, peak, end, and duration of it in extensive areas using vegetation indices (Zhang et al., 2003; Ivits et al., 2013; van Leeuwen et al., 2013). Combiningphenology and vegetation indices allows informing about terrestrial ecosystem dynamics, aiding in the characterization of their structure, dominant composition, and providing the necessary information for their classification from a functional perspective. Studies of primary productivity and its phenology based on satellite vegetation indices enable continuous monitoring thereof, presenting the additional advantage of rapid updates as technological development progresses (Comer et al., 2003; Zeng et al., 2022; Ferchichi et al., 2022).

This chapter aims to show the application of phenological metrics obtained from satellite sensors in characterizing natural vegetation dynamics in Temperate forests of South America. This work focuses on phenological variables representing the temporality, quantity, and trend of primary productivity over 20 years observed through the EVI. First, a description of the phenological variables is provided to achieve this. Secondly, a selection of these variables is used to describe the behavior of sclerophyll, spinal, deciduous, conifer and evergreen forests in the temperate region in central and southern Chile.

14.2 Remote Sensing Phenology

Phenology is the study of periodic events associated with the biological cycles of living organisms, including their relationship with seasonal or interannual variations in climate. In the case of plant primary productivity, it corresponds to the events that account for this process over time, such as the start date, end, or maximum of photosynthetic or growth activity during the year. Given that through satellite sensors, it is possible to observe the behavior of primary productivity through spectral combinations or vegetation indices (mainly NDVI and EVI), phenology can be characterized through satellite images. This allows for monitoring seasonal and interannual vegetation changes at local, regional, or global scales (http://remotesensing.usgs.gov).

The events associated with phenology observed from satellite images (referred to as remote sensing phenology) are obtained from the analysis of time series representing the behavior of the primary productivity in terms of vegetation index values over a period of at least 2 years of observation (Fig. 14.1). The spatial and temporal resolution of the series may vary depending on the satellite sensor used, with common resolutions being 1.1 km, 500 m, 250 m, and 10 m, associated with AVHRR, MODIS sensors, and Sentinel 2 satellites. Temporally, it is possible to construct series with extensions of 46, 21, and 6 years, respectively, for the same sensors.

From a time series that represents vegetation index values, it is possible to recognize the following phenological events associated with vegetation primary productivity (Jönsson & Eklundh, 2002, 2004), which are related to temporality (occurrence) and the productivity values recorded in it. Both types are described below.

Temporal Phenological Metrics (Fig. 14.2a):

- Start of season (SOS): Date in Julian day when productivity value in terms of vegetation index has increased by x% from the previous minimum value observed

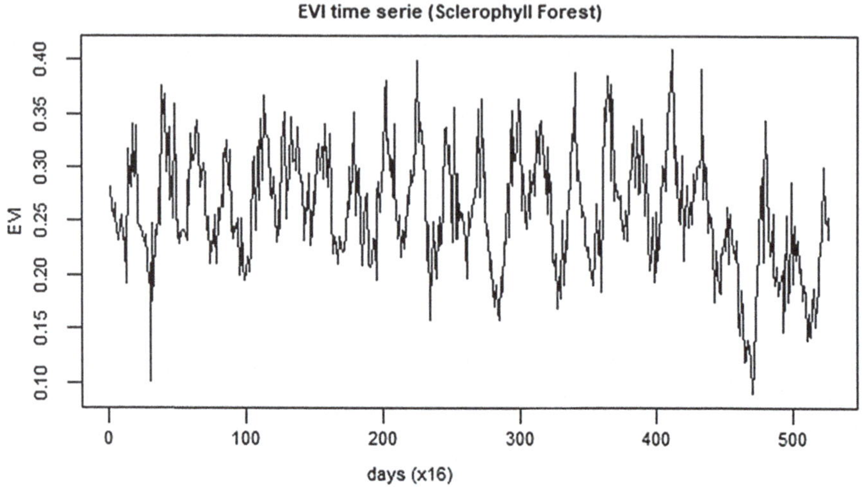

Fig. 14.1 Example of a time series for the EVI covering years 2002 and 2022. The series consists of 23 values per year, corresponding to an average observation every 16 days. In total, each series has 529 observations. Author's elaboration

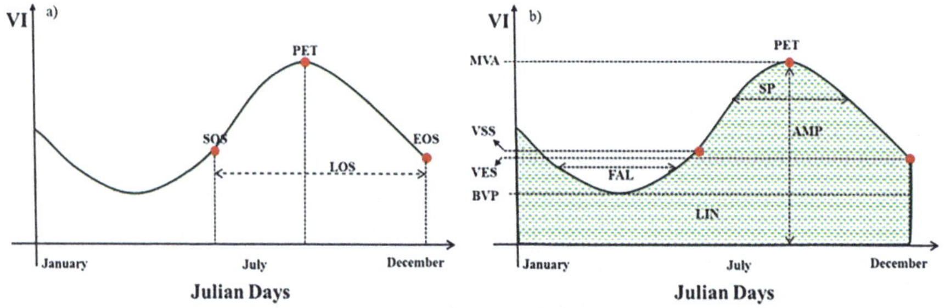

Fig. 14.2 Different phenological variables associated with a temporal series of VI vegetation index values, considering a one-year period.: (**a**) temporal phenological metrics, (**b**) Vegetation Index-based productivity phenological metrics. (Author's elaboration)

before the maximum value in the previous spring. The increment value "x" depends on each species group or vegetation community, with 20–40% being used for Mediterranean and Temperate forests.

- End of season (EOS): The date in Julian day when productivity has reached y% of the minimum value after the spring maximum. Like to the previous case, y% for Mediterranean and Temperate forests can vary between 20% and 40%.
- Length of season (LOS): The number of Julian days between the start and end of the growing season (SOS-EOS).
- Peak of time (PET). Date in Julian day when the maximum vegetation index value occurs within the growth season given by SOS and EOS.

SOS indicates the point from which productivity accelerates due to improved environmental conditions that conditioning vegetation development, while EOS indicates the time when a period of low productivity or senescence begins. LOS indicates the number of days the growing season has lasted, and PET is the day when the decrease in productivity-related activity begins.

Vegetation Index-Based Productivity Phenological Metrics (Fig. 14.2b):

- Start of season productivity value (VSS): The vegetation index value at the start of the growing season (SOS).
- End of season productivity value (VES): The vegetation index value at the end of the growing season (EOS).
- Maximum productivity value (MVA): the maximum vegetation index value between the start (SOS) and end (EOS) of the season.
- Base productivity value (BVP): the average value obtained between the vegetation index value at the start (SOS) and end (EOS) of the season.
- Productivity amplitude (AMP): The difference between the maximum vegetation index value (MVA) and the base value (BVP) for the season.
- Seasonal integral (LIN): The sum of vegetation index values under the accumulated curve between the start (SOS) and end (EOS) of the season.
- Spring productivity (SP): The annual average of the vegetation index for the spring season (September 21 to December 21).
- Autumn productivity (FAL): The annual average of the vegetation index for the autumn season (March 21 to June 21).

The different productivity indicators obtained from the time series associated with vegetation indices account for short- and long-term changes in ecosystem activity. These indicators are sensitive to intra- and inter-annual variations in climate (Sepúlveda et al., 2018). However, the internal ecosystem mechanisms and processes that produce these events are still difficult to elucidate directly from satellite information (Dronova & Taddeo, 2022).

The values of the phenological metrics are obtained by adjusting a function that relates the vegetation index behavior to time. The first step for this adjustment corresponds to an analysis of the integrity of the time series, where anomalous and missing values are identified, which can be imputed following the methodology proposed by (Moritz & Bartz-Beielstein, 2017). Then, a model is fitted to the cleaned series that relates the vegetation index values to time, for which it is possible to use functions such as adaptive Savitzky-Golay, asymmetric Gaussian, double logistic, or non-parametric fits with different degrees of complexity (Jönsson & Eklundh, 2002). Table 14.1 shows an example of values for some phenological variables obtained from the TIMESAT 3.3 program, including the above functions (Jönsson & Eklundh, 2004). Unlike global functions, for example based on Fourier transformations, local functions are capable of capturing changes in seasonal synchronization between different years, a property that makes them suitable for studying vegetation dynamics and phenology.

Table 14.1 Example of values for phenological variables for five phenological periods obtained from EVI time series processed by the TimeSat 3.3 program

Season	SOS	EOS	LOS	BVP	PET	AMP	LIN	FAL	SP
1	260	474	214	0.365	379	0.177	7.383	0.428	0.472
2	280	509	229	0,399	384	0.154	8.202	0.498	0.490
3	276	539	263	0.386	436	0.167	9.679	0.530	0.526
4	229	532	302	0.354	362	0.206	10.530	0.478	0.482
5	265	498	233	0.394	358	0.125	8.228	0.461	0.511

Author's elaboration

14.3 Study Cases in Southern Temperate Forests

To represent vegetation dynamics through phenological variables, a case study is presented below, encompassing the Mediterranean and Temperate regions of Chile between latitudes 32° S and 44° S and longitudes 70.8° W and 74.8° W. The region's climate is temperate Mediterranean (Csb2), characterized by marked dry and warm summer seasons (Peel et al., 2007). Mean temperatures in warmer and colder months are 29.4 °C and 14.3 °C, respectively, and mean annual rainfall ranges between 300 and 2000 mm per year. Due to extreme drought in the study area during the last decade, total yearly rainfall decreased by 25–45% (Garreaud et al., 2020).

The vegetation is characterized by sclerophyll, deciduous, evergreen, and conifer forests, grasslands, and shrublands, developed under a marked dry and warm summer season influenced by mountain physiography (Luebert & Pliscoff, 2018). This study case describes phenology for natural vegetation, including principally woodland growth forms, and excludes highly artificial ecosystems based on land cover maps developed by Zhao et al. (2016) and CONAF (2021).

The spatial resolution defined for the calculation of phenological variables is 6.25 ha (250 × 250 m), corresponding to the resolution of the MODIS satellite product MOD13Q1v061 "MOD13Q1.061 Terra Vegetation Indices 16-Day Global 250 m" obtained from the MODIS sensor (Didan, 2015). Furthermore, the temporal extension is considered to span 20 years from 2002 to 2022. The temporal granularity is 16 days for EVI vegetation index data series and 12 months for the phenological variables derived from them. The variables presented in Fig. 14.3 correspond to: (a) Start of Season (SOS), (b) End of Season (EOS), (c) Peak Time (PET) and Seasonal Integral (LIN); these show an example of spatial representation of the annual average phenological variables for a deciduous forest in the Maule region (SOS, EOS, and LIN).

Additionally, images representing the temporal trend of phenology for each selected variable were obtained. This trend was calculated through a simple linear regression model, where the slope and significance level of each pixel were stored.

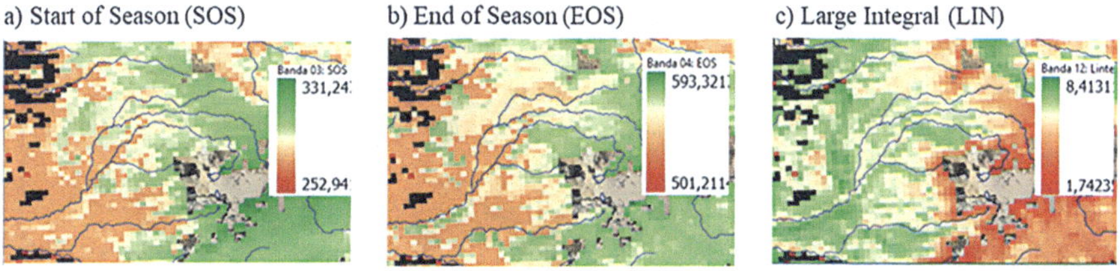

Fig. 14.3 Example of a spatial representation of phenological variables for deciduous forest in the Maule mountain range: (**a**) Start of Season (SOS) in Julian days, (**b**) End of Season (EOS) in Julian days, and (**c**) Large Integral (LIN) in accumulated EVI values. The images represent the annual mean value for a 20 year. (Author's elaboration)

14.4 Results and Discussion

The vegetation productivity observed through the MODIS sensor confirms the north-south gradient, which is presented across the 12° of latitude covering the study area (Fig. 14.4). The EVI index takes values between 0.1 and 1 when there is photosynthetic activity or the presence of active and structured vegetation; values between 0 and 0.1 are associated with bare soils and less than 0 with water bodies, ice, or urban cover (Zhengxing et al., 2003). For the study area, it is possible to observe that between the parallels 32° S and 34° S, natural vegetation reaches average annual EVI values lower than 0.22. This behavior is mainly associated with sclerophyll forests and shrublands that develop under low precipitation (300 mm/year) and long annual drought periods (8 months), which have intensified over the last 13 years (Miranda et al., 2020). This situation contrasts with the EVI values presented south of parallel 38° S (average annual >0.40), which is characteristic of coniferous and evergreen forests that develop under shorter dry periods (4 months) and precipitation amounts ranging from 1000 to 2300 mm/year. An intermediate situation is represented by deciduous forests located on the foothills of the Andes Mountain range (parallels 36° S–39° S), where productivity reaches EVI values between 0.30 and 0.40 and under annual precipitation values averaging around 700 mm/year.

Regarding the inter and intra-annual variability of the EVI index (Fig. 14.5), it is observed that the study region, presents different amplitude values, which are associated with the variability of vegetation productivity and its relationship with climate. High amplitude values are mainly present in deciduous and shrubland forests due to the natural loss of much of their photosynthetic coverage in winter (EVI amplitude = 0.40). The opposite, lower amplitude, is presented by coniferous and evergreen forests (EVI amplitude = 0.10), whose characteristic is the conservation of foliage throughout the year, which is also associated with a more extended period of productivity. Due to the presence of clouds, forests located south of parallel 39°S present more noise in their signal, which can sometimes be a problem for calculating phenological variables. These responses associated with different vegetation

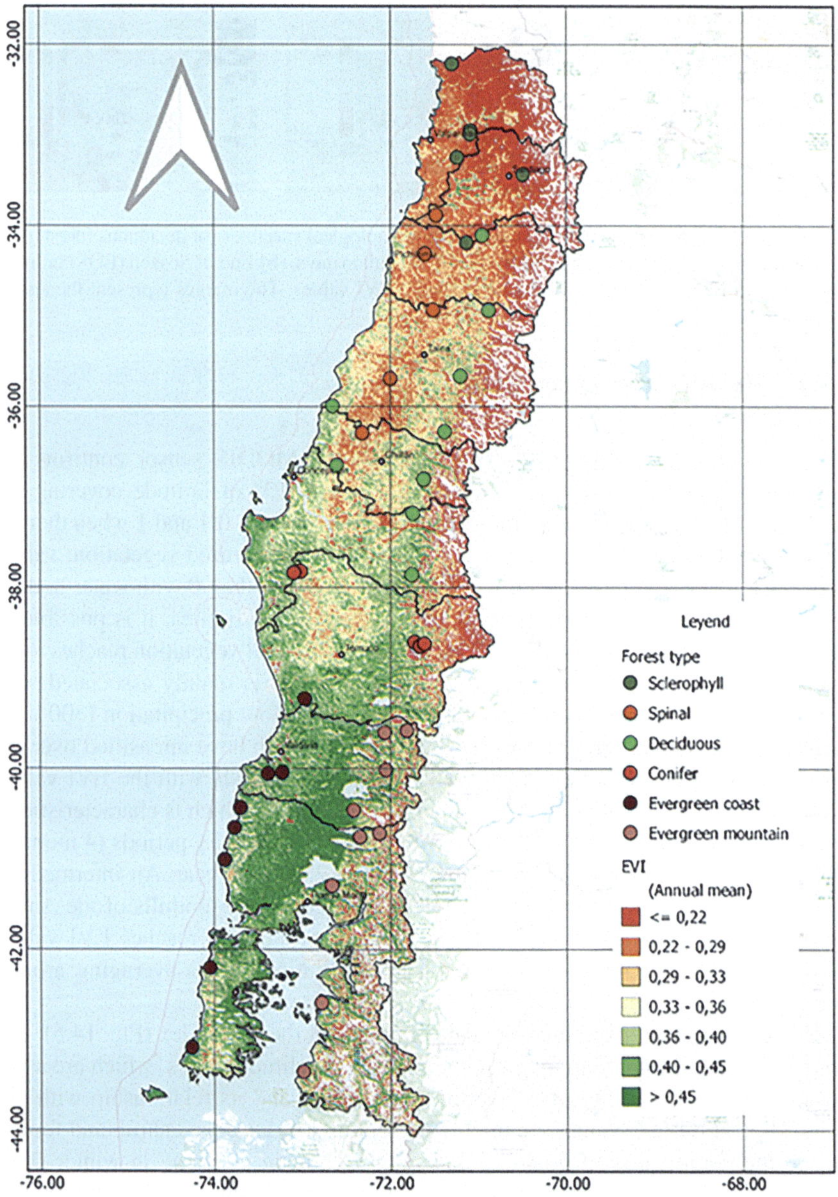

Fig. 14.4 EVI annual average for the study period. (Author's elaboration)

types have already been observed in other studies using other global (van Leeuwen et al., 2013) and local capture sensors in this study region (Glade et al., 2016).

Regarding the phenological variables (Fig. 14.6 and Table 14.2), SOS shows its productive onset through a marked latitudinal gradient (Fig. 14.6a). This is

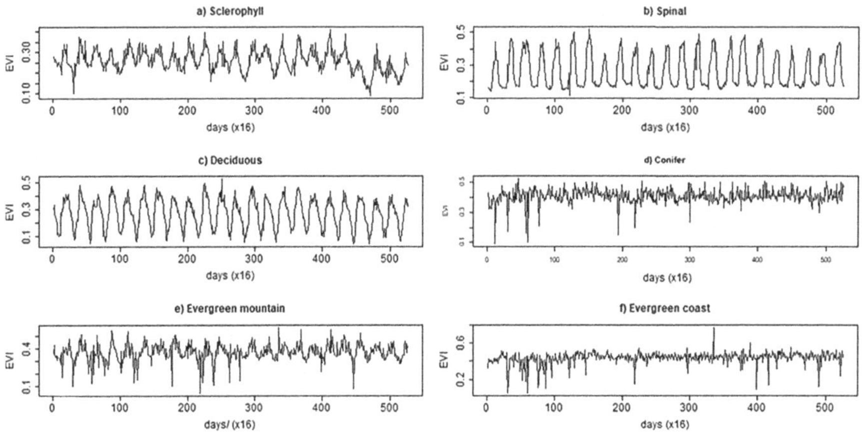

Fig. 14.5 EVI signals for forests: (**a**) Sclerophyll, (**b**) Shrubland, (**c**) Deciduous, (**d**) Conifers, (**e**) Mountain evergreen, and (**f**) Coastal evergreen. (Author's elaboration)

characterized by an average start of activity around the first half of July (Julian day 203) for shrublands, the first week of August (Julian day 255) for sclerophyll forests, the first week of September (Julian day 284) for conifer and evergreen forests, and the beginning of October (Julian day 303) for deciduous forests. A similar behavior in terms of gradient is presented by End of Season (EOS) (Fig. 14.6b). On the other hand, the day of maximum productive activity represented by the PET variable occurs 2.6 months after the start of the season (SOS) for shrublands, three months for sclerophyll and deciduous forests, and between 3.5 and 4 months for coniferous and evergreen forests (Fig. 14.6c). The sclerophyll and evergreen forests stand out for their more significant variability, mainly due to the structural variability they present throughout their distribution (Bannister & Donoso, 2013).

Furthermore, it is possible to observe a marked latitudinal gradient for the accumulated productivity (LIN), which dramatically differentiates the behavior of forests located north and south of the study region. Accumulated values fluctuating between 1 and 3.5 are observed for shrublands, 3.5 and 6 for sclerophyll forests, and 6 and 16 for deciduous and evergreen forests (Figure 14.6d). It is noteworthy that the accumulated values of the EVI, in this case, can be 16 times higher between one type of forest and another, which is clearly reflected in its structural and seasonal dynamics.

Regarding the trend of primary productivity represented by LIN, Fig. 14.7 shows those values that were significant for the linear regression applied to the 20-year observation series. Three behaviors are highlighted: (a) sectors with negative slope (decrease in accumulated productivity), which are mainly found between parallels 32° and 36° S, associated with sclerophyll and shrublands; (b) sectors with positive slope (increase in accumulated productivity), mainly associated with vegetation located on the limit of tree vegetation between parallels 32° and 38° S; and (c) zero slope representing no changes mainly located towards the south of the study region.

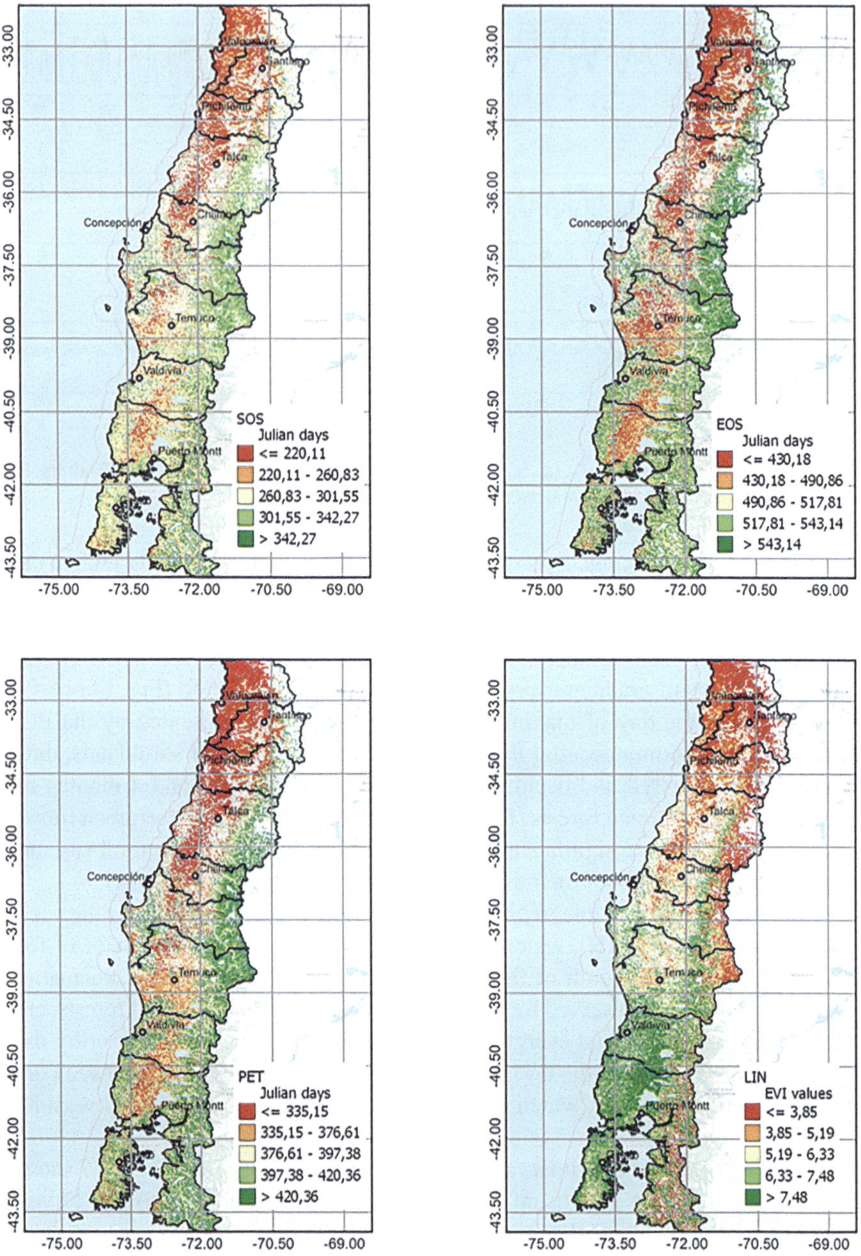

Fig. 14.6 Phenological variables: (**a**) Start of Season (SOS), (**b**) End of Season (EOS), (**c**) Peak Time (PET) and (**d**) accumulated integral (LIN). The first three variables are measured in Julian days (1–365) and the last in EVI values

Table 14.2 Mean and standard deviation of phenological variables start of season (SOS), end of season (EOS), day of maximum activity (PET), and accumulated productivity integral (LIN) for different types of forests associated with the study area. The values of the first three variables correspond to Julian days (1–365). The LIN variable is represented in EVI vegetation index values (−1, 1)

	SOS		EOS		PET		LIN	
Forest type	Mean	Std. Dev.	Mean	Std. Dev.	Mean	Std. Dev.	Mean	Std. Dev.
Sclerophyll	254.92	30.22	469.57	40.19	350.72	31.88	4.24	1.03
Spinal	203.42	10.12	349.44	15.44	282.91	11.06	3.64	0.17
Deciduous	302.84	12.92	525.22	19.24	395.02	15.38	6.62	1.35
Conifer	288.56	28.62	518.17	19.25	396.15	21.80	6.83	0.68
Evergreen mountain	280.41	46.92	511.55	56.56	398.71	54.61	6.55	1.15
Evergreen coast	283.72	33.50	528.83	47.66	410.89	48.83	6.92	1.48

Author's elaboration

It is important to note that a series of isolated pixels with positive and negative slopes are randomly distributed throughout the study region. Their trend values may be due to anthropogenic processes independent of the climate, such as forest management, land use change, among others.

These results coincide with observations in recent years regarding the effect of the last heatwaves and prolonged drought, which began in 2010 (Garreaud et al., 2017). The peaks reached in 2019–2020 generated widespread browning of vegetation between the Coquimbo and O'Higgins regions (Miranda et al., 2020, 2023). It should also be noted that productivity declines may be associated with the effects of forest fires, especially those occurring in 2017 and 2022.

14.5 Conclusion

The increase in droughts and heatwaves and the change in the hydrological cycle due to climate change are extensively and recurrently affecting ecosystem productivity, especially terrestrials.

Monitoring these changes is an urgent task that can be extensively addressed using satellite remote sensing. Based on the spectral data they capture, metrics representing the phenological behavior of primary productivity can be obtained.

The use of phenological metrics from satellite images allows for the detection of changes associated with the onset, end, and duration of productivity over time and the quantification of its accumulated value.

Furthermore, it allows differentiation between vegetation structures (forest types) and observation of their trends over time. In our study case, which was associated with the Temperate region of South America, differences in satellite phenological variables associated with different types of forests were observed, located

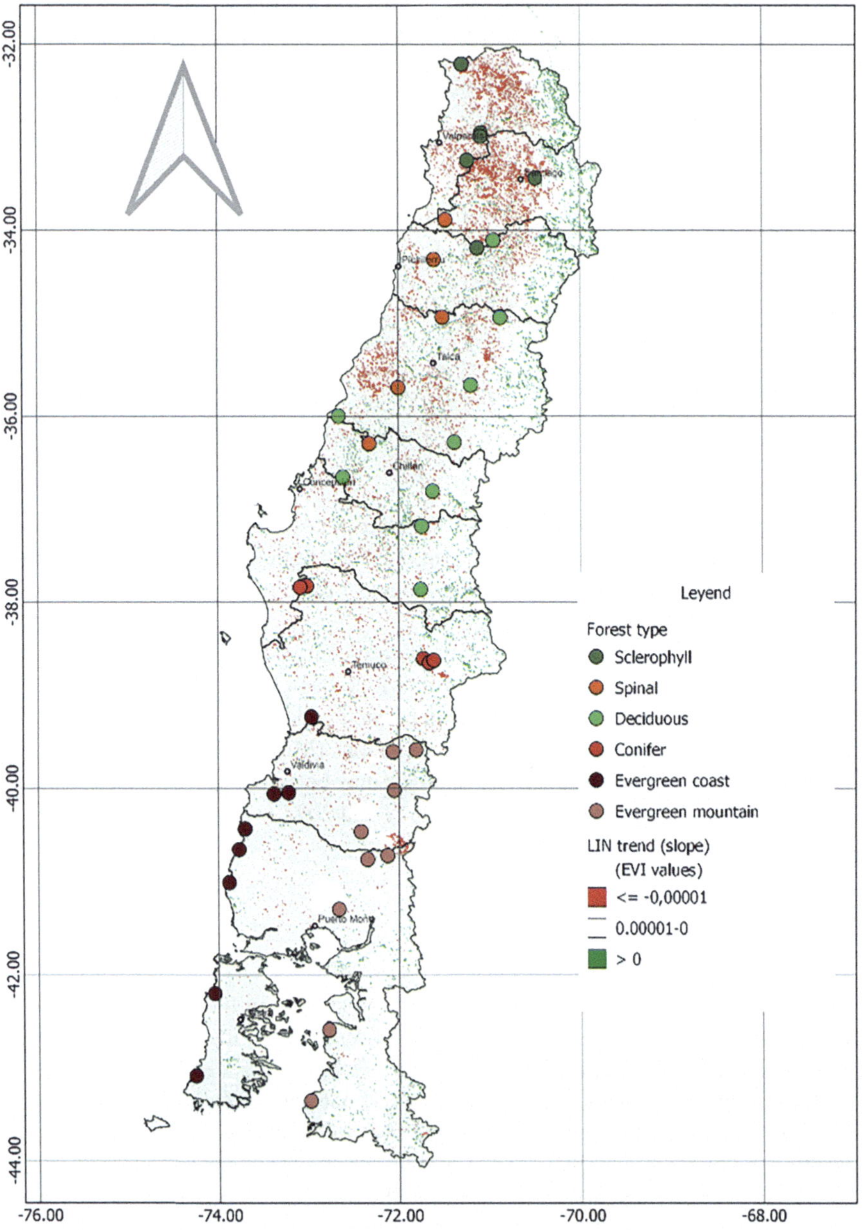

Fig. 14.7 Trend of primary productivity represented by the phenological variable LIN. Red-colored pixels represent a negative trend, while green-colored pixels represent a positive trend for the LIN variable for a 20-year observation period. (Author's elaboration)

under different environmental conditions regarding water availability and temperature (latitudinal gradient).

Moreover, through the measurement of satellite phenological variables over a long period, negative and positive trends were observed depending on the location and characteristics of the vegetation. The most unfavorable condition regarding productivity trend is observed in the northern zone of the study region, where there is an accumulation of pixels with a negative significant trend.

This information reinforces what has already been observed in the region associated with widespread browning events due to droughts and heatwaves occurring in the last 5 years in this region. The use of phenological metrics obtained by satellite sensors is presented as key tool for monitoring ecosystem changes occurring at different spatial and temporal scales. Their improvement over time is key to providing more precise information for decision making and planning.

References

Bannister, J. R., & Donoso, P. J. (2013). Forest typification to characterize the structure and composition of old-growth evergreen forests on Chiloe Island, North Patagonia (Chile). *Forests, 4*(4), 1087–1105. https://doi.org/10.3390/f4041087

Berdugo, M., Vidiella, B., Solé, R. V., & Maestre, F. T. (2022). Ecological mechanisms underlying aridity thresholds in global drylands. *Functional Ecology, 36*(1), 4–23. https://doi.org/10.1111/1365-2435.13962

Comer, P., Faber-Langendoen, D., Evans, R., Gawler, S., Josse, C., Kittel, G., Menard, S., Pyne, M., Reid, M., Schulz, K., Snow, K., & Teague, J. (2003). *Ecological systems of the United States: A working classification of U.S. Terrestrial Systems*. Nature Serve.

CONAF, 2021. Catastro de los Recursos Vegetacionales Nativos de Chile al año 2020. *Departamento de Monitoreo de Ecosistemas Forestales*. 76 p

Didan, K. (2015). MOD13Q1 MODIS/Terra vegetation indices 16-day L3 Global 250m SIN Grid V006 [data set]. In NASA EOSDIS Land Processes Distributed Active Archive Center.

Dronova, I., & Taddeo, S. (2022). Remote sensing of phenology: Towards the comprehensive indicators of plant community dynamics from species to regional scales. *Journal of Ecology, 110*(7), 1460–1484. https://doi.org/10.1111/1365-2745.13897

Ferchichi, A., Abbes, A. B., Barra, V., & Farah, I. R. (2022). Forecasting vegetation indices from spatio-temporal remotely sensed data using deep learning-based approaches: A systematic literature review. *Ecological Informatics, 68*, 101552. https://doi.org/10.1016/j.ecoinf.2022.101552

Garreaud, R. D., Álvarez-Garreton, C., Barichivich, J., Boisier, J. P., Christie, D., Galleguillos, M., LeQuesne, C., McPhee, J., y Zambrano-Bigiarini, M. (2017). The 2010–2015 megadrought in central Chile: impacts on regional hydroclimate and vegetation. *Hydrol. Earth Syst. Sci., 21*, 6307–6327, https://doi.org/10.5194/hess-21-6307-2017

Garreaud, R. D., Boisier, J. P., Rondanelli, R., Montecinos, A., Sepúlveda, H. H., & Veloso-Aguila, D. (2020). The Central Chile Mega Drought (2010–2018): A climate dynamics perspective. *International Journal of Climatology, 40*(1), 421–439. https://doi.org/10.1002/joc.6219

Glade, F. E., Miranda, M. D., Meza, F. J., & van Leeuwen, W. J. D. (2016). Productivity and phenological responses of natural vegetation to present and future inter-annual climate variability across semi-arid river basins in Chile. *Environmental Monitoring and Assessment, 188*(12), 676. https://doi.org/10.1007/s10661-016-5675-7

Gordo, O., & Sanz, J. J. (2010). Impact of climate change on plant phenology in Mediterranean ecosystems. *Global Change Biology, 16*(3), 1082–1106. https://doi.org/10.1111/j.1365-2486.2009.02084.x

Hernández, A., Miranda, M. D., Arellano, E. C., & Dobbs, C. (2016). Landscape trajectories and their effect on fragmentation for a Mediterranean semi-arid ecosystem in Central Chile. *Journal of Arid Environments, 127*, 74–81. https://doi.org/10.1016/j.jaridenv.2015.10.004

Huang, X., Xiao, J., & Ma, M. (2019). Evaluating the performance of satellite-derived vegetation indices for estimating gross primary productivity using FLUXNET observations across the globe. *Remote Sensing, 11*(15). https://doi.org/10.3390/rs11151823

Ivits, E., Cherlet, M., Mehl, W., & Sommer, S. (2013). Ecosystem functional units characterized by satellite observed phenology and productivity gradients: A case study for Europe. *Ecological Indicators, 27*, 17–28. https://doi.org/10.1016/j.ecolind.2012.11.010

Jönsson, P., & Eklundh, L. (2002). Seasonality extraction by function fitting to time-series of satellite sensor data. *IEEE Transactions on Geoscience and Remote Sensing, 40*(8), 1824–1832. https://doi.org/10.1109/TGRS.2002.802519

Jönsson, P., & Eklundh, L. (2004). TIMESAT—A program for analyzing time-series of satellite sensor data. *Computers & Geosciences, 30*(8), 833–845. https://doi.org/10.1016/j.cageo.2004.05.006

Kariyeva, J., Leeuwen, W. J. D., Woodhouse, C., & a. (2012). Impacts of climate gradients on the vegetation phenology of major land use types in Central Asia (1981–2008). *Frontiers of Earth Science, 6*(2), 206–225. https://doi.org/10.1007/s11707-012-0315-1

Kwork, R. (2018). Ecology's remote-sensing revolution. *Nature, 556*, 137–138.

Luebert, F., & Pliscoff Varas, P. A. (2018). Sinopsis bioclimática y vegetacional de Chile. (2 ed.).

Malhi, Y., Franklin, J., Seddon, N., Solan, M., Turner, M. G., Field, C. B., & Knowlton, N. (2020). Climate change and ecosystems: Threats, opportunities and solutions. *Philosophical Transactions of the Royal Society B: Biological Sciences, 375*(1794), 20190104. https://doi.org/10.1098/rstb.2019.0104

Miranda, A., Lara, A., Altamirano, A., Di Bella, C., González, M. E., & Julio Camarero, J. (2020). Forest browning trends in response to drought in a highly threatened mediterranean landscape of South America. *Ecological Indicators, 115*(August 2019), 106401. https://doi.org/10.1016/j.ecolind.2020.106401

Miranda, A., Syphard, A. D., Berdugo, M., Carrasco, J., Gómez-González, S., Ovalle, J. F., Delpiano, C. A., Vargas, S., Squeo, F. A., Miranda, M. D., Dobbs, C., Mentler, R., Lara, A., & Garreaud, R. (2023). Widespread synchronous decline of Mediterranean-type forest driven by accelerated aridity. *Nature Plants., 9*, 1810–1817. https://doi.org/10.1038/s41477-023-01541-7

Moritz, S., & Bartz-Beielstein, T. (2017). Impute TS: Time series missing value imputation in R. *R Journal, 9*(1), 207–218. https://doi.org/10.32614/rj-2017-009

Peel, M. C., Finlayson, B. L., & Mcmahon, T. A. (2007). Updated world map of the Köppen-Geiger climate classification. *Hydrology and Earth System Sciences, 11*, 1633–1644.

Root-Bernstein, M., Guerrero-Gatica, M., Piña, L., Bonacic, C., Svenning, J.-C., & Jaksic, F. M. (2016). Rewilding-inspired transhumance for the restoration of semiarid silvopastoral systems in Chile. *Regional Environmental Change., 17*, 1381–1396. https://doi.org/10.1007/s10113-016-0981-8

Sepúlveda, M., Bown, H. E., Miranda, M. D., & Fernández, B. (2018). Impact of rainfall frequency and intensity on inter- and intra-annual satellite-derived EVI vegetation productivity of an Acacia caven shrubland community in Central Chile. *Plant Ecology, 219*(10), 1209–1223. https://doi.org/10.1007/s11258-018-0873-8

Smith-Ramírez, C., Grez, A., Galleguillos, M., Cerda, C., Ocampo-Melgar, A., Miranda, M. D., Muñoz, A. A., Rendón-Funes, A., Díaz, I., Cifuentes, C., Alaniz, A., Seguel, O., Ovalle, J., Montenegro, G., Saldes-Cortés, A., Martínez-Harms, M. J., Armesto, J. J., & Vita, A. (2023). Ecosystem services of Chilean sclerophyllous forests and shrublands on the verge of collapse: A review. *Journal of Arid Environments, 211*(May 2022), 104927. https://doi.org/10.1016/j.jaridenv.2022.104927

Ustin, S. L., & Middleton, E. M. (2021). Current and near-term advances in earth observation for ecological applications. *Ecological Processes, 10*(1), 1. https://doi.org/10.1186/s13717-020-00255-4

van Leeuwen, W., Hartfield, K., Miranda, M., & Meza, F. (2013). Trends and ENSO/AAO driven variability in NDVI derived productivity and phenology alongside the Andes Mountains. *Remote Sensing, 5*(3), 1177–1203. https://doi.org/10.3390/rs5031177

Zeng, Y., Hao, D., Huete, A., et al. (2022). Optical vegetation indices for monitoring terrestrial ecosystems globally. *Nature Reviews Earth and Environment, 3*, 477–493. https://doi.org/10.1038/s43017-022-00298-5

Zhang, X., Friedl, M. A., Schaaf, C. B., Strahler, A. H., Hodges, J. C. F., Gao, F., Reed, B. C., & Huete, A. (2003). Monitoring vegetation phenology using MODIS. *Remote Sensing of Environment, 84*(3), 471–475. https://doi.org/10.1016/S0034-4257(02)00135-9

Zhao, Y., Feng, D., Yua, L., Wang, X., Chen, Y., Hernández, H.J., Galleguillos, M., Estades, C., Biging, G., Radke, J. Gong, P. (2016). Detailed dynamic land cover mapping of Chile: accuracy improvement by integrating multi-seasonal land cover data. *Remote Sensing of Environment 183*, 170–185

Zhengxing, W., Chuang, L., & Alfredo, H. (2003). From AVHRR-NDVI to MODIS-EVI: Advances in vegetation index research. *Acta Ecologica Sinica, 23*(5).

Chapter 15
Sentinel Hydrological Processes: Informing Drought Management Strategies in Chile

Aurora Gaxiola, Camilo del Río, Marcelo D. Miranda, and Francisco Suárez

Abstract This chapter explores the role of hydrological processes as sentinel indicators of climate change, emphasizing the time of appearance of anomalies in fog, groundwater, and vegetation dynamics in Mediterranean and semi-arid regions of Chile. By analyzing case studies on the spatial-temporal variability of fog, groundwater storage, and primary productivity, we explore how these variables signal changes in the hydrological cycle under increasing climatic stress. Groundwater availability and fog frequency offer crucial insights into water resource dynamics, while changes in phenology reveal vegetation responses to prolonged drought and heat waves. The findings of these original studies contribute to improving drought management strategies and highlight the importance of early detection of hydrological anomalies and ecosystem resilience in regions highly vulnerable to climate variability.

Keywords Drought mitigation · Fog · Groundwater · Phenology · Time of appearance anomalies

A. Gaxiola (✉)
Facultad de Ciencias Biológicas, Pontificia Universidad Católica de Chile, Santiago, Chile
e-mail: agaxiola@uc.cl

C. del Río
Instituto de Geografía, Pontificia Universidad Católica de Chile, Santiago, Chile

M. D. Miranda
Facultad de Agronomía y Sistemas Naturales, Pontificia Universidad Católica de Chile, Santiago, Chile

F. Suárez
Escuela de Ingeniería, Pontificia Universidad Católica de Chile, Santiago, Chile

D. Rivera, G. Donoso (eds.), *Droughts in Chile*, Global Issues in Water Policy 31,
https://doi.org/10.1007/978-3-031-85040-0_15

15.1 Introduction

In the face of increasing anthropogenic climate change, the concept of sentinel ecosystems emerges as a valuable tool to understand and anticipate the repercussions of environmental changes on our planet's habitats. Primarily associated with the study of glaciers (O'Gorman et al. 2014; Mark & Fernández, 2017), lakes (Schindler, 2009; Williamson et al., 2009; Echeverría et al., 2019) and wetlands (Tarnocai, 2006; Mitsch et al., 2013), sentinel ecosystems have proven to be indicators par excellence of climate changes by reflecting alterations in their biotic and abiotic characteristics. In other words, these ecosystems serve as sensitive indicators of long-term trends, establishing correlations with climatic variations and providing crucial insights into the potential consequences of such variations. For example, lakes assimilate and "store" information about alterations in the biogeochemical dynamics of their surrounding catchment areas, evidence that represents centuries of information (Fuentealba et al., 2020). On the other hand, wetlands, particularly peatlands, have emerged as exceptional sentinels, given the responsiveness of their biota to climatic conditions (Mataloni, 2016). The resilience of peatlands, exemplified in their ability to accurately reflect climatic dynamics over long extended periods, is evident in the late Holocene climate reconstruction (Van Bellen et al., 2016).

However, what is often analyzed in sentinel ecosystems are elements of biogeochemical cycles, i.e. fluxes and pools (Mueller et al., 2009; Dangles et al., 2017; Cantonati et al., 2022). For example, sediments in lakes and peatlands allow the identification of changes in fluxes in the carbon cycle (e.g. sequestration and productivity; Molot & Dillon, 1996; Lolu et al., 2020). In glacier ice, on the other hand, it is possible to analyze changes in the carbon pool from the analysis of atmospheric CO_2 trapped in ice bubbles (Thompson, 2010). Also, studies have shown how historical changes in vegetation cover within a watershed alter water fluxes and reservoirs (Istanbulluoglu & Bras, 2005; Karssenberg et al., 2017), making of vegetation key to understanding the hydrological cycle (Allen et al., 2015). Therefore, we propose the consideration of the hydrological cycle as a sentinel process, particularly relevant as water scarcity is an urgent and critical global challenge for countries such as Chile where desertification is a constant threat.

The concept of sentinel ecosystems or processes represents complexities as it requires not only a multidisciplinary approach but also the scientific capacity to generate, store and analyze long-term data that allow not only the identification of patterns but also anomalies (O'Gorman et al., 2014). Such complexities can be theoretically addressed through concepts such as environmental thresholds (Folke et al., 2004) and tipping points (Scheffer et al., 2009) as these allow for correlating long-term data with ecological processes and deciphering the resilience and vulnerability of ecosystems to climate change (Barnosky et al., 2012). Indeed, studies that have used these theoretical frameworks have allowed us to identify diverse ecological and environmental issues (Heinze et al., 2021) and have sparked debates about the environmental challenges that lie ahead (Barnosky et al., 2012). However, environmental thresholds or tipping points within sentinel processes can present technical challenges due to their complex and multifaceted nature (Tabor & Aguirre, 2004), and actions to prevent in the face of climate change are urgent (Marquet et al., 2019). The challenge for the scientific community, therefore, is to develop the capacity to understand and comprehend the historical temporal dynamics of social-ecological systems, and at the same time identify changes, however subtle, that may initially indicate trends toward critical thresholds or tipping points, offering valuable clues about the trajectory of ecosystem change and the urgency of proactive intervention measures (Scheffer et al., 2009; Barnosky et al., 2012; Heinze et al., 2021, but Hillebrand et al., 2023).

A key question for local governments is when we can expect to see the local effects of climate change and when we should change our management and conservation policies (e.g., Haasnoot et al., 2013). In the face of this uncertainty, time-of-emergence (ToE) analyses become relevant (Henson et al., 2017). ToE analyses emerge as a valuable complementary tool, as these analyses can be applied to long-term databases of environmental variables (Ascott et al., 2022; Rojas et al., 2019; Farías et al., 2021; Alaniz et al., 2024). However, the obstacle faced by many researchers is access to this type of data. Additionally, the ability to integrate the information is key, as it is not only important to identify an anomaly or the time of emergence, but also the sequence of occurrence in which these changes happen (Giorgi & Bi, 2009; King et al., 2015). In this sense, for this work, we consider the use of the concept of sentinel variables of the hydrological cycle. The premise of this work is that analysis of these "sentinel" variables highlights their inherent value in elucidating the intricate connections between climate, ecosystems, and human well-being. By identifying and analysing these variables, we aim to highlight the interconnection of key sentinel processes and their broader implications for water availability, soil health, biodiversity, and the well-being of all species. Ultimately, our compelling goal is to present a multidisciplinary scientific approach that can inform adaptive management strategies and policy interventions aimed at mitigating the detrimental effects of climate change in Chile.

Therefore, in this chapter, we present three short case studies on sentinel variables of the hydrological cycle in order to promote a discussion on the importance of knowing the spatio-temporal changes of water resources (fog and groundwater) and biological resources (phenology/primary productivity). In the first instance, del Río et al. presents in Sect. 15.2.1 the case of coastal fog as a sentinel variable by analyzing changes in its spatial distribution. This approach improves our understanding of the role of fog water in the hydrological cycle and provides a sentinel baseline of fog spatiotemporal patterns. A relevant aspect considering that in central and northern Chile, fog is an important water input for terrestrial coastal ecosystems and people (Larraín et al., 2002; Cereceda et al., 2008; Osses et al., 2017; Moat et al., 2021). In this case study, the main tools were the processing of data obtained by the Geostationary Operational Environmental Satellites Satellite (GOES-16) over a five-year time window and the analysis and processing of satellite images with the Google Earth Engine platform. The approaches allowed for modeling the spatiotemporal distributions (daily, annual, and interannual) of low clouds and fog from Arica to El Maule. Undoubtedly, understanding spatiotemporal dynamics and the factors that determine these helps to dimension the elements of the hydrological cycle in hyper-arid and semi-arid regions, such as those of northern and central Chile (Garreaud et al., 2008, 2017). Additionally, it allows us to understand potential historical and present changes in the ecological and evolutionary dynamics of coastal vegetation (Armesto et al., 1993; Armesto & Vidiella, 1993; Rundel et al., 1991; del Val et al., 2006; Latorre et al., 2011; Koch et al., 2019). This becomes important when remembering that vegetation and soil are key elements of the water cycle. Finally, having the ability to identify patterns and anomalies in the spatiotemporal dynamics of fog is essential to inform public policy on coastal vegetation conservation strategies and urban development programs. The case study of del Río et al. also contributes to the consideration of fog as an option to mitigate the effects of drought, as it is established as a complementary water source, both for human use and for the maintenance, conservation, and restoration of ecosystems.

Suárez (Sect. 15.2.2) presents a long-term analysis of groundwater availability as a sentinel of the hydrological cycle. It discusses the importance of groundwater as a key component of the hydrological cycle and is particularly important for combating droughts. It is key to remember that droughts are one of the natural disasters that cause the most damage, as these impact both ecosystems and society for extended periods (Bahinipati, 2020; Barría et al., 2021; Singh et al., 2021). Therefore, indices of groundwater availability are required to identify trends and anomalies in the availability of this variable in the hydrological cycle. The methodology of this study therefore consisted of calculating the standardized groundwater storage index (SGWI; Halder et al., 2020) and generating a time series of over 20 years for Chile. This was done to visualize anomalies in groundwater storage variations and correlate them with periods of drought and assess drought severity. Suárez's case study highlights the fact that groundwater exploitation has allowed for the sustenance of agriculture and human consumption during periods of drought (Arumí et al., 2013; Herrera et al., 2023). This contrasts with the case presented by del Río et al., as fog still does not yet play a relevant role in mitigating water scarcity, probably because atmospheric water harvesting technologies still require further development. However, both studies emphasize that knowledge of the spatio-temporal patterns of both sentinel variables is still limited (Suárez et al., 2021). The chapter on groundwater emphasizes the importance of generating and parameterizing standardized indices that allow for the identification of drought risk in groundwater. In other words, to generate the knowledge and analytical capacity that enable us to identify critical values of groundwater availability at which not only human activities such as agriculture could collapse, but also native vegetation, which is the key regulator of the biogeochemical cycles of water and carbon.

In the last study case by Miranda et al. (Sect. 15.2.3) on spatiotemporal changes in primary productivity based on the analysis of temporal trends in leaf phenology. The theoretical perspective and results of this study complement the view of the hydrological cycle as a sentinel process, as monitoring vegetation dynamics has become an essential strategy to detect changes in the hydrological and other biogeochemical cycles (Kariyeva et al., 2012). Changes in vegetation structure and plant metabolic capacity can be considered as the "canary in the coal mine" as vegetation-cover loss above 50% can lead terrestrial ecosystems (mainly semi-arid) to desertification and collapse (Tabor & Aguirre, 2004). In some ecosystems, plants use groundwater (Dawson & Pate, 1996) and atmospheric sources in the form of fog (Dawson, 1998; Weathers, 1999) as their main water source, so changes in these sources can have detrimental effects on the primary productivity of ecosystems (Latorre et al., 2011). In other words, since the climate-precipitation-vegetation relationship determines the distribution and dynamics of many terrestrial ecosystems (Rustad, 2008), small changes in conditions can affect the dynamic equilibrium of ecosystems and their associated functions (Malhi et al., 2020). The work of Miranda et al. is primarily based on the analysis of a 20-year time series of vegetation indices; Normalized Difference Vegetation Index (NDVI) and Enhanced Vegetation Index (EVI), (Huang et al., 2019). In particular, the results allow us to identify the impact at the physiological level of the latest heat waves and prolonged drought that had the most dramatic effect in Chile between 2019 and 2020.

15.2 Case Studies

15.2.1 Fog and Low Clouds as Sentinels Against Drought[1]

On the coast of northern and central Chile (between 18° S and 34° S), there is a low cloud cover of the stratocumulus type (Sc). These clouds almost permanently cover the southeastern Pacific Ocean and, on the mainland, meet the Coastal Cordillera, producing advective fog, locally called camanchaca (Cereceda et al., 2002; del Río et al., 2018, 2021). Fog is an important component of the local climate, providing humidity and water to the environment, and promoting biodiversity (Rundel & Mahu, 1976). From a hydrological point of view, the role that fog plays in the water balance is unclear, as it is atmospheric water that rarely precipitates to the ground. However, fog interacts with the surface and vegetation, contributing to enhance physical and biological processes that are relevant in the context of droughts.

From the physical point of view, the presence of fog attenuates direct solar radiation (Osses et al., 2017), one of the most relevant elements in evaporation. In turn, the presence of fog decreases air temperature and increases water vapor in the atmosphere, decreasing air vapor pressure, which also contributes to attenuate evaporation. From an ecological point of view, fog structures and regulates spatial patterns of vegetation in coastal ecosystems (Borthagaray et al., 2010; Stanton et al., 2013). Likewise, in hyper-arid to semiarid regions, such as northern and central Chile, the presence of fog, and its contribution of water and nutrients, acquires special importance and has been shown to positively impact productivity (Wang et al., 2019).

[1]The authors would like to thank Felipe Lobos and Klaus Keim for their contributions to the work required to complete this this section.

Given the relevance of the presence of fog and low clouds on climate and ecosystems, understanding their spatiotemporal dynamics (daily, annual, and interannual) can be crucial to understanding the effects that drought could have on different territories. For example, the mega drought (Garreaud et al., 2020), associated with the effects of climate change and which has been affecting the central zone of Chile for more than a decade, has produced a loss of productivity in ecosystems and a "browning" of vegetation that is lower in coastal sectors where the presence of fog and low clouds is regular (Miranda et al., 2020). Likewise, the constant presence of fog along the coast of northern and central Chile has been mainly studied as a potential complementary water source (Cereceda et al., 2008; Larraín et al., 2002; del Río et al., 2018; Montecinos et al., 2018; Carvajal et al., 2022). Thus, fog is also presented as an option to attenuate or mitigate the effects of drought.

In this sense, understanding seasonal fog cycles and spatial-temporal variability is key to identify changes in fog supply and its climatic, ecological, and hydrological cycle implications.

15.2.1.1 Methods

The identification and spatiotemporal characterization of fog and low clouds frequency is performed by processing data obtained from the Geostationary Operational Environmental Satellites (GOES-16), available from August 2017 to 2022.The temporal resolution of the GOES-16 satellite is 10 min and its spatial resolution is 2 km. These resolutions allow an accurate characterization of the spatiotemporal distribution of the haze, including its daily and seasonal cycles. The algorithm for fog and low cloud imaging is described in del Río et al. (2021) and distinguishes between a daytime and a nighttime process. The daytime algorithm uses the change detection technique, i.e., it determines the differences between a completely cloud-free image with each available image in the analyzed period by setting a threshold of 0.15 radiance. Then, high clouds are differentiated from low clouds (and/or fog) by a thermal threshold of 273 K, where clouds above this threshold are classified as low clouds. On the other hand, for the nighttime algorithm, two thermal bands are processed, the 3.8 μm shortwave infrared band and the 10.9 μm longwave infrared band. To differentiate low clouds from high clouds or bare ground, a threshold of −2 K is used (Lee et al., 1997). Thus, values below this threshold were classified as low cloudiness and those above the threshold as cloud-free. Then, pixels classified as fog or low clouds are tested by a filter, which distinguishes and classifies whether the cloud top temperature, estimated by the bright temperature at the 10.9 μm wavelength, is above or below the 273 K threshold (Torregrosa et al., 2015). The result of both nighttime and daytime algorithms is integrated into monthly frequency averages where 0% corresponds to fog-free skies and 100% is fully overcast within a given period, e.g., 1 month. These results are grouped into 4 frequency ranges: 3–10%, 10–20%, 2–40%, and greater than 40%. For the analysis and processing of satellite images, the Google Earth Engine platform (Gorelick et al., 2017) was used, downloading the available images, both for day and night, from August 2017 to May 2022.

15.2.1.2 Analysis of Results and Conclusions

The result of the analysis shows marked differences in the spatio-temporal distribution of fog frequency from Arica to Maule. Concerning the spatial distribution patterns, we found that from latitude 28° S to the north, the lowest fog frequency occurs during summer and autumn (Figs. 15.1 and 15.2) and the highest in the spring period (Figs. 15.3 and 15.4).

From latitude 28° S to the south, there is a change in the pattern, with the highest frequency occurring in summer and the lowest in winter. This spatiotemporal variability is related to the north-south displacement of the Pacific anticyclone and the subsidence of large-scale warm air masses, greater in central Chile during the summer, and greater in the north during the winter (del Río et al., 2021). At the same time, another regional factor that determines latitudinal variation is the sea surface temperature, which presents a thermal gradient from south to north, with variations associated with the annual solar cycle (higher temperatures in summer and lower in winter). The interaction of these physical mechanisms, subsidence, and sea surface temperature, results in a strong thermal inversion, which favors the formation of fog and low clouds, preventing the humid marine air from mixing with the dry air of the troposphere (Muñoz et al., 2011). From a longitudinal point of view, the frequency

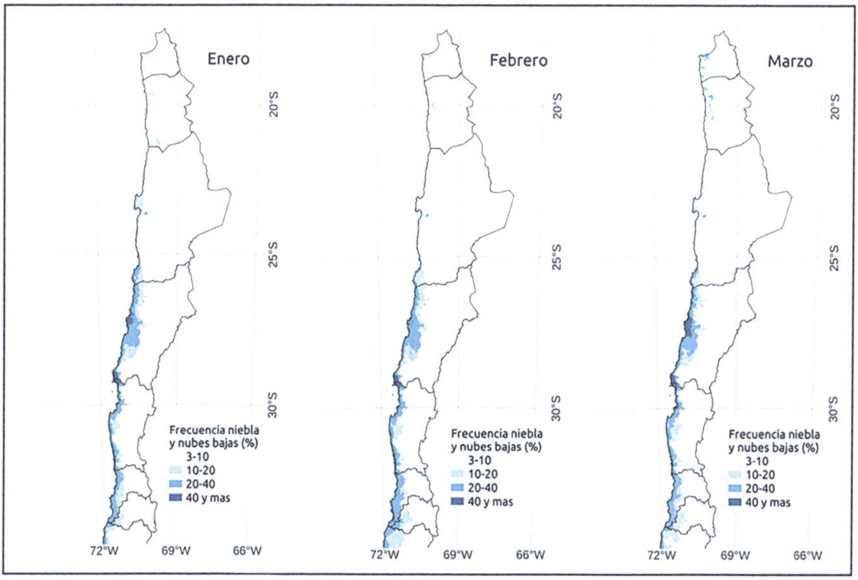

Fig. 15.1 Frequency of fog and low clouds during the summer between 2017 and 2022

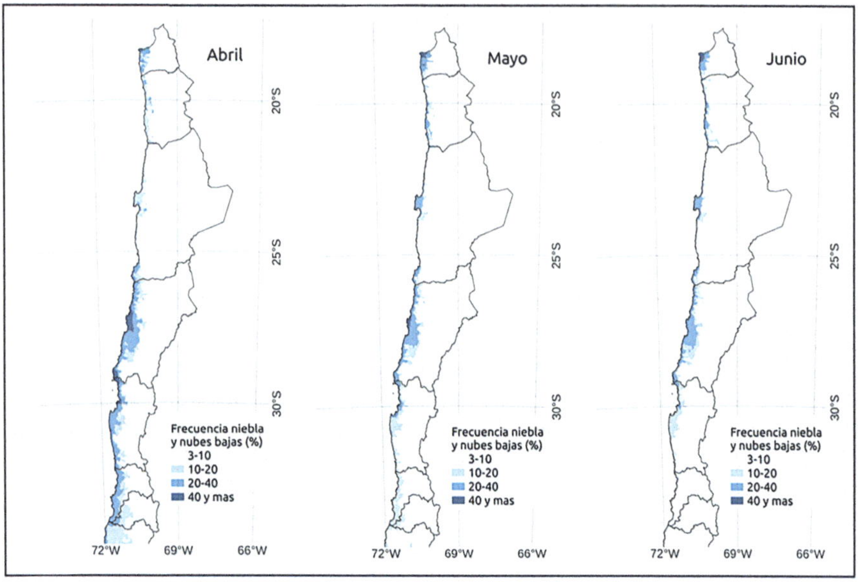

Fig. 15.2 Frequency of fog and low clouds during the autumn (2017–2022)

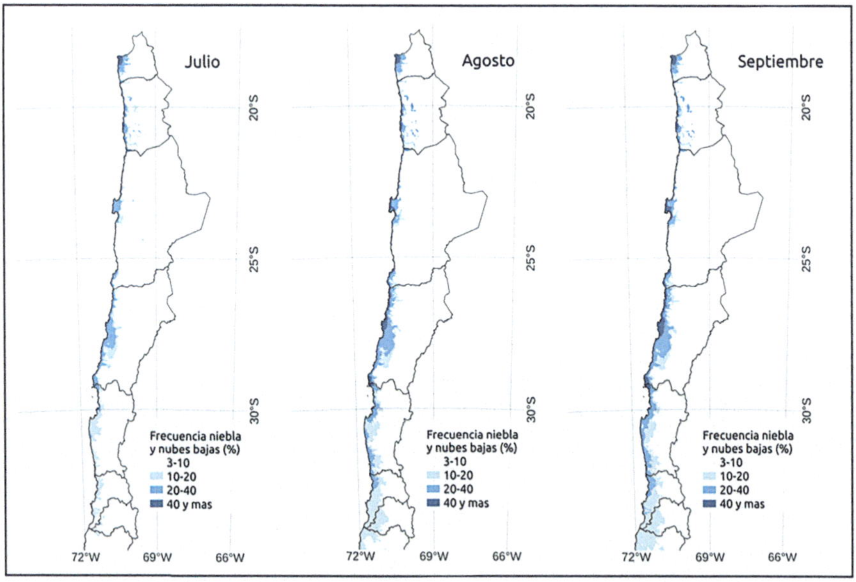

Fig. 15.3 Frequency of fog and low clouds during the winter between 2017 and 2022

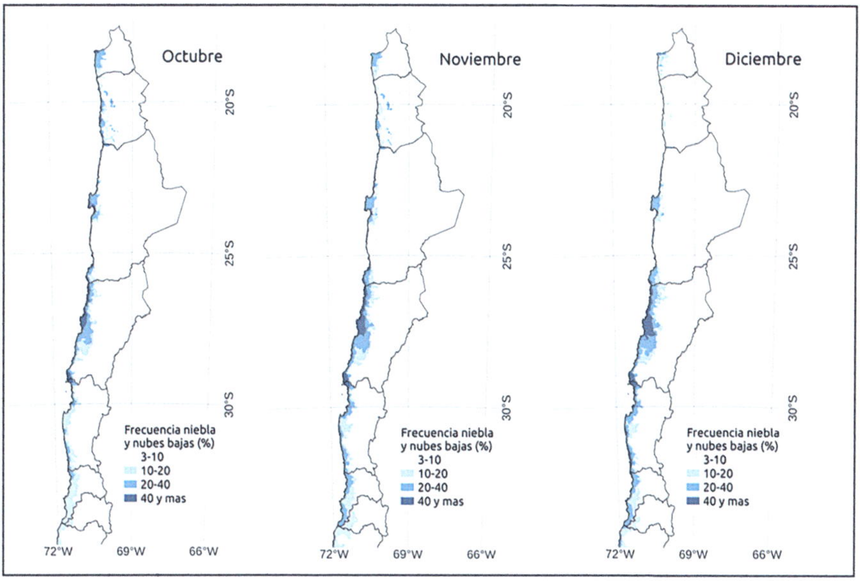

Fig. 15.4 Frequency of fog and low clouds during the spring between 2017 and 2022

of fog presents a non-homogeneous gradient of decrease, i.e., the highest frequency of fog occurs in the territories near the ocean and decreases towards the Andes (Fig. 15.5).

However, it is only in certain places that these low clouds interact with the surface (summit and high slopes), generating fog that can be used by vegetation. In a country like Chile, this is key, since this means that longitudinal fog patterns are determined mainly by topography, which determines the inland entrance of fog (del Río et al., 2021). In other words, at higher altitudes, fog is intercepted by the mountains and does not enter inland. On the contrary, in lowland areas, the valleys act as fog corridors from the coast to the interior. This is key to dealing with drought because it suggests that coastal fog and low coastal clouds can help mitigate the effects of climate change on coastal ecosystems and those present in inland valleys. Therefore, the frequency of fog and low clouds serves as a sentinel alert of the effects of drought and makes it possible to determine which territories may be more or less impacted.

Understanding the presence and spatiotemporal dynamics of fog can be of great importance in the face of drought, since its presence attenuates its effects, and presents itself as a complementary source of water. In this sense, in central Chile the presence of fog and low clouds is greater in summer, when precipitation is scarce, producing a complementary cycle of water supply between winter precipitation and the presence of fog in summer. Similarly, the greatest presence of fog and low clouds in the extreme north of Chile occurs in winter and spring, complementing the scarce high Andean rains that occur during the summer.

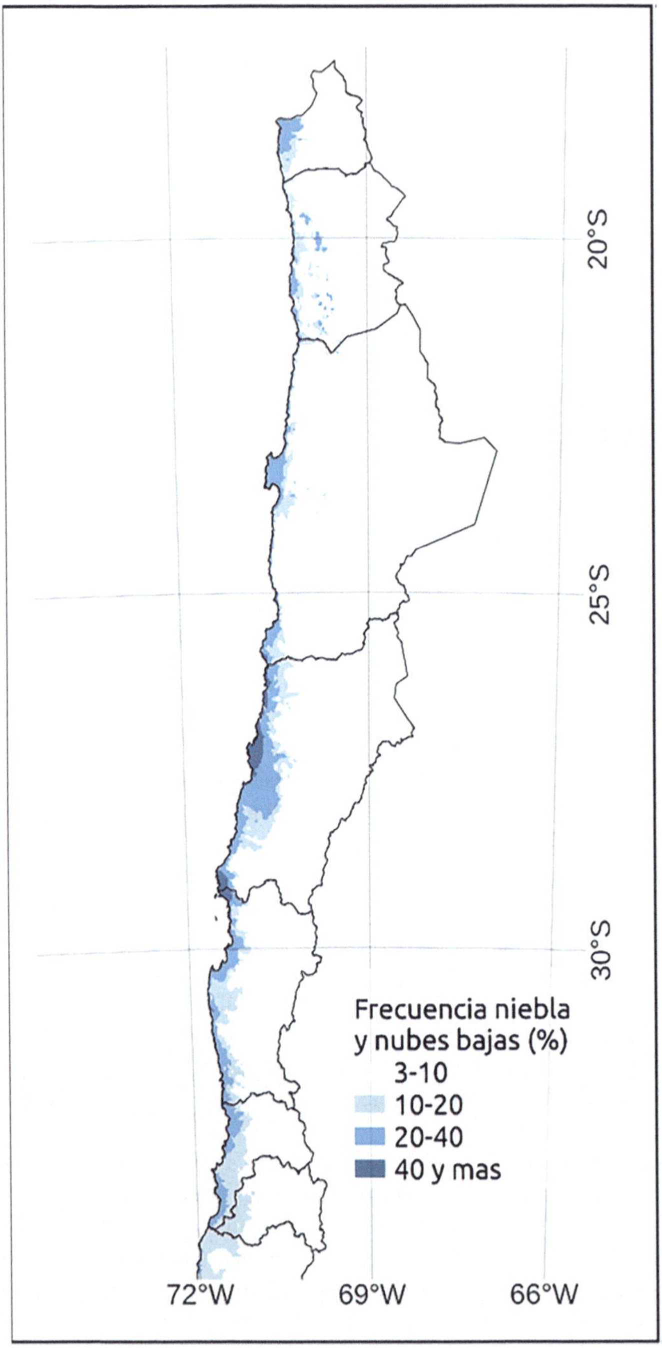

Fig. 15.5 Frequency of fog and low clouds between 2017 and 2022

15.2.2 *Observing Groundwater Anomalies Using the Standardized Groundwater Storage Index (SGWI) Sentinels of Drought Effects*

Groundwater is a key component of the hydrological cycle, as it stores water that would otherwise flow rapidly to the ocean or evaporate if remains as surface water. Groundwater can represent the primary water source for ecosystems and human communities. Unfortunately, it has been managed as if it were a renewable resource, without considering that droughts strongly impact the management of this underground resource (Petersen-Perlman et al., 2022). Groundwater has been used to mitigate water scarcity in the face of drought events driven by current climate change. Indeed, droughts are one of the natural disasters that cause the most damage, despite being slow-evolving natural processes; public and private institutions have not been able to allocate resources to respond to this threat (Venton, 2018). Consequently, worldwide it is common to not find robust legislation supporting water management under drought conditions. This lack of legislation generates water, energy, and food insecurity at the regional level (Wilhite & Glantz, 1985; Calow et al., 2010). Therefore, it is relevant to search for indices that allow identification groundwater-related drought risk (Funk & Shukla, 2020; Ali et al., 2022; Babre et al., 2022; Fankhauser et al., 2022).

Efforts have been made in Chile to identify and manage droughts (e.g., see Chaps. 4, 5, 6 and 7 of this book). Considering that the last decade (2010–2021) was the driest in the main Mediterranean climate regions, with average precipitation deficits ranging from −2 to −33%, in southwestern Australia and central Chile, respectively (Garreaud et al., 2020), there is still much work to be done. Under this drought context, groundwater exploitation in Chile has sustained agriculture and human consumption (Arumí et al., 2013; Herrera et al., 2023), even when the hydrogeological knowledge of the country is still limited (Suárez et al., 2021). Consequently, it is relevant to visualize anomalies in groundwater storage, with the purpose of managing the groundwater resources in the context of drought, and as a sentinel variable of the hydrological cycle.

15.2.2.1 Methods

Similar to how indices like the Standardized Precipitation Index (SPI; McKee et al., 1993) and the Standardized Precipitation Evapotranspiration Index (SPEI; Vicente-Serrano et al., 2010) use precipitation and evaporation data to measure droughts, this section recommends the use of the Standardized Groundwater Storage Index outlined in Eq. 15.1 (SGWI; Halder et al., 2020).

$$SGWI = \frac{x_i - x_m}{\sigma} \tag{15.1}$$

where x_i is the groundwater storage of the month i of the selected territorial unit, x_m is the average of the groundwater storage in the entire period of analysis in the territorial unit and σ is the standard deviation of all data in the territorial unit. Therefore, the SGWI enables the visualization of anomalies in groundwater storage variations (GWS) and links them to drought periods.

To estimate the time series of SGWI in Chile, the GWS data from the Global Land Data Assimilation System (GLDAS-2.2; Li et al., 2019) were obtained through the Google Earth Engine (GEE; Gorelick et al., 2017) platform. GLDAS-2.2 uses meteorological information from the European Centre for Medium-Range Weather Forecasts (ECMWF; https://www.ecmwf.int/) and incorporates data on anomalies in total terrestrial water storage obtained through the Gravity Recovery and Climate Experiment (GRACE; Watkins et al., 2015) from 2003 to the present. The GWS data were obtained at a daily temporal resolution and a spatial scale of 1°. Daily information was aggregated to the monthly scale from January 2003 to December 2021, and the spatial resolution was defined using the administrative regions of Chile.

15.2.2.2 Analysis of Results and Conclusions

Our results demonstrate that the SGWI can be utilized as an indicator of drought conditions, as it quantifies the deviation of groundwater storage from its long-term average storage. Figure 15.6 depicts the monthly time series of the SDWI over the last two decades for three distinct geographical zones of the country. In the Arica and Parinacota Region, a series of positive and negative anomalies is observed without a clear trend, whereas in the Metropolitan Region, a clear negative anomaly is evident between the years 2010–2018, consistent with the widely documented mega-drought in the central zone of the country (Garreaud et al., 2020). In the Magallanes Region, a positive anomaly is observed until 2011, followed by a significant negative anomaly occurring between 2011 and 2014. Negative anomalies have been observed again in recent years in this region. The information presented in Fig. 15.6 allows for a clear differentiation of the severity and duration of hydrogeological drought conditions in different zones of the country, where negative SGWI values indicate groundwater storage lower than the average, suggesting the onset of a drought in a particular area.

To delve into the trend of groundwater storage anomalies, Fig. 15.7 illustrates the monthly storage in the different regions of the country. On one hand, the northern zone has lower water storage (arid zone) and exhibits lower variability. On the other hand, the regions in the central and southern zones have higher water storage (e.g., La Araucanía and Los Ríos, averaging a storage of 1200 mm), as well as high variability (from O′ Higgins to Biobío) ranging between 250 and 1000 mm. Our results regarding variability strongly suggest that regions of the country with higher variability in groundwater storage potentially face greater risk, as they are areas that concentrate national agro-industry. Additionally, the data suggest that in the northern zone, groundwater storage and precipitation positively correlate, which may

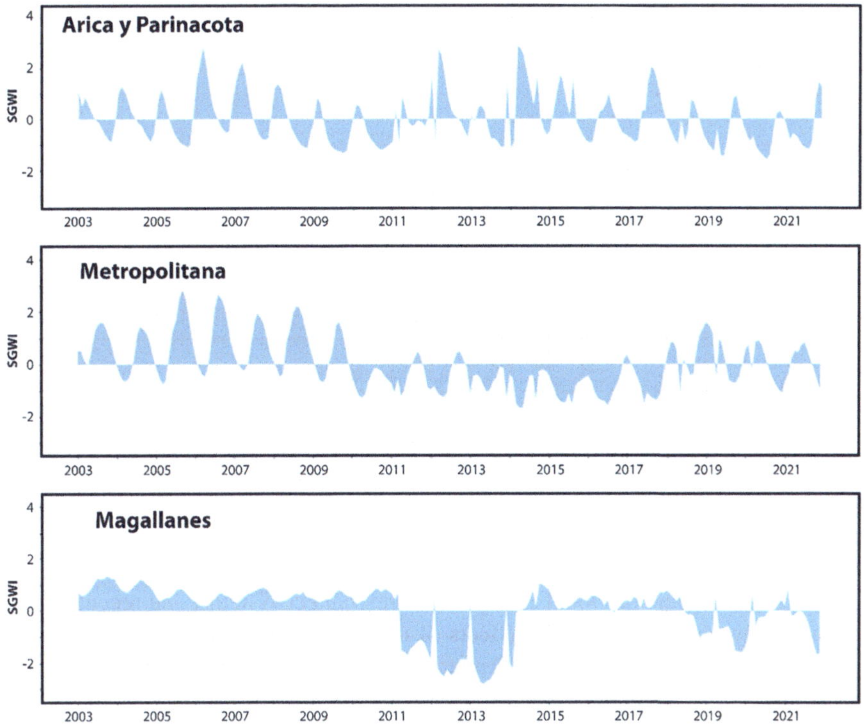

Fig. 15.6 Standardized Groundwater Storage Index (SGWI) between 2003 and 2021 in three regions of Chile: Arica y Parinacota (above); Regi6n Metropolitana (middle); y Regi6n de Magallanes (bottom)

alleviate drought conditions or exacerbate them in case of precipitation scarcity or extreme rainfall events. Therefore, monitoring storage variations could be employed to assess groundwater risk during drought events, making it a valuable sentinel variable for evaluating the severity and duration of groundwater droughts.

15.2.3 Phenology of Forest Productivity as a Sentinel Against Drought[2]

Vegetation-related spatiotemporal dynamics play a fundamental role in understanding the changes that occur in the structure and functioning of ecosystems in response to natural and anthropogenic disturbances (Berdugo et al., 2022; Miranda et al., 2023). Monitoring vegetation dynamics has become an essential strategy for

[2]The authors would like to thank Valentina Bravo for their contributions to the work required to complete this this section

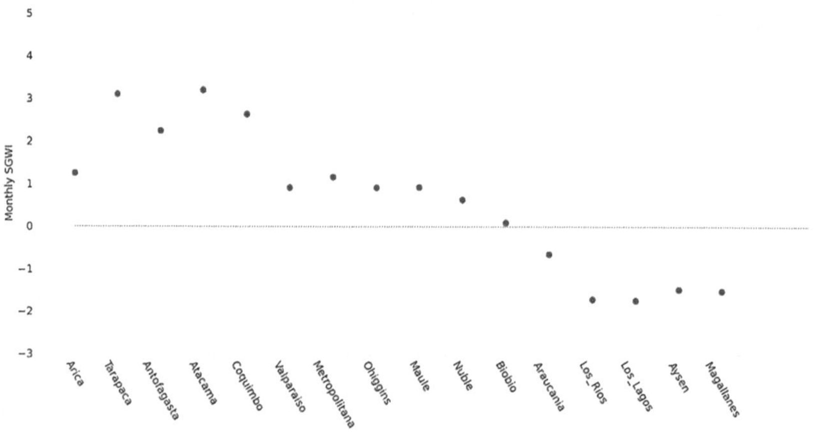

Fig. 15.7 Descriptive statistics of monthly groundwater storage (GWS) in Chile between 2003 and 2021 (above); Predominance of anomalies in groundwater storage (SGWI) in Chile between 2003 and 2021 (below)

detecting the impacts of global change on terrestrial ecosystems, and recently, this has been key for identifying the effects of disasters such as fire and droughts on vegetation (Kariyeva et al., 2012). Primary productivity can be estimated from time series based on vegetation indices derived from satellite images, such as the Normalized Difference Vegetation Index (NDVI) and Enhanced Vegetation Index, among others (Huang et al., 2019). These allow for continuous monitoring of ecosystems at different spatial scales (Malhi et al., 2020), and satellite image analysis facilitates obtaining information on temporal variations in forests at a large scale, which in turn facilitates the study of their spatiotemporal trajectories (Ustin & Middleton, 2021).

Satellite phenology, which involves the use of satellite spectral data to track seasonal variations in land surfaces, provides consistent estimates of primary productivity, in addition to temporal events associated with the onset, peak, end, and duration of primary productivity over large areas using vegetation indices (Zhang et al., 2003; Ivits et al., 2013; van Leeuwen et al., 2013). This combination of studying phenology through vegetation indices allows for informing on vegetation dynamics, enabling to characterize its structure, and dominant composition, and to provide the necessary information for its classification from a functional perspective, and understanding changes associated with extreme events and potential collapse thresholds (Gray & Ewers, 2021). In scenarios of prolonged drought such as the one experienced in central Chile since 2010, the use of these indices is key, and makes it possible to relate changes in water sources (see previous case studies) and the effect these could have on vegetation dynamics.

15.2.3.1 Methods

The study area encompasses the Mediterranean and Temperate regions of Chile between latitude 32.0° S and 44.0° S and longitude 70.8° W to 74.8° W, including the administrative regions from Coquimbo to Los Lagos. The spatial resolution defined for the analysis is 6.25 hectares (250 × 250 m), corresponding to the resolution of the MOD13Q1v061 satellite product "MOD13Q1.061 Terra Vegetation Indices 16-Day Global 250 m" obtained from the MODIS sensor (Didan, 2015). Furthermore, the temporal extent covers 20 years from 2002 to 2022, with a temporal granularity according to the type of product to be obtained, 16 days for productivity data (EVI) and 12 months for phenological variables. Within the study area, the area of analysis was delimited as the surface dominated by forests, shrublands, and grasslands (Zhao et al., 2016).

Events associated with satellite phenology (remote sensing phenology) are derived from time series analysis representing the behavior of primary productivity in vegetation index values (for this study, EVI), over at least 2 years of data capture. The temporal resolution of the time series may vary depending on the satellite sensor used. This study is 16 days, according to the MOD13Q1.061 product described above (more information in the Phenology Book Chapter). These data can be obtained by fitting a function that relates the behavior of the vegetation index (EVI) to time using a polynomial approach or a non-parametric fit with different degrees of complexity. The fitting of functions and calculating phenological variables are performed using the Timesat 3.3 program (Jönsson & Eklundh, 2004) and the R data analysis program (R Core Team, 2020).

15.2.3.2 Analysis of Results and Conclusions

Monitoring primary productivity through phenology analysis is particularly crucial during events such as droughts and heatwaves as it allows for understanding ecosystems' state and sensitivity. In this context, using vegetation indices obtained from satellite sensors emerges as an efficient alternative. Our results show a marked latitudinal gradient in vegetation primary productivity represented by the large integral (LIN) variable, with accumulated values fluctuating between 1 and 3.5 for scrubland, 3.5 and 6 for sclerophyll forest, and between 6 and 16 for deciduous and evergreen temperate forests (Fig. 15.8). Productivity in crop areas reaches values close to forests, especially between the Valparaíso and Maule regions (Fig. 15.8), highlighting the importance of water use for maintaining native and agroforestry ecosystems. On the other hand, productivity start of season (SOS) stands out, showing an activity onset in a latitudinal vegetation gradient associated with shrublands and forests between latitudes 32°–38° S, with a limit marked by Nothofagus deciduous forests in the mountains and radiata pine and eucalyptus forest plantations on the coast (Fig. 15.8). Agricultural areas exhibit a continuous SOS period occurring at the end of each year. In other words, there are changes in the onset of photosynthetic activity across the study region.

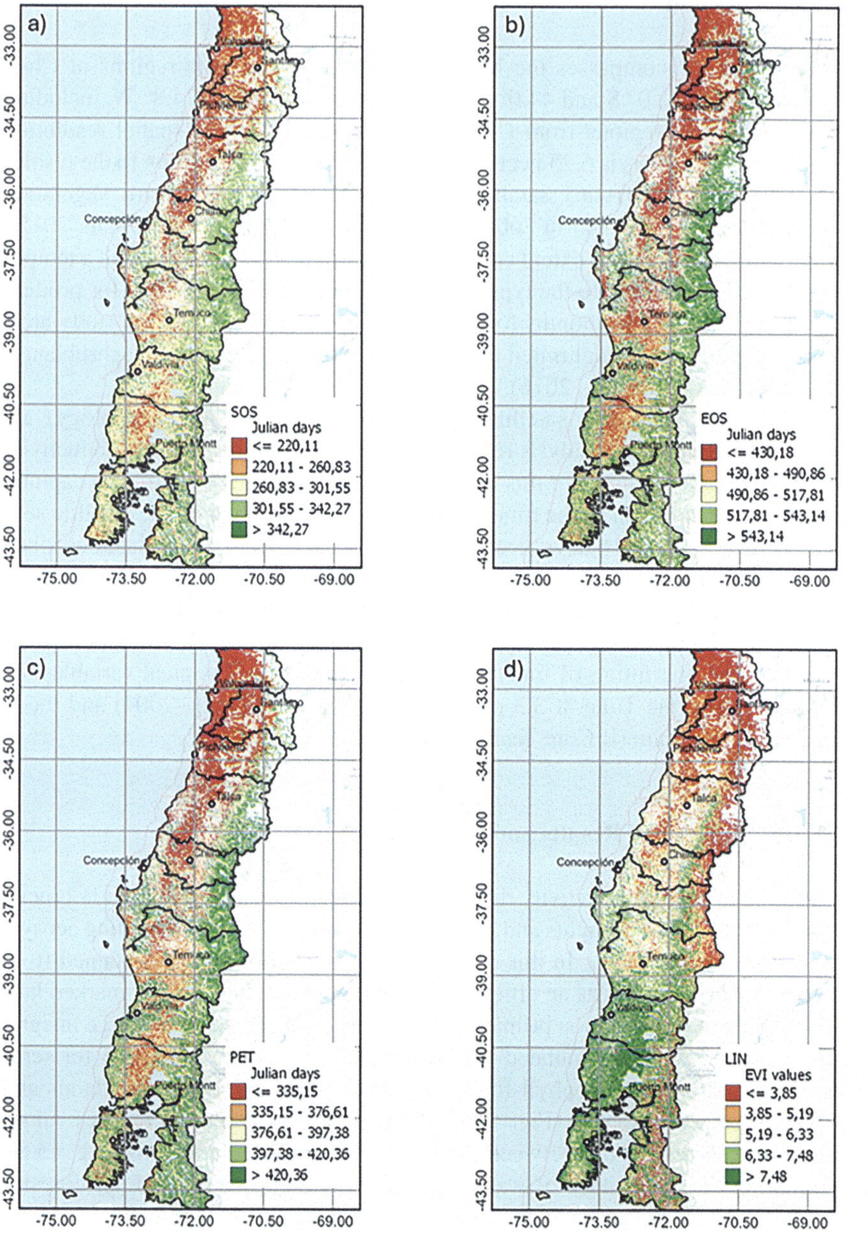

Fig. 15.8 Phenological variables: (**a**) start of the season (SOS), (**b**) end of the season (EOS), (**c**) Peak values (PET), (**d**) large integral (LIN). Lin is represented in EVI values and a, b, and c are represented in Julian Days (1 to 365)

Regarding the trend of primary productivity represented by the LIN variable over the last 20 years, Fig. 15.9 shows significant values for linear regression between the dependent LIN variable and time represented by 20 years of observation. Therefore, this allows us to identify the importance of land management and governance since some sectors with a negative slope show a decrease in

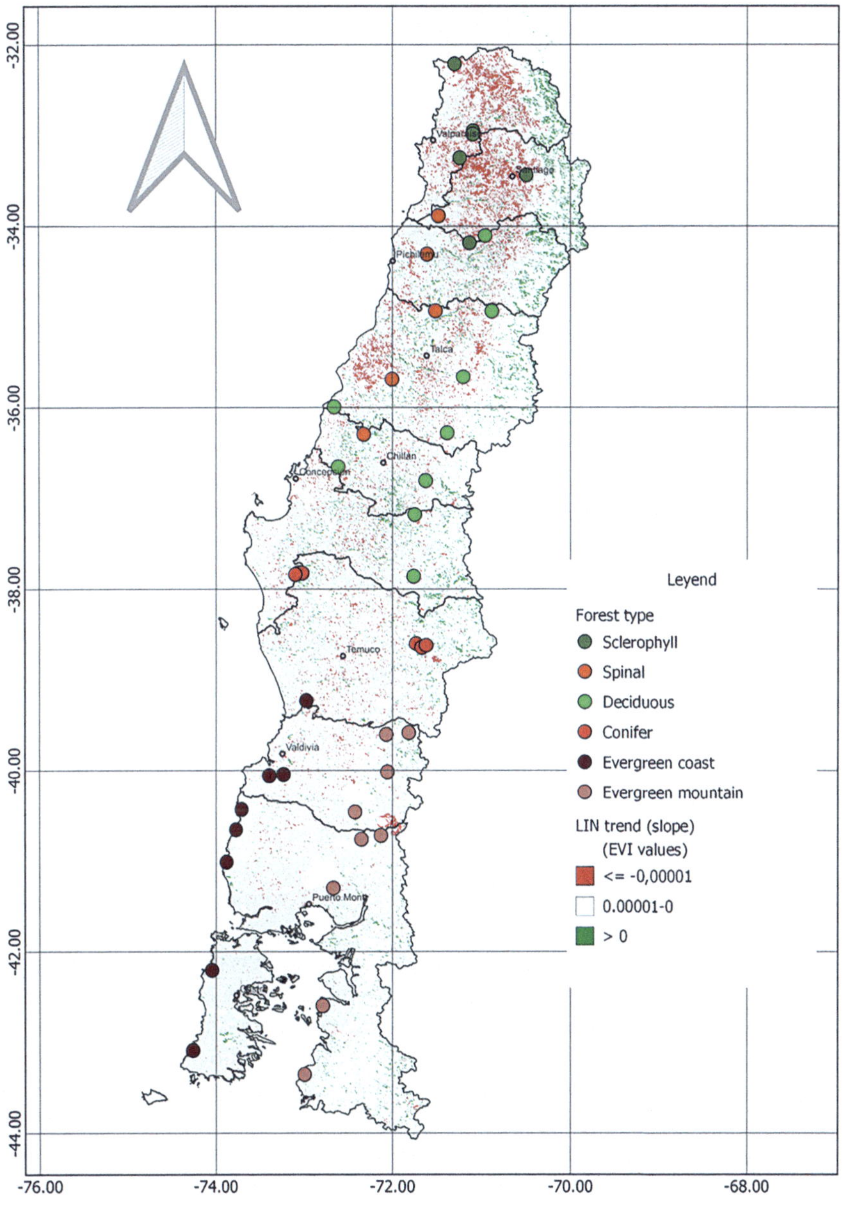

Fig. 15.9 The trend of primary productivity is represented by the LIN variable. Red pixels represent negative trends, and green pixels positive trends for a period of 20 years

productivity over the last 20 years (Fig. 15.9). On the other hand, there are sectors with a positive slope where productivity has increased over the last 20 years (Fig. 15.9). Likewise, some areas do not show phenological changes, and these are "subsidized" sectors, agricultural areas from the central valley of the region up to latitude 38° S (Fig. 15.9). These results coincide with recent observations regarding the effect of the last heatwaves and prolonged drought, which began in 2010 (Garreaud et al., 2017). The peaks reached in 2019–2020 caused widespread browning of vegetation between the Coquimbo and O'Higgins regions (Miranda et al., 2023). It should also be noted that decreases in productivity may be associated with the effects of forest fires, especially those occurring in 2017 and 2022.

Detailed knowledge about vegetation dynamics is essential for planning conservation and ecosystem restoration actions and identifying tipping points. The spatiotemporal patterns of leaf phenology serve as a sentinel in the face of global events such as climate change. Understanding these dynamics facilitates the implementation of effective and sustainable measures to protect and preserve natural resources in central Chile and beyond.

15.3 Concluding Remarks

The studies presented in this chapter offer valuable contributions to drought management and prevention in Chile by providing a deeper understanding of key hydrological and ecological processes in the country. The analysis of coastal fog, groundwater, and leaf phenology sheds light on critical aspects of the hydrological cycle and the response of ecosystems to changing climatic conditions. For example, the study of coastal fog reveals its potential as a supplementary water source, which could have significant implications for alleviating water scarcity in coastal regions. Furthermore, research on groundwater availability highlights its importance in sustaining agriculture and human consumption during periods of drought, emphasizing the need for continuous monitoring and the generation of standardized indices to assess the risk of drought in these groundwater resources. Finally, the analysis of leaf phenology underscores the role of vegetation in the hydrological cycle and highlights its vulnerability to anthropogenic climatic change, emphasizing the importance of conservation and restoration of plant cover as a preventive measure against drought.

Despite the significant contributions of these studies, it is important to recognise their limitations and opportunities for improvement in drought management and prevention in Chile. A key limitation is the lack of continuous monitoring of ecosystems and critical variables, which hinders the early identification of significant changes and the effectiveness of management interventions. In addition, the absence of analyses such as emergency timing represents a gap in our understanding of the timing and magnitude of climate change effects on ecosystems. Addressing these limitations requires greater investment in long-term monitoring of infrastructure and data analysis, as well as the integration of multidisciplinary approaches that incorporate analyses such as time of emergence to assess the vulnerability and resilience of ecosystems to drought and other climate change impacts. Multidisciplinary studies are essential for understanding the complexity of natural and social systems and developing effective adaptation and mitigation strategies to address the challenge of climate change.

References

Alaniz, A. J., Marquet, P. A., Carvajal, M. A., Vergara, P. M., Moreira-Arce, D., Muzzio, M. A., & Keith, D. A. (2024). Perspectives on the timing of ecosystem collapse in a changing climate. *Conservation Biology, e14247.*

Ali, S., Liu, D., Fu, Q., Cheema, M. J. M., Pal, S. C., Arshad, A., Pham, Q. B., & Zhang, L. (2022). Constructing high-resolution groundwater drought at spatio-temporal scale using GRACE satellite data based on machine learning in the Indus Basin. *Journal of Hydrology, 612*, 128295.

Allen, C. D., Breshears, D. D., & McDowell, N. G. (2015). On underestimation of global vulnerability to tree mortality and forest die-off from hotter drought in the Anthropocene. *Ecosphere, 6*(8), 1–55.

Armesto, J. J., & Vidiella, P. E. (1993). Plant life-forms and biogeographic relations of the flora of Lagunillas (30|circS) in the fog-free Pacific coastal desert. *Annals of the Missouri Botanical Garden*, 499–511.

Armesto, J. J., Vidiella, P. E., & Gutiérrez, J. R. (1993). Plant communities of the fog-free coastal desert of Chile: plant strategies in a fluctuating environment. *Revista Chilena de Historia Natural, 66*, 271–282.

Arumí, J. L., Rivera, D., Holzapfel, E., & Munoz, E. (2013). Effect of drought on groundwater in a Chilean irrigated valley. *Proceedings of the Institution of Civil Engineers: Water Management, 166*(5), 231–241. Thomas Telford Ltd.

Ascott, M. J., Macdonald, D. M. J., Sandwidi, W. J. P., Black, E., Verhoef, A., Zongo, G., Tirogo, J., & Cook, P. (2022). Time of emergence of impacts of climate change on groundwater levels in sub-Saharan Africa. *Journal of Hydrology, 612*, 128107.

Babre, A., Kalvāns, A., Avotniece, Z., Retiķe, I., Bikše, J., Popovs, K., Jemeljanova, M., Zelenkevičs, A., & Dēliņa, A. (2022). The use of predefined drought indices for the assessment of groundwater drought episodes in the Baltic States over the period 1989–2018. *Journal of Hydrology: Regional Studies, 40*, 101049.

Bahinipati, C. S. (2020). Assessing the costs of droughts in rural India: a comparison of economic and non-economic loss and damage. *Current Science, 118*(11), 1832.

Barnosky, A. D., Hadly, E. A., Bascompte, J., Berlow, E. L., Brown, J. H., Fortelius, M., Getz, W. M., Harte, J., Hastings, A., Marquet, P. A., & Martinez, N. D. (2012). Approaching a state shift in Earth's biosphere. *Nature, 486*(7401), 52–58.

Barría, P., Chadwick, C., Ocampo-Melgar, A., Galleguillos, M., Garreaud, R., Díaz-Vasconcellos, R., Poblete, D., Rubio-Álvarez, E., & Poblete-Caballero, D. (2021). Water management or megadrought: what caused the Chilean Aculeo Lake drying? *Regional Environmental Change, 21*, 1–15.

Berdugo, M., Vidiella, B., Solé, R. V., & Maestre, F. T. (2022). Ecological mechanisms underlying aridity thresholds in global drylands. *Functional Ecology, 36*(1), 4–23. https://doi.org/10.1111/1365-2435.13962

Borthagaray, A. I., Fuentes, M. A., & Marquet, P. A. (2010). Vegetation pattern formation in a fog-dependent ecosystem. *Journal of Theoretical Biology, 265*(1), 18–26.

Calow, R. C., MacDonald, A. M., Nicol, A. L., & Robins, N. S. (2010). Ground water security and drought in Africa: linking availability, access, and demand. *Groundwater, 48*(2), 246–256.

Cantonati, M., Lichtenwöhrer, K., Leonhardt, G., Seifert, L., Mustoni, A., Hotzy, R., Schubert, E., Blattner, L., Bilous, O., Lotz, A., & Poschlod, B. (2022). Using springs as sentinels of climate change in nature parks north and south of the alps: A critical evaluation of methodological aspects and recommendations for long-term monitoring. *Water, 14*(18), 2843.

Carvajal, D., Mora-Carreño, M., Sandoval, C., & Espinoza, S. (2022). Assessing fog water collection in the coastal mountain range of Antofagasta, Chile. *Journal of Arid Environments, 198*, 104679.

Cereceda, P., Larraín, H., Osses, P., Farías, M., & Egaña, I. (2008). The spatial and temporal variability of fog and its relation to fog oases in the Atacama Desert, Chile. *Journal Atmospheric Research, 87*, 312–323.

Cereceda, P., Osses, P., Larrain, H., Farías, M., Pinto, R., & Schemenauer, R. S. (2002). Advective, orographic and radiation fog in the Tarapacá region, Chile. *Journal Atmospheric Research, 64*, 261–271.

Dangles, O., Rabatel, A., Kraemer, M., Zeballos, G., Soruco, A., Jacobsen, D., & Anthelme, F. (2017). Ecosystem sentinels for climate change? Evidence of wetland cover changes over the last 30 years in the tropical Andes. *PLoS One, 12*(5), e0175814.

Dawson, T. E. (1998). Fog in the California redwood forest: ecosystem inputs and use by plants. *Oecologia, 117*, 476–485.

Dawson, T. E., & Pate, J. S. (1996). Seasonal water uptake and movement in root systems of Australian phraeatophytic plants of dimorphic root morphology: a stable isotope investigation. *Oecologia, 107*, 13–20.

del Río, C., Garcia, J. L., Osses, P., Zanetta, N., Lambert, F., Rivera, D., et al. (2018). ENSO influence on coastal fog-water yield in the Atacama Desert, Chile. *Aerosol and Air Quality Research, 18*(1), 127–144.

del Río, C., Lobos-Roco, F., Latorre, C., et al. (2021). Spatial distribution and interannual variability of coastal fog and low clouds cover in the hyperarid Atacama Desert and implications for past and present Tillandsia landbeckii ecosystems. *Plant Systematics and Evolution, 307*, 58.

del Val, E. K., Armesto, J. J., Barbosa, O., Christie, D. A., Gutiérrez, A. G., Jones, C. G., Marquet, P. A., & Weathers, K. C. (2006). Rain forest islands in the Chilean semiarid region: fog-dependency, ecosystem persistence and tree regeneration. *Ecosystems, 9*, 598–608.

Didan, K. (2015). MOD13Q1 MODIS/Terra Vegetation Indices 16-Day L3 Global 250m SIN Grid V006 [Data set]. In NASA EOSDIS Land Processes Distributed Active Archive Center.

Echeverría, S., Hausner, M. B., Bambach, N., Vicuña, S., & Suárez, F. (2019). Modeling present and future ice covers in two Antarctic lakes. *Journal of Glaciology*, 1–14.

Fankhauser, K., Macharia, D., Coyle, J., Kathuni, S., McNally, A., Slinski, K., & Thomas, E. (2022). Estimating groundwater use and demand in arid Kenya through assimilation of satellite data and in-situ sensors with machine learning toward drought early action. *Science of the Total Environment, 831*, 154453.

Farías, A. A., Armas, C., Gaxiola, A., Cea, A. P., Luis Cortes, J., López, R. P., Casanoves, F., Holmgren, M., Meserve, P. L., Gutiérrez, J. R., & Kelt, D. A. (2021). Species interactions across trophic levels mediate rainfall effects on dryland vegetation dynamics. *Ecological Monographs, 91*(2), e01441.

Folke, C., Carpenter, S., Walker, B., Scheffer, M., Elmqvist, T., Gunderson, L., & Holling, C. S. (2004). Regime shifts, resilience, and biodiversity in ecosystem management. *Annual Review of Ecology, Evolution, and Systematics, 35*, 557–581.

Fuentealba, M., Latorre, C., Frugone-Álvarez, M., Sarricolea, P., Giralt, S., Contreras-Lopez, M., Prego, R., Bernárdez, P., & Valero-Garcés, B. (2020). A combined approach to establishing the timing and magnitude of anthropogenic nutrient alteration in a mediterranean coastal lake-watershed system. *Scientific Reports, 10*(1), 5864.

Funk, C., & Shukla, S. (2020). *Drought early warning and forecasting: Theory and practice*. Elsevier Inc.

Garreaud, R., Barichivich, J., Christie, D. A., & Maldonado, A. (2008). Interannual variability of the coastal fog at Fray Jorge relict forests in semiarid Chile. *Journal of Geophysical Research: Biogeosciences, 113*(G4).

Garreaud, R. D., Álvarez-Garretón, C., Barichivich, J., Pablo Boisier, J., Christie, D., Galleguillos, M., LeQuesne, C., McPhee, J., & Zambrano-Bigiarini, M. (2017). The 2010–2015 mega-drought in central Chile: impacts on regional hydroclimate and vegetation. *Hydrology and Earth System Sciences, 21*(12), 6307–6327.

Garreaud, R. D., Boisier, J. P., Rondanelli, R., Montecinos, A., Sepúlveda, H. H., & Veloso-Aguila, D. (2020). The central Chile mega drought (2010–2018): a climate dynamics perspective. *International Journal of Climatology, 40*(1), 421–439.

Giorgi, F., & Bi, X. (2009). Time of emergence (TOE) of GHG-forced precipitation change hot-spots. *Geophysical Research Letters, 36*(6).

Gorelick, N., Hancher, M., Dixon, M., Ilyushchenko, S., Thau, D., & Moore, R. (2017). Google Earth Engine: Planetary-scale geospatial analysis for everyone. *Remote Sensing of Environment, 202*, 18–27.

Gray, R. E., & Ewers, R. M. (2021). Monitoring forest phenology in a changing world. *Forests, 12*(3), 297.

Haasnoot, M., Kwakkel, J. H., Walker, W. E., & Ter Maat, J. (2013). Dynamic adaptive policy pathways: A method for crafting robust decisions for a deeply uncertain world. *Global Environmental Change, 23*(2), 485–498.

Halder, S., Roy, M. B., & Roy, P. K. (2020). Analysis of groundwater level trend and groundwater drought using Standard Groundwater Level Index: a case study of an eastern river basin of West Bengal, India. *SN Applied Sciences, 2*, 1–24.

Heinze, C., Blenckner, T., Martins, H., Rusiecka, D., Döscher, R., Gehlen, M., et al. (2021). The quiet crossing of ocean tipping points. *Proceedings of the National Academy of Sciences, 118*(9), e2008478118.

Henson, S. A., Beaulieu, C., Ilyina, T., John, J. G., Long, M., Séférian, R., et al. (2017). Rapid emergence of climate change in environmental drivers of marine ecosystems. *Nature Communications, 8*(1), 14682.

Herrera, C., Urrutia, J., Gamboa, C., Salgado, X., Godfrey, L., Rivas, A., Jódar, J., Custodio, E., León, C., Sigl, V., Delgado, K., & Arriagada, E. (2023). Evaluation of the impact of the intensive exploitation of groundwater and the mega-drought based on the hydrochemical and isotopic composition of the waters of the Chacabuco-Polpaico basin in central Chile. *Science of the Total Environment, 165055.*

Hillebrand, H., Kuczynski, L., Kunze, C., Rillo, M. C., & Dajka, J. C. (2023). Thresholds and tipping points are tempting but not necessarily suitable concepts to address anthropogenic biodiversity change—an intervention. *Marine Biodiversity, 53*(3), 43. https://doi.org/10.1017/jog.2019.78

Huang, X., Xiao, J., & Ma, M. (2019). Evaluating the performance of satellite-derived vegetation indices for estimating gross primary productivity using FLUXNET observations across the globe. *Remote Sensing, 11*(15). https://doi.org/10.3390/rs11151823

Istanbulluoglu, E., & Bras, R. L. (2005). Vegetation-modulated landscape evolution: Effects of vegetation on landscape processes, drainage density, and topography. *Journal of Geophysical Research: Earth Surface, 110*(F2).

Ivits, E., Cherlet, M., Mehl, W., & Sommer, S. (2013). Ecosystem functional units characterized by satellite observed phenology and productivity gradients: A case study for Europe. *Ecological Indicators, 27*, 17–28. https://doi.org/10.1016/j.ecolind.2012.11.010

Jönsson, P., & Eklundh, L. (2004). TIMESAT—a program for analyzing time-series of satellite sensor data. *Computers & Geosciences, 30*(8), 833–845. https://doi.org/10.1016/j.cageo.2004.05.006

Kariyeva, J., Leeuwen, W. J. D., Woodhouse, C., & a. (2012). Impacts of climate gradients on the vegetation phenology of major land use types in Central Asia (1981–2008). *Frontiers of Earth Science, 6*(2), 206–225. https://doi.org/10.1007/s11707-012-0315-1

Karssenberg, D., Bierkens, M. F., & Rietkerk, M. (2017). Catastrophic shifts in semiarid vegetation-soil systems may unfold rapidly or slowly. *The American Naturalist, 190*(6), E145–E155.

King, A. D., Donat, M. G., Fischer, E. M., Hawkins, E., Alexander, L. V., Karoly, D. J., Dittus, A. J., Lewis, S. C., & Perkins, S. E. (2015). The timing of anthropogenic emergence in simulated climate extremes. *Environmental Research Letters, 10*(9), 094015.

Koch, M. A., Kleinpeter, D., Auer, E., Siegmund, A., del Rio, C., Osses, P., García, J., Marzol, M. V., Zizka, G., & Kiefer, C. (2019). Living at the dry limits. Ecological genetics of Tillandsia lomas in the Chilean Atacama Desert. *Plant Systematics and Evolution, 305*, 1041–1053.

Larraín, H., Velásquez, F., Cereceda, P., Espejo, R., Pinto, R., Osses, P., & Schemenauer, R. S. (2002). Fog measurements at the site "Falda Verde" north of Chañaral compared with other fog stations of Chile. *Journal of Atmospheric Research, 64*, 273–284.

Latorre, C., González, A. L., Quade, J., Fariña, J. M., Pinto, R., & Marquet, P. A. (2011). Establishment and formation of fog-dependent Tillandsia landbeckii dunes in the Atacama Desert: Evidence from radiocarbon and stable isotopes. *Journal of Geophysical Research: Biogeosciences, 116*(G3).

Lee, T., Turk, J., & Richardson, K. (1997). Stratus and fog products using GOES-8-9 3.9 μm Data. *Weather and Forecasting, 12*, 664–677.

Li, B., Rodell, M., Kumar, S., Beaudoing, H. K., Getirana, A., Zaitchik, B. F., Goncalves, L. G., Cossetin, C., Bhanja, S., Mukherjee, A., Tian, S., Tangdamrongsub, N., Long, D., Nanteza, J., Lee, J., Policelli, F., Goni, I. B., Daira, D., Bila, M., de Lannoy, G., Mocko, D., Steele-Dunne, S. C., Save, H., & Bettadpur, S. (2019). Global GRACE data assimilation for groundwater and drought monitoring: Advances and challenges. *Water Resources Research, 55*(9), 7564–7586. https://doi.org/10.1029/2018wr024618

Lolu, A. J., Ahluwalia, A. S., Sidhu, M. C., Reshi, Z. A., & Mandotra, S. K. (2020). Carbon sequestration and storage by wetlands: Implications in the climate change scenario. In *Restoration of wetland ecosystem: A trajectory towards a sustainable environment* (pp. 45–58). Springer.

Malhi, Y., Franklin, J., Seddon, N., Solan, M., Turner, M. G., Field, C. B., & Knowlton, N. (2020). Climate change and ecosystems: Threats, opportunities and solutions. *Philosophical Transactions of the Royal Society, B: Biological Sciences, 375*(1794). https://doi.org/10.1098/rstb.2019.0104

Mark, B. G., & Fernández, A. (2017). The significance of mountain glaciers as sentinels of climate and environmental change. *Geography Compass, 11*(6), e12318.

Marquet P. A., A. Altamirano, M. T. K. Arroyo, M. Fernández, S. Gelcich, K. Górski, E. Habit, A. Lara, A. Maass, A. Pauchard, P. Pliscoff, H. Samaniego, C. Smith-Ramírez (editores) (2019). Biodiversidad y cambio climático en Chile: Evidencia científica para la toma de decisiones. Informe de la mesa de Biodiversidad. Comité Científico COP25; Ministerio de Ciencia, Tecnología, Conocimiento e Innovación. https://cdn.digital.gob.cl/filer_public/d2/ce/d2ce6fb0-272d-4f6c-aa95-7dd275c32b6b/libro-biodiversidad.pdf

Mataloni, G. (2016). Diversity patterns of plankton communities in Tierra del Fuego peat bog pools as sentinels of climate change. *Biodiversity, 17*(1–2), 26–33.

McKee, T. B., Doesken, N. J., & Kleist, J. (1993, January). The relationship of drought frequency and duration to time scales. *Proceedings of the 8th Conference on Applied Climatology* (Vol. 17, No. 22, pp. 179–183).

Miranda, A., Lara, A., Altamirano, A., Di Bella, C., González, M. E., & Camarero, J. J. (2020). Forest browning trends in response to drought in a highly threatened mediterranean landscape of South America. *Ecological Indicators, 115, 106401*(2020) ISSN 1470-160X.

Miranda, A., Syphard, A. D., Berdugo, M., Carrasco, J., Gómez-González, S., Ovalle, J. F., Delpiano, C. A., Vargas, S., Squeo, F. A., Miranda, M. D., Dobbs, C., Mentler, R., Lara, A., & Garreaud, R. (2023). Widespread synchronous decline of Mediterranean-type forest driven by accelerated aridity. *Nature Plants*. https://doi.org/10.1038/s41477-023-01541-7

Mitsch, W. J., Bernal, B., Nahlik, A. M., Mander, Ü., Zhang, L., Anderson, C. J., et al. (2013). Wetlands, carbon, and climate change. *Landscape Ecology, 28*, 583–597.

Moat, J., Orellana-Garcia, A., Tovar, C., Arakaki, M., Arana, C., Cano, A., et al. (2021). Seeing through the clouds–Mapping desert fog oasis ecosystems using 20 years of MODIS imagery over Peru and Chile. *International Journal of Applied Earth Observation and Geoinformation, 103*, 102468.

Molot, L. A., & Dillon, P. J. (1996). Storage of terrestrial carbon in boreal lake sediments and evasion to the atmosphere. *Global Biogeochemical Cycles, 10*(3), 483–492.

Montecinos, S., Cereceda, P., & Rivera, D. (2018). Fog collection and its relationship with local meteorological variables in a semiarid zone in Chile. *Atmosfera, 31*(2), 143–153.

Mueller, D. R., Van Hove, P., Antoniades, D., Jeffries, M. O., & Vincent, W. F. (2009). High Arctic lakes as sentinel ecosystems: Cascading regime shifts in climate, ice cover, and mixing. *Limnology and Oceanography, 54*(6part2), 2371–2385.

Muñoz, R. C., Zamora, R. A. & Rutllant, J. A., (2011). The coastal boundary layer at the eastern margin of the southeast Pacific (23.4 S, 70.4 W): Cloudiness-conditioned climatology. *Journal of climate, 24*(4), pp. 1013–1033. https://journals.ametsoc.org/view/journals/clim/24/4/2010jcli3714.1.xml.

O'Gorman, E. J., Benstead, J. P., Cross, W. F., Friberg, N., Hood, J. M., Johnson, P. W., Sigurdsson, B. D., & Woodward, G. (2014). Climate change and geothermal ecosystems: natural laboratories, sentinel systems, and future refugia. *Global Change Biology, 20*(11), 3291–3299.

Osses P.; Escobar R.; del Río C. Vargas C, Guerra N. (2017). El clima desértico costero con nublados abundantes del desierto de Atacama y su relación con los recursos naturales energía solar y agua de niebla. Caso de estudio Alto Patache, Región de Tarapacá, Chile. Revista de Geografía Norte Grande, N°68: 33–48. doi:https://doi.org/10.4067/S0718-34022017000300033.

Petersen-Perlman, J. D., Aguilar-Barajas, I., & Megdal, S. B. (2022). Drought and groundwater management: Interconnections, challenges, and policyresponses. *Current Opinion in Environmental Science & Health, 28*, 100364.

R Core Team. (2020). *R: A language and environment for statistical computing*. R Foundation for Statistical Computing. https://www.R-project.org/

Rojas, M., Lambert, F., Ramirez-Villegas, J., & Challinor, A. J. (2019). Emergence of robust precipitation changes across crop production areas in the 21st century. *Proceedings of the National Academy of Sciences, 116*(14), 6673–6678.

Rundel, P. W., & Mahu, M. (1976). Community structure and diversity in a coastal fog desert in northern Chile. *Flora, 165*(6), 493–505.

Rundel, P. W., Dillon, M. O., Palma, B., Mooney, H. A., Gulmon, S. L., & Ehleringer, J. R. (1991). The phytogeography and ecology of the coastal Atacama and Peruvian deserts. *Aliso: A Journal of Systematic and Floristic Botany, 13*(1), 1–49.

Rustad, L. E. (2008). The response of terrestrial ecosystems to global climate change: towards an integrated approach. *Science of the Total Environment, 404*(2–3), 222–223.

Scheffer, M., Bascompte, J., Brock, W. A., Brovkin, V., Carpenter, S. R., Dakos, V., et al. (2009). Early-warning signals for critical transitions. *Nature, 461*(7260), 53–59.

Schindler, D. W. (2009). Lakes as sentinels and integrators for the effects of climate change on watersheds, airsheds, and landscapes. *Limnology and Oceanography, 54*(6part2), 2349–2358.

Singh, C., Jain, G., Sukhwani, V., & Shaw, R. (2021). Losses and damages associated with slow-onset events: urban drought and water insecurity in Asia. *Current Opinion in Environmental Sustainability, 50*, 72–86.

Stanton, D. E., Negret, B. S., Armesto, J. J., & Hedin, L. O. (2013). Forest patch symmetry depends on direction of limiting resource delivery. *Ecosphere, 4*(5), 1–12.

Suárez, F., Leray, S., & Sanzana, P. (2021). Groundwater resources. *Water Resources of Chile*, 93–127.

Tabor, G. M., & Aguirre, A. A. (2004). Ecosystem health and sentinel species: adding an ecological element to the proverbial "canary in the mineshaft". *EcoHealth, 1*(3), 226–228.

Tarnocai, C. (2006). The effect of climate change on carbon in Canadian peatlands. *Global and Planetary Change, 53*(4), 222–232.

Thompson, L. G. (2010). Climate change: The evidence and our options. *The Behavior Analyst, 33*, 153–170.

Torregrosa, A., Combs, C., & Peters, J. (2015). GOES-derived fog and low cloud indices for coastal north and central California ecological analyses. *Earth and Space Science, 3*, 46–67.

Ustin, S. L., & Middleton, E. M. (2021). Current and near-term advances in Earth observation for ecological applications. *Ecological Processes, 10*(Issue 1). https://doi.org/10.1186/s13717-020-00255-4

Van Bellen, S., Mauquoy, D., Hughes, P. D., Roland, T. P., Daley, T. J., Loader, N. J., et al. (2016). Late-Holocene climate dynamics recorded in the peat bogs of Tierra del Fuego, South America. *The Holocene, 26*(3), 489–501.

van Leeuwen, W., Hartfield, K., Miranda, M., & Meza, F. (2013). Trends and ENSO/AAO Driven Variability in NDVI Derived Productivity and Phenology alongside the Andes Mountains. *Remote Sensing, 5*(3), 1177–1203. https://doi.org/10.3390/rs5031177

Venton, C. C. (2018). *Economics of resilience to drought* (p. 43). USAID.

Vicente-Serrano, S. M., Beguería, S., & López-Moreno, J. I. (2010). A multiscalar drought index sensitive to global warming: the standardized precipitation evapotranspiration index. *Journal of Climate, 23*(7), 1696–1718.

Wang, L., Kaseke, K. F., Ravi, S., Jiao, W., Mushi, R., Shuuya, T., & Maggs-Kölling, G. (2019). Convergent vegetation fog and dew water use in the Namib Desert. *Ecohydrology, 12*(7), e2130.

Watkins, M. M., Wiese, D. N., Yuan, D. N., Boening, C., & Landerer, F. W. (2015). Improved methods for observing Earth's time variable mass distribution with GRACE using spherical cap mascons. *Journal of Geophysical Research: Solid Earth, 120*(4), 2648–2671. https://doi.org/10.1002/2014jb011547

Weathers, K. C. (1999). The importance of cloud and fog in the maintenance of ecosystems. *Trends in Ecology & Evolution, 14*(6), 214–215.

Wilhite, D. A., & Glantz, M. H. (1985). Understanding the drought phenomenon: the role of definitions. *Water International, 10*, 111–120.

Williamson, C. E., Saros, J. E., Vincent, W. F., & Smol, J. P. (2009). Lakes and reservoirs as sentinels, integrators, and regulators of climate change. *Limnology and Oceanography, 54*(6part2), 2273–2282.

Zhang, X., Friedl, M. A., Schaaf, C. B., Strahler, A. H., Hodges, J. C. F., Gao, F., Reed, B. C., & Huete, A. (2003). Monitoring vegetation phenology using MODIS. *Remote Sensing of Environment, 84*(3), 471–475. https://doi.org/10.1016/S0034-4257(02)00135-9

Zhao, Y., Feng, D., Yu, L., Wang, X., Chen, Y., Bai, Y., Hernández, H. J., Galleguillos, M., Estades, C., Biging, G. S., Radke, J. D., & Gong, P. (2016). Detailed dynamic land cover mapping of Chile: Accuracy improvement by integrating multi-temporal data. *Remote Sensing of Environment, 183*, 170–185. https://doi.org/10.1016/j.rse.2016.05.016

Chapter 16
Contingent Agreements to Reduce the Impacts of Droughts on Human Water Consumption

Sebastián Vicuña, Antonia Ávila, Jorge Gironás, Óscar Melo, Daniela Rivera, and María Molinos

Abstract One of the expected impacts of climate change is the alteration of precipitation and temperatures, modifying hydrological regimes and water availability. Faced with these changes, it is essential to adopt measures to guarantee the quantity, quality and continuity of drinking water supply to the population. This chapter proposes an instrument that consists of the design and implementation of pre-agreement contracts for the temporary transfer of water volumes from agriculture to water utilities. The application of the instrument is presented using the Aconcagua River basin and the Valparaíso-Viña del Mar conurbation in Chile as a case study. Through

S. Vicuña (✉)
Department of Hydraulics and Environmental Engineering, Pontificia Universidad Católica de Chile, Santiago, Chile

Centro de Cambio Global UC, Pontificia Universidad Católica de Chile, Santiago, Chile

Centro de Investigación para la Gestión Integrada del Riesgo de Desastre (CIGIDEN), Santiago, Chile

Center for Water Law and Management, Pontificia Universidad Católica de Chile, Santiago, Chile
e-mail: svicuna@uc.cl

A. Ávila
Master in Engineering Sciences, Pontificia Universidad Católica de Chile, Santiago, Chile

J. Gironás
Department of Hydraulics and Environmental Engineering, Pontificia Universidad Católica de Chile, Santiago, Chile

Centro de Cambio Global UC, Pontificia Universidad Católica de Chile, Santiago, Chile

Centro de Investigación para la Gestión Integrada del Riesgo de Desastre (CIGIDEN), Santiago, Chile
Centro de Desarrollo Urbano Sustentable (CEDEUS), Santiago, Chile

D. Rivera, G. Donoso (eds.), *Droughts in Chile*, Global Issues in Water Policy 31,
https://doi.org/10.1007/978-3-031-85040-0_16

Ó. Melo
Centro de Cambio Global UC, Pontificia Universidad Católica de Chile, Santiago, Chile

Center for Water Law and Management, Pontificia Universidad Católica de Chile, Santiago, Chile

Department of Agricultural Economics, Pontificia Universidad Católica de Chile, Santiago, Chile

D. Rivera
Faculty of Law and the Center for Water Law and Management, Pontificia Universidad Católica de Chile, Santiago, RM - Santiago, Chile

M. Molinos

Center for Water Law and Management, Pontificia Universidad Católica de Chile, Santiago, Chile

Institute of Sustainable Processes, University of Valladolid, Valladolid, Spain

simulation models, the transfers, and economic compensations of a potential application of the instrument are characterized and evaluated considering different future climate scenarios, drought triggers for activation of pre-agreements, compensation scenarios based on human basic water consumption thresholds, and opportunity cost of water used in the agricultural sector.

Keywords Drinking water supply · Water transfer agreements · Water reallocation · Aconcagua · Valparaíso-Viña del Mar

16.1 Introduction

One of the expected impacts of climate change in the world in general, and in Chile in particular, is an alteration in precipitation and temperature levels, which would result in new hydrological regimes, modifying water availability over time (Arnell & Gosling, 2013). These alterations could compromise the capacity of hydrological systems to meet multiple water demands, such as ecosystem, human consumption, and production, and increase the risks associated with extreme events related to this element. One way to address these risks is through structured planning that allows the needs of the different actors, their interdependence, and the growing competition for the use of the resource due to reduced availability to be coupled and articulated. This type of approach would make it possible to improve water security and reduce negative impacts on society, the economy, and the environment.

One of the key areas of water vulnerability to climate change is the population's access to drinking water (Center for Climate and Resilience Research et al., 2021). Given the preeminence of this use, it is vital to guarantee the quantity, quality and continuity of its supply, avoiding failures that may have harmful or irreversible consequences. In the case of Chile, the water utilities in charge of providing water in the concessioned operational territories[1] are currently facing lower water availability, changes in the seasonality of water flows, algae bloom due to the increase in water temperature, worsening turbidity levels, salinization of coastal water reservoirs, among others (Superintendencia de Servicios Sanitarios, 2020). To ensure the supply of drinking water, water utilities must conduct long-term planning to reduce vulnerability through the implementation of adaptation measures. The planning of drinking water supply in concessioned areas is done through development programs or plans. These are investment programs that water utilities must develop over a horizon of up to 15 years, and which must be updated every 5 years. It is important to note that these plans must be approved by the water regulator, Superintendencia de Servicios Sanitarios (SIIS). In these plans, the companies commit to take a series of actions to ensure the quality and continuity of the drinking water service.

There are two types of adaptation measures to avoid failures in a drinking water supply system: structural and nonstructural (Ricalde et al., 2022). The first refers to the construction or repair of infrastructure, increasing the capacity of the system to produce, treat, and distribute water, either in the short or long term. This can be achieved, for example, with new reservoirs, reuse of wastewater, reduction of system losses, and desalination plants, among others. The latter modifies the system from a management perspective, conducting actions without necessarily altering its physical configuration. For example, using financial tools in water resource systems, financial concessions for drought management portfolios, leasing, or transfer of water rights (WR), or tools for demand management through tariffs.

With some exceptions, such as the case of England (Environment Agency, 2017), the development programs or plans of water companies do not explicitly consider climate change scenarios in the design of adaptation measures; they consider the premise that water demand may vary, but that the climate remains with the same characteristics as in the past. In the context of a changing climate, this leads to unplanned reactive measures, which is far from being an optimal solution from an environmental, social, and economic (non-sustainable) perspective. In the scenario of severe drought that has persisted for the last 14 years in central Chile, the need to take measures to ensure the drinking water supply for cities such as Santiago or Valparaiso has generated tensions between water companies and other relevant users of water resources, such as farmers.

Based on the results of previous works (Vicuña et al., 2018; Ricalde et al., 2022), the objective of this chapter is to design a public policy instrument to reduce the impacts of droughts on human water consumption in a context of deep uncertainties and interests of multiple actors. To this end, three specific objectives were considered: (a) to propose changes or adjustments in the elaboration of programs or development plans of concessioned water utilities to better prepare for and face the effects of drought on water availability; (b) to evaluate the proposed instrument for the case

[1] More information about how the Chilean water supply and sanitation system works can be reviewed at Molinos-Senante (2018).

of the urban area of Valparaíso-Viña del Mar and; (c) to generate recommendations for the implementation of the aforementioned policy instrument.

16.2 Central Aspects of the Proposal

Given the uncertainty regarding the magnitude of future changes in water availability due to climate change, the growing costs of achieving higher levels of water security, especially in the face of a low anticipation and planning capacity, and the willingness of some holders of WR to agree on temporary transfers of water volumes among themselves in conditions of severe drought, a favorable scenario arises for thinking about and implementing flexible instruments that make such agreements feasible and formalized.

Given the central role of the development programs or plans in the provision of drinking water and sanitation services by the water utility and in the supervision exercised in this respect by the Superintendency of Sanitary Services (SISS, by its Spanish acronym)—the regulatory body of water provision and sanitation services in Chile—, it is considered convenient and opportune to incorporate flexible instruments for water demand management. In this context, this study proposes and evaluates the implementation of pre-agreements for the temporary transfer of water volumes between sanitary service providers and other holders of WR that are part of water users' organizations.[2] These pre-agreements should contemplate adequate compensation (for the cost associated with the transfer of water) and should be outlined as a complement to other structural measures and investments that must necessarily be conducted by the sanitation company to ensure the supply of drinking water in its operational territory, especially in drought conditions.

16.3 Case Study

This research focuses on the Aconcagua River basin, located in central Chile, specifically in the Valparaíso region (see Fig. 16.1). Its extension is 7340 km^2 and its main river is the Aconcagua (DGA, 2004). This river is formed by the junction of the Juncal and Blanco rivers and flows into the bay of Concón, with a length of 142 km from the junction with the Blanco River (DGA, 2004).

This basin is responsible for supplying the main cities of the Valparaíso region, which are Viña del Mar and Valparaíso. It also covers about 45% of the Valparaíso

[2]These correspond to entities, regulated in the Water Code, formed by two or more persons with water rights in the same river, aquifer, canal, reservoir or other hydraulic works of common intake and distribution. Their objective is to manage and distribute water among their members according to their rights of use, to administer the common works, and to resolve conflicts between one or more users, or between them and the organization.

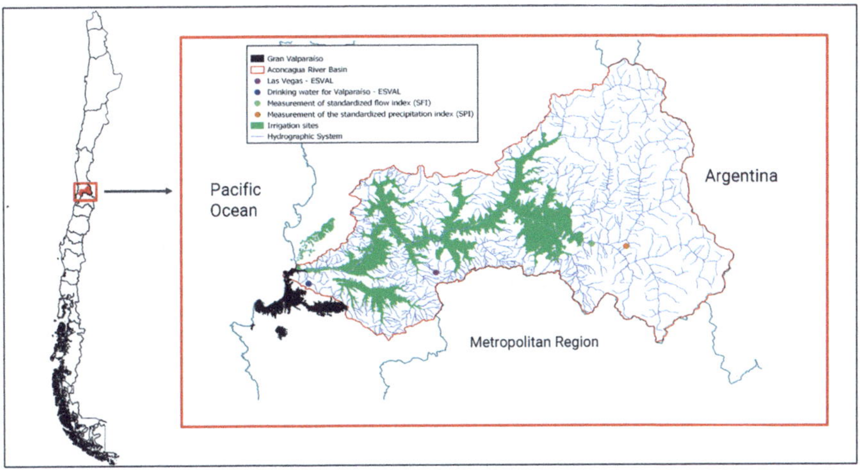

Fig. 16.1 Representation of the study area. (Source: Prepared by the authors)

region. In population and economic terms, the Valparaíso region is the second most populated in the country, with 10% of the total population (INE, 2017) and is the third in generating the highest GDP in the country with 7.7% (OBSERVA, 2022). In sectoral terms, the activity that stands out the most is the manufacturing industry, followed by personal services and transportation, information, and communications (DGA, 2020). Specifically, in the basin, there are urban potable water (water consumption of 2.84 m^3/s), mining (water consumption of 1.34 m^3/s), agricultural (water consumption of 25.79 m^3/s), industrial (water consumption of 2.63 m^3/s), and hydroelectric generation activities (water consumption of 33.67 m^3/s) (DGA, 2020). ESVAL is the water utility in charge of water supply for a large part of the population in the urban region of Valparaíso-Viña del Mar. Figure 16.1 presents the Aconcagua River basin, the areas where irrigation sites are located, the urban area supplied by ESVAL, and other points of interest that are used later in the analysis methodology.

16.4 Analysis Methodology

To evaluate the effect that this pre-agreement between the water utility and other water users in the Aconcagua River basin could have, an analysis methodology is developed that considers three key steps: development of future climate scenarios; activation, and execution of the agreement and finally, economic analysis of compensation. Figure 16.2 shows a summary of the methodology presented in the analysis of the implementation's impact of this adaptation strategy.

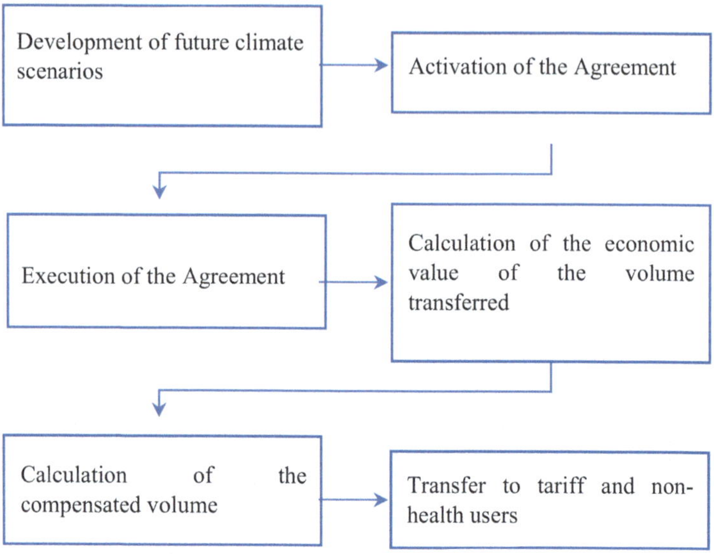

Fig. 16.2 Methodology summary. (Elaborated by the authors)

16.4.1 Development of Future Climate Scenarios

To address future climate uncertainty, two greenhouse gas emission scenarios based on the latest IPCC report were considered. An optimistic scenario, known as SSP1-2.6, and a pessimistic scenario, known as SSP5-8.5 (IPCC, 2023). There are a large number of global climate models (GCMs) that use these scenarios to create possible trajectories of future climate series. However, not all of them are well suited to the study area. To determine the best models to work with, two bias indices were used. The first is the mean annual bias (SPA, by its Spanish acronym), which measures the average annual precipitation or temperature ($\overline{x}$) of model j at station i and compares it to the same value calculated for station i (see Eq. 16.1). The second is the coefficient of variation bias (SCV, by its Spanish acronym), which measures the ratio between the coefficient of variation $\left(\frac{s}{\overline{x}}\right)$ of model j at station i and the coefficient of variation at the corresponding station (see Eq. 16.2). In the case of both Eqs. (16.1) and (16.2), the closer the value is to zero, the better the model fits.

$$SPA_i = \frac{\overline{x}_{ji} - \overline{x}_i}{\overline{x}_i} \tag{16.1}$$

$$SCV_i = \frac{\frac{s_{ji}}{\overline{x}_{ji}} - \frac{s_i}{\overline{x}_i}}{\frac{s_i}{\overline{x}_i}} \tag{16.2}$$

On the other hand, given the substantial number of weather stations in the study area, an aggregate indicator of both measures of bias was constructed for each *k* variable (precipitation and temperature). Finally, by averaging the indices for both variables, the values used to select the best-fitting models are obtained (Eq. 16.3).

$$MR = \frac{\sum_k \sum_i SPA_{ki} + \sum_k \sum_i SCV_{ki}}{2} \tag{16.3}$$

Using this methodology, 10 GCMs are finally considered for the analysis NESM3, KIOST-ESM, CanESM5, KACE-1-0-G, CMCC-CM2-SR5, CMCC-ESM2, ACCESS_CM2, ACCESS_ESM1-5, CAMS-CSM1-0 y CESM2-WACCM.[3] The study period for the analysis is from 1980 to 2061.

16.4.2 Activation and Execution of the Agreement

To simulate the activation and execution of the agreement, WEAP[4] (Yates et al., 2005) and MODFLOW[5] (Harbaugh, 2005) models that exist in the Strategic Water Management Plan (PEGH, by its Spanish acronym) of the Aconcagua River basin are used (DGA, 2020). The advantage of these models is that they incorporate those channels through which water is redistributed to meet the drinking water production needs of ESVAL, the water utility for the Valparaíso-Viña del Mar conurbation. This allows us to analyze the amount of water that will need to be transferred in the different climate scenarios modeled.

The activation of the agreement is conducted considering drought indicators. Considering the methodology of the General Directorate of Water (Dirección General de Aguas—DGA) (DGA, 2022), two possible drought indices are constructed: standardized precipitation index (SPI) and standardized flow index (SFI). The SPI (SFI) makes it possible to identify periods with precipitation (flow rates) significantly lower than the normal or expected condition. Both indexes are simple to calculate and are monitored periodically, which ensures an initiative-taking follow-up to trigger time for the set of actions to be considered. Following the recommendations of DGA (2022), an accumulation period is used in the calculation. Some issues to keep in mind are:

[3] The acronyms correspond to the acronyms of global climate models (or GCMs). There are about 30 models developed by different research institutions and meteorological services in developed countries.

[4] WEAP is a software developed by the Stockholm Environment Institute (SEI) that allows the modeling of water systems under a graphical environment and a user-friendly interface.

[5] MODFLOW is a finite difference flow modeler developed by the U.S. Geological Survey, which consists of a source code that solves the groundwater flow equation through interactions. It is used in hydrogeology to simulate groundwater flow in any aquifer.

- The index is tracked monthly, but the agreement is triggered semi-annually, to ease the application of the instrument, at the beginning of April and at the beginning of October. The former marks the beginning of the hydrological year and the latter the end of the rainy season and the beginning of the snowmelt period. These dates can be reconsidered in another hydrological context.
- Selection of base meteorological station: WEAP-MODFLOW models are built based on meteorological series distributed in the basin. In the case of the Aconcagua, data from the Chacabuquito fluviometric station and meteorological records from the upper part of the basin were used.
- The historical period can be used for the definition of precipitation normality conditions; in the case analyzed, the historical period used is from 1980 to 2018. To add precision, that process can be updated every 10 years.
- To follow in line with the procedures established by the DGA, it was decided that the indexes to be used are SPI accumulated to 12 months (SPI-12) and SFI accumulated to 6 months (SFI-6). In addition, a sensitivity analysis of the value of the drought trigger (threshold) was performed. For this purpose, following Mckee et al. (1993) it was decided to consider three possible triggers with values equal to −1.04, −1.54 and −2.04, which define the occurrence of moderate, severe, and extreme drought, respectively.

When the defined threshold is exceeded in one of the trigger months (April or October), a temporary transfer of water volumes from WR holders (in this case, mainly farmers in the different sections of the basin) to the water utility is triggered. For the study case, this implies applying the legal priority of personal and domestic water use (article 5 bis of the Water Code) to operate its drinking water production infrastructure.

As a result of the activation of the agreement, a record is left of the volumes of water available for the water utility and the changes in availability for other users. It is important to note that no increase in groundwater extraction by groundwater rights holders is allowed to compensate for the reduction in the volume of surface water, or vice versa.

16.4.3 Economic Analysis of Compensation

The economic analysis seeks to evaluate the costs of the agreements and the characterization of their possible water supply. For this purpose, the economic demand for water in the study area is evaluated to represent the opportunity cost generated by the transfer of volumes between farmers and water utilities. The aim is to identify the opportunity cost of water in the agricultural sector in the context of drought.

The opportunity cost is presented as a range of possible values. The value of water irrigating a set of crops was estimated as the economic productivity of water; that is, the ratio between the gross margin generated by each crop and the water consumed by that crop.

The water value of the set of crops present in the Aconcagua Basin is obtained through a weighted average of them, according to the current surface area of each one, based on data from the last Agricultural Census and the Fruit Census.[6,7] In addition, a range of minimum and maximum water values is reported, derived from a grouping of the crops into five groups and verifying, within each one, which are the minimum and maximum gross margins. These margins were then assigned to all crops in the same group, in the logic of potential substitution between similar crops and considering data availability. Crops were grouped into cereals, forage crops, vegetables, vines, and fruit trees. In addition, all the communes belonging to the province of San Felipe, i.e., San Felipe, Catemu, Llaillay, Panquehue, and Santa María, were included. Finally, taking the average of the value calculated for each commune in the province, the value of water to be used is 0.85 UDS/m^3, where this result was sensitized by ±0.5 USD/m^3.

The economic cost of the compensation is applied in two possible scenarios in relation to the amount of water consumed by households and other customers within the concessioned area. One considers total compensation and the other, partial compensation. Partial compensation uses three consumption thresholds per customer (10, 15, and 20 m^3/month per customer) above which compensation must be made. The first of these values corresponds to the volume of water classified as optimal access by the World Health Organization (WHO) for the human right to water, which has also been established by the Supreme Court in some judgments pronounced in cases of lack of access to drinking water (100 L/person/day; Howard et al., 2020). The other two values are based on the limit that has generally been used in practice and the one that is configured as the legal ceiling of the drinking water subsidy (15 and 20 m^3/month per household, respectively; article 2 letter b Law No. 18,778 of 1989; Gómez-Lobo & Contreras, 2003).

Figure 16.3 shows a summary of the methodology presented in the analysis of the implementation's impact of this adaptation strategy. The following section presents the results of the analysis.

16.5 Results and Discussion

This section presents the results of the application of the proposed methodology for the case study of the Aconcagua River basin and the water supply needs that ESVAL must meet the needs of personal and domestic use (human consumption) of Valparaíso-Viña del Mar.

First, the different climatic conditions that can be identified in the Aconcagua River basin are described. Based on information obtained through bias correction

[6] https://www.ine.gob.cl/estadisticas/economia/agricultura-agroindustria-y-pesca/censos-agropecuarios

[7] https://bibliotecadigital.odepa.gob.cl/bitstream/handle/20.500.12650/70428/catastroFruticolaValpo2020.pdf

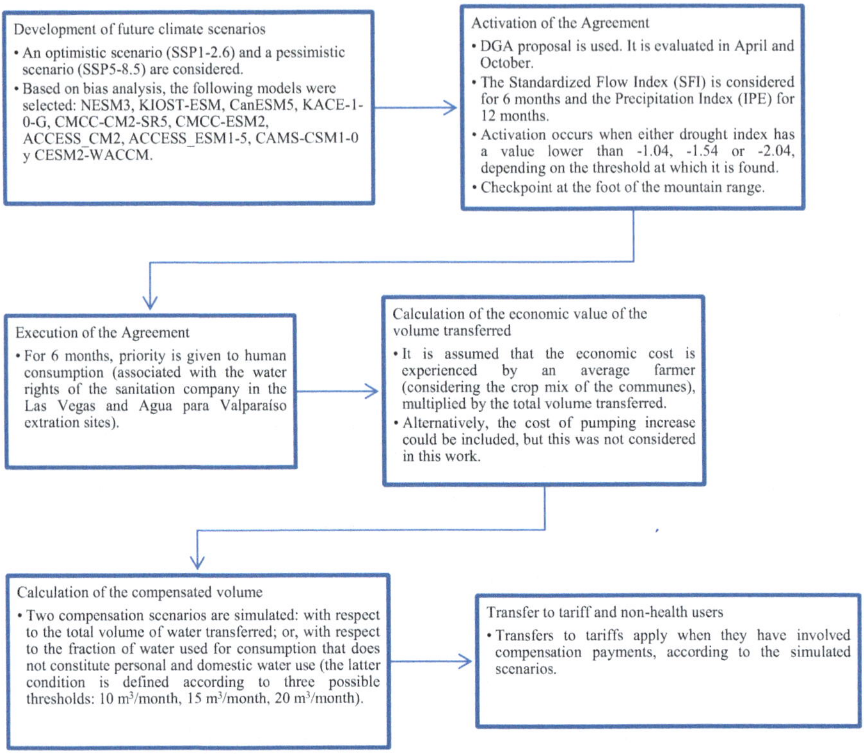

Fig. 16.3 Analysis methodology. (Prepared by the authors)

with the UQM[8] method (Aedo et al., 2021), applied for each of the selected GCMs according to the methodology described above. The expected changes in terms of precipitation and flow in the study area can be seen in Fig. 16.4.

Applying the methodology described above, it is possible to obtain the number of times a severe drought condition would be declared in the future, and, therefore, the agreement would be activated in the different projected climate scenarios. It is important to note that the study horizon ranges from 2019 to 2061. Figure 16.5a, b show the number of activations of the agreement versus the change in average flow for each of the climate scenarios considered. Three thresholds are considered to distinguish between moderate, severe, and extreme drought conditions.

A first finding, evidenced from Figs. 16.4 and 16.5 is that some models are activated a greater number of times even though they are not those with lower flow availability; for example, the CMCC_ESM2 model in both scenarios. This is because the SPI and SFI indices of these models turn out to be more critical in the months in which it was defined that the contract would be activated, i.e., April and

[8]This corresponds to the Unbiased Quantile Mapping method (UQM), which seeks to preserve both the trend of the quantiles and the changes in the standard deviation and mean of the historical observed series in the modeled series.

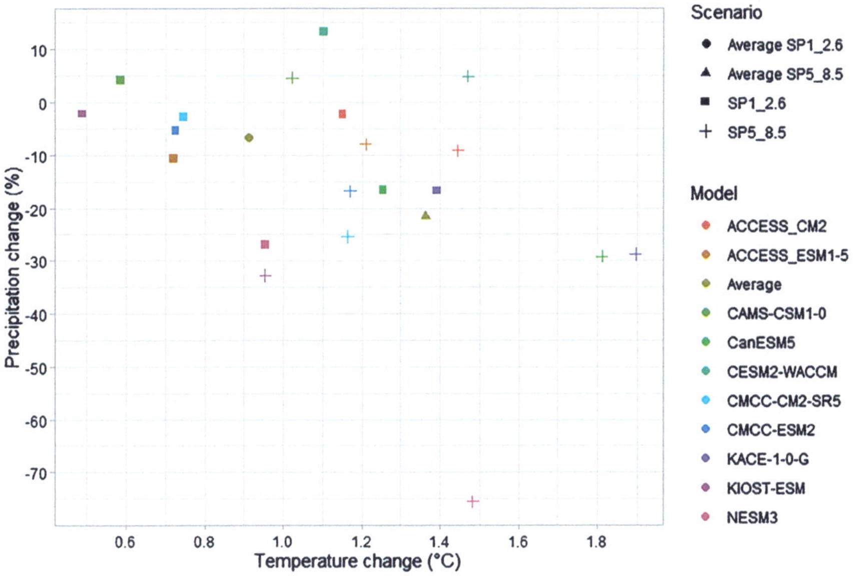

Fig. 16.4 Change in climate comparing the period 1980–2018 with 2020–2058. (Elaborated by the authors)

October. A second finding is that even though there could be an increase in flow, the agreement is still activated. This leads to the conclusion that there is a base situation that would cause shortages in future periods, even though the situation is better than the current one. Finally, it is also possible to appreciate that in the pessimistic scenario there is a higher number of activations, but at the same time, there is a greater dispersion compared to the optimistic scenario.

To effectively demonstrate the implications of activating the agreement, an example is provided using a model and scenario (NESM3 and SSP5-8.5) during a specific hydrological period (June 2041 to July 2041). Figure 16.6 shows how water volume varies for the ESVAL and irrigation sites in the climate scenario and model indicated. In October 2041, the agreement would be activated, which leads to what is shown in Fig. 16.6a, b. It can be observed that the volume of water received by ESVAL and the irrigation sites before October 2041 is almost identical.

To estimate the volume of water transferred to be compensated, we considered the drinking water consumption data by rank, provided by SISS, using ESVAL reports (SIFAC II report).[9] The monthly average of water consumption data by range (0–10, 11–15, 16–20, etc. in m^3 per month) was considered for the period from January 2019 to December 2021 and for Gran Valparaíso, which includes the municipalities of Viña del Mar, Valparaíso, Concón, Quilpué, and Villa Alemana. The percentages obtained for the monthly average are shown in Fig. 16.7. This percentage represents, on the one hand, non-residential water consumption and, on the

[9] See https://www.siss.gob.cl/586/w3-article-7247.html.

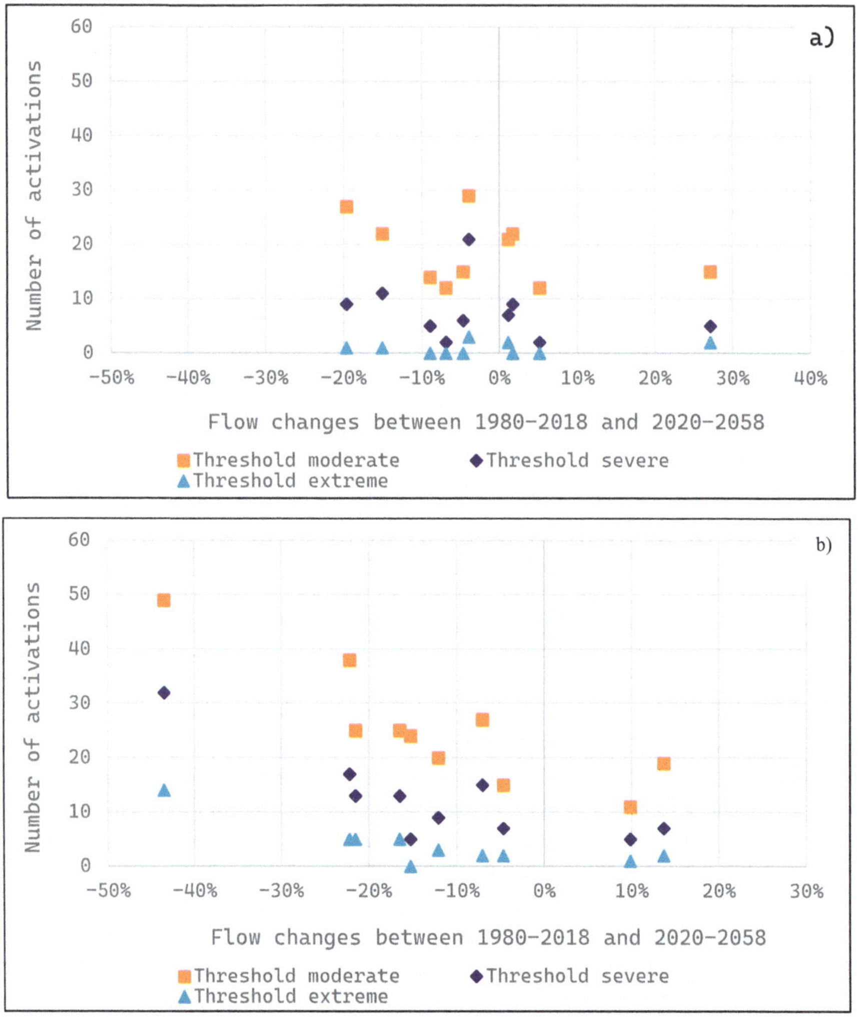

Fig. 16.5 Number of activations for different drought thresholds versus change in flow in different climate scenarios for: (**a**) favorable scenario (SSP1-2.6) and (**b**) pessimistic scenario (SSP5-8.5). (Elaborated by the authors)

other hand, water consumption over the limit in cubic meters per month at the monthly level.

As mentioned above, in the last stage of the calculation of the economic cost of compensation, three values were considered to calculate the amount to be compensated, which are 0.35, 0.85, and 1.35 USD/m^3. These values should be considered as a reference range, since, among other things, the current crop composition of the basin was not considered in detail.

The final compensation amount was calculated with Eq. (16.4).

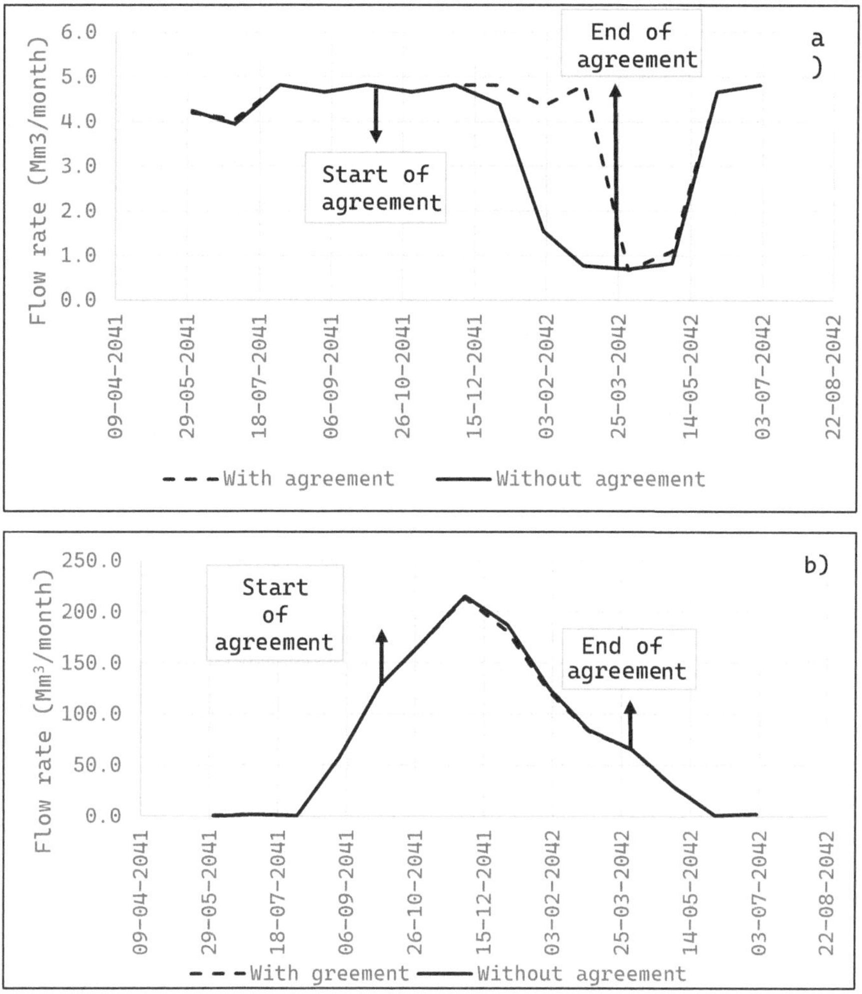

Fig. 16.6 Example of agreement activation for the NESM3 model and SSP585 scenario, period June 2041 to July 2042. Note: (**a**) water flow received by ESVAL; (**b**) flow received by irrigation sites. (Elaborated by the authors)

$$\begin{aligned} \textit{Amount to be compensated} &= \textit{Percentage to be compensated} \\ &\quad * \textit{Volume of water received} * \textit{Value to be compensated} \end{aligned} \tag{16.4}$$

The results obtained for the amount to be compensated for each percentage of water, for the SSP5-8.5 scenario and the −1.04 threshold, are plotted in Figs. 16.8 and 16.9. It is important to note that in all cases the entire study horizon was considered.

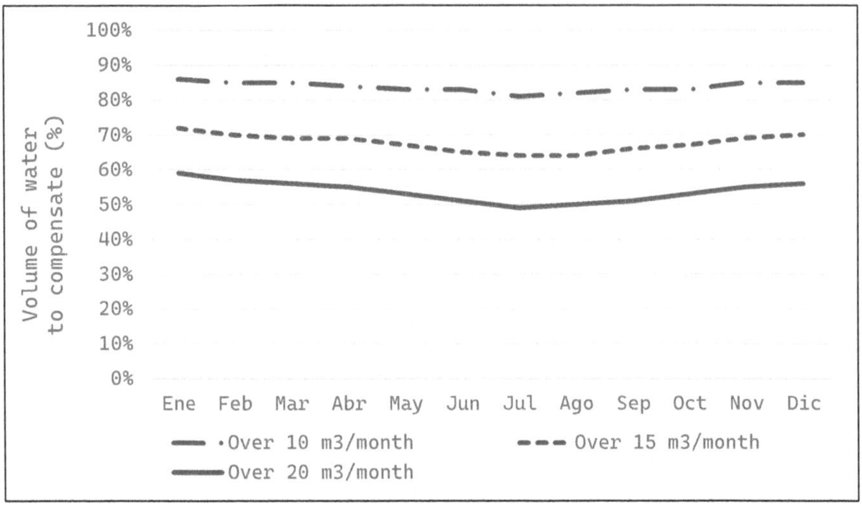

Fig. 16.7 Volume of water to be compensated, considering different thresholds for personal and domestic use or human consumption. (Elaborated by the authors)

Considering a household in Greater Valparaíso of four people consuming 18 m^3/month, and the tariff data shown on ESVAL's web page,[10] we obtain that the value of the bill is approximately 42.44 USD/month. Finally, taking this number to an annual value, the result is 509.28 USD/customer/year. Comparing this value with the values shown in Fig. 16.9, the higher value represents 0.61% of the reference annual bill.

16.6 Implementation and Feasibility of the Proposal

This section presents an analysis of the several aspects to be considered at the time of implementing the proposed instrument, especially its regulatory background and some aspects of its design.

16.6.1 Regulatory Aspects

The design and implementation of an instrument such as the one proposed is characterized by a proactive approach to risk management. Under this vision, and in anticipation of the obligation to adopt water redistribution agreements arising from the declaration of scarcity zones for the surveillance boards of a given basin

[10] https://www.esval.cl/personas/necesito-ayuda/tus-boletas-y-cobros/tarifas/

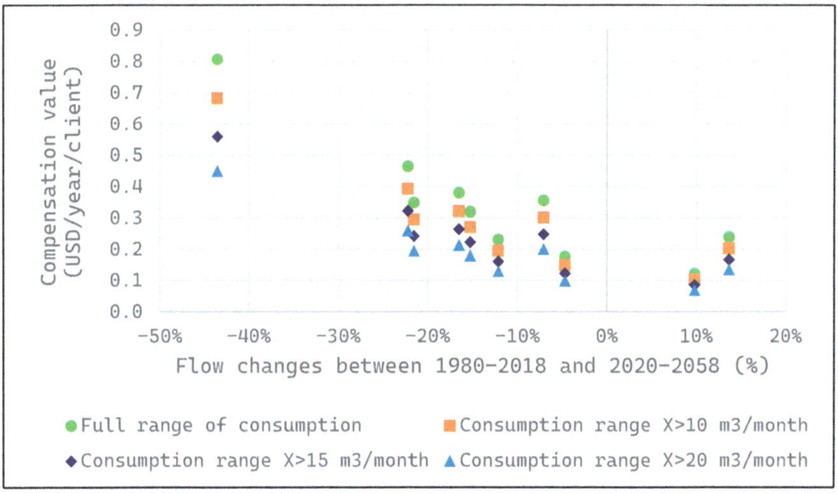

Fig. 16.8 Compensation value for each consumption range. Note: Considering SSP5-8.5 scenario, moderate drought threshold (−1.04) and water value of 0.35 USD/m³. (Elaborated by the authors)

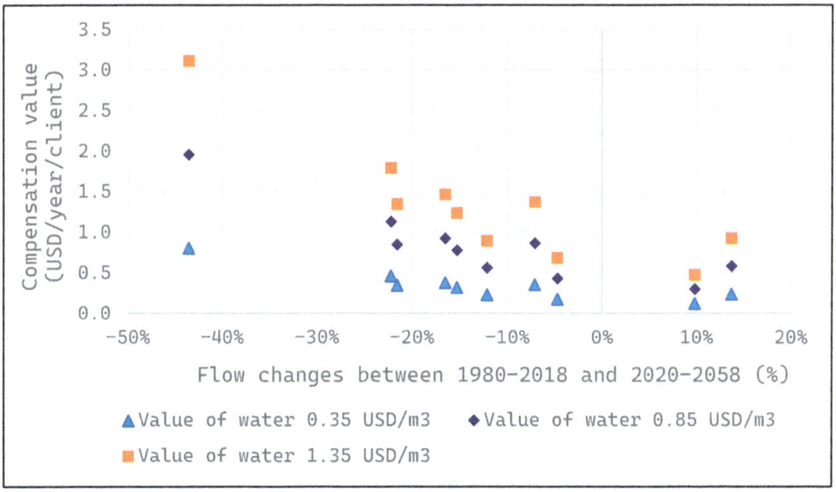

Fig. 16.9 Compensation value for each water value. Note: Considering SSP5-8.5 scenario, moderate drought threshold (−1.04) and the entire consumption range. (Elaborated by the authors)

(Article 314 of the Water Code), the holders of WR that make up such organizations (particularly water utilities, farmers, industrial water users, among others) may enter into a pre-agreement and/or agreement for the temporary transfer of water volumes. This would operate or become effective under specifically defined circumstances and conditions regarding, for example, the triggering drought thresholds, the compensation to be paid by the water utility to the actor that transfers its volumes, and the operating conditions to be maintained by this provider, among others. It is assumed that the design of this agreement is defined in the context of an infrastructure decision-making process at the basin scale to ensure that investments are made to maintain the quality and continuity of the system's operation in a situation in which there is no drought as defined in the instrument design.

From the legal point of view, the feasibility or viability of this proposal is justified by the following fundamental pillars of the regulations in force in Chile:

(a) Consecration of the human right to water. According to the Water Code, modified by Law 21.435, which entered into force on April 6, 2022, access to drinking water and sanitation are essential and inalienable human rights and must be guaranteed by the State (Article 5 of the Water Code). The object of the human right to water is personal and domestic use; that is, the use of water is limited to those basic and essential needs for life.
(b) Prioritization of human consumption, domestic subsistence use, and sanitation in the entire system regulated by the Water Code. Also, because of the legal reform, the Water Code provides that, among the various functions of water, human consumption, domestic subsistence use, and sanitation will always prevail, either when granting or limiting the exercise of WR (Rivera et al., 2020). In turn, it is established that domestic subsistence use refers to the use that a person or family makes of the water they extract to satisfy their needs for drinking, personal hygiene, watering their animals, and growing subsistence horticultural products (Article 5 bis of the Water Code). This is related to the different compensation scenarios.

One of the manifestations of the prioritization of the subsistence function occurs in circumstances of severe drought, which, previously qualified by the DGA, may lead to the declaration of a water shortage zone, for a maximum period of 1 year, which may be extended successively. During the period of validity of such declaration, a series of special rules for the redistribution of water, contained in article 314 of the Water Code, apply. In what is of special interest for this study:

(a) The DGA may require from the surveillance boards a redistribution agreement, which must ensure the prevalence of human consumption, sanitation, or subsistence domestic use.
(b) Associations of canal owners or water communities whose members include sanitation service providers must take the necessary measures to ensure that they receive the necessary flows to guarantee human consumption, sanitation, or subsistence domestic use.
(c) Any holder of WR who, because of this redistribution agreement, receives less water than that which would correspond to them according to the existing availability, shall be entitled to be compensated by whomsoever may correspond. This compensation shall not apply if such a lower proportion is a consequence of the application of the prioritization of human consumption, sanitation, and subsistence domestic use.

(d) Continuity and quality requirements to be met by sanitation service providers. The water utility must ensure the continuity and quality of their services, which may only be affected by reasons of force majeure. The SISS may require providers to enter into contracts to guarantee the provision of raw water to produce potable water when its lack affects the continuity of the service; if the lack of provision of raw water is due to force majeure, and the providers are obliged to enter into contracts for its provision, new tariffs must be set to include the higher costs, if any (article 35 DFL 382, General Law of Sanitary Services and 97 Decree 1.199).

These conditions or budgets must be endorsed and justified in the development program of each provider, an essential instrument for the operation and materialization of the water provision and sanitation service. It corresponds to the investment plan for a given time horizon, the purpose of which is to allow the provider to replace, extend, and expand its facilities to meet the requirements of the service demand (Article 53 letter k) DFL 382, General Law of Sanitary Services. Both the SISS and the provider may promote the modification of the program. When this instrument is updated, the water utility must prove that they have in a domain or in use the rights to use surface or groundwater to meet the demand in the next 5 years, as foreseen in the same program (articles 155 and following Decree 1.199). According to the full cost recovery principle considered in the Chilean water regulatory model, the economic costs associated with the development plans are incorporated into the water tariffs.

To the extent that the proposed instrument can be considered in the development plan or program of water utilities, as a pre-agreement, there is no need to make significant regulatory changes for its implementation. The above is without prejudice to definitions that go beyond the objectives of this instrument, and that could require regulatory clarifications, such as the volume of water associated with human consumption (personal and domestic use of water), and other basic assumptions for the application of this prioritization, such as the operating conditions to be met by the water utility.

16.6.2 Considerations to Be Observed in the Design of the Instrument

In addition to the regulatory aspects, it is important to consider the following notions when designing this instrument:

(a) In recent years, different mechanisms of water redistribution from agriculture to urban use have been applied, formally or informally, in several basins (before the latest modifications of the Water Code). This experience should be considered in this discussion:

- To a significant extent, these redistribution mechanisms have materialized due to the drought, but the over-granting of WR that exist in some areas has also played a role in their origin.

- The implementation of these agreements or redistributions has been different in the case of Aconcagua-Valparaíso than in the case of Maipo-Santiago. In the latter, the agreements have been easier to obtain. One of the reasons that explains this difference is the fact that in Aconcagua the population is relatively downstream in the basin, unlike Maipo, where the population is upstream, which evidences the importance of taking measures at the basin scale and involving all relevant stakeholders.

(b) In the eventual implementation of the instrument, it seems important to:

- Encourage savings in water consumption in cities that can be concretized through different tools (educational campaigns, rationing, progressive drinking water tariffs, among others).
- Consider that the application of this instrument responds to the logic of short term measures and not to strategic long term planning. Without prejudice to this, long-term structural measures associated with investment decisions should be designed and implemented. Specially to adapt to drought conditions not covered in the instrument.
- Define precisely and quantify what "human consumption" means at the time of applying the instrument and estimate the potential compensation to be paid for the transfer or temporary assignment of water volumes from one actor to another.
- Deepen the study of the feasibility of the instrument in the case of groundwater, where the application may be more complex because the governance and control/management capacity of this water source is more challenging since traditionally individual or isolated extractions have prevailed in this field, without greater coordination between the respective holders of water rights.

Given the importance of water resource management at the basin level, it is important to incorporate in the discussion and analysis other water users, including the environment, which could be affected by this new water distribution. In this sense, and considering the human right to water, rural sanitation services have a special incidence. Other aspects that could be incorporated as the instrument begins to be applied are related to water quality and the refinement of the cost of the pre-agreement, which could include not only the benefit lost by farmers in each season but also the social cost of ceasing agricultural activity.

16.6.3 Complementary Nature of the Proposed Instrument

This proposal should be understood as a complement to other tools that must necessarily be adopted to address adaptation to droughts and climate change. In other words, it does not seek to ignore the importance of other actions to address this challenge, but rather to be one more element that can contribute to its solution.

On the other hand, the quantitative analysis conducted to quantify the compensations is illustrative and, in no case, seeks to establish the definitive amount of them.

At the same time and given the lack of detailed and updated information on income, costs, and cultivated area, a wide range of values has been proposed, which are merely referential.

Acknowledgments We would like to thank Sebastián Aedo, Valentina Strappa, and Juan de Dios Guzmán, who collaborated in the development of this work. We would also like to thank the Centro de Políticas Públicas UC, which allowed us to carry out a similar work that served as a starting point for this work. We would like to thank funding via ANID project FSEQ210018, FONDECYT project 1211846 and ANID project ATE 220055.

References

Aedo, S., Chadwick, C., Gironas, J., & González-Leiva, F. (2021). ClimQMBC is a new package to bias correct climatic variables while preserving raw GCM changes in the mean and standard deviation for R, Python, and MATLAB. In *AGU Fall Meeting Abstracts* (Vol. 2021, pp. A55Q-1631).

Arnell, N. W., & Gosling, S. N. (2013). The impacts of climate change on river flow regimes at the global scale. *Journal of Hydrology, 486*, 351–364. https://doi.org/10.1016/j.jhydrol.2013.02.010

Center for Climate and Resilience Research, Álamos, N., Monsalve, T., Billi, M., Lefort, I., Allendes, A., Navea, J., Calvo, R., & Urquiza, A. (2021). *Vulnerabilidad hídrica territorial: Marco Analítico y Aplicaciones.* https://www.cr2.cl/wp-content/uploads/2021/11/Alamos-Monsalve-et-al-2021-Vulnerabilidad-h%C3%ADdrica-territorial.pdf

Environment Agency. (2017). *Water resources planning: How water companies ensure a secure supply of water for homes and businesses.* https://www.gov.uk/government/publications/water-resources-planning-managing-supply-and-demand/water-resources-planning-how-water-companies-ensure-a-secure-supply-of-water-for-homes-and-businesses

Gómez-Lobo, A., & Contreras, D. (2003). Water subsidy policies: A comparison of the Chilean and Colombian schemes. *World Bank Economic Review, 17*(3), 391–407. https://doi.org/10.1093/WBER/LHG028

Harbaugh, A. W. (2005). *MODFLOW-2005, the US Geological Survey modular ground-water model: The ground-water flow process* (Vol. 6). US Department of the Interior, US Geological Survey.

Howard, G., Bartram, J., Williams, A., Overbo, A., Fuente, D., & Geere, J.-A. (2020). *Domestic water quantity, service level and health* (2nd ed.). World Health Organization, 60 pp.

Instituto Nacional de Estadística. (2017). *CENSO 2017*. http://resultados.censo2017.cl/

McKee, T. B., Doesken, N. J., & Kleist, J. (1993). The relationship of drought frequency and duration to time scales. In *8th Conference on Applied Climatology*, Anaheim, 17-22 January 1993 (pp. 179–184).

Ministerio de Obras Públicas, Dirección General de Aguas. (2020). *Plan estratégico de gestión hídrica en la cuenca del Aconcagua: informe final.* https://snia.mop.gob.cl/repositoriodga/handle/20.500.13000/125414?locale-attribute=es#:~:text=El%20objetivo%20general%20de%20este%20estudio%20es%20proponer,que%20aseguren%20su%20abastecimiento%20en%20cantidad%20y%20calidad

Ministerio de Obras Públicas, Dirección General de Aguas. Chile. (2004). *DIAGNOSTICO Y CLASIFICACION DE LOS CURSOS Y CUERPOS DE AGUA SEGUN OBJETIVOS DE CALIDAD CUENCA DEL RIO ACONCAGUA.* https://mma.gob.cl/wp-content/uploads/2017/12/Aconcagua.pdf

Ministerio de Obras Públicas, Dirección General de Aguas. Chile. (2022). Deja sin efecto la resolución d.g.a. n° 1.674 (exenta), de 12 de junio de 2012, y establece criterios que determinan el carácter de severa sequía, de conformidad a lo dispuesto en el artículo 314 del código de aguas. Recovered from: https://www.bcn.cl/leychile/navegar?idNorma=1177291&idVersion=2024-01-10&idParte=

Ministerio de Obras Públicas, Superintendencia de Servicios Sanitarios. (2020). *Informe de Gestión del sector Sanitario*. https://www.siss.gob.cl/586/articles-19431_recurso_1.pdf

Molinos-Senante, M. (2018). Urban water management. In G. Donoso (Ed.), *Water policy in Chile. Global issues in water policy* (Vol. 21). Springer. https://doi.org/10.1007/978-3-319-76702-4_9

Observatorio del Sistema Nacional de Ciencia, Tecnología, Conocimiento e Innovación. (2022). *PIB total y per cápita regional*. https://observa.minciencia.gob.cl/indicadores/contexto-socioeconomico/pib-per-capita-regional

Panel intergubernamental de expertos sobre el Cambio Climático. (2023). *Synthesis report of the IPCC sixth assessment report (AR6)*. https://report.ipcc.ch/ar6syr/pdf/IPCC_AR6_SYR_LongerReport.pdf

Ricalde, I., Vicuña, S., Melo, O., Tomlinson, J. E., Harou, J. J., & Characklis, G. (2022). Assessing tradeoffs in the design of climate change adaptation strategies for water utilities in Chile. *Journal of Environmental Management, 302*, 114035.

Rivera, D., Molinos, M., & Donoso, G. (2020). *Priorización del uso del agua para consumo humano*. Temas de la Agenda Pública, 15(131), Centro de Políticas Públicas UC, Santiago, 15 pp

Vicuña, S., Gil, M., Melo, O., Donoso, G., & Merino, P. (2018). Water option contracts for climate change adaptation in Santiago, Chile. *Water International, 43*(2), 237–256.

Yates, D., Sieber, J., Purkey, D., Huber-Lee, A., & WEAP21, À. (2005). A demand, priority, and preference driven water planning model: Part 1, model characteristics. *Water International, 30*(4), 487–500.

Chapter 17
Water Reuse in Chile: An Alternative to Face the Drought

Eduardo Leiva, Ignacio T. Vargas, and Rodrigo Labatut

Abstract In recent years, drought events have become increasingly common, causing serious environmental disasters. The drought causes a significant reduction in the availability of both surface and groundwater resources, which can lead to different effects, such as a reduction in agricultural productivity, reduced energy generation, an increase in water treatment costs, a marked deterioration of natural habitats, and an impact on economic and social activities. Due to drought and water scarcity, several regions of the world will face unprecedented pressures exacerbating the current imbalance between water demand and regional availability. In Chile, drought events are increasingly common, so it is necessary to evaluate changes in the management strategies of water resources to address this problem. A powerful strategy to face to the drought is water reuse. The reuse of water offers a supply alternative for various beneficial uses, reducing pressures on better quality water resources. However, and although this approach has been studied during several decades, implementing suitable and environmentally acceptable water reuse strategies still represents a significant challenge. This chapter provides an overview of water reuse strategies applied in Chile, focusing on greywater and wastewater reuse, along with

E. Leiva (✉)
Departamento de Ingeniería Hidráulica y Ambiental, Pontificia Universidad Católica de Chile, Santiago, Chile

Facultad de Química y de Farmacia, Departamento de Química Inorgánica, Pontificia Universidad Católica de Chile, Santiago, Chile

Center for Water Law and Management, Pontificia Universidad Católica de Chile, Santiago, Chile
e-mail: ealeiva@uc.cl

I. T. Vargas
Departamento de Ingeniería Hidráulica y Ambiental, Pontificia Universidad Católica de Chile, Santiago, Chile

Centro de Desarrollo Urbano Sustentable (CEDEUS), Santiago, Chile

R. Labatut
Departamento de Ingeniería Hidráulica y Ambiental, Pontificia Universidad Católica de Chile, Santiago, Chile

D. Rivera, G. Donoso (eds.), *Droughts in Chile*, Global Issues in Water Policy 31,
https://doi.org/10.1007/978-3-031-85040-0_17

the key factors, policies, and regulations necessary for the sustainable implementation and widespread adoption of these strategies. The potential uses of and the risks associated with reclaimed water reuse are discussed. In addition, the political, regulatory, economic, environmental, technological, and social challenges related to water reuse in the Chilean context are reviewed.

Keywords Water reuse · Greywater · Drought · Water scarcity

17.1 Introduction

Water scarcity is a problem of great concern in the world. In the coming years, water resources in various regions are expected to face unprecedented pressures due to continued population growth, changes in consumption habits and climate change (Liu et al., 2017; Kummu et al., 2016). In particular, the phenomenon of climate change has led to an increase in the frequency and intensity of droughts in many parts of the world (Cook et al., 2018). According to the World Resources Institute (WRI), four billion people, mainly distributed in the regions of Middle East, North Africa, and South Asia, live under highly water-stressed conditions for at least 1 month a year (WRI, 2023). In eastern US and central Europe, summers have become warmer and dryer, affecting groundwater levels and surface water reservoirs (Nahrstedt et al., 2020). This situation generates regional water conflicts, limits economic development, affects social welfare, decreases food availability and causes environmental degradation.

Since the 1980s, global water demand has been increasing at an approximate annual rate of 1%. It is projected that by the year 2050, demand will increase by 20–30% compared to the current level, widening the gap between the demand for clean water and the available supply (UN-Water, 2010; Uhlenbrook & Connor, 2019). Globally, there is enough freshwater availability to meet the world's demand; however, there are periodic spatial and temporal variations in demand and water availability, leading to scenarios of uncertainty regarding the actual resource availability and causing marked periods of water scarcity in different parts of the world at specific times of the year. Various studies have assessed water scarcity on a global scale, but due to the resolution of these studies, water scarcity at the local or intra-basin level has not been adequately examined, underestimating the impact of water scarcity (Vorosmarty et al., 2000; Alcamo & Henrichs, 2002; Oki & Kanae, 2006; Oki et al., 2001; Ndehedehe et al., 2023). At the local level, variability in water availability can be even greater, and the effects of drought can be exacerbated, leading to much greater impacts than previously reported. Additionally, the growing water scarcity is compounded by historical inequalities in access to water experienced by certain sectors of the population (Deshpande et al., 2020). Particularly, the most vulnerable populations are disproportionately affected by the availability of this resource, limiting their quality of life, economic prospects, and in many cases, their subsistence.

The global climate crisis has imposed severe water stress on semi-arid climates. Climate change projections for these zones have revealed a dramatic reduction in rainfall, affecting water availability, ecosystems, and cities. Chile is facing a high level of water stress. WRI ranks Chile as the 16th most water-stressed regions in the world (WRI, 2023), resulting from a megadrought that has lasted nearly 14 years, involving 8 out of its 15 continental regions (CR2, 2023). Projections for Chilean basins indicate that the ratio between the future total water demand and the available renewable surface water is expected to reach a value superior to 80% by 2040 (Maddocks et al., 2015). On the other hand, recently the longest and most severe heat wave in the middle of the winter, significantly impacting the snowpack levels of the Andes mountains, which constitute the freshwater reservoir for the summer months (CR2, 2023). The distribution of water resources across the country exhibits notable disparities (Aitken et al., 2016; Rodríguez et al., 2020a). This discrepancy is particularly alarming in rural regions, where approximately 300,000 rural households in Chile lack the essential infrastructure required for accessing safe drinking water (Amulén, 2019). Chile has a 99.97% coverage of access to safe drinking water in urban areas, but only 72% in rural areas, which strongly restricts the development of these areas (Blanco & Donoso, 2016).

Chile has an average runoff of 53,953 m^3 per person per year, a value significantly higher than the global average of 6600 m^3 per person per year (Valdés-Pineda et al., 2014). This positions Chile as a privileged country in terms of water resources, however, there is a significant inequality regarding water availability, population, and water demand for various uses across the country. The central and northern regions of Chile represent a critically important sector for the country due to the growing agricultural and mining activities (Aitken et al., 2016; Rehner et al., 2014; Oyarzún & Oyarzún, 2011). However, drought events in recent years have hit these regions the hardest, leading to serious local conflicts over water resources (Budds, 2009). Meteorological drought has translated into hydrological drought and reduced aquifer recharge, resulting in conflicts among water users, environmental impacts on ecosystems, and economic hardships for various industries (Aitken et al., 2016; Oyarzún & Oyarzún, 2011; Moskvitch, 2012). These issues have elevated the importance of freshwater availability and the search for alternative water sources. In response to this challenge, water reuse options have emerged as an attractive alternative, especially for alleviating pressures on higher-quality water resources and addressing problems that arise when the available water supply is already stretched and cannot meet the growing demand of a community.

17.2 Water Reuse: Global Context

A powerful alternative to address drought is water reuse. Water reuse has been practiced by humanity for thousands of years. There are reports of wastewater reuse for crop irrigation dating back to the Bronze Age (3200–1100 BC) (Angelakis & Gikas, 2014; Angelakis & Snyder 2015). However, despite the inherent benefits that these practices may have brought, many were not carried out properly and did not take into account modern water treatment processes. Water treatment and sanitation have become more commonplace only in the last century, leading to significant improvements in quality of treated waters and the impact of wastewater discharge.

As previously mentioned, water scarcity and drought has generated an urgent need for technologies that allow for the proper management of water resources while maximizing their primary and secondary uses. Water reuse is a practice that enables the more efficient use of water resources for a specific purpose and can be classified into direct water reuse and indirect water reuse (Dishman et al., 1989). Direct water reuse involves treating water with the ultimate goal of reusing it for a predetermined secondary purpose (Dishman et al., 1989). In contrast, indirect water reuse involves the use of water by specific localities downstream from other communities, effectively reusing the wastewater from the upstream population for various purposes (Dishman et al., 1989). The fundamental difference lies in the approach to the process.

The potential uses of recycled water can vary widely, including urban reuse, groundwater recharge/non-potable reuse, agricultural reuse, environmental reuse, industrial reuse, and potable reuse (USEPA, 2012). However, these uses and the benefits they bring will depend on various factors, many of which are influenced by the advancements, culture, and legal framework of each country. These factors include the level of treatment, the condition of water resources, environmental status, public willingness, guidelines, and regulations (USEPA, 2012). Reusing water reduces the strain on freshwater resources and helps alleviate issues that arise when the available water supply is overcommitted and cannot meet increasing demand.

Among the key benefits of water reuse are contributions to sustainable development, water conservation, a stable water supply, and numerous environmental, economic, and health benefits for individuals (Shoushtarian & Negahban-Azar, 2020). However, alongside these benefits, there are also significant challenges that must be overcome to sustainably implement these practices over time (Shoushtarian & Negahban-Azar, 2020). The main challenges include political, legal and economic concerns, technical issues, social considerations, as well as the potential environmental impacts of their implementation (Shoushtarian & Negahban-Azar, 2020). Therefore, to efficiently implement these practices, it is essential to strike a balance that allows for maximizing benefits while also mitigating potential negative impacts.

Originally, water reuse was oriented towards using treated wastewater to improve the management of water resources in response to the increasing population living in cities (Salgot & Folch, 2018). However, due to differences in water quality, the concept of water reuse also expanded to include the potential reuse of greywater, which would require less treatment, and the reuse of industrial water.

17.3 Water Reuse in Chile

En Chile, despite the existence of areas with serious water supply issues, the concept of water reuse is relatively new when compared to other countries like the United States, Mexico, Israel, or Australia, where water reuse has been practiced for many years (Bravo, 2011). Water reuse practices in recent decades in Chile have been primarily focused on indirect reuse and small-scale approaches in decentralized systems. However, in February 2018, in Chile, Law No. 21,075 was enacted, which regulates the collection, reuse, and disposal of greywater (Diario Oficial de la República de Chile, 2018). This instrument has become a legal incentive to formalize systems and greywater reuse practices in urban and rural areas. Thus, greywater reuse has gained attention primarily due to its lower treatment costs and simpler treatment technology (Rodríguez et al., 2021, 2022a; Van de Walle et al., 2023; Al Arni et al., 2022; Gómez-Monsalve et al., 2022). The separation of greywater from general sewage and its reuse is now one of the most popular and promising approaches for sustainable and integral water management in scenarios of water scarcity (Vuppaladadiyam et al., 2019). However, the treatment and reuse processes will depend on the quality of the greywater and, ultimately, the water treatment technologies implemented.

Over the years, various water and wastewater treatment technologies have been implemented in Chile, many of which are applicable for water reuse. Conventional water reuse processes, such as membrane processes, adsorption processes using commercial sorbents, oxidation processes, artificial wetlands, and biological treatment systems, are the most common. However, new technological applications are required to enhance water/wastewater treatment processes and their sustainability. Thus, future technological applications for water reuse must address several challenges, including the need to improve efficiency to achieve the required water quality levels, reduce energy costs, and minimize failures to ensure public health protection. Indeed, the development of new technologies can provide better alternatives for the sustainable provision of safe, clean, and adequate water supply, as it would allow mitigating the undesired consequences of conventional water treatment processes, such as the release of contaminants, a decrease in purification efficiency, and high energy requirements.

17.3.1 Greywater Reuse

Greywater includes domestic wastewater from showers, bathtubs, dishwashers, kitchen sinks, and laundry, excluding the toilet for sanitary reasons. Greywater represents about 65% of the total water consumption in a household (Batisha, 2020). Thus, its reutilization could result in a significant reduction of the total household drinking water budget. Due to its high organic matter content, nutrients, fats, surfactants, and pathogens, greywater requires treatment before being used. Depending on the intended use, the treated water's quality standards vary according to the criteria established in different countries. In Chile, the legal framework and social acceptability have evolved within the last 5 years, allowing the possibility of greywater reuse (Diario Oficial de la República de Chile, 2018; Amaris et al., 2021). The Chilean regulation (i.e., Chilean Law N° 21,075) considers ornamental, recreational, and urban uses for greywater. The parameters used to assess the quality of treated greywater are the 5-day Biochemical Oxygen Demand (BOD_5), Total Suspended Solids (TSS), Fecal Colloids (CF), Turbidity, and residual free chlorine.

While TSS and CFs are addressed with primary treatment and disinfection, respectively, BOD5 and turbidity, associated with the dissolved organic matter in greywater, must undergo secondary treatment (generally biological).

Conventional treatment technologies such as activated sludge and anaerobic digestion use space and energy and are complex to operate. Thus, new technologies are required to meet the needs of decentralized water systems. These systems manage and treat small volumes of wastewater (i.e., less than 10 m^3/day) originating in single households or communities on a smaller scale than centralized systems. This reduces transport costs and extensive infrastructure requirements, allowing the development of more compact, sustainable, and resilient systems (Capodaglio et al., 2017; Garrido-Baserba et al., 2018). Also, when combined with source separation, decentralization enables the direct recovery of nutrients, heat, and energy contained in the greywater, which contributes to the revalorization of waste material and sustainable development (Besson et al., 2021; Hadengue et al., 2022; Larsen et al., 2016). Thus, Tailored or fit-for-purpose technologies are essential to address the different quality standards required depending on the objective and local regulations, completing the requested quality standards for a high range of influent greywater qualities and avoiding unnecessary treatment for more efficient use of the available resources (Capodaglio, 2021; Van de Walle et al., 2023).

According to the Chilean water quality standard for greywater reuse, irrigation of ornamental green spaces is the most accessible application for treated domestic greywater, saving drinking water for human consumption or other critical uses with higher sanitary requirements. Under a Mediterranean climate, the water consumption of certain species can range from 2.6 to 9 L/m^2 per day (Schweitzer & Erell, 2014). In a semi-arid climate, depending on the green space, consumption can reach up to 60 L/m^2 per day (Reyes et al., 2016), a value that is ten times the reference of evapotranspiration. Even though using plants resistant to drought will reduce water consumption for irrigation, this will probably not be enough. Thus, greywater emerges as an alternative to sustain green spaces in the context of climate change and water scarcity. A study of greywater generation in Chile indicated that a four-person household generates about 356 L daily (Cruz, 2017). Thus, considering this number, a quick estimation suggests that a building with 100 occupants would generate almost 9 m^3 per day, allowing the irrigation from 150 to 1000 m^2 for a semi-arid or Mediterranean climate, respectively. This further underscores the potential of greywater reuse.

17.3.2 Reuse of Municipal Wastewater and Industrial Waters

Regions where surface and ground waters are under high stress, greywater resources are limited, and/or desalination is not a feasible option, have resorted to domestic wastewaters as an additional/alternative source of water to satisfy their needs. The most common scenario of treated wastewater reuse is the *de facto* reuse (indirect water reuse), where downstream communities use surface water (e.g., rivers) for drinking, irrigation, or other purposes, after it has been subjected to upstream discharges from conventional wastewater treatment plants (WWTP) (AWWA, 2016). In addition, non-potable reuse refers to recycled or reclaimed water that is not used for drinking but is safe to use for irrigation or industrial processes (e.g., landscape irrigation, cooling water, groundwater recharge, habitat restoration, toilet flushing, car wash, fountains) (IWA, 2023). In contrast, wastewater reclamation projects involve treating the water to an extent that enables it to comply with the regulations for drinking, crops irrigation, or industrial purposes (Shahid et al., 2022). The projects can be divided into indirect potable reuse (IPR) applications, where the treated wastewater is introduced into an environmental buffer (e.g., groundwater aquifer or a surface water reservoir, lake, or river) before entering the drinking water treatment plant (DWTP); and, direct potable reuse (DPR) applications, where the treated wastewater is directly introduced into the DWTP (AWWA, 2016).

Wastewater reclamation facilities rely on a multiple barrier treatment approach, where a train consisting of a sequence of bio-, physico-, and even, thermo-chemical processes may be required. The specific treatment steps used are, however, dependent upon both the final use of the treated water and the specific quality standards established for each application. Thus, similarly to the reuse of greywater, a "fit-for-purpose" approach to wastewater reuse is more desirable, as it can reduce capital and operational costs (Larsen et al., 2016). In a typical IPR or DPR wastewater reclamation facility, advanced treatment steps follow the effluent of a conventional WWTP. These steps include ozonation, adsorption with activated carbon, chemical disinfectants, UV radiation, advanced oxidation processes (AOPs) and membrane filtration steps, among others (NWRI, 2015; Rizzo et al., 2020). These train sequences are capable of removing pathogens as well as most contaminants of emerging concern (CECs) (Kim et al., 2018).

Israel is, by far, the world leader in water reuse, reusing nearly 90% of its produced water, mostly for irrigation of agriculture crops (Fluence Corporation, 2021). In the US, 43 out of 50 states are practicing water reuse (IWA, 2023), with an estimated 2.53 million m^3/day of wastewater reclaimed for irrigation in 2015 (Ritter, 2021). California, Florida, Arizona, and Texas account for 95% of the wastewater reuse for irrigation in the US (Ritter, 2021). In Florida, over 55% of the reuse is used for landscape irrigation and 11% for agriculture irrigation, whereas in California, only 18% was used for landscape irrigation and 29% of the reuse was used for agriculture irrigation (IWA, 2023). In Europe, more than 40,000 million m^3 of wastewater is treated in the EU every year, but only 2.4% is further treated to be reused (WISE, 2023). The leader countries in terms of percent of reclaimed water are the Mediterranean, where 12% in Italy and Spain, 60% in Malta, and 90% in Cyprus (Interreg, 2023). In Chile, there are currently 297 WWTP with different treatment systems, which operate on around 1284 million m^3 of wastewater per year, which corresponds to an average flow of 40.7 m^3/s. From this amount, 96% of treated wastewater is discharged to surface, continental or marine water bodies, and only 4% of treated water is reused directly and in a planned manner for different uses (SISS, 2018). An initiative led by Fundación Chile in the region of Coquimbo, has implemented six rural, pilot-scale wastewater reuse plants since 2019. These plants supply treated wastewater for the development of artificial wetlands as well as for the irrigation of ornamental plants, recreational parks, and natural areas (Fundación Chile, 2023). A key concept demonstrated by this initiative, is that low cost, simple wastewater treatment technologies can be effective and efficient enough to produce water with sufficient quality for such purposes, decreasing the demand for less available and costly potable water.

Considering that bacteria concentrations in untreated wastewater can vary between 10^4 and 10^8 CFU/100 mL (López et al., 2019), microbial composition is one of the main health hazards in wastewater reuse, and thus one of the most important parameters to regulate (Schellenberg et al., 2020). The use of untreated, raw wastewater in agriculture can lead to the production of contaminated food and to the spread of contamination in the environment through water leaching and runoff (Farhadkhani et al., 2020). Thus, regardless of the final use, if not adequately treated, wastewater reuse can represent a source of microbiological health risk to the population as a whole (Bonetta et al., 2022). Fortunately, Conventional WWTP can remove up to 98% of pathogens, such as *E. coli*, during the biological process stage; and up to 99.999% can be achieved by a final disinfection process (Raboni et al., 2016). However, conventional domestic WWTP are not very effective at removing CECs, such as pesticides, pharmaceuticals, microplastics, endocrine disrupting compounds, and personal care products, or antibiotic resistant bacteria (ARB) and antibiotic resistance genes (ARGs) (Fluence Corporation, 2021). Thus, if limits for these constituents are to be established by authorities for wastewater reuse, then conventional WWTP will certainly require upgrade their facilities, adding advanced treatment steps, as mentioned above, to meet more stringent water quality standards.

Today, wastewater reclamation is a technically and economically viable strategy to produce water with the required quality to meet the standards for a variety of uses, with applications ranging from crops irrigation, groundwater/aquifer recharge, surface water augmentation, to direct human consumption. However, the public is still skeptical about the health risk of reuse projects, which is especially true if conventional WWTP/DWTP, particularly that new CECs are becoming more ubiquitous in today's domestic wastewaters (Fluence Corporation, 2021; Xue et al., 2023). New, updated water reuse regulations, enhanced, continuous effluent monitoring, education, and outreach will all be required to convince the public that adequately treated wastewater can be safe for irrigating crops and even for human consumption.

17.3.3 Regulations and Guidelines

Regulations and guidelines are critical for designing and implementing water reuse systems (Watson et al., 2017; Shoushtarian & Negahban-Azar, 2020; Lane et al., 2017; Rodríguez et al., 2022a, 2022b). The regulations include standards that establish the requirements for water treatment processes, a determination of applicable treatment systems, criteria and water quality standards, effluent monitoring and control processes and frequency, necessary signage, operating standards, permitted uses for treated water, among other guidelines (Shoushtarian & Negahban-Azar, 2020; Rodríguez et al., 2022b; Hespanhol & Prost, 1994; United States Environmental Protection Agency et al., 2004). Often, the development of these regulations can be controversial as it may limit the applicability of water reuse systems in specific contexts. For this reason, the development and implementation of these regulations are a challenge. Despite the fact that water reuse has been practiced for several years, regulations are still quite limited. In the case of wastewater reuse, it is still an emerging practice in many countries, and several of them do not have specific regulations.

In the case of Chile, the political constitution of Chile does not include a specific section associated with water reuse, beyond establishing a constitutional guarantee of environmental protection (Article 19, No. 8) (Leiva et al., 2020). Similarly, Chilean environmental regulations do not contain specific and detailed rules for the implementation of treatment systems aimed at water reuse (Leiva et al., 2020). These regulations only establish general guidelines for water/wastewater treatment projects that must undergo the Environmental Impact Assessment System (SEIA, *by the Spanish acronym of Sistema de Evaluación de Iimpacto Ambiental*) and should eventually consider measures of compensation, reparation, or mitigation. Nevertheless, as mentioned earlier, in Chile, there is Law No. 21.075, enacted in February 2018, which regulates the collection, reuse, and disposal of greywater (Diario Oficial de la República de Chile, 2018). This legal instrument aims to regularize the reuse of greywater in Chile and promote these practices throughout the country. This regulation applies exclusively to the reuse of greywater and does not cover the reuse of blackwater or of black water or wastewater, establishing a differentiation and separation between the two. The norm applies to those who voluntarily install greywater reuse systems, excluding buildings with specific characteristics that currently have a mandatory requirement to construct greywater reuse systems according to the General Ordinance of Urbanism and Construction (Decreto 10, 2020).

Several considerations for the implementation of greywater reuse systems in Chile are established in Law N°. 21.075 (Diario Oficial de la República de Chile, 2018). In this regard, concerning governance, it will be the Regional Ministerial Secretariats of Health that must approve and authorize greywater reuse projects implemented throughout the country. Additionally, they will be responsible for monitoring and ensuring compliance with the provisions of the regulation of Law N°. 21,075 (Leiva et al., 2020). On the other hand, entities such as the Superintendency of Sanitary Services (SISS, *by the spanish acronym of Superintendencia de servicios sanitarios*), the Superintendency of Environment (SMA, *by the spanish acronym of Superintendencia de medio ambiente*), or the Ministry of Housing and Urban Planning (MINVU, *by the spanish acronym of Ministerio de Vivienda y Urbanismo*) may have involvement in specific aspects of the implementation of greywater recycling systems, such as environmental impact statements and studies. Similarly, sanitation companies may only have involvement in domestic greywater recycling systems, not in rural areas (Leiva et al., 2020). Unfortunately, the law does not establish a public body or entity to provide support for the implementation of greywater recycling systems, which can evidently be a significant barrier. Despite these obstacles, Law N° 21,075 is a substantial initiative that can contribute to regulating greywater treatment and reuse systems and, at the same time, encourage their expansion in Chile.

17.4 Future Challenges: Drivers and Barriers

The implementation of water reuse systems brings with it a wide variety of future challenges. To successfully implement these systems and encourage the adoption of water reuse practices, it is necessary to consider both the drivers that promote their development and the barriers that can obstruct it (Lee & Jepson, 2020). A balance between drivers and barriers will determine the success of the adoption and widespread use of water reuse practices.

17.4.1 Political and Regulatory Factors

Political factors can play a significant role in promoting or restricting the implementation of water reuse systems. As mentioned earlier, in Chile, Law N° 21,075 serves as a public instrument that can either favor, enhance, or limit the implementation of greywater reuse systems. Another example can be found in Australian cities like Sydney and Melbourne, where government-introduced water reuse policies have encouraged the use of recycled water (Rogers et al., 2015; Bischel et al., 2012). Similarly, in 2009 in California, USA, the California Recycled Water Policy was launched (Lee & Tan, 2016). As a result of this strategy, water recycling practices in Northern California have increased, leading to the establishment and implementation of standards and guidelines to promote water recycling in municipalities (Lee & Tan, 2016). In Chile, the General Ordinance of Urbanism and Construction is another instrument that promotes greywater reuse in the country (Decreto 10, 2020). This ordinance mandates the installation of greywater reuse systems in buildings with a surface area equal to or greater than 5000 m^2, with the exception of residential buildings, where implementation is voluntary (Decreto 10, 2020).

Although policy and regulatory factors can serve as drivers for the implementation of water reuse systems, the absence of standardized frameworks and guidelines to guide and facilitate their implementation has been identified in the literature as one of the major obstacles to urban water reuse (Lee & Jepson, 2020). Moreover, legal provisions pertaining to water quality parameters and their relationship to approved uses of treated water and water rights in each country can be extremely important for promoting water reuse strategies and practices, especially when faced with scenarios of water scarcity (Watson et al., 2017; Lane et al., 2017; Bischel et al., 2012). At the same time, cooperation between water supply facilities and wastewater treatment facilities is a key requirement to promote the integration and implementation of water reuse systems and practices (Bischel et al., 2012). However, in the case of Chile, effective coordination is also strongly required between public and private stakeholders to efficiently and effectively promote successful water reuse strategies (Lane et al., 2017; Lee & Tan, 2016). In Chile and Latin America, centralized systems have more advanced infrastructure for implementing water reuse practices, while the greatest water needs are often found in rural areas where decentralized systems have limited infrastructure (Rodríguez et al., 2022b; Desbureaux & Rodella, 2019). Finally, in Chile, legal factors can become barriers to adopting water reuse practices. Insufficient institutional support or cooperation among government agencies, primarily due to the lack of clear governance in Law No. 21.075, can be a significant barrier. Moreover, the presence of unclear and specific regulations can often restrict the adoption of water reuse practices. For this reason, having a regulatory framework with clear standards and guidelines is crucial.

17.4.2 Financial and Economic Factors

Different studies have reported that the feasibility of water reuse projects in rural and urban areas will be influenced by funding for these projects, government subsidies, tax incentives, and the associated implementation and operational costs (Van Rensburg, 2016; Bichai et al., 2018; Stathatou et al., 2017; Sgroi et al., 2018). Economic factors can pose a significant barrier due to the high capital costs associated with these technologies, compared to the water savings achieved. Rodríguez et al. (2020b) showed that pilot systems for greywater treatment and reuse in rural areas of Chile may not be economically feasible since the investment costs exceed US$ 5200, an amount that is not covered by the water savings. Even so, this study suggests that other benefits such as thermal regulation and improved quality of life could make these projects feasible if a social evaluation is considered, rather than purely an economic one.

From an economic perspective, the costs of operation and the lack of funding for water sampling, water analysis, and treatment facilities may restrict the implementation of water reuse practices (Rodríguez et al., 2020b; Hughes, 2013; Friedler, 2008; Miller, 2006). In fact, the investment costs of water reuse systems can make their implementation prohibitive, making government subsidies necessary to promote their widespread adoption. Instances of this can be observed in Abu Dhabi and Dubai (UAE), where the adoption of water reuse has been promoted and encouraged by tax incentives and price discounts (Bichai et al., 2018). Since the existing costs of development and adoption of water reuse technologies and practices are significantly greater than the returns realized through fresh water savings, the introduction of subsidies and tax reductions is a significant driver to promote the implementation and adoption of water reuse technologies and practices (Xue et al., 2023; Rodríguez et al., 2020b; Amaris et al., 2020; Beveridge et al., 2017; Grundmann & Maaß, 2017). However, the challenges associated with adapting household sewage networks reduce the willingness to pay for water reuse systems, which emerges as another significant obstacle to the implementation and adoption of water reuse technologies and practices (Watson et al., 2017; Bint et al., 2019; Paola et al., 2018).

17.4.3 Environmental Factors

Environmental factors are another crucial factor for implementing water reuse systems. Water scarcity acts as a significant catalyst for the development and implementation of these technologies. In many regions of the world, drought has reached critical levels, driving the need and search for alternative water sources (Munné et al., 2023; Ishii et al., 2015; Gosling & Arnell, 2016; Kunz et al., 2015; Qin et al., 2019; Schewe et al., 2014). On the other hand, climate change, the widening gap between water supply and demand, as well as natural disasters, have reduced the water security of cities and underscore the need for the development of strategies for more sustainable water resource management, such as water reuse (Miller, 2006; Beveridge et al., 2017; Bint et al., 2019; Jiang et al., 2018). Similarly, other environmental factors that influence and promote the adoption of water reuse practices include the protection of available freshwater sources and the increase in CECs in natural waters (Anderson, 2003; Chen & Wang, 2009).

In terms of barriers to the implementation of water reuse systems, environmental factors may not be the primary determinants. However, a key factor is the timing of drought impacts (Watson et al., 2017). It has been reported that during severe drought events, the need for and promotion of these practices become evident (Watson et al., 2017). However, when short-term periods of increased water availability occur, interest from authorities and individuals for these systems diminishes. Therefore, the environmental focus of water reuse becomes a reactive measure in drought scenarios rather than a proactive measure to reduce future drought impacts.

17.4.4 Technical and Technological Factors

Another key aspect for the implementation of water reuse systems is the technical capacity that each country has to implement effective water treatment systems, both centralized and decentralized. It is essential to have facilities adapted to integrate water reuse systems, for example, in municipal wastewater treatment. Furthermore, water treatment technologies need to be adaptable to the quality of the influent water, all while remaining cost-effective, in order to ensure the long-term feasibility and sustainability of water reuse practices (Bichai et al., 2018; Maniam et al., 2022). For example, the application of advanced nanotechnology to traditional water purification processes provides opportunities to develop more cost-effective and environmentally acceptable water reuse technologies (Eslamian et al., 2016; Qu et al., 2013). Membrane processes, such as microfiltration, ultrafiltration, and nanofiltration for advanced water purification, as well as nanofiltration and reverse osmosis for desalination and water reuse, have been extensively studied (Wintgens et al., 2005; Yang et al., 2020; Fernández-Medrano et al., 2022; Mendret et al., 2019; Kaya & Dayanir, 2020). However, these technological advancements must be aligned with their applicability and economic viability in water scarcity scenarios.

As mentioned earlier, from a technological perspective, insufficient infrastructure and wastewater treatment facilities are one of the main barriers to the successful adoption of water reuse practices (Bichai et al., 2018; Carr & Potter, 2013). In the case of Chile, the availability of treatment infrastructure in urban areas is quite high, with wastewater treatment coverage exceeding 99% (Villamar et al., 2018). However, in rural areas, the situation is different, and the infrastructure is often insufficient and lacks adequate facilities for wastewater treatment and water reuse. Moreover, wastewater treatment in these areas is more complex due to their locations, the lower availability of operator training, and constrained resources. In this context, several barriers to adopting water reuse occur in rural areas where, precisely, the needs are greater. Various difficulties, such as high operating costs, poor maintenance, inadequate infrastructure, low levels of technical training, and limited operation and monitoring of treatment systems, hinder progress in these regions (Diario Oficial de la República de Chile, 2018; Mizyed, 2013; Murei et al., 2022). These barriers can become insurmountable obstacles for rural areas and require a more targeted design of water reuse systems.

17.4.5 Social and Perception Factors

Finally, one of the most critical aspects for the successful adoption of water reuse practices and technologies is public perception. Gross et al. (2015) reported that in different societies, there is a psychological barrier that generates a sense of rejection towards the use of recycled water. This perception arises because people believe that recycled water is dirty or that the source from which it comes is not suitable due to health risks. It's important to note that this perception factor is associated with the end use of the water, with water for drinking purposes generating the greatest resistance, as opposed to other uses such as irrigation or toilet flushing, which are more readily accepted.

Amaris et al. (2020), through a comprehensive survey using advanced discrete choice models, sought to understand water use and reuse preferences among specific groups of people in Santiago, Chile. They found that, in decreasing order of preference, there is a general acceptance of using high-quality treated greywater for toilet flushing, laundry, garden irrigation, handwashing, and use in showers/bathtubs. However, the results also showed that there is no preference for using this water for drinking, confirming that social perception of water reuse for consumption purposes is rather negative. In this regard, before implementing water reuse technologies and projects, efforts should be made to educate the population about the advantages of water reuse and the benefits of its adoption. Public awareness and understanding can play a significant role in overcoming the psychological barriers associated with water reuse.

An educated and informed population is a significant driver for water reuse. Several studies have shown that an informed public with a higher environmental awareness is more likely to accept water reuse practices (Bichai et al., 2018; Bint et al., 2019; Poortvliet et al., 2018; Hartley, 2006; Smith et al., 2018; Wilcox et al., 2016; Ricart & Rico, 2019). Likewise, it has been observed that this perception is more positive in populations more affected by drought (Ormerod & Scott, 2013; Šteflová et al., 2018). Indeed, public perception regarding the water scarcity scenario can be a driver or a barrier to the adoption of water reuse practices. Positive public perception is strongly influenced by water scarcity scenarios and the population's environmental awareness and responsibility, while negative public perception is rooted in a lack of knowledge about water treatment processes and quality, as well as potential risks (Smith et al., 2018; Ormerod & Scott, 2013).

Negative public perception is also supported by a lack of trust, skepticism about its quality and benefits, as well as cultural factors (Lee & Jepson, 2020; Nkhoma et al., 2021). In some countries, cultural factors related to religion restrict and impede the adoption of water reuse practices. A clear example of this can be seen in Jakarta, where individuals, driven by their religious beliefs, limit their interaction with reclaimed water (Priadi et al., 2017).

Finally, the balance between all these drivers and barriers will determine the viability, sustainability, and successful adoption of water reuse systems and practices.

17.5 Conclusions

In this chapter, water reuse issues and challenges were examined, in order to determine the key aspects for adopting these practices as an alternative to address drought.

In Chile, the growing drought and water scarcity have had a negative impact on economic development and the well-being of communities, driving the search for alternative water sources that can help alleviate the pressure on the scarce available water resources and contribute to the water security of communities. The potential benefits of water reuse are numerous and include water conservation, improvements in the water supply, as well as various environmental, economic, and health benefits. On the other hand, the need for cities to transition to more sustainable practices and a circular economy aligned with the use of natural resources further emphasizes the importance of water reuse. However, while it is evident that water reuse can significantly enhance the management of water resources, the successful implementation of these systems involves overcoming various challenges, such as technological and economic aspects, institutional/regulatory issues, and social and community perception. Different successful experiences have been developed in Chile, including the reuse of greywater, wastewater, and industrial waters. In this context, the enactment of Chilean Law N° 21,075, which regulates the collection, reuse, and disposal of greywater, has been a significant boost for the widespread adoption of water reuse practices in the country. However, to successfully implement and adopt water reuse practices, it is necessary to achieve a balance that allows for the enhancement of the drivers that promote its development and the reduction of the barriers that obstruct it. In this context, the primary limitations are associated with the lack of economic incentives to adopt these practices, which hinders the implementation of reuse systems, particularly decentralized systems. On the other hand, the current skepticism surrounding the benefits of water reuse (wastewater and greywater) is another significant obstacle and requires a necessary boost aligned with methods of public participation and education. Finally, to maximize the relationship between drivers and barriers, it is necessary for initiatives to promote the reuse of wastewater to be aligned among different institutions and decision-makers, including state authorities, sanitation companies, civil society, and private entities, in order to enhance these practices in the face of the growing drought scenarios projected for the future and contribute to the sustainable management of water resources.

Acknowledgements The authors gratefully acknowledge the research support provided by FSEQ210018 project E.L acknowledges support from ANID/FONDECYT Regular 1241833 (2024–2028). I.V. acknowledges support from CEDEUS (ANID/FONDAP 1522A0002) and ANID/ FONDECYT 1201134, and EL, IV and R.L. acknowledges support from Technological Consortium for the Management of Water Resources (CoTH2O-CORFO, 20CTECGH-145896).

References

Aitken, D., Rivera, D., Godoy-Faúndez, A., & Holzapfel, E. (2016). Water scarcity and the impact of the mining and agricultural sectors in Chile. *Sustainability, 8*(2), 128.

Al Arni, S., Elwaheidi, M., Salih, A. A., Ghernaout, D., & Matouq, M. (2022). Greywater reuse: An assessment of the Jordanian experience in rural communities. *Water Science and Technology, 85*(6), 1952–1963.

Alcamo, J., & Henrichs, T. (2002). Critical regions: A model-based estimation of world water resources sensitive to global changes. *Aquatic Sciences, 64*, 352–362.

Amaris, G., Dawson, R., Gironás, J., Hess, S., & de Dios Ortúzar, J. (2020). Understanding the preferences for different types of urban greywater uses and the impact of qualitative attributes. *Water Research, 184*, 116007.

Amaris, G., Dawson, R., Gironás, J., Hess, S., & de Dios Ortúzar, J. (2021). From mathematical models to policy design: Predicting greywater reuse scheme effectiveness and water reclamation benefits based on individuals' preferences. *Sustainable Cities and Society, 74*, 103132.

Amulén, F. (2019). *Pobres de agua. Radiografía del agua rural de Chile: Visualización de un problema oculto*. Fundación Amulén. https://cambioglobal.uc.cl/wp-content/uploads/2023/09/Informe_Final_PobresdeAgua_FundAmulen.pdf

Anderson, J. (2003). The environmental benefits of water recycling and reuse. *Water Science and Technology: Water Supply, 3*(4), 1–10.

Angelakis, A. N., & Gikas, P. (2014). Water reuse: Overview of current practices and trends in the world with emphasis on EU states. *Water Utility Journal, 8*(67), e78.

Angelakis, A. N., & Snyder, S. A. (2015). Wastewater treatment and reuse: Past, present, and future. *Water, 7*(9), 4887–4895.

AWWA. (2016). *Potable reuse 101 an innovative and sustainable water supply solution*. https://www.awwa.org/Portals/0/AWWA/ETS/Resources/PotableReuse101.pdf?ver=2018-12-12-182505-710

Batisha, A. F. (2020). Greywater in Egypt: The sustainable future of non-conventional water resources. *Environmental Science and Pollution Research, 27*(28), 35428–35438.

Besson, M., Berger, S., Tiruta-Barna, L., Paul, E., & Spérandio, M. (2021). Environmental assessment of urine, black and grey water separation for resource recovery in a new district compared to centralized wastewater resources recovery plant. *Journal of Cleaner Production, 301*, 126868.

Beveridge, R., Moss, T., & Naumann, M. (2017). Sociospatial understanding of water politics: Tracing the multidimensionality of water reuse. *Water Alternatives, 10*(1), 22–40.

Bichai, F., Grindle, A. K., & Murthy, S. L. (2018). Addressing barriers in the water-recycling innovation system to reach water security in arid countries. *Journal of Cleaner Production, 171*, S97–S109.

Bint, L., Garnett, A., Siggins, A., & Jaques, R. (2019). Alternative water sources in New Zealand's commercial buildings. *Water Supply, 19*(2), 371–381.

Bischel, H. N., Simon, G. L., Frisby, T. M., & Luthy, R. G. (2012). Management experiences and trends for water reuse implementation in northern California. *Environmental Science & Technology, 46*(1), 180–188.

Blanco, E., & Donoso, G. (2016). Agua potable rural: desafíos para la provisión sustentable del recurso. *Actas de Derecho de Agua*, 63–79.

Bonetta, S., Pignata, C., Gasparro, E., Richiardi, L., Bonetta, S., & Carraro, E. (2022). Impact of wastewater treatment plants on microbiological contamination for evaluating the risks of wastewater reuse. *Environmental Sciences Europe, 34*(1), 20.

Bravo, M. B. (2011). *Contexto legal: reutilización de aguas grises, SustentaBiT* (11th ed., pp. 34–38). Cámara Chilena de la Construcción. https://biblioteca.cchc.cl/datafiles/23333-2.pdf

Budds, J. (2009). Contested H2O: Science, policy and politics in water resources management in Chile. *Geoforum, 40*(3), 418–430.

Capodaglio, A. G. (2021). Fit-for-purpose urban wastewater reuse: Analysis of issues and available technologies for sustainable multiple barrier approaches. *Critical Reviews in Environmental Science and Technology, 51*(15), 1619–1666.

Capodaglio, A. G., Callegari, A., Cecconet, D., & Molognoni, D. (2017). Sustainability of decentralized wastewater treatment technologies. *Water Practice and Technology, 12*(2), 463–477.

Carr, G., & Potter, R. B. (2013). Towards effective water reuse: Drivers, challenges and strategies shaping the organisational management of reclaimed water in Jordan. *The Geographical Journal, 179*(1), 61–73.

Chen, R., & Wang, X. C. (2009). Cost–benefit evaluation of a decentralized water system for wastewater reuse and environmental protection. *Water Science and Technology, 59*(8), 1515–1522.

Cook, B. I., Mankin, J. S., & Anchukaitis, K. J. (2018). Climate change and drought: From past to future. *Current Climate Change Reports, 4*, 164–179.

CR2. (2023). *La megasequía 2010-2019: una lección para el futuro*. https://www.cr2.cl/megasequia/

Cruz, P. (2017, May). *Propuesta de reglamento sanitario para la reutilización de aguas grises*. Ministerio de Salud. Seminario Aguas Grises un Recurso Latente. AIDIS, Santiago.

Decreto 10. (2020). *Modifica Decreto Supremo N° 47, de Vivienda y Urbanismo, de 1992, Ordenanza general de Urbanismo y Construcciones, en lo relativo a establecer los trámites y requisitos de los permisos de loteo y edificación que incorporen sistemas de Reutilización de aguas grises*. https://www.bcn.cl/leychile/navegar?i=1153728&f=2222-02-02

Desbureaux, S., & Rodella, A. S. (2019). Drought in the city: The economic impact of water scarcity in Latin American metropolitan areas. *World Development, 114*, 13–27.

Deshpande, A., Miller-Petrie, M. K., Lindstedt, P. A., Baumann, M. M., Johnson, K. B., Blacker, B. F., et al. (2020). Mapping geographical inequalities in access to drinking water and sanitation facilities in low-income and middle-income countries, 2000–17. *The Lancet Global Health, 8*(9), e1162–e1185.

Diario Oficial de la República de Chile. (2018). *Ley N° 21.075: Regula la Recolección, Reutilización y Disposición de Aguas Grises* [Law No 21.075: Regulates the collection, reuse and disposal of greywater, Authors Translation]. Biblioteca del Congreso Nacional de Chile.

Dishman, C. M., Sherrard, J. H., & Rebhun, M. (1989). Gaining support for direct potable water reuse. *Journal of Professional Issues in Engineering, 115*(2), 154–161.

Eslamian, S., Amininezhad, S. M., Amininejad, S. M., & Adamowski, J. (2016). Application of nanotechnology in water reuse. In *Urban water reuse handbook* (pp. 883–888). CRC Press.

Farhadkhani, M., Nikaeen, M., Hadi, M., Gholipour, S., & Yadegarfar, G. (2020). Campylobacter risk for the consumers of wastewater-irrigated vegetables based on field experiments. *Chemosphere, 251*, 126408.

Fernández-Medrano, V., Cuartas-Uribe, B., Bes-Piá, M. A., & Mendoza-Roca, J. A. (2022). Application of nanofiltration and reverse osmosis membranes for tannery wastewater reuse. *Water, 14*(13), 2035.

Fluence Corporation. (2021). *Israel leads world in water recycling*. Fluence Corporation. https://www.fluencecorp.com/israel-leads-world-in-water-recycling/

Friedler, E. (2008). The water saving potential and the socio-economic feasibility of greywater reuse within the urban sector–Israel as a case study. *International Journal of Environmental Studies, 65*(1), 57–69.

Fundación Chile. (2023). *Cinco nuevos sistemas de reúso de aguas residuales tratadas enfrentan la sequía de la región de Coquimbo*. Available online: https://fch.cl/noticias/cinco-nuevos-sistemas-de-reuso-de-aguas-residuales-tratadas-enfrentan-la-sequia-de-la-region-de-coquimbo/

Garrido-Baserba, M., Vinardell, S., Molinos-Senante, M., Rosso, D., & Poch, M. (2018). The economics of wastewater treatment decentralization: A techno-economic evaluation. *Environmental Science & Technology, 52*(15), 8965–8976.

Gómez-Monsalve, M., Domínguez, I. C., Yan, X., Ward, S., & Oviedo-Ocaña, E. R. (2022). Environmental performance of a hybrid rainwater harvesting and greywater reuse system: A case study on a high water consumption household in Colombia. *Journal of Cleaner Production, 345*, 131125.

Gosling, S. N., & Arnell, N. W. (2016). A global assessment of the impact of climate change on water scarcity. *Climatic Change, 134*, 371–385.

Gross, A., Maimon, A., Alfiya, Y., & Friedler, E. (2015). *Greywater reuse*. CRC Press.

Grundmann, P., & Maaß, O. (2017). Wastewater reuse to cope with water and nutrient scarcity in agriculture—A case study for Braunschweig in Germany. In *Competition for water resources* (pp. 352–365). Elsevier.

Hadengue, B., Morgenroth, E., Larsen, T. A., & Baldini, L. (2022). Performance and dynamics of active greywater heat recovery in buildings. *Applied Energy, 305*, 117677.

Hartley, T. W. (2006). Public perception and participation in water reuse. *Desalination, 187*(1–3), 115–126.

Hespanhol, I., & Prost, A. M. E. (1994). WHO guidelines and national standards for reuse and water quality. *Water Research, 28*(1), 119–124.

Hughes, S. (2013). Authority structures and service reform in multilevel urban governance: The case of wastewater recycling in California and Australia. *Urban Affairs Review, 49*(3), 381–407.

Interreg. (2023). *The new regulation aims to encourage and facilitate water reuse in the EU*. https://interreg-baltic.eu/project-posts/the-new-regulation-aims-to-encourage-and-facilitate-water-reuse-in-the-eu/

Ishii, S. K., Boyer, T. H., Cornwell, D. A., & Via, S. H. (2015). Public perceptions of direct potable reuse in four US cities. *Journal-American Water Works Association, 107*(11), E559–E570.

IWA. (2023). *Water reuse in the United States: A trend on the rise*. https://iwa-network.org/learn/water-reuse-in-the-united-states-a-trend-on-the-rise/

Jiang, D., Fischer, M., Huang, Z., & Kunz, N. (2018). Identifying drivers of China's provincial wastewater reuse outcomes using qualitative comparative analysis. *Journal of Industrial Ecology, 22*(2), 369–376.

Kaya, Y., & Dayanir, S. (2020). Application of nanofiltration and reverse osmosis for treatment and reuse of laundry wastewater. *Journal of Environmental Health Science and Engineering, 18*, 699–709.

Kim, S., Chu, K. H., Al-Hamadani, Y. A., Park, C. M., Jang, M., Kim, D. H., et al. (2018). Removal of contaminants of emerging concern by membranes in water and wastewater: A review. *Chemical Engineering Journal, 335*, 896–914.

Kummu, M., Guillaume, J. H., de Moel, H., Eisner, S., Flörke, M., Porkka, M., et al. (2016). The world's road to water scarcity: Shortage and stress in the 20th century and pathways towards sustainability. *Scientific Reports, 6*(1), 38495.

Kunz, N. C., Fischer, M., Ingold, K., & Hering, J. G. (2015). Why do some water utilities recycle more than others? A qualitative comparative analysis in New South Wales, Australia. *Environmental Science & Technology, 49*(14), 8287–8296.

Lane, R., Bettini, Y., McCallum, T., & Head, B. W. (2017). The interaction of risk allocation and governance arrangements in innovative urban stormwater and recycling projects. *Landscape and Urban Planning, 164*, 37–48.

Larsen, T. A., Hoffmann, S., Lüthi, C., Truffer, B., & Maurer, M. (2016). Emerging solutions to the water challenges of an urbanizing world. *Science, 352*(6288), 928–933.

Lee, K., & Jepson, W. (2020). Drivers and barriers to urban water reuse: A systematic review. *Water Security, 11*, 100073.

Lee, H., & Tan, T. P. (2016). Singapore's experience with reclaimed water: NEWater. *International Journal of Water Resources Development, 32*(4), 611–621.

Leiva, E., Sánchez, R., Serrano, J., Schneider, N., Alvial, C., & Rodríguez, C. (2020). Reutilización de aguas grises en Chile: propuesta de implementación en comunidades rurales como alternativa de mitigación para la escasez hídrica. *Propuestas para Chile*, 173.

Liu, J., Yang, H., Gosling, S. N., Kummu, M., Flörke, M., Pfister, S., et al. (2017). Water scarcity assessments in the past, present, and future. *Earth's Future, 5*(6), 545–559.

López, A., Rodríguez-Chueca, J., Mosteo, R., Gómez, J., Rubio, E., Goñi, P., & Ormad, M. P. (2019). How does urban wastewater treatment affect the microbial quality of treated wastewater? *Process Safety and Environmental Protection, 130*, 22–30.

Maddocks, A., Young, R. S., & Reig, P. (2015). *Ranking the world's most water-stressed countries in 2040*. World Resources Institute. https://www.wri.org/insights/ranking-worlds-most-water-stressed-countries-2040

Maniam, G., Zakaria, N. A., Leo, C. P., Vassilev, V., Blay, K. B., Behzadian, K., & Poh, P. E. (2022). An assessment of technological development and applications of decentralized water reuse: A critical review and conceptual framework. *Wiley Interdisciplinary Reviews: Water, 9*(3), e1588.

Mendret, J., Azais, A., Favier, T., & Brosillon, S. (2019). Urban wastewater reuse using a coupling between nanofiltration and ozonation: Techno-economic assessment. *Chemical Engineering Research and Design, 145*, 19–28.

Miller, G. W. (2006). Integrated concepts in water reuse: Managing global water needs. *Desalination, 187*(1–3), 65–75.

Mizyed, N. R. (2013). Challenges to treated wastewater reuse in arid and semi-arid areas. *Environmental Science & Policy, 25*, 186–195.

Moskvitch, K. (2012). *War for water in Chile's Atacama desert: Vines or mines*. BBC News. http://www.bbc.com/news/business-17423097

Munné, A., Solà, C., Ejarque, E., Sanchís, J., Serra, P., Corbella, I., et al. (2023). Indirect potable water reuse to face drought events in Barcelona city. Setting a monitoring procedure to protect aquatic ecosystems and to ensure a safe drinking water supply. *Science of the Total Environment, 866*, 161339.

Murei, A., Mogane, B., Mothiba, D. P., Mochware, O. T. W., Sekgobela, J. M., Mudau, M., et al. (2022). Barriers to water and sanitation safety plans in rural areas of South Africa—A case study in the Vhembe District, Limpopo Province. *Water, 14*(8), 1244.

Nahrstedt, A., Gaba, A., Zimmermann, B., Jentzsch, T., Kroemer, K., Tiemann, Y., et al. (2020). Reuse of municipal wastewater for different purposes based on a modular treatment concept. *Journal of Water Reuse and Desalination, 10*(4), 301–316.

Ndehedehe, C. E., Ferreira, V. G., Adeyeri, O. E., Correa, F. M., Usman, M., Oussou, F. E., et al. (2023). Global assessment of drought characteristics in the Anthropocene. *Resources, Environment and Sustainability, 12*, 100105.

Nkhoma, P. R., Alsharif, K., Ananga, E., Eduful, M., & Acheampong, M. (2021). Recycled water reuse: What factors affect public acceptance? *Environmental Conservation, 48*(4), 278–286.

NWRI. (2015). *Framework for direct potable reuse*. National Water Research Institute.

Oki, T., & Kanae, S. (2006). Global hydrological cycles and world water resources. *Science, 313*(5790), 1068–1072.

Oki, T., Agata, Y., Kanae, S., Saruhashi, T., Yang, D., & Musiake, K. (2001). Global assessment of current water resources using total runoff integrating pathways. *Hydrological Sciences Journal, 46*(6), 983–995.

Ormerod, K. J., & Scott, C. A. (2013). Drinking wastewater: Public trust in potable reuse. *Science, Technology, & Human Values, 38*(3), 351–373.

Oyarzún, J., & Oyarzún, R. (2011). Sustainable development threats, inter-sector conflicts and environmental policy requirements in the arid, mining rich, northern Chile territory. *Sustainable Development, 19*(4), 263–274.

Paola, V., Mustafa, A. A., & Giacomo, Z. (2018). Willingness to pay for recreational benefit evaluation in a wastewater reuse project. Analysis of a case study. *Water, 10*(7), 922.

Poortvliet, P. M., Sanders, L., Weijma, J., & De Vries, J. R. (2018). Acceptance of new sanitation: The role of end-users' pro-environmental personal norms and risk and benefit perceptions. *Water Research, 131*, 90–99.

Priadi, C. R., Suleeman, E., Darmajanti, L., Novriaty, S., Suwartha, N., Resnawati, R., et al. (2017). Water recycling opportunity in the business sectors of Greater Jakarta, Indonesia. *International Journal of Technology, 8*(6), 1031–1039.

Qin, Y., Mueller, N. D., Siebert, S., Jackson, R. B., AghaKouchak, A., Zimmerman, J. B., et al. (2019). Flexibility and intensity of global water use. *Nature Sustainability, 2*(6), 515–523.

Qu, X., Brame, J., Li, Q., & Alvarez, P. J. (2013). Nanotechnology for a safe and sustainable water supply: Enabling integrated water treatment and reuse. *Accounts of Chemical Research, 46*(3), 834–843.

Raboni, M., Gavasci, R., & Torretta, V. (2016). Assessment of the fate of Escherichia coli in different stages of wastewater treatment plants. *Water, Air, & Soil Pollution, 227*(12), 455.

Rehner, J., Baeza, S. A., & Barton, J. R. (2014). Chile's resource-based export boom and its outcomes: Regional specialization, export stability and economic growth. *Geoforum, 56*, 35–45.

Reyes, R., Bustamante, W., Gironás, J., Pastén, P. A., Rojas, V., Suárez, F., et al. (2016). Effect of substrate depth and roof layers on green roof temperature and water requirements in a semi-arid climate. *Ecological Engineering, 97*, 624–632.

Ricart, S., & Rico, A. M. (2019). Assessing technical and social driving factors of water reuse in agriculture: A review on risks, regulation and the yuck factor. *Agricultural Water Management, 217*, 426–439.

Ritter, W. (2021). State regulations and guidelines for wastewater reuse for irrigation in the US. *Water, 13*(20), 2818.

Rizzo, L., Gernjak, W., Krzeminski, P., Malato, S., McArdell, C. S., Perez, J. A. S., et al. (2020). Best available technologies and treatment trains to address current challenges in urban wastewater reuse for irrigation of crops in EU countries. *Science of the Total Environment, 710*, 136312.

Rodríguez, C., Sánchez, R., Lozano-Parra, J., Rebolledo, N., Schneider, N., Serrano, J., & Leiva, E. (2020a). Water balance assessment in schools and households of rural areas of Coquimbo region, north-central Chile: Potential for greywater reuse. *Water, 12*(10), 2915.

Rodríguez, C., Sánchez, R., Rebolledo, N., Schneider, N., Serrano, J., & Leiva, E. (2020b). Cost–benefit evaluation of decentralized greywater reuse systems in rural public schools in Chile. *Water, 12*(12), 3468.

Rodríguez, C., Sánchez, R., Rebolledo, N., Schneider, N., Serrano, J., & Leiva, E. (2021). Life cycle assessment of greywater treatment systems for water-reuse management in rural areas. *Science of the Total Environment, 795*, 148687.

Rodríguez, C., Carrasco, F., Sánchez, R., Rebolledo, N., Schneider, N., Serrano, J., & Leiva, E. (2022a). Performance and treatment assessment of a pilot-scale decentralized greywater reuse system in rural schools of north-central Chile. *Ecological Engineering, 174*, 106460.

Rodríguez, C., García, B., Pinto, C., Sánchez, R., Serrano, J., & Leiva, E. (2022b). Water context in Latin America and the Caribbean: Distribution, regulations and prospects for water reuse and reclamation. *Water, 14*(21), 3589.

Rogers, B. C., Brown, R. R., De Haan, F. J., & Deletic, A. (2015). Analysis of institutional work on innovation trajectories in water infrastructure systems of Melbourne, Australia. *Environmental Innovation and Societal Transitions, 15*, 42–64.

Salgot, M., & Folch, M. (2018). Wastewater treatment and water reuse. *Current Opinion in Environmental Science & Health, 2*, 64–74.

Schellenberg, T., Subramanian, V., Ganeshan, G., Tompkins, D., & Pradeep, R. (2020). Wastewater discharge standards in the evolving context of urban sustainability—The case of India. *Frontiers in Environmental Science, 8*, 30.

Schewe, J., Heinke, J., Gerten, D., Haddeland, I., Arnell, N. W., Clark, D. B., et al. (2014). Multimodel assessment of water scarcity under climate change. *Proceedings of the National Academy of Sciences of the United States of America, 111*(9), 3245–3250.

Schweitzer, O., & Erell, E. (2014). Evaluation of the energy performance and irrigation requirements of extensive green roofs in a water-scarce Mediterranean climate. *Energy and Buildings, 68*, 25–32.

Sgroi, M., Vagliasindi, F. G., & Roccaro, P. (2018). Feasibility, sustainability and circular economy concepts in water reuse. *Current Opinion in Environmental Science & Health, 2*, 20–25.

Shahid, M. K., Kashif, A., Pathak, P., Choi, Y., & Rout, P. R. (2022). Water reclamation, recycle, and reuse. In *Clean energy and resource recovery* (pp. 39–50). Elsevier.

Shoushtarian, F., & Negahban-Azar, M. (2020). Worldwide regulations and guidelines for agricultural water reuse: A critical review. *Water, 12*(4), 971.

SISS. (2018). *Informe de gestión del sector sanitario*. http://www.siss.gob.cl/586/articles-17722_recurso_1.pdf

Smith, H. M., Brouwer, S., Jeffrey, P., & Frijns, J. (2018). Public responses to water reuse—Understanding the evidence. *Journal of Environmental Management, 207*, 43–50.

Stathatou, P. M., Kampragou, E., Grigoropoulou, H., Assimacopoulos, D., Karavitis, C., & Gironás, J. (2017). Creating an enabling environment for WR&R implementation. *Water Science and Technology, 76*(6), 1555–1564.

Šteflová, M., Koop, S., Elelman, R., Vinyoles, J., & Van Leeuwen, K. (2018). Governing non-potable water-reuse to alleviate water stress: The case of Sabadell, Spain. *Water, 10*(6), 73.

Uhlenbrook, S., & Connor, R. (2019). *The United Nations world water development report 2019: Leaving no one behind*. https://www.unwater.org/publications/un-world-water-development-report-2019.

United States Environmental Protection Agency, Office of Wastewater Management, Municipal Support Division, National Risk Management Research Laboratory (US), & Technology Transfer, & Support Division. (2004). *Guidelines for water reuse*. US Environmental Protection Agency.

UN-Water. (2010). Climate change adaptation: The pivotal role of water. In *Policy brief*. UN-Water.

USEPA (US Environmental Protection Agency). (2012). *Guidelines for water reuse*. US Environmental Protection Agency.

Valdés-Pineda, R., Pizarro, R., García-Chevesich, P., Valdés, J. B., Olivares, C., Vera, M., et al. (2014). Water governance in Chile: Availability, management and climate change. *Journal of Hydrology, 519*, 2538–2567.

Van de Walle, A., Kim, M., Alam, M. K., Wang, X., Wu, D., Dash, S. R., et al. (2023). Greywater reuse as a key enabler for improving urban wastewater management. *Environmental Science and Ecotechnology, 16*, 100277.

Van Rensburg, P. (2016). Overcoming global water reuse barriers: The Windhoek experience. *International Journal of Water Resources Development, 32*(4), 622–636.

Villamar, C. A., Vera-Puerto, I., Rivera, D., & De la Hoz, F. (2018). Reuse and recycling of livestock and municipal wastewater in Chilean agriculture: A preliminary assessment. *Water, 10*(6), 817.

Vorosmarty, C. J., Green, P., Salisbury, J., & Lammers, R. B. (2000). Global water resources: Vulnerability from climate change and population growth. *Science, 289*(5477), 284–288.

Vuppaladadiyam, A. K., Merayo, N., Prinsen, P., Luque, R., Blanco, A., & Zhao, M. (2019). A review on greywater reuse: Quality, risks, barriers and global scenarios. *Reviews in Environmental Science and Bio/Technology, 18*, 77–99.

Watson, R., Mukheibir, P., & Mitchell, C. (2017). Local recycled water in Sydney: A policy and regulatory tug-of-war. *Journal of Cleaner Production, 148*, 583–594.

Wilcox, J., Nasiri, F., Bell, S., & Rahaman, M. S. (2016). Urban water reuse: A triple bottom line assessment framework and review. *Sustainable Cities and Society, 27*, 448–456.

Wintgens, T., Melin, T., Schäfer, A., Khan, S., Muston, M., Bixio, D., & Thoeye, C. (2005). The role of membrane processes in municipal wastewater reclamation and reuse. *Desalination, 178*(1–3), 1–11.

WISE. (2023). *Water reuse*. https://water.europa.eu/freshwater/europe-freshwater/water-reuse

WRI. (2023). *25 countries, housing one-quarter of the population, face extremely high water stress*. https://www.wri.org/insights/highest-water-stressed-countries

Xue, B., Guo, X., Cao, J., Yang, S., Qiu, Z., Wang, J., & Shen, Z. (2023). The occurrence, ecological risk, and control of disinfection by-products from intensified wastewater disinfection during the COVID-19 pandemic. *Science of the Total Environment, 900*, 165602.

Yang, J., Monnot, M., Ercolei, L., & Moulin, P. (2020). Membrane-based processes used in municipal wastewater treatment for water reuse: State-of-the-art and performance analysis. *Membranes, 10*(6), 131.

Chapter 18
Desalinization in Times of Drought

María Molinos-Senante, Daniela Rivera, Guillermo Donoso, and Óscar Melo

Abstract Several factors are contributing to the increased adoption of desalination. Population growth has significantly impacted areas experiencing water stress, while drought conditions, partly influenced by climate change, are simultaneously affecting many regions that previously considered their traditional water supplies to be sufficient. Meanwhile, the cost of desalinated water has been decreasing as technology evolves and the cost of other sources increases. In Chile, there are currently 22 desalination plants in operation, 6 under construction, 3 with environmental approval, and 12 under preliminary evaluation. The current capacity of the operating plants in Chile is 8200 L/s. However, despite this boom in desalination, its orderly and sustainable economic and environmental development presents significant challenges. In this context, the main regulatory, legal, environmental, social and economic challenges for the sustainable development of desalination in Chile are reviewed, outlining some ideas to advance in them.

Keywords Desalinization · Challenges · Complementary water sources · Chile

M. Molinos-Senante (✉)
Institute of Sustainable Processes, University of Valladolid, Valladolid, Spain

Center for Water Law and Management, Pontificia Universidad Católica de Chile, Santiago, Chile
e-mail: mmolinos@uc.cl

D. Rivera
Faculty of Law and the Center for Water Law and Management, Pontificia Universidad Católica de Chile, Santiago, RM - Santiago, Chile

G. Donoso
Faculty of Agronomy and Natural Systems and the Center for Water Law and Management, Pontificia Universidad Católica de Chile, Santiago, RM - Santiago, Chile

Ó. Melo
Center for Water Law and Management, Pontificia Universidad Católica de Chile, Santiago, Chile

Department of Agricultural Economics, Pontificia Universidad Católica de Chile, Santiago, Chile

D. Rivera, G. Donoso (eds.), *Droughts in Chile*, Global Issues in Water Policy 31,
https://doi.org/10.1007/978-3-031-85040-0_18

18.1 Introduction

Water distribution across the Earth's surface is markedly uneven, with merely 3% of surface water classified as fresh, while the vast majority, or 97%, is contained within oceans. Regarding freshwater reserves, glaciers hold 69%, groundwater sources account for 30%, and a minor fraction of less than 1% is found in lakes, rivers, and swamps (USGS, 2019). Thus, most of the water on the earth's surface is in oceans which, due to its high content of salts cannot be directly used for human consumption or agricultural purposes. Nevertheless, considering that seawater desalination is the process by which the dissolved mineral salts in water are removed, desalinated water is an alternative water resource that can play a key role in alleviating water scarcity in coastal areas.

The first seawater distillation plant was built by the British government in Aden (Yemen) in 1869 to supply the ships of the Empire with fresh water in route to India (Khordagui, 1999). From this anecdotal desalination plant, currently there are about 20,000 desalination plants in the world, with a total capacity of approximately 100 million cubic meters per day, which can treat seawater (60%) or brackish water (40%) (International Desalination Association, 2022) (Fig. 18.1). Over time, desalination has transitioned from a novel to a critical alternative water source on a global scale. Consequently, numerous countries, particularly those grappling with water scarcity, have integrated desalination into their overarching water management strategies (Asli et al., 2023).

Several factors are converging that explain the boom in desalination. Population growth in many places with water stress, including parts of China, India, South Africa, and the United States, especially in Arizona and California. Moreover, drought, partly driven by a changing climate, is occurring simultaneously in many regions that until not long ago thought their supplies were abundant (Robbins, 2019). Meanwhile, the cost of desalinated water has been decreasing as technology

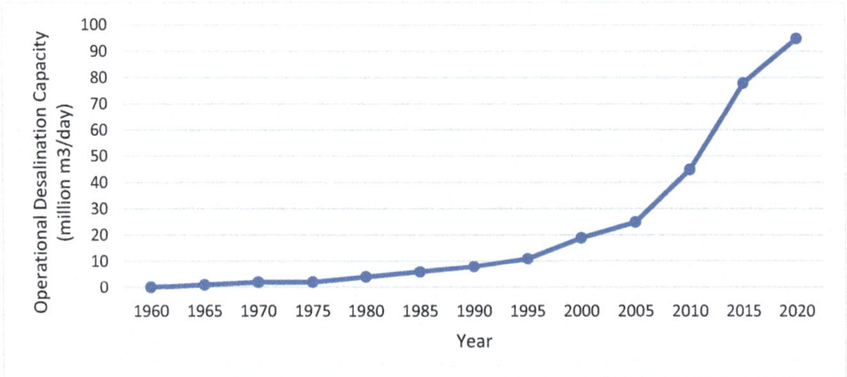

Fig. 18.1 Evolution of the operational desalination capacity worldwide. (Source: Own elaboration based on Angelakis et al., 2021)

evolves and the cost of other sources increases. In the last three decades, the cost of desalination has reduced by more than half (Shokri & Sanavi Fard, 2023). However, a boom in desalination does not mean that everywhere with access to the sea has found a new source of fresh water to address water scarcity. There are a series of elements and factors that must be considered for their development and implementation to be sustainable. Section 18.2 of this chapter presents a brief overview of desalination in Chile. Subsequently, Sects. 18.3 and 18.4 discuss the main challenges for a sustainable desalination in Chile from a regulatory and environmental perspective, respectively. Section 18.5 focuses on social issues, whereas Sect. 18.6 addresses the economics of desalination. The last section concludes with some key messages.

18.2 Desalination in Chile

With a coastline length of 6435 km, Chile presents vast potential for the development of desalination facilities. Actually, the first "desalination plant" located in Las Salinas dates back to 1892 (Aguas Antofagasta, 2024). Nevertheless, it was in the 1990s when the mining industry began to explore the use of seawater as an alternative source for industrial processes (B2B Media Group, 2022). However, the oldest desalination plants among those currently in operation supply the urban water sector: the Arica Plant began operating in 1998 (Herrera-León et al., 2019), and the North Antofagasta Desalination Plant started its operations in 2003 (Herrera-León et al., 2019). Since those years, there has been an increase in the number of industrial desalination plants across the country with emphasis in the North area, where most of the mining activity takes place.

The first national registry of seawater desalination plants and projects, with a capacity greater than 20 L/s, conducted by ACADES (2023), indicates that in 2023 Chile had 31 facilities, in operation (22), under construction (6), or with environmental approval (3), and 12 in preliminary evaluation.[1] The 22 desalination plants currently in operation have a capacity of 8200 L/s, which will increase to 25,000 L/s when the 31 identified plants are operating. Regionally, 13 of the 22 operating desalination plants (59%) are in the Antofagasta Region, 5 of the 22 (23%) in Atacama, 2 (9%) in Valparaiso, and the Biobio and Magallanes regions each have one desalination plant. While there is a predominance of desalination in the northern part of the country (Antofagasta and Atacama), it is also significant to note the presence of desalination plants in the southern zone of Chile.

Regarding the use of desalinated water, 10 of the 22 operational desalination plants (45%), including those located in Valparaiso, Biobio, and Magallanes, produce water for industrial use with a total capacity of 410 L/s. Three of the 22 facilities (14%) provide water for domestic purposes with a total capacity of 1578 L/s.

[1] Available at: https://acades.cl/wp-content/uploads/2023/03/Catastro-plantas-ACADES.pdf.

The rest of the desalination plants (9 of 22, 41%) are used in the mining industry with a capacity of 6291 L/s. There is a marked predominance of the mining sector in the production and use of desalinated water. In relation to the desalination plants in the construction phase, 4 of the 6 are associated with mining with a total capacity of 2560 L/s, one of them for potable water (634 L/s), and another facility will be multipurpose with a capacity of 1000 L/s.

Considering that the projection for desalination plants in Chile amounts to 15 new facilities with a total capacity of 22,297 L/s, desalination is being considered a relevant solution to address the country's water deficit (Vicuña et al., 2022). However, despite the increase in recent years in the number and capacity of desalination plants in the country, along with future projections, Chile faces significant gaps and challenges for the sustainable regulation of desalination, which will be addressed in the following section.

18.3 Challenges and Gaps for the Sustainable Regulation of Desalination in Chile

Chile's desalination regulatory and institutional framework has challenges and gaps that must be addressed for sustainable project development and operation. These are outlined below.

18.3.1 Lack of Complete and Updated Information on Water Supply/Demand in Chile

Desalination can be considered an additional water source to those currently used by the main water users. This implies identifying uncertainties regarding water supply and demand and understanding how they interact with desalination. In turn, this understanding is multi-scale, as it requires knowledge at a global scale (such as the effects of climate change), but also at a local scale (for example, consumer preferences). In this respect, one of the most noted challenges in water matters in Chile is the absence of information and monitoring systems to correctly and completely measure water availability, in volume and quality, as well as its consumption, which is the basis for adequate management (Rivera et al., 2021).

This lack of understanding may result in proposing desalination investments that fail to achieve the anticipated benefits or do not represent the most cost-effective or efficient solution for enhancing water security. The selection of an appropriate site for the desalination plant is essential to guarantee both uninterrupted operations and the environmental and social sustainability of its processes (García-Bartolomei et al., 2022). Planning desalination infrastructure to combat droughts is important to prevent these facilities from becoming obsolete after the droughts end. It is therefore crucial to identify any information gaps or uncertainties that may pose a risk to the development of these types of investments.

On one hand, there are uncertainties associated with changes in water consumption. These are given by changes in population, cultivated area, development of mining projects, and technological changes. The general trend in the country has been towards a reduction in demand per water potable urban client. This trend, combined with demographic changes in each city, leads to fluctuations in total consumption. Conversely, agricultural conditions vary by basin; however, there has been a general shift towards fruit cultivation, replacing traditional crops. In the mining sector, a noticeable trend is emerging to meet water requirements with seawater, which will ultimately reduce the pressure on continental water resources. Conversely, the national water supply has experienced substantial changes and is projected to evolve further due to climate change. In the central-southern regions of the country, anticipated reductions in precipitation and the retreat of snowlines are expected to impact both surface flows and groundwater reserves. The variability of water availability resulting from climate change remains less understood.

It is therefore essential to reduce information gaps by creating mechanisms to generate and disseminate information, analyse the benefits, and conduct research tailored to the specific characteristics of each territorial region. This process should also address the relationship between water supply and demand (Aitken et al., 2022). In this regard, the implementation and maintenance of an updated National Unified Water Information System, as proposed by the Mesa Nacional del Agua (2022), is recommended. From an articulative and coordinating perspective, the General Directorate of Water (DGA) should lead and be responsible for this initiative, with adequate economic resources (public budget) allocated to ensure its success.

18.3.2 Fragmented Institutional Framework, Lack of Integrated Water Resources Management, and Absence of a Planning Instrument

Chile's water framework includes both centralized state institutions and independent Water User Organizations (WUOs). Centralized bodies are state administration components. The DGA is a key institution with multiple powers and responsibilities. There are challenges impacting the effective and efficient exercise of centralized water management in Chile. According to the Banco Mundial (2011, 2013) and the Mesa Nacional del Agua (2022), the most notable challenges and issues include:

1. Absence of national and catchment-level hydrological planning and objectives,
2. Weak institutional presence of the DGA,
3. Fragmented powers and functions without effective coordination,
4. Insufficient funding, and
5. Lack of stakeholder participation in water resource management.

WUOs are responsible for water management at the local level (Vergara & Rivera, 2018). They have existed in Chilean water resource management since colonial times, even though the first law that regulates them was passed in 1908 (Obando, 2009). A significant proportion of these WUOs has not updated their capacity to meet actual water management challenges (DGA, 2018; Montginoul et al., 2016; Rinaudo & Donoso, 2019).

The division of powers and functions among various public and private entities, combined with a lack of coordination, has hindered the development of an effective, transparent, and secure institutional framework. This fragmentation contributes to the challenges in establishing a sustainable desalination scenario. The institutions involved do not always act in a coordinated manner or consider water management from a comprehensive perspective that includes the entire hydrographic basin and the various stakeholders and sectors (Rivera et al., 2021).

There is an absence of integrated water resources management in Chile, as each water source is typically treated in isolation rather than as part of a cohesive system. Furthermore, Chile lacks a national water plan that serves as a guiding and binding instrument, developed with standards of participation, information dissemination, and inclusion (Aitken et al., 2022). This plan should be properly aligned with the strategic water management plans that have been established and are continuing to be formulated across various basins within the country (Rivera et al., 2021).

The fragmented institutional and regulatory framework, coupled with a lack of clear direction and control in water management and weak institutional capacities, adversely impacts the sustainable development of desalination. The solutions that have emerged thus far tend to be specific to individual companies or sectors, particularly within the mining industry. There is currently no regulation specifying the development of shared desalinated water supply systems both at the national and local levels (Aitken et al., 2022).

These institutional weaknesses result in significant regulatory disparities, especially in project execution based on their purposes: mining, potable water, or multipurpose use. Additionally, the permit system is extensive, covering maritime concessions, environmental permits, major hydraulic works (sometimes issued by the DGA), easements, and more. There is a notable delay in obtaining these permits. From the decision to initiate a medium to large-scale desalination project (exceeding 200 L/s) until the production of desalinated water, it typically takes over 6 years, on average (ACADES, 2021).

Given this scenario of dispersion and insufficient institutional coordination regarding water issues, to advance in desalinization it is necessary to:

1. Promote effective institutional coordination, for which two alternatives are, at present, being considered: creating a Sub-secretariat of Water Resources (Bill 14,446-09) and establishing a National Water Agency. These options have emerged during recent constitutional processes in the country (Rivera et al., 2022; Irarrázabal et al., 2023).
2. Create a national water planning framework. In Chile, water policy and national plans are not regulated, so their inclusion should be encouraged. A comprehensive

water planning framework is proposed to ensure long-term water security. Water planning, which includes the consideration of natural and supplementary sources like desalination, aims to ensure the sustainable management and utilization of water resources. This approach also seeks to secure the state's long-term water supply in light of climate change projections.[2] Additionally, this comprehensive water planning framework must include mechanisms for monitoring the plan and provisions for its periodic updates. The creation of national or territorial-scale programs and policies for the planning and execution of desalination projects are necessary for their development and implementation coordinated with other alternative water resources, such as the reuse of treated wastewater (Voutchkov, 2022).
3. Ensure that the Strategic Plans for Water Resources in Basins include analyses and references on shared desalination infrastructure.[3]
4. Incorporate the study of desalination into the Water Resources Climate Change Adaptation Plan, to be conducted by the Ministry of the Environment (Vicuña et al., 2022).
5. Establish connections with territorial planning instruments to align desalination plant locations with existing criteria and guidelines, considering the current social context (Aitken et al., 2022). Likewise, location criteria should be established to better integrate desalination plants with these instruments or to propose amendments to the Urban Planning and Construction Law or coastal legislation, defining priority areas for this industry (García-Bartolomei et al., 2022).

18.3.3 Absence of Specific and Systemic Regulations on Desalination and the Use of Seawater in General, Linked to the Legal Framework and Overall Water Management

As Voutchkov (2022) points out, in most countries where desalination is being implemented, there is no solid and comprehensive institutional, legal, and policy framework on the subject. The main institutional challenges that hinder the development of desalination are the lack of plans and policies for the implementation of desalination projects nationwide, and complex environmental regulations and

[2] For example, hydrological water management plans in Spain give greater prominence to desalination, thus ensuring supply and balancing demands and environmental requirements. In turn, the National Urban Water and Desalination Plan in Australia helped major cities secure their water supplies and reduce their dependence on natural continental sources by supporting infrastructure and research projects in desalination, water recycling, and rainwater collection and reuse.

[3] The 2022 reform of the Water Code establishes that the DGA must develop a Strategic Plans for Water Resources for each of the 101 basins in the country. The regular updating of these plans should be implemented in accordance with articles 293 bis of the Water Code and 13 of the Framework Law on Climate Change.

policies. Chile is no exception to this global trend, as there is no regulation that systematically, comprehensively, and orderly governs the use of seawater in general, nor specifically addresses desalination. Moreover, there is no territorial or strategic planning related to water use. Thus, desalination in Chile has developed without a comprehensive regulation, relying only on sector-specific or, in other words, fragmented regulation. The same view is shared by the national association of desalination in Chile, ACADES, who stated that the existing scattered regulations were not designed to enhance the development of this alternative source, and the bills in progress continue to regulate this issue only partially, focusing only on some of its aspects and stages (ACADES, 2021).

The main legal and regulatory discussions on desalination have focused on the legal classification of desalinated water and the role of environmental authorization issues.

18.3.3.1 Legal Classification or Nature of Desalinated Water

Chile's Water Code encompasses both terrestrial and maritime waters; however, its regulations are solely applicable to terrestrial waters. The Code acknowledges rainwater, flowing and stationary surface waters, groundwater, and glaciers, over which the establishment of water rights is expressly prohibited.

The high seas are a common good; due to its relative abundance and non-competitive characteristics, the high seas do not require regulation. The territorial sea is defined in Chile as a national good of public use. Conversely, it is essential to clarify the legal status of desalinated water. In this context, various authors present differing opinions on the matter. Plaza (2017) contends that general legislation does not definitively address this matter, as the public order norms delineated in the Political Constitution supersede those in the Civil Code. This legal framework precludes individuals from obtaining ownership of assets designated by law as national goods for public use, encompassing all waters without exception. Consequently, Plaza (2017) argues that desalinated waters are national goods for public use. This situation has significant implications, such that desalinated waters cannot be owned by private individuals.

Conversely, Celume and Goldenberg (2021) contend that seawater is a common resource accessible to all populations, and thus does not necessitate a special legislative regime for its extraction and use (free access). Once extracted from the sea, regardless of whether desalination has taken place, the water becomes the property of the individual who extracted it and can be freely traded and utilized (Celume & Goldenberg, 2021). Vicuña et al. (2022) propose a similar approach, suggesting that desalinated water is distinct from seawater. They argue that, in the absence of regulations specifying that desalinated water remains a national asset for public use, it does not possess this legal status. Consequently, it would be owned by the concession holder, who could then engage in commercial transactions involving the water.

Thus, a comprehensive law governing the use of seawater within territorial seas and the desalination process is required. This legislation should encompass various aspects including the extraction of seawater, its processing and treatment, the characteristics and management of desalinated water, and the transfer or conveyance of desalinated water to its intended destinations or uses, among other considerations. This regulation must also include a brief reference to desalination of water other than seawater (inland brackish water), as it is a practice that is also developed in Chile.

In implementing these regulatory changes, it is essential to ensure alignment with water institutions and management frameworks, including the national water plan and the Strategic Plans for Water Resources in Basins, among others.

18.3.3.2 Role of Environmental Authorization

The Environmental Impact Assessment System (in Spanish, SEIA) is an essential tool to mitigate negative environmental impacts on marine flora and fauna throughout all phases of a desalination project, including construction, operation, closure, and dismantling. Article 3 of the Regulation of the Environmental Impact Assessment System (Decree 40), specifically section o), mandates that submarine outfalls be subjected to the SEIA. Consequently, desalination projects with submarine outfalls must undergo the SEIA process. However, the regulation does not address seawater capture systems. Therefore, it is crucial to regulate the inclusion of desalination projects in the SEIA from a comprehensive perspective, encompassing all elements of the installation that may potentially cause environmental impacts throughout the project's life cycle.

In Chile, the SEIA performs environmental impact assessments either through an Environmental Impact Declaration or Environmental Impact Study. This depends on whether a project affects factors listed in article 11 of Law 19,300. Due to insufficient data on the marine environment affected by desalination projects, it is unclear which assessment method applies. Although not legally required, most large-scale desalination plants in Chile have environmental qualification resolutions from SEIA. This often happens because they are evaluated as part of larger projects like mining and power generation or through other unrelated causes. The environmental assessment has been restricted to direct impacts and there is no baseline to assess negative effects of brine; requirements depend on the purpose of the desalination plant, such as mining, energy generation, and drinking water (Alvez et al., 2020).

Clear criteria are required for these evaluations, considering factors such as plant size (discharge flow) and presence of conservation areas or sensitive species. The Ministry of the Environment should define thresholds and requirements for these assessments, based on the magnitude of the desalination plant and the flow of extraction or rejection involved (Skewes, 2017).

Santibáñez and Santibáñez (2020) identify several environmental factors that must be considered in the environmental impact assessment of desalination plants. These include: (a) the effects of seawater intake on the marine ecosystem, seabed, currents, and water quality; (b) the design and layout of aqueducts; (c) emissions of gases such as nitrogen oxide; (d) noise pollution; (e) chemical processes involved in desalination and waste disposal; (f) impacts on coastal soil; (g) the substantial energy demands of the desalination process, which represent a primary operating cost and contribute to subsequent emissions; (h) the effects of brine disposal in the coastal marine area, whether through discharge into the sea, infiltration, or evaporation in land-based ponds; and (i) the impact of projects on nearby communities.

18.4 Challenges for Environmentally Sustainable Desalination in Chile

The challenges for environmentally sustainable desalination in Chile, as in other countries, cover multiple aspects. Some of the main challenges in the Chilean context are as follows:

18.4.1 Energy Efficiency and Use of Renewable Energy

Desalination plants are one of the most energy-intensive alternative water sources (Stokes & Horvath, 2006; Østergaard et al., 2014; Goh et al., 2017; Šteflová et al., 2021). Energy costs represent between one third and half of the total cost of desalination (CCC, 2004). Therefore, the water-energy nexus is key to providing sustainability for both water and energy, as the security of both is fundamentally linked (Goh et al., 2017; Hamiche et al., 2016; Shahzad et al., 2017). This implies the need to consider long-term energy potential before investing in desalination.

The energy demand for desalination has significantly decreased in recent decades, reaching approximately 3 kWh/m^3 or less (Voutchkov, 2018; Zarzo, 2021). These decreases result from new technologies with more efficient desalination processes. Nevertheless, worldwide, the energy source currently used for desalination is predominantly non-renewable, with a high carbon footprint. Pistocchi et al. (2020) indicate that the expected growth of desalination would lead to a projected 180% increase in carbon emissions by 2040, if renewable sources are not considered. In this context, renewable energy represents a sustainable solution to the shortage of energy and water (Goh et al., 2017). On the other hand, energy recovery devices can be used to exploit the pressure of the rejected brine, with the potential to recover up to 50% of the consumed energy (Goh et al., 2017). In the case of Chile this technology is implemented in the Antofagasta desalination plant, with an energy recovery rate of 30%.

According to López and Del Campo (2021), the electric energy demand of the main Chilean desalination plants operating in 2021 was 34.2 GWh/year (see Table 18.1).

Table 18.1 Electrical energy use in main Chilean desalination plants (2021)

Desalination plant	Use	Capacity (L/s)	Electrical demand KWh/year
Tocopilla	Drinking water	75	386,316
Spence	Mining	1000	5,150,880
La Chimba	Drinking water	850	4,378,248
Coloso	Mining	2500	12,877,200
Manto Verde	Mining	120	618,106
Totoralillo	Multi-use	600	3,090,528
Candelaria	Mining	500	2,575,440
Atacama region	Drinking water	450	2,317,896
Other		550	2,832,984
TOTAL		6645	34,227,598

Source: Own elaboration based on López and Del Campo (2021)

18.4.2 Environmental Impacts Associated with Seawater Abstraction and Brine Discharge

According to Díaz (2022), the main aspects to consider in desalination projects from an environmental viewpoint are:

1. Salinity of brine discharge. The salinity of the brine (the reject stream from the desalination process) is approximately twice that of seawater (66 g/L).
2. Presence of chemicals in brine discharge. For the pretreatment of seawater, before the reverse osmosis process itself, chemical reagents are added to the water, which then become part of the brine discharged into the sea.
3. Seawater intake. During this process, along with seawater, phytoplankton-zooplankton and other larger organisms can be captured if the intake systems are not adequate.
4. Construction of facilities. During the construction phase of desalination plants, environmental impacts can also be generated associated with the resuspension of sediments and underwater noise.

SEIA is envisioned as a key instrument for preventing negative environmental impacts on marine flora and fauna in the context of the construction and operation of desalination plants. As presented previously, it is necessary to have precise and unequivocal thresholds that establish the conditions for the construction and operation of desalination plants. Critical aspects to consider include modelling of brine discharge to determine the potential impacts of brine and determine the impacts of seawater intake. Based on the development of a baseline, it is important to define design criteria that minimize environmental impact. Additionally, desalination plant projects should incorporate a comprehensive environmental monitoring program to implement protection guidelines for sensitive areas against the salinity effects generated by brine discharges.

18.4.3 Potential Greenhouse Gas Emissions

The operation of desalination plants involves significant energy consumption. This energy demand can result in substantial greenhouse gas (GHG) emissions if it is generated from non-renewable sources. Internationally, it is estimated that the carbon footprint of desalination plants using reverse osmosis is between 0.4 and 6.7 kgCO2eq/m^3 (Cornejo et al., 2014). The carbon footprint of the operation of desalination plants depends on the energy consumption of each facility (see Table 18.1); and the GHG emission factor, that is the kg of CO2eq emitted per kWh of energy produced, which is determined by the mix of electric power production from which the plant is supplied.

The average emission factor for the National Electric System (SEN) in 2021 in Chile was 0.3907 tCO2eq/MWh (Ministry of Energy, 2022). Hence, assuming that all the desalination plants listed in Table 18.1 were supplied through the SEN, their annual carbon footprint is estimated at 13,361.94 tCO2eq. The total GHG emissions in Chile in 2021 were 85,700,000 tCO2eq/year (DatosMacro, 2022). Thus, the electricity consumption of desalination plants in Chile corresponds to approximately 0.016% of the total GHG emissions. Although this figure may appear small, in the context of greenhouse gas mitigation needs and a potential rise in the number of desalination plants, renewable energy sources, particularly solar energy, offer an opportunity to reduce the carbon footprint of the water desalination process.

18.4.4 Earthquake and Tsunami Risks

Desalination plants are predominantly situated along coastal regions and are thus susceptible to tsunami events. Although tsunamis are rare, they are highly destructive natural phenomena. Notably, Chile is situated within a seismically active zone, where the Nazca plate subducts beneath the South American plate. Historical and instrumental data document the magnitude and intensity of earthquakes that have occasionally triggered tsunamis, with records spanning at least five centuries for the region (NGDC/WDS, 2019).

The effects of earthquakes and tsunamis on infrastructure are numerous due to dynamic and hydrodynamic forces, although in Chile the application of seismic standards has been effective in mitigating this risk. Specifically, for drinking water systems, the main impacts have been summarized in the PAHO report (1998), which could affect desalination plants and, therefore, should be considered during their construction and operation:

1. Total or partial destruction of intakes, transport structures, treatment, storage, and distribution facilities.
2. Breakages in delivery and distribution pipelines and damage to connections between pipes or tanks, resulting in water loss.
3. Interruption of electricity, communications, and access routes.
4. Changes in water quality.

In this context, it is crucial to evaluate the tsunami risk ex-ante to the construction of desalination plants to adopt appropriate mitigation measures, which, according to the definition of risk, may be associated with reducing exposure and/or vulnerability. Various studies (e.g., Tada et al., 2013; Aguirre et al., 2018; González et al., 2020) have proposed different methods to evaluate tsunami disaster risk. These approaches should be thoroughly analysed to develop a unified risk management methodology.

18.5 Social Challenges for Sustainable Desalination

The success of desalination projects relies on economic, environmental viability, and support from society and local communities affected by the project (Etale et al., 2020). Public opposition to essential infrastructure projects like desalination plants indicates a failure in policy communication, potentially leading to insufficient future funding and poor oversight, endangering vital services (Haddad et al., 2018). Thus, before developing a desalination project, decision-makers should assess the perception of potential consumers of desalinated water (Villar-Navascués et al., 2020).

Studies on people's perceptions of desalination fall into three categories. The first includes studies on general attitudes towards desalination as a water supply and, therefore, these studies are not linked to a specific desalination project. The second focuses on studies that analyze public opinions on actual future desalination projects; and the third examines attitudes in places where desalination systems have already been established (Friedler et al., 2006).

To determine which perceptions matter, previous studies categorize people into two groups: (a) people who consume (or will consume) desalinated water; and, (b) people who live near the coastal edge and/or desalination infrastructure. Most studies suggest that residents view desalination positively as a solution to potential shortages in drinking water supply (Gibson et al., 2015; Heck et al., 2016a, 2016b). Conversely, people residing on the coastal edge may oppose desalination plants due to their beliefs about the biological, ecological, and environmental impacts of the plants (King et al., 2012; Heck et al., 2016a). This dichotomy has been observed in the case of the Chilean "Aconcagua" desalination project located in the Puchuncaví bay with a maximum desalination capacity of 1000 L/s. Although the project has had an environmental qualification resolution since 2020, it has not been exempted from strong opposition by fishermen from the bay itself and Quintero who argue that the operation of the desalination plant will impact coastal ecosystems and fishing. On the other hand, the company responsible for the project reached early agreements with several rural drinking water systems located in Olmué and Limache, whose leaders consider desalinated water as an essential water source to ensure the supply of drinking water in their municipalities (Fundación Terram, 2023).

Access to the coastal edge and the landscape impacts of desalination plants are another concern for local communities (Haddad et al., 2018). Residents sometimes oppose projects related to natural resource management in their vicinity, even though they would otherwise generally support them. This corresponds to the well-known "not-in-my-back-yard" opposition, which has been widely used to explore whether proximity to a new development influences the perception of local residents. The sociodemographic characteristics of local communities can also affect the acceptance or rejection of desalination projects (Haddad et al., 2018). Psychological factors such as beliefs and perceptions about the impacts of desalination have also been identified as influential in people's attitudes towards desalination (Gibson et al., 2015; Heck et al., 2016b). Additionally, institutional variables, such as people's trust in the entity (public or private) responsible for or promoting the construction and/or operation of the desalination plant, can influence the attitude of local communities (Hynes et al., 2014; Potts et al., 2016).

In this context, instruments such as social life cycle assessment can be valuable for identifying the social impacts on local communities. According to Nazaran et al. (2021), three key aspects to consider are: (a) safety and health; (b) community engagement; and (c) local employment. Additionally, qualitative methods, such as semi-structured interviews with key actors, can be applied to better understand the attitude of local communities towards desalination (Liu et al., 2022). According to GHD (2008), the main objectives of a social impact management plan are as follows:

1. Manage social impacts throughout the design, construction, and operational phases of a desalination project.
2. Ensure that new information about social impacts is considered in its assessment, its significance, and the required management actions.
3. Periodically evaluate the effectiveness of mitigation, improvement, and follow-up measures.
4. Flexibility and adaptability to identify unexpected impacts and manage them appropriately.
5. Foster stronger working relationships between the community and the project proponent.

18.6 Economics of Desalination

Financial and economic factors are important for both the implementation and operation of desalination plants, and they can influence their viability as an alternative to other water sources in a context of drought. Desalination projects concentrate on Capital Expenditure (CAPEX), the costs associated with investment, obtaining easements, permits, and the selection of corrosion-resistant materials. The operation expenditure (OPEX) is proportional to energy consumption. Cooley et al. (2006) state that the cost corresponding to energy consumption in desalination plants using reverse osmosis represents between 44% and 50% of the total.

Table 18.2 presents, as a reference, a comparative quantification between the technologies applied in desalination processes and their costs. The total cost of producing desalinated water using Multistage Flash Distillation (MSF) technology ranges from USD$1.02/m³ to USD$1.74/m³, with an average of USD$1.44/m³. The technology of Multiple Effect Distillation (MED) presents an average production cost of USD$1.44/m³ with a range from USD$0.75/m³ to USD$1.85/m³. Seawater Reverse Osmosis (RO) today surpasses thermal technologies, representing 66% of the global desalination capacity installed. The costs of this technology have significantly decreased in recent years, reaching an average of USD$0.64/m³, with a range from USD$0.36/m³ to USD$0.92/m³ (World Bank, 2019; Zhu et al., 2019; Zarzo, 2021).

Pumping and transporting water necessitates significant infrastructure investments and entails substantial energy consumption, which adversely affects operational costs. Due to these expenses, the cost of desalinated water can be higher compared to other sources. However, this is contingent upon various factors, including the distance to the application site, the price of energy, and the availability of alternative resources.

Orphanopoulus (2021) conducted an estimation of the total cost of desalination for a specific seawater desalination plant in Chile. This assessment referenced the actual costs of the Atacama plant and considered a capacity of 300 L/s, marine works of 300 L/s, energy consumption ranging from 25 to 30 MW, construction of a medium voltage line to connect the plant with the existing grid, and the necessary plant facilities. These facilities include seawater intake and brine discharge systems, diffusers designed to meet marine conditions, physical and chemical pretreatment systems, a reverse osmosis desalination system, a post-treatment system, an accumulation unit for the produced water, and complementary facilities such as urbanization, lighting, air conditioning, and warehouses. The estimated total cost, encompassing both CAPEX and OPEX, was calculated to be USD$1.17/m³.

The economic value for the supply of desalinated seawater to the mining sector is clear (USD$43.89/m³), estimated based on its apparent water productivity for mining (Donoso, 2021). This conclusion does not apply to agricultural users, a sector that presents the lowest economic water productivity in Chile, and lower than the

Table 18.2 Desalination costs according to technology

Desalination technology	Capital expenditure USD$/m³	Energy consumption kWh/m³	Operational expenditure USD$/m³	Global trend	Energy source
Ro	700–900	2–2.8	0.45–0.92	Growing	Electrical
Electrodialysis	–	16–19	0.58	Stable	Electrical
Med	900–2300	3.4–4	0.75–1.85	Decreasing	Caloric/electrical
MFD	800–4100	5–8	1.10–1.50	Decreasing	Caloric/electrical

Source: Own elaboration based on Cooley et al. (2006), National Water Commission (2008), and World Bank (2019)

average economic water productivity of the sector in Latin America and the Caribbean of USD\$1.01/m^3 (World Bank, 2019; Mekonnen & Hoekstra, 2011). The economic water productivity of agriculture in Chile varies between USD\$0.2/m^3 for cereal production and USD\$0.85/m^3 for fruit production (Banco Central de Chile, 2019; DGA, 2016; Donoso, 2021). Considering the values of economic productivity, the higher value agricultural sectors would be willing to pay USD\$0.8/m^3. Therefore, desalinated seawater for agriculture, even if provided through an efficient network, would require subsidies. These results for Chile are consistent with Martínez-Álvarez et al. (2019), who concluded that the cost of desalinated water for irrigation in Spain in 2017 was between USD\$0.66 m^3 and USD\$0.85/m^3. This cost is, in most cases, high compared to the crop value, so a subsidy would be required. In fact, these authors find that farmers have systematically been unwilling to pay the full costs of desalinated seawater, leading the Spanish Government to apply direct subsidies as an extraordinary measure to mitigate the effects of water scarcity.

Despite these cases, desalinated seawater is used in some areas worldwide to irrigate high-value crops and improve the socioeconomic resilience of the region, but innovative and integrated water management strategies, including subsidies, are needed for greater acceptance (Martínez-Álvarez et al., 2019; Shalsi et al., 2019). The use of desalinated seawater for irrigating high-value crops is present in southeast Spain, the Canary Islands, and Israel. Martínez-Álvarez et al. (2019) concluded that such use of desalinated seawater is promising in coastal regions with water scarcity where highly profitable crops can be cultivated, although innovative and integrated water management strategies are required to support greater acceptance.

18.7 Conclusions

Chile's unique geography, extensive coastline, and growing water scarcity issues, exacerbated by frequent droughts, highlight the importance of desalination as a strategic solution for addressing the nation's water demands. Over the years, desalination has evolved from Chile's first plant in 1892 to a robust sector, with 31 facilities currently operational or in development. This growth underscores desalination's pivotal role, particularly in the mining industry, to offset limited freshwater resources. However, achieving sustainable desalination practices requires overcoming regulatory, environmental, social, and economic challenges.

The effective integration of desalination into Chile's water management strategy relies on a deep understanding of water supply-demand dynamics, yet there is a recognized lack of comprehensive data and monitoring. A National Unified Water Information System could enhance data-driven decisions, enabling Chile to meet both immediate and future water needs influenced by climate change, demographic shifts, and agriculture.

A fragmented governance framework and the Water Code's lack of desalination-specific regulations also impede Chile's ability to maximize desalination's potential. Developing a cohesive national water planning framework is essential to incorporate desalinated water into broader water resource management strategies, ensuring alignment with local needs and climate change projections. A comprehensive desalination law that addresses seawater extraction, processing, desalinated water management, and its distribution is necessary. Moreover, regulatory guidelines should include inland desalination (such as for brackish water) to adapt to the diverse applications of desalination across Chile.

SEIA is vital in addressing desalination projects' environmental impacts on marine ecosystems. Yet, clearer criteria are needed, covering aspects like plant size, discharge volumes, and ecological sensitivity, to ensure robust assessments. Community involvement is another critical component; as desalination projects impact local areas, protocols promoting social acceptance and community participation should be established to foster local support. At the same time, desalination costs, which are closely linked to the energy issue, must be carefully analysed, including the possibility of public-private partnerships.

By addressing these considerations holistically, Chile can establish an effective regulatory framework for desalination that not only mitigates water scarcity and drought impacts but also aligns with environmental conservation and community well-being. This integrated approach will support a sustainable, resilient water management system for Chile's evolving needs.

18.8 Normative

DFL 1, fija texto del Código de Aguas, 1981.

Ley 19.300, aprueba ley sobre bases generales del medio ambiente, 1994.

Decreto 40, aprueba reglamento del Sistema de Evaluación de Impacto Ambiental, 2013.

Acknowledgements This work was developed within the framework of the ANID FSEQ210018, and Consorcio Tecnológico Del Agua CotH2O—Gestión Integrada de Recursos Hídricos 20CTECGH-145896 projects, whose support and financing we would like to thank.

References

ACADES. (2021). *Desalinización. Consideraciones generales para una nueva regulación institucional y legal* (p. 3). ACADES. https://www.acades.cl/wp-content/uploads/2022/03/Ej_Publicacion2.pdf

ACADES. (2023). *Primer catastro de proyectos y plantas desalinizadoras en Chile*. Asociación Chilena de Desalación y Reúso. https://www.acades.cl/wp-content/uploads/2023/03/Catastro-plantas-ACADES.pdf

Aguas Antofagasta. (2024). *Inicios de la Desalinización en Chile*. http://www3.aguasantofagasta.cl/DESALAR/chile.html

Aguirre, P., Vásquez, J., de la Llera, J. C., González, J., & González, G. (2018). Earthquake damage assessment for deterministic scenarios in Iquique, Chile. *Natural Hazards, 92*(3), 1433–1461.

Aitken, D., Saavedra, B., Figueroa, A., López, G., Becerra, L., Henríquez, A., Mosella, C., & Cid, F. (2022). *Cinco nichos socio-tecnológicos críticos para una minería sostenible en la región andina*. Centro de Estudios del Cobre y la Minería y Cooperación Alemana para el Desarrollo.

Alvez, A., Aitken, D., Rivera, D., Vergara, M., McIntyre, N., & Concha, F. (2020). At the crossroads: Can desalination be a suitable public policy solution to address water scarcity in Chile's mining zones. *Journal of Environmental Management, 258*, 110039.

Angelakis, A. N., Valipour, M., Choo, K.-H., Ahmed, A. T., Baba, A., Kumar, R., Toor, G. S., et al. (2021). Desalination: From ancient to present and future. *Water, 13*(16), 2222.

Asli, B. Z., McIntyre, N., Djordjevic, S., Farmani, R., & Pagliero, L. (2023). A closer look at the history of the desalination industry: The evolution of the practice of desalination through the course of time. *Water Supply, 23*(6), 2517–2526.

B2B Media Group. (2022). *Catastro de Plantas Desalinizadoras y Sistemas de Impulsión de Agua de Mar.* https://www.b2bmg.cl/catastro-de-plantas-desalinizadoras-en-chile/

Banco Central de Chile. (2019). *Bases de datos estadísticas*. https://si3.bcentral.cl/Siete/secure/cuadros/arboles.aspx?idCuadro=CCNN2013_G_VE

Banco Mundial. (2011). *Diagnóstico de la gestión de los recursos hídricos*. Banco MundialSantiago.

Banco Mundial. (2013). *Estudio para el mejoramiento del marco institucional para la gestión del agua*. Banco Mundial.

CCC (California Coastal Commission). (2004). *Seawater desalination and the California Coastal Act*. http://www.coastal.ca.gov/

Celume, T., & Goldenberg, J. L. (2021). El libre acceso a las aguas del mar: una aproximación a su calificación jurídica y económica. *Revista de Derecho Administrativo Económico, 34*, 131–156.

Cooley, H., Gleick, P., & Wolff, G. (2006). *Desalination, with a grain of salt: A California perspective*. Pacific Institute for Studies in Development, Environment, and Security. https://pacinst.org/wp-content/uploads/2006/06/desalination-grain-of-salt-a-cali-perspective-june-2006.pdf

Cornejo, P. K., Santana, M. V. E., Hokanson, D. R., Mihelcic, J. R., & Zhang, Q. (2014). Carbon footprint of water reuse and desalination: A review of greenhouse gas emissions and estimation tools. *Journal of Water Reuse and Desalination, 4*(4), 238–252.

DatosMacro. (2022). *Emisiones de gases de efecto invernadero en Chile*. https://datosmacro.expansion.com/energia-y-medio-ambiente/emisiones-co2/chile#:~:text=Las%20emisiones%20de%20CO2%20en,de%20menos%20a%20m%C3%A1s%20contaminantes

DGA. (2016). *Atlas del Agua Chile*. Dirección General de Aguas, Ministerio Obras Públicas. http://www.dga.cl/DGADocumentos/Atlas2016parte1-17marzo2016b.pdf

DGA. (2018). *Diagnóstico nacional de organizaciones de usuarios*. DGA. https://bibliotecadigital.ciren.cl/bitstreams/ce09c1fa-693d-4b57-8a43-91ae857ab43e/download

Díaz, H. (2022). *¿Tiene impacto la desalinización?* Desalación y medio ambiente marino. https://www.acades.cl/wp-content/uploads/2022/06/Humberto-Diaz.pdf

Donoso, G. (2021). Economics of water resources. In B. Fernández & J. Gironás (Eds.), *Water resources of Chile* (pp. 335–346). Springer. https://doi.org/10.1007/978-3-030-56901-3_18

Etale, A., Fielding, K., Schäfer, A. I., & Siegrist, M. (2020). Recycled and desalinated water: Consumers' associations, and the influence of affect and disgust on willingness to use. *Journal of Environmental Management, 261*, 110217.

Friedler, E., Lahav, O., Jizhaki, H., & Lahav, T. (2006). Study of urban population attitudes towards various wastewater reuse options: Israel as a case study. *Journal of Environmental Management, 81*(4), 360–370.

Fundación Terram. (2023). *La "cara y el sello" de la planta desalinizadora*. https://www.terram.cl/2023/06/el-cara-y-sello-de-la-planta-desalinizadora-mientras-en-puchuncavi-y-quintero-protestan-en-limache-y-olmue-celebran/

García-Bartolomei, E., Vásquez, V., Rebolledo, G., Vivallo, A., Acuña-Ruz, T., Rebolledo, J., Orrego, R., & Barra, R. (2022). Defining priority areas for the sustainable development of the desalination industry in Chile: A GIS multi-criteria analysis approach. *Sustainability, 14*, 7772.

GHD. (2008). *Report for the proposed southern seawater desalination project: Social impact assessment* (Perth, Western Australia).
Gibson, F. L., Tapsuwan, S., Walker, I., & Randrema, E. (2015). Drivers of an urban community's acceptance of a large desalination scheme for drinking water. *Journal of Hydrology, 528*, 38–44.
Goh, P. S., Matsuura, T., Ismail, A. F., & Ng, B. C. (2017). The water-energy nexus: Solutions towards energy-efficient desalination. *Energy Technology, 5*(8), 1136–1155.
González, J., González, G., Aránguiz, R., Catalán, P. A., & Cienfuegos, R. (2020). A hybrid deterministic and stochastic approach for tsunami hazard assessment in Iquique, Chile. *Natural Hazards, 100*(1), 231–254.
Haddad, B., Heck, N., Paytan, A., & Potts, D. (2018). Social issues and public acceptance of seawater desalination plants. In *Sustainable desalination handbook: Plant selection, design and implementation* (pp. 505–525). Elsevier.
Hamiche, A. M., Stambouli, A. B., & Flazi, S. (2016). A review of the water-energy nexus. *Renewable and Sustainable Energy Reviews, 65*, 319–331.
Heck, N., Paytan, A., Potts, D. C., & Haddad, B. (2016a). Coastal residents' literacy about seawater desalination and its impacts on marine ecosystems in California. *Marine Policy, 68*, 178–186.
Heck, N., Paytan, A., Potts, D. C., & Haddad, B. (2016b). Predictors of local support for a seawater desalination plant in a small coastal community. *Environmental Science and Policy, 66*, 101–111.
Herrera-León, S., Cruz, C., Kraslawski, A., & Cisternas, L. A. (2019). Current situation and major challenges of desalination in Chile. *Desalination and Water Treatment, 171*, 93–104.
Hynes, S., Norton, D., & Corless, R. (2014). Investigating societal attitudes towards the marine environment of Ireland. *Marine Policy, 47*, 57–65.
International Desalination Association. (2022). *Desalination and water reuse by the numbers.* https://idadesal.org/
Irarrázabal, R., Alburquenque, W., Baltra, M. L., Fermandois, A., Helfmann, C., Ibáñez, J. E., Ovalle, A., & Rivera, D. (2023). *Comentario a las normas aprobadas en general por la Comisión Experta. Medio ambiente, recursos naturales y desarrollo sostenible.* Foro Constitucional UC. https://foroconstitucional.uc.cl/wp-content/uploads/2023/04/Medio-ambiente-recursos-naturales-y-desarrollo-sostenible.pdf
Khordagui, H. K. (1999). Desalination. In *Environmental geology. Encyclopedia of earth science.* Springer. https://doi.org/10.1007/1-4020-4494-1_78
King, T. J., Ooi, D., Cary, J., Fisher, A., Schibeci, R., & Murphy, K. (2012). *Public perceptions of, and responses to, desalination in Australia* (Alfred Deakin Research Institute Working Paper Series). Deakin University
Liu, T.-K., Ye, J.-A., & Sheu, H.-Y. (2022). Exploring the social acceptability for the desalination plant project: Perceptions from the stakeholders. *Desalination, 532*, 15757.
López, W., & Del Campo, M. E. (2021). *Consumo de energía e impacto de la descarga de salmuera de plantas desalinizadoras.* https://www.acades.cl/wp-content/uploads/2022/04/Comite-Tecnico-Primera-Sesion-02.08.2021.pdf
Martínez-Álvarez, V., Maestre-Valero, J. F., González-Ortega, M. J., Gallego-Elvira, B., & Martin-Gorriz, B. (2019). Characterization of the agricultural supply of desalinated seawater in southeastern Spain. *Water, 11*, 1233.
Mekonnen, M. M., & Hoekstra, A. Y. (2011). National water footprint accounts: The green, blue and grey water footprint of production and consumption, UNESCO-IHE. *Value of Water Research Report Series,* No. 50
Mesa Nacional del Agua. (2022). *Informe final.* Mesa Nacional del Agua. http://derechoygestionaguas.uc.cl/es/documentos/new/507-2022-mesa-nacional-del-agua-informe-final/file
Ministry of Energy. (2022). *Factors of greenhouse gas emissions in Chile.* http://energiaabierta.cl/visualizaciones/factor-de-emision-sic-sing/

Montginoul, M., Rinaudo, J.-D., Brozović, N., & Donoso, G. (2016). Controlling groundwater exploitation through economic instruments: Current practices, challenges and innovative approaches. In A. J. Jakeman, O. Barreteau, R. J. Hunt, J.-D. Rinaudo, & A. Ross (Eds.), *Integrated groundwater management: Concepts, approaches and challenges* (pp. 551–581). Springer. https://doi.org/10.1007/978-3-319-23576-9_22

National Water Commission. (2008). *Emerging trends in desalination: A review*. Commonwealth of Australia.

Nazaran, I. S., Ghani, L. A., Muhammad, Z., & Saputra, J. (2021). Social impact of small seawater desalination plant of local community: A social life cycle assessment. In *Proceedings of the International Conference on Industrial Engineering and Operations Management,* Monterrey, Mexico, November 3–5, 2021.

NGDC/WDS. (2019). *National Geophysical Data Center, World Data Service. Global Historical Tsunami Database*. National Geophysical Data Center.

Obando, I. M. (2009). Estructura y jurisdicción de las organizaciones de usuarios de aguas en chile durante el siglo XIX. *Revista de Derecho de la Pontificia Universidad Católica de Valparaíso* n.32 Valparaíso jun. 2009.

Orphanopoulus, D. (2021). *Desalinización en Chile hoy: ¿La Agricultura Mirando al Mar? Seminario Deasalación ¿La Agricultura Mirando al Mar?* Santiago.

Østergaard, P. A., Lund, H., & Mathiesen, B. V. (2014). Energy system impacts of desalination in Jordan. *International Journal of Sustainable Energy Planning and Management, 1*, 29–40. https://doi.org/10.5278/ijsepm.2014.1.3

Pistocchi, A., Bleninger, T., Breyer, C., Caldera, U., Dorati, C., Ganora, D., et al. (2020). Can seawater desalination be a win-win fix to our water cycle? *Water Research, 182*, 115906.

Plaza, R. (2017). ¿Es necesario legislar sobre el uso del agua de mar y su desalinización? El marco jurídico actual de las aguas desaladas y el análisis de los proyectos de ley en curso. *Revista de Derecho Ambiental, 7*, 60–93.

Potts, T., Pita, C., O'Higgins, T., & Mee, L. (2016). Who cares? European attitudes towards marine and coastal environments. *Marine Policy, 72*, 59–66.

Rinaudo, J.-D., & Donoso, G. (2019). State, market or community failure? Untangling the determinants of groundwater depletion in Copiapó (Chile). *International Journal of Water Resources Development, 35*(2), 283–304. https://doi.org/10.1080/07900627.2017.1417116

Rivera, D., Alburquenque, W., Cortés, S., Donoso, G., Melo, O., Molinos, M., Parodi, A., Reyes, F., Soto, R., and Vicuña, S. (2021). *Aguas y nueva Constitución. Perspectivas y propuestas*. Foro Constitucional UC. https://derechoygestionaguas.uc.cl/es/documentos/new/495-foro-constitucional-comision-de-aguas-documento/file

Rivera, D., Donoso, G., Molinos, M. and Vicuña, S. (2022). *Comentarios a la propuesta de norma constitucional que crea la Agencia Nacional del Agua, contenida en el 3° Informe de la Comisión de Sistemas de Justicia*. Foro Constitucional UC. https://foroconstitucional.uc.cl/wp-content/uploads/2022/04/Minuta-Agencia-Nacional-del-Agua.pdf

Robbins, J. (2019). *As water scarcity increases, desalination plants are on the rise, Yale Environment 360*. https://e360.yale.edu/features/as-water-scarcity-increases-desalination-plants-are-on-the-rise

Santibáñez, F., & Santibáñez, C. (2020). *Análisis de las soluciones posibles a la escasez hídrica del Norte chileno. El caso de la desalación de agua marina*, Santiago

Shahzad, M. W., Burhana, M., Ang, L., & Nga, K. C. (2017). Energy-water-environment nexus under pinning future desalination sustainability. *Desalination, 413*, 52–64.

Shalsi, S., Ordens, C. M., Curtis, A., & Simmons, C. T. (2019). Can collective action address the "tragedy of the commons" in groundwater management? Insights from an Australian case study. *Hydrogeology Journal, 27*(7), 2471–2483.

Shokri, A., & Sanavi Fard, M. (2023). Techno-economic assessment of water desalination: Future outlooks and challenges. *Process Safety and Environmental Protection, 169*, 564–578.

Skewes, F. (2017). Autorización ambiental para actividades de desalinización de agua de mar. *Revista de Derecho Ambiental, 7*, 35–59.

Šteflová, M., Koop, S. H. A., Fragkou, M. C., & Mees, H. (2021). Desalinated drinking-water provision in water-stressed regions: Challenges of consumer-perception and environmental impact lessons from Antofagasta, Chile. *International Journal of Water Resources Development, 38*(5), 742–765.

Stokes, J., & Horvath, A. (2006). Life cycle energy assessment of alternative water supply systems. *The International Journal of Life Cycle Assessment, 11*(5), 335–343.

Tada, K., Komine, H., & Murakami, S. (2013). Prediction of time to reduce tsunami sediment salinity by rainfall after the Great East Japan earthquake. Coupled phenomena in environmental geotechnics: From theoretical and experimental research to practical applications. *Proceedings of the International Symposium, ISSMGE TC, 215*, 671–674.

USGS. (2019). *The distribution of water on, in, and above the Earth*. https://www.usgs.gov/media/images/distribution-water-and-above-earth

Vergara, A., & Rivera, D. (2018). Legal and institutional framework of water resources. In G. Donoso (Ed.), *Water policy in Chile* (pp. 67–85). Springer.

Vicuña, S., Daniele, L., Farías, L., González, H., Marquet, P. A., Palma-Behnke, R., Stehr, A., Urquiza, A., Wagemann, E., Arenas-Herrera, M. J., Bórquez, R., Cornejo-Ponce, L., Delgado, V., Etcheberry, G., Fragkou, M. C., Fuster, R., Gelcich, S., Melo, O., Monsalve, T., … Winckler, P. (2022). *Desalinización: Oportunidades y desafíos para abordar la inseguridad hídrica en Chile* (p. 188). Comité Asesor Ministerial Científico sobre Cambio Climático; Ministerio de Ciencia, Tecnología, Conocimiento e Innovación. https://www.minciencia.gob.cl/uploads/filer_public/e0/48/e048f6f8-ca73-4d90-b108-8ca5976565c1/desalinizacion__oportunidades_y_desafios_para_abordar_la_inseguridad_hidrica_en_chile.pdf

Villar-Navascués, R., Ricart, S., Gil-Guirado, S., Rico-Amorós, A. M., & Arahuetes, A. (2020). Why (not) desalination? Exploring driving factors from irrigation communities' perception in South-East Spain. *Water, 12*(9), 2408.

Voutchkov, N. (2018). *Total cost of water production. A PowerPoint presentation to the Korea University.*

Voutchkov, N. (2022). Desalinated water. In M. Qadir, V. Smakhtin, S. Koo-Oshima, & E. Guenter (Eds.), *Unconventional water resources* (pp. 233–254). Springer.

World Bank. (2019). *The role of desalination in an increasingly water-scarce world*. International Bank for Reconstruction and Development.

Zarzo, D. (2021). *La Desalación Agro en el Mundo. Seminario Desalación ¿La Agricultura Mirando al Mar?* Ministerio de Agricultura.

Zhu, Z., Peng, D., & Wang, H. (2019). Seawater desalination in China: An overview. *Journal of Water Reuse and Desalination, 9*(2), 115–132.

Chapter 19
Evaluation of Multipurpose Reservoir Operating Policies at Basin and Electric Power System Scale

Antonia Durán, Marcel Favereau, Álvaro Lorca, Sebastián Vicuña, Óscar Melo, and Matías Negrete-Pincetic

Abstract The impacts of climate change such as hydrological droughts have motivated significant changes in reservoir management strategies. The operation of multipurpose reservoirs is vital for efficiently utilizing stored water resources, serving various needs such as electricity generation and agricultural irrigation. Despite substantial efforts to support decision-making within each economic sector, there is a notable lack of comprehensive integration across these sectors in concurrent analy-

A. Durán
Department of Hydraulic and Environmental Engineering, Pontificia Universidad Católica de Chile, Santiago, Chile

Department of Industrial and Systems Engineering, Pontificia Universidad Católica de Chile, Santiago, Chile

M. Favereau (✉)
Department of Industrial and Systems Engineering, Pontificia Universidad Católica de Chile, Santiago, Chile

Department of Engineering Sciences, Universidad Andrés Bello, Santiago, Chile
e-mail: marcel.favereau@unab.cl

Á. Lorca
Department of Industrial and Systems Engineering, Pontificia Universidad Católica de Chile, Santiago, Chile

Department of Electrical Engineering, Pontificia Universidad Católica de Chile, Santiago, Chile

S. Vicuña
Department of Hydraulic and Environmental Engineering, Pontificia Universidad Católica de Chile, Santiago, Chile

Ó. Melo
Department of Agricultural Economics, Pontificia Universidad Católica de Chile, Santiago, Chile

M. Negrete-Pincetic
Department of Electrical Engineering, Pontificia Universidad Católica de Chile, Santiago, Chile

D. Rivera, G. Donoso (eds.), *Droughts in Chile*, Global Issues in Water Policy 31,
https://doi.org/10.1007/978-3-031-85040-0_19

ses. To address this gap, we propose an integrated approach that combines a large-scale hydrothermal scheduling model with a basin-scale water resources model to comprehensively analyze the operations of both the power and agricultural systems. This approach enables the assessment of operational policies for multipurpose reservoirs and their performance at both local and regional scales under diverse hydrological scenarios. An essential modification analyzed here is the prioritization of water allocation to agricultural users. We rigorously evaluate the impacts of this modification across various hydrological conditions, using a prominent Chilean basin as a case study. Applying this methodology to Laja Lake, the largest Chilean multipurpose reservoir with significant hydroelectric capacity and extensive agricultural areas, we analyze the situation in 2025 and find that hydrological variations directly affect both the electrical and agricultural aspects of performance. During drought conditions, there is a noticeable increase in thermal generation, costs, emission intensity, and water deficits. Furthermore, the adjusted policy reveals intricate trade-offs between emissions from the power sector and agricultural water deficits. In drier scenarios, improving agricultural reliability incurs minimal additional operational costs and reduces power sector emissions, supporting the adoption of a policy aligned with net-zero objectives.

Keywords Hydrological droughts · Multipurpose reservoirs · Reservoir management strategies · Hydropower and irrigation · Sectoral integration in reservoir management

19.1 Introduction

Reservoirs represent hydraulic infrastructure of substantial capacity, designed, and managed to fulfill many purposes. Their construction and operational oversight aim to strike a delicate equilibrium among various functions, encompassing municipal and industrial irrigation, hydroelectric power generation, and flood mitigation (Labadie, 2004). This multifaceted utility gives rise to the concept of multipurpose reservoirs, where a web of norms, regulations, legal statutes, and contractual obligations govern the allocation of water resources. These intricate agreements, forged between entities such as electric utilities, water supply agencies, and water rights holders, delineate the distribution of water volumes among stakeholders through operational rules (Mower & Miranda, 2013).

The significance of multipurpose reservoir operations assumes heightened prominence when viewed through the lens of the water-energy-food nexus paradigm. This holistic framework endeavors to interconnect critical economic activities linked to water resources while striving to balance potential conflicts between various sectors (Smajgl et al., 2016; FAO, 2014). One method that allows for coordinating multiple users involves a standard operating policy, which determines the water release from the reservoir based on inflows, demand, and reservoir storage level (Bolouri-Yazdeli et al., 2014). Notably, the vertical segmentation of stored water, classified into distinct levels based on waterline elevations from bottom to surface, assumes conceptual importance. These delineated zones encompass the dead storage zone, buffer zone, conservation zone, flood control zone, and spill zone, each playing a pivotal role in shaping the operational dynamics of reservoirs (Jain, 1993).

The emergence of unusual hydroclimatic phenomena, characterized by increased frequency and heightened intensity, is becoming more prevalent because of climate change and its associated effects (Loucks, 2000; Jazmin et al., 2016; Kim et al., 2021). These shifting climatic patterns have necessitated a fundamental reassessment of conventional reservoir planning strategies. Consequently, there is a need for a harmonized and all-encompassing approach that considers the interconnectedness of water and power systems (Hightower & Pierce, 2009). This approach must operate within a regulatory and institutional framework that assumes the responsibility for the effective and reliable governance of these vital resources.

Hydropower operation is generally planned as part of an electric grid programming process, which is usually modeled using optimization techniques such as the hydrothermal scheduling problem (HTSP) (Labadie, 2004). This problem is applied in large-scale systems, where generation is determined by thermal, hydroelectric, solar, wind, and other generation sources. Its main objective is to minimize operational costs over a medium-term planning horizon (typically ranging from 2 to 5 years), while its primary decisions include the generation of power plants in each period, the amount of water to be used from each reservoir, and the water flow through the hydrological network of the system. An important result of this model is the valuation of stored water, which is represented by the dual variables associated with the reservoir balance constraints in the optimization problem and the cost of avoided thermal generation in subsequent periods. Its resolution requires algorithms tailored to this type of problem, and for this purpose several methods have been developed. Given that hydro inflows are typically considered random variables, one of the most widely used methods is stochastic dual dynamic programming (SDDP) (Pereira & Pinto, 1991), which solves multi-stage stochastic optimization problems where the present cost and a future cost function of another subproblem are minimized at each stage, where the latter is iteratively approximated using Benders cuts (Benders, 1962). One limitation of this method is that it requires convexity to find an optimal solution and obtain the future value of water.

In order to integrate other sectors into the hydrothermal scheduling problem and thus represent multipurpose reservoirs with more detail, several approaches have been developed incorporating agriculture into their modeling. For example, Pereira-Bonvallet et al. (2016) present a suboptimal approach based on the SDDP method for coordinating a hydrothermal system with non-convex irrigation constraints. This approach is applied to the Central Interconnected System (SIC) in Chile (since 2017 named as National Electric System), where operation rules that coordinate water uses between the electricity and agriculture sectors are modeled and compared with a mixed-integer linear programming (MILP) approach. The study results demonstrate that the presented approach determines feasible solutions close to the optimal solution with costs similar to those determined in the optimal condition by MILP. However, the volumes of stored water tend to be lower in the same situation. Additionally, Gonzalez et al. (2016) consider a hydrothermal scheduling problem using an implicit stochastic version of SDDP to obtain the marginal value of hydropower generation. This value is subsequently used in the decision problem of the farmer, an optimization model called Positive Mathematical Programming (PMP), which assigns available land and water to possible crops, thereby maximizing net benefit. In this way, demand functions are obtained to develop efficient operation schemes. From this application, the authors provide an economically optimal operation in which the total benefits of water use in agriculture and energy are higher than allocation by rights or contractual agreements. Later, Matus et al. (2019) present a methodology that implements the operational agreement of the Laja reservoir in 2017 into a long-term planning model known as Long-term Planning (PLP), which is based on the SDDP method and is used by the Chilean power system operator. The results showed that the simultaneous use of irrigation rights and generation allows for feasible and economically viable solutions, and the agreement demonstrates a sustainable recovery of the Laja Lake volume. Finally, Navarro et al. (2023) develop a medium-term stochastic multistage programming model for the hydrothermal scheduling problem in the Chilean national power system. This model considers the effects of short-term operational constraints, such as short-term storage, reserve requirements, irrigation demand, and uncertainty in the hydro inflows of the power system's hydrological network, represented by a vector autoregressive model. Additionally, this research explicitly models an approximation of the agreements between irrigation and generation of Chile's multipurpose reservoirs: Maule and Laja Lakes. However, this representation is a linear approximation of the agreements, as their mathematical implementation requires non-convex functions incompatible with the SDDP method.

On the other hand, there are water resources models (WRMs), which can represent the watershed-scale hydrological system of a multipurpose reservoir, reproducing the operation of the reservoir and the water transfers among the different users who utilize the resource throughout the basin. A variety of models exist in literature. One is the model developed by Tomlinson et al. (2020), a water resources simulator called Pywr, a free and open-source Python library. Pywr's structure consists of a network of nodes connected by directed links, where constraints and attributes are stored. Pywr solves dynamic problems in discrete periods, performing resource reallocation at each time step through a linear programming algorithm based on costs or penalties. It incorporates functionalities that allow the addition of boundary conditions, such as input data series or state functions. Another characteristic is the ability to store simulation results in each period for subsequent processing, observation, or preservation. Additionally, Pywr offers a flexible interface to integrate multi-objective optimization problems, allowing to fully leverage its capacity to simulate multiple scenarios.

Pywr has been used in different countries, although not extensively. For instance, Khadem et al. (2021) modeled the water system from the Kielder reservoir to London in England, obtaining the spatial and temporal distribution of water in the system and then incorporating the reservoir volume values into a flow temperature model. In Basheer et al. (2021), Pywr was used to simulate the Nile River system in Egypt, with the authors extending one of its functionalities to temporarily add water and electricity metrics for integration into an Egyptian economic model. In Chile, Ricalde et al. (2022) modeled the upper part of the Maipo basin, including demand sites such as agriculture and associated rivers. Through a multi-objective evolutionary algorithm, they simulated the flow passing through each node in different hydrological scenarios and then estimated metrics to evaluate system performance. Lastly, Jander et al. (2023) considered the water network in the Maule basin in Chile to study the water-energy-food nexus. This research included physical, productive, and regulatory characteristics and employed a multi-objective evolutionary algorithm to obtain portfolios of non-inferior adaptation strategies to climate change. An interesting aspect of this research is the validation of the WRM's ability to model the non-convex functions of operational agreements, thereby opening the possibility for future studies to utilize this tool to simulate such constraints that other models cannot. However, one limitation of this model is that it only represents hydroelectric generation at the basin level and not the entire power system.

Thus, hydrothermal scheduling is a problem that primarily allows the modeling of large-scale power systems and has been extended to integrate agricultural systems in terms of operational constraints and contractual agreements that coordinate both sectors. However, these methods have not mathematically represented basin-level water interactions to characterize the network's water transfer. Moreover, SDDP's need for convex constraints or continuous decision variables limits its use in modeling non-convex irrigation agreements. Although a method has been developed in the literature to address this issue, known as SDDiP (Stochastic Dual Dynamic Integer Programming) (Zou et al., 2019), it has not been applied to large-scale instances (Lorca et al., 2020). Additionally, there have been no apparent efforts in literature to analyze different operation agreements of multipurpose reservoirs in the medium-term and study their effects on HTSP planning models.

Furthermore, water resources simulation models have not incorporated the power system to a magnitude encompassing different generation plants and transmission lines, among other elements. Instead, research has mainly been oriented toward simulating a specific basin's water and agricultural network. Additionally, hydrothermal scheduling models have not attempted to implement integrative methodological approaches with water resources models (Lorca et al., 2020); instead, they have been developed as separate approaches. This gap highlights the opportunity to apply a different methodological approach that enables the analysis of the economic effects of decisions made in one sector to the other. Consequently, the contributions of this research are as follows:

1. An integrated medium-term approach is proposed, based on a hydrothermal scheduling model and a water resources model to represent and evaluate policies defining the operation of multipurpose reservoirs primarily serving agriculture and hydroelectricity in various hydrological scenarios.
2. A sensitivity analysis is conducted on the inflows of the hydrological system encompassing the electricity and agricultural systems to evaluate the operational agreement in four hydrological scenarios of wet, medium, dry, and very dry conditions.

3. A modification is made to the water resources management policy of a multipurpose reservoir, wherein the priority for the agricultural sector is increased, and the performance of this change is assessed in the four hydrological scenarios.

This work is developed through an application on Laja Lake in Chile, a multipurpose reservoir with a storage capacity of over 5500 hm^3, salient for its contribution to hydroelectric power generation and agricultural irrigation. Downstream of the reservoir, run-of-river power plants have a total installed capacity exceeding 1100 MW, accounting for 15% of the installed hydroelectric capacity (Galetovic & Muñoz, 2009; Pereira-Bonvallet et al., 2016; Matus et al., 2019). Furthermore, the reservoir supports three main irrigation channels, sustaining approximately 117,000 ha of agricultural land, representing 6% of the agricultural areas contributing to the economic growth of the Biobío region.

The remainder of this chapter is organized as follows. Section 19.2 presents the integration methodology and an in-depth explanation of the hydrothermal scheduling and water resources simulation models. In Sect. 19.3, the case study and the results of the computational experiments are exposed. Finally, Sect. 19.4 presents a discussion and concluding remarks.

19.2 Proposed Hydrothermal Scheduling and Water Resources Simulation Models

19.2.1 Multistage Stochastic Optimization Model of Hydrothermal Scheduling

In this section, the proposed hydrothermal scheduling model is presented. First the multistage stochastic version is presented. It is important to mention that the decisions of this model are all continuous variables related to the storage of the reservoirs, the movement of water along the waterways of the water network, and the generation decisions of hydro, thermal, and renewable power plants.

To correctly model the water system, we consider three types of nodes. First, there are storage nodes that allow the storage of water (i.e., reservoirs). Then, there are pass nodes that only allow deciding water movements along the water network. Finally, there are generation nodes, which would be those water nodes with an electric generator: reservoir or run-of-the-river. It is important to note that, although most reservoirs have a generator, these were separated from their respective reservoirs for simplicity of notation and modeling. Nodes ($\mathcal{N}$) are composed by reservoirs ($\mathcal{N}^{a}$), intakes ($\mathcal{N}^{p} \setminus \mathcal{N}^{g}$), hydroelectric plants ($\mathcal{N}^{g}$), and irrigation nodes ($\mathcal{K}$). Inflows are composed of natural flows ($\boldsymbol{\alpha}$), generable flows ($\mathbf{q}$), non-generable flows ($\mathbf{y}$), seepage ($\boldsymbol{\nu}$), and the irrigation (non-generable) flows that preserve feasibility in the water balance constraints related to irrigation zones.

It is important to highlight that the hydro generation nodes also allow water movement, but this is not its only objective. The waterways connecting these nodes are divided into two types. On the one hand, there are the so-called generable waterways, which consist of waterways whose water flow passes through an electric generator to reach another node in the network. On the other hand, there are the non-generable waterways, which are those flows that do not go through a generator and only move water between one network node and another (without generating electricity), i.e., it can be considered as spillage in the case of reservoirs.

The formulation of the proposed hydrothermal scheduling problem (Model 19.1) is as follows:

$$\min \sum_{t\in\mathcal{T}} \sum_{b\in\mathcal{R}(t)} \left(\sum_{g\in\mathcal{G}} C^{G}_{gtb} p_{gtb} + \sum_{i\in\mathcal{B}} \left(C^{LS}_{ib} p^{LS}_{itb} + C^{OG}_{ib} p^{OG}_{itb} \right) \right) \beta'_{tb} \tag{19.1a}$$

s.t.

$$w_{et} = w_{e,t-1} + \sum_{b\in\mathcal{R}(t)} \beta_{tb} \big(\alpha_{et} - \upsilon_{et} + \sum_{u\in\mathcal{I}^{G}(e)} q_{(u,e),t,b} + \sum_{u\in\mathcal{I}^{N}(e)} y_{(u,e),t,b} - \sum_{u\in\mathcal{O}^{G}(e)} q_{(e,u),t,b} + \sum_{u\in\mathcal{O}^{N}(e)} q_{(e,u),t,b} \big) \ \forall e\in\mathcal{N}, t\in\mathcal{T}. \tag{19.1b}$$

$$\upsilon_{et} = \lambda^{int}_{e} + \lambda^{slope}_{e} w_{e,t-1} \quad \forall e\in\mathcal{N}^{a}, t\in\mathcal{T} \tag{19.1c}$$

$$\sum_{u\in\mathcal{I}^{G}(n)} q_{(u,n),t,b} + \sum_{u\in\mathcal{I}^{N}(n)} y_{(u,n),t,b} + \alpha_{nt} = \sum_{u\in\mathcal{O}^{\mathcal{G}}(n)} q_{(n,u),t,b} + \sum_{u\in\mathcal{O}^{\mathcal{N}}(n)} y_{(n,u),t,b} \ \forall n\in\mathcal{N}^{p}, t\in\mathcal{T}, b\in\bar{\mathcal{R}}(t) \tag{19.1d}$$

$$\sum_{g\in\mathcal{G}(i)} p_{gtb} + \sum_{j\in\mathcal{N}^{g}(i)} p^{h}_{jtb} + \sum_{l\in\mathcal{L}^{out}(i)} f_{ltb} - \sum_{l\in\mathcal{L}^{in}(i)} f_{ltb} + p^{LS}_{itb} - p^{OG}_{itb} = p^{d}_{itb} \forall i\in\mathcal{B}, t\in\mathcal{T}, b\in\bar{\mathcal{R}}(t) \tag{19.1e}$$

$$f_{ltb} = B_{lt} \left(\theta_{s(l)tb} - \theta_{r(l)tb} \right) \ \forall l\in\mathcal{L}, t\in\mathcal{T}, b\in\bar{\mathcal{R}}(t) \tag{19.1f}$$

$$-\overline{f}_{lt} \le f_{ltb} \le \overline{f}_{lt} \quad \forall l\in\mathcal{L}, t\in\mathcal{T}, b\in\bar{\mathcal{R}}(t) \tag{19.1g}$$

$$p^{h}_{jtb} = \delta_{j} \sum_{u\in\mathcal{O}^{G}(j)} q_{(j,u),t,b} \quad \forall j\in\mathcal{N}^{g}, t\in\mathcal{T}, b\in\bar{\mathcal{R}}(t) \tag{19.1h}$$

$$0 \le \sum_{n\in\mathcal{N}^{\mathcal{T}}(u)} q_{(u,n),t,b} \le \bar{q}_{u} \quad \forall u\in\mathcal{N}^{g}, t\in\mathcal{T}, b\in\bar{\mathcal{R}}(t) \tag{19.1i}$$

$$\mathcal{V}^{end} \le \sum_{e\in\mathcal{N}^{a}} \left(\varphi_{e} w_{eT} \right) \tag{19.1j}$$

$$0 \le q_{(u,n),t,b} \le \bar{q}_{(u,n)} \quad \forall u\in\mathcal{N}, n\in\mathcal{N}^{G}(u), t\in\mathcal{T}, b\in\bar{\mathcal{R}}(t) \tag{19.1k}$$

$$0 \le y_{(u,n),t,b} \le \bar{y}_{(u,n)} \quad \forall u\in\mathcal{N}, n\in\mathcal{N}^{N}(u), t\in\mathcal{T}, b\in\bar{\mathcal{R}}(t) \tag{19.1l}$$

$$0 \le p^{h}_{ntb} \le \bar{p}^{h}_{nt} \quad \forall n\in\mathcal{N}^{G}, t\in\mathcal{T}, b\in\bar{\mathcal{R}}(t) \tag{19.1m}$$

$$\underline{w}_{et} \le w_{et} \le \overline{w}_{et} \quad \forall e \in \mathcal{N}^{a}, t \in \mathcal{T} \tag{19.1n}$$

$$0 \le p_{gtb} \le \overline{p}_{gt} \quad g \in \mathcal{G} \setminus \mathcal{G}^{V}, t \in \mathcal{T}, b \in \bar{\mathcal{R}}(t) \tag{19.1o}$$

$$0 \le p_{gtb} \le \overline{p}_{gt} \xi_{gtb} \quad g \in \mathcal{G}^{V}, t \in \mathcal{T}, b \in \bar{\mathcal{R}}(t) \tag{19.1p}$$

where the objective function (Eq. 19.1a) seeks to minimize the expected value (with respect to the random variable of hydro inflows, $\boldsymbol{\alpha}$) of the total generation operation costs throughout the planning horizon, considering the variable costs of the power plants along with the load shedding and over-generation. Constraints (Eq. 19.1b) represent the dynamics of the reservoirs by coupling spatial and temporal decisions, considering both seepage, inflow, and outflow of both generable and non-generable water of the waterways. Constraints (Eq. 19.1c) model the seepage (slow escape of water through a porous material or small holes) of the reservoirs with a linear approximation regarding the storage variable $\mathbf{w}$, and constraints (Eq. 19.1d) represent the mass balance in all pass nodes (passing and generating). Constraints (Eq. 19.1e) represent the electricity demand-supply constraints. Equations (19.1f) and (19.1g) represent the flow equations in the transmission lines under a DC power flow approximation and the maximum transmission capacity of the lines, respectively. Constraints (Eq. 19.1h) and (Eq. 19.1i) define the power of the hydroelectric plants and the maximum generable flow in each hydroelectric plant, respectively. Constraints (Eq. 19.1j) define the terminating conditions of the model based on a linear function that defines the value of water at the end of the time horizon obtained from an exogenous long-term hydrothermal scheduling model. Finally, constraints (Eq. 19.1k)–(Eq. 19.1p) represent the bounds of each decision variable.

The Model 19.1 lacks some additional constraints. In particular, the constraints

$$\alpha_{nt} = m_{nt} + g_{nt} + \zeta_{nt} \qquad \forall n \in \mathcal{N}, t \in \mathcal{T} \tag{19.1q}$$

$$\zeta_{nt} = \sum_{k \in \mathcal{N}} \sum_{l=1}^{L} \Phi_{nk}^{l} \zeta_{k,t-l} + \epsilon_{nt} \quad \forall n \in \mathcal{N}, t \in \mathcal{T} \tag{19.1r}$$

correspond to a vector autoregressive (VAR) model of order L that represents the uncertain hydro inflows, considering an additive decomposition to the inflows time series, preserving linearity. This decomposition separates the water inflow of the node n additively at the stage t into a trend (m_{nt}), a seasonality (g_{nt}), and a residual component (ζ_{nt}). Then, since this residual component is typically stationary (and distributed as Gaussian), it can be modeled with an autoregressive model that allows capturing the temporal relationship between the data. In addition, to incorporate the spatial correlations, a VAR derivation was considered. With this, a water inflow in a node n and a stage t of the model is defined as $\alpha_{nt} = m_{nt} + g_{nt} + \zeta_{nt}$. Trend and seasonality were fitted using a Theil-Sen estimator and a cosine-sine model, respectively, based on the works of Favereau et al. (2022) and Favereau et al. (2023). In addition, Φ_{nk}^{l} is the covariance between the water nodes n at time t and k l steps before, and ϵ_{nt} is an independent and identically distributed (iid) random variable (considered as a multivariate Gaussian in this work).

There exist two main basins in Chile concerning hydroelectric generation and agriculture: Laja and Maule. Although the normative and operational description is presented in detail in Sect. 19.3.1, we consider a linear approximation of these highly nonlinear and nonconvex constraints for Model 19.1.

19.2.2 Water Resources Model

This section presents the water resources simulation model based on Pywr, the library developed by Tomlinson et al. (2020), which is used to study the flows of the water network in an agricultural area where the reservoir resources are shared with a large-scale power system. Compared to hydrothermal scheduling, an attractive characteristic of this model is that it incorporates integer variables to represent irrigation agreements.

Pywr is an open-source Python dynamic modeling library for water resource networks, whose main structure is a directed graph of nodes (vertices) connected by links (edges). The nodes are the components in which the data and constraints are assigned to reallocate the water resources when solving it. Within the set of nodes, the main ones are the base of the model: input, output, storage, and link. The water resources must enter through an input node and leave through an output node, where the state variable associated with them is the flow that goes through the nodes. In this way, the set of feasible routes is defined as any combination between the input and output nodes. Loops can be formed in the network, leading to more complex network structures, such as bi-directional links between nodes. Every route in the network must comply that every node must be on at least one path, all routes must start with an input node, and all routes must end with an output node.

The input, output, and link nodes share properties, such as minimum flow, maximum flow, and cost. Maximum and minimum flow properties constrain the flow through the node, while the cost is not necessarily monetized. Rather, it is a penalty incurred due to the amount of net flow through it. On the other hand, storage nodes differ from the rest of the nodes since they are made up of at least one input node and one output node, in addition to having properties of both minimum and maximum volumes. The state variable associated with this type of node is the current volume stored in the reservoir, whose update occurs at the end of the period. Storage nodes can also have costs, which are applied to the underlying input and output nodes.

To solve the problem of determining the allocation of water resources, Pywr considers a path-based allocation algorithm. There is also an edge-based allocation algorithm declared in Tomlinson et al. (2020) as an option to solve the problem, but it was observed that the former has marginally better performance than the latter. Both approaches have similar mathematical structures aiming to achieve each node's properties. The path-based allocation algorithm is described by Model 19.2:

$$\min \sum_{p \in \mathcal{P}} c_p x_p \tag{19.2a}$$

s.t.

$$c_p = \sum_{n \in p} c_n \quad \forall p \in \mathcal{P} \tag{19.2b}$$

$$A_n^{\mathrm{f}} = \sum\nolimits_{p \in \mathcal{P} | n \in p} x_p \quad \forall n \in \mathcal{N} \tag{19.2c}$$

$$a_n^{\mathrm{f}} \leq A_n^{\mathrm{f}} \leq b_n^{\mathrm{f}} \quad \forall n \in \mathcal{N} \tag{19.2d}$$

$$A_d = \sum\nolimits_{p \in \mathcal{P} | o_c \in p} x_p - \sum\nolimits_{p \in \mathcal{P} | i_c \in p} \frac{x_p}{f_{i_c}} = 0 \quad \forall (o_c, i_c) \in \mathcal{C} \tag{19.2e}$$

$$A_n^{\mathrm{s}} = \sum\nolimits_{p \in \mathcal{P} | n \in p \forall n \in O_n} x_p - \sum\nolimits_{p \in \mathcal{P} | n \in p \forall n \in I_n} x_p \quad \forall n \in \mathcal{S} \tag{19.2f}$$

$$a_n^{\mathrm{s}} \leq A_n^{\mathrm{s}} \leq b_n^{\mathrm{s}} \quad \forall n \in S \tag{19.2g}$$

$$a_n^{\mathrm{s}} = \frac{V_n^{\min} - V_n}{\Delta t} \quad \forall n \in \mathcal{S} \tag{19.2h}$$

$$b_n^{\mathrm{s}} = \frac{V_n^{\max} - V_n}{\Delta t} \quad \forall n \in \mathcal{S} \tag{19.2i}$$

where the objective function (Eq. 19.2a) consists of minimizing the total cost of the flows x_p of each path (or route) $p \in \mathcal{P}$ within the network, where the cost of each path p corresponds to the sum over all the costs associated with the nodes n within the path (Eq. 19.2b). Concerning the other constraints, Eqs. (19.2c) and (19.2d) ensure that the total flow of a certain node n (A_n^{f}) must be equal to the sum over all the paths that contain that node, where a_n^{f} and b_n^{f} correspond to the minimum and maximum flow, respectively. A domain is designated to each node to group common nodes such that the same resource flows through paths between nodes in the same domain. In Eq. (19.2e), the flow between domains must equal 0 to ensure inter-domain mass balance, where o_c and i_c are the output and input node for each of the inter-domain connections, respectively. Also, here $\mathcal{C}$ and f_{i_c} are the set of all edges and a conversion factor, respectively. In the constraints Eq. (19.2f), the net flow in a reservoir node (A_n^{s}) corresponds to the difference between the total flow of input (I_n) and output (O_n) nodes. Additionally, A_n^{s} is constrained by the minimum (a_n^{s}) and maximum (b_n^{s}) flows in constraints Eq. (19.2g). These limits are defined by Eqs. (19.2h) and (19.2i), where the minimum ($V_n^{\min}$) and maximum ($V_n^{\max}$) volumes are combined with the current one (V_n).

19.2.3 Integration of the Hydrothermal Scheduling and Water Resources Simulation Models

To integrate the hydrothermal scheduling model with the water resources simulation model, two considerations must be taken. First, the policies obtained from both models will be evaluated on the same hydro inflows, represented by four exceedance probabilities (detailed in Sect. 19.3.2), based on the statistical history of inflow observations. These four scenarios correspond to exceedance probabilities of 25% (wet), 50% (medium), 75% (dry), and 95% (very dry) of hydro inflows, based on the statistical history of inflow observations. Both models consider the same water network, with the difference that the water resources simulation model uses the actual irrigation agreements, and the hydrothermal scheduling model uses a linear approximation. Actual irrigation agreements are presented in Sect. 19.3.1. Second, the results of both models will be mainly related to the intensity of CO_2 emissions, the power system's operational cost, and the agricultural system's water deficit. Figure 19.1 depicts a scheme that summarizes this integration, which will be explained in more detail in Sect. 19.3.1.

The intensity of the emissions of the power system in stage t (ϱ_t), is estimated as the sum of the products between (a) the total generation produced by each technology g in stage t ($\varsigma_{gt}^{\text{gen}}$), and (b) the emission factor of the total life cycle of the technology g (ς_g^{lc}), divided by the total demand of the stage t of the power system (ς_t^{dem}). This expression is measured in kg CO_2 eq/MWh and summarized as

$$\varrho_t = \frac{\sum_{g \in \mathcal{G}} \varsigma_{gt}^{\text{gen}} \cdot \varsigma_g^{\text{lc}}}{\varsigma_t^{\text{dem}}} \quad \forall t \in \mathcal{T}. \tag{19.3}$$

In particular, let us define

$$\varsigma_t^{\text{dem}} = \sum_{i \in \mathcal{B}} \sum_{b \in \mathcal{R}(t)} p_{itb}^{\text{d}} \quad \forall t \in \mathcal{T} \tag{19.3a}$$

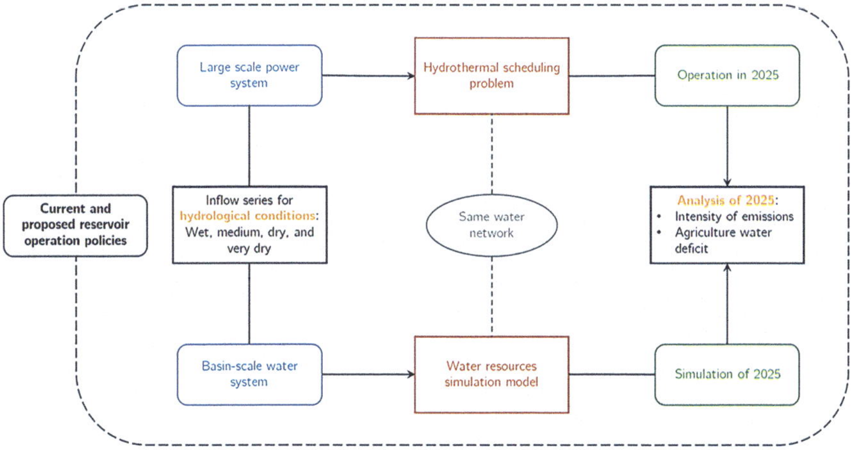

Fig. 19.1 Scheme of integration of the hydrothermal scheduling and the water resources simulation models. (Author's own elaboration)

$$\varsigma_{gt}^{\text{gen}} = \begin{cases} \sum_{b \in \bar{\mathcal{R}}(t)} p_{gtb} & \text{if } g \notin \mathcal{G}^{\text{V}} \\ \sum_{b \in \bar{\mathcal{R}}(t)} p_{gtb}^{\text{h}} & \text{if } g \in \mathcal{G}^{\text{V}} \end{cases} \quad \forall g \in \mathcal{G}, t \in \mathcal{T} \tag{19.3b}$$

The emission factors of the total life cycle of greenhouse gas (GHG) emissions for the different generation technologies were obtained from the database of Nicholson and Heath (2021), from which the maximum values of the total life cycle of GHGs were considered. Table 19.1 summarizes these values.

One criterion to evaluate the performance of the agricultural system is the water deficit in agricultural extraction site a in stage t ($\varsigma_{gt}^{\text{def}}$). This deficit corresponds to the difference between the consumptive water right of the agricultural irrigator a in the stage t ($\varsigma_{at}^{\text{right}}$) and the flow assigned to the agricultural extraction site a in the stage t ($\varsigma_{at}^{\text{ext}}$), measured in hm^3, as shown by

$$\varsigma_{at}^{\text{def}} = \varsigma_{at}^{\text{right}} - \varsigma_{at}^{\text{ext}} \quad \forall a \in \mathcal{A}, t \in \mathcal{T}. \tag{19.4}$$

In this way, the total deficit in stage t ($\varsigma_t^{\text{totdef}}$), measured in hm^3, can be summarized by

$$\varsigma_t^{\text{totdef}} = \sum_{a \in \mathcal{A}^{*}} \varsigma_{at}^{\text{def}} \quad t \in \mathcal{T}. \tag{19.5}$$

One last indicator that will be used to analyze the performance of the water resources simulation model will be the agricultural system's reliability (ς^{rel}), a performance indicator proposed by Hashimoto et al. (1982). This criterion is defined as the relative time in which failures do not occur in the system. For this application, a failure in stage t ($\varsigma_t^{\text{fail}} = \sum_{a \in \mathcal{A}} f_{at}$) is the event in which water deficits occur, where f_{at} is mathematically defined as

Table 19.1 Total life cycle maximum emissions factors for power generation technologies

Generation technology	Maximum emission factor ($\text{kg } CO_2$ eq/MWh)
Biomass	1300
Solar-PV	226
Geothermal	245
Hydroelectric (reservoir)	43
Hydroelectric (run-of-river)	75
Wind	81
Natural gas	988
Diesel	1170
Coal	1689

Nicholson and Heath (2021)

$$f_{at} = \begin{cases} 1 & \text{if } \varsigma_{gt}^{\text{def}} > 0 \\ 0 & \text{if } \varsigma_{gt}^{\text{def}} = 0 \end{cases} \quad \forall a \in \mathcal{A}, t \in \mathcal{T}. \tag{19.6}$$

In this way, the following equation defines the reliability of the agricultural system.

$$\varsigma^{\text{rel}} = 1 - \frac{\sum_{t \in \mathcal{T}} \varsigma_t^{\text{fail}}}{|\mathcal{T}|}. \tag{19.7}$$

19.3 Computational Experiments

In this section, we analyze the results of extensive computational experiments to assess the performance of both power and agriculture systems based on the hydrothermal scheduling and water resources simulation models. Both models are implemented in an Intel®Xeon®CPU E5-2630 v4 with 64 GB RAM server. The stochastic hydrothermal scheduling model was implemented with the SDDP.jl package (Dowson & Kapelevich, 2021) in the Julia language (JuMP environment), and solved with Gurobi 0.9.5.

19.3.1 Case Study

The case study corresponds to the Chilean electric power system with current data obtained from the power system operator and the projected future information based on a carbon neutrality scenario of the Long-term Energy Planning study (for 2023–2027) of the Ministry of Energy (Barría et al., 2021). The case study considers 64 (monthly) stages, 228 buses, 331 transmission lines, 1019 power plants, 210 hydro nodes, and 410 hydro connections. It is important to highlight that this representation considers the entire real hydraulic network: the same number of reservoirs, hydro connections, and hydro nodes of seven basins.

The Chilean power system contains 12 reservoirs, 2 of which are operated in accordance with the decisions of the system operators and the agreements that define the availability of water delivered to run-of-river plants and to neighboring local farmers. One of these agreements is the one promulgated in 1958 between the Hydraulic Works Department (DOH) and Chile's National Electricity Company (ENDESA, 1958), which defines the management of Laja Lake, which provides water for hydroelectric generation and irrigation for agriculture. The last update of the agreement was promulgated by ENDESA (2017), which contemplates modifications to carry out a better distribution of water resources. In this agreement, different extraction rules are defined in four operational strata of the lake (Fig. 19.2):

- Upper storage: The maximum volume for irrigation and generation is 25% and 65% of the volume at the beginning of the season, respectively. The remaining 10% is allocated to the lake. In addition, a maximum extraction for generation is established as 1200 hm^3.
- Intermediate storage: The maximum volume for irrigation and generation is 40% of the volume at the beginning of the season for each, while the remaining 20% goes to the lake.
- Transition storage: The maximum extraction volume for irrigation is 40% of the volume at the beginning of the season and 5% for generation. The remaining 55% is used to improve the lake's condition.
- Lower storage: Here, 570 hm^3 are exclusively devoted to irrigation and 30 hm^3 for mixed-use, while the maximum volume for generation will be equal to 5% of the volume at the beginning of the season.

The irrigation season is defined between December 1 and April 30, which is the period wherein the volume extractions are specified in the agreement. For this reason, the volume at the beginning of the season corresponds to the water level observed or calculated in Laja Lake on November 30 of each year, from which the extraction volumes for irrigation and electricity generation are estimated.

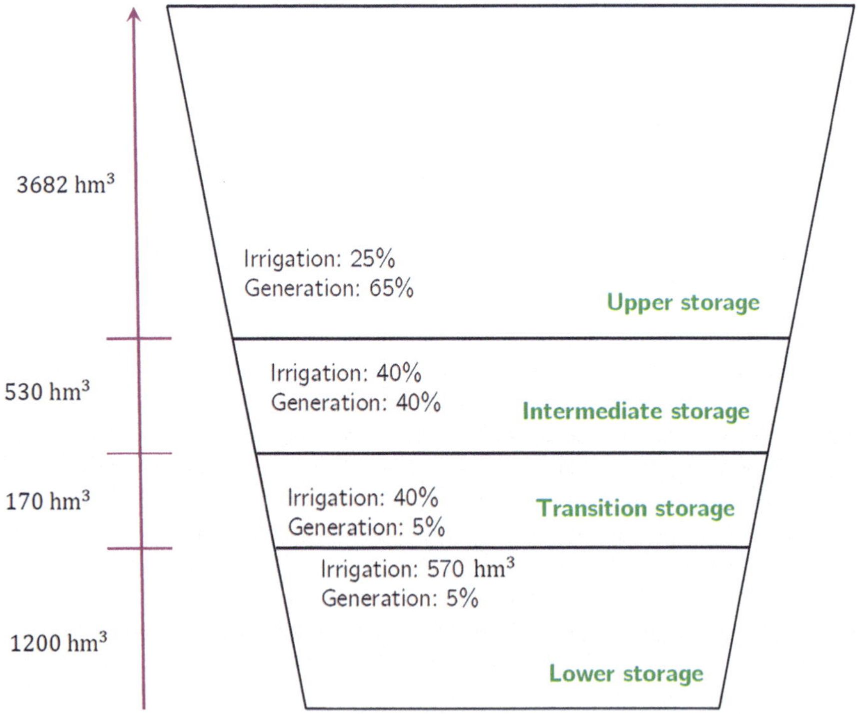

Fig. 19.2 Operational strata of Laja Lake, based on ENDESA agreement of 2017

Mathematically, the operating rules form a piecewise linear function where a constant amount of volume is added to a percentage of the available volume of the respective storage zone. This percentage varies and is distributed between irrigation, generation, and the lake in each reservoir storage zone. The expressions for irrigation ($V_{\text{irrig}}^{\max}$) and generation ($V_{\text{gen}}^{\max}$) are summarized in the Eqs. (19.8a) and (19.8b), respectively, based on $V_{30_{\text{nov}}}$ (which corresponds to the initial volume at the beginning of the season that is stipulated in the agreement between the DOH and ENDESA).

$$V_{\text{irrig}}^{\max}\left(V_{30_{\text{nov}}}\right)=\begin{cases} 600 & \text{if} \quad V_{30_{\text{nov}}} \le 1200\text{hm}^3 \\ 600+0.4\left(V_{30_{\text{nov}}}-1200\right) & \text{if} \quad 1200 \le V_{30_{\text{nov}}} \le 1370\text{hm}^3 \\ 668+0.4\left(V_{30_{\text{nov}}}-1370\right) & \text{if} \quad 1370 \le V_{30_{\text{nov}}} \le 1900\text{hm}^3 \\ 880+0.25\left(V_{30_{\text{nov}}}-1900\right) & \text{if} \quad 1900 \le V_{30_{\text{nov}}} \le 5582\text{hm}^3 \end{cases} \tag{19.8a}$$

$$V_{\text{gen}}^{\max}\left(V_{30_{\text{nov}}}\right)=\begin{cases} 0.05 \cdot V_{30_{\text{nov}}} & \text{if} \quad V_{30_{\text{nov}}} \le 1200\text{hm}^3 \\ 60+0.05\left(V_{30_{\text{nov}}}-1200\right) & \text{if} \quad 1200 \le V_{30_{\text{nov}}} \le 1370\text{hm}^3 \\ 68.5+0.4\left(V_{30_{\text{nov}}}-1370\right) & \text{if} \quad 1370 \le V_{30_{\text{nov}}} \le 1900\text{hm}^3 \\ 280.5+0.65\left(V_{30_{\text{nov}}}-1900\right) & \text{if} \quad 1900 \le V_{30_{\text{nov}}} \le 5582\text{hm}^3 \end{cases} \tag{19.8b}$$

According to ENDESA (1958), some water rights are established in the Laja river basin for irrigators in the area according to:

- First irrigators: Its maximum irrigation demand is established at 90 m^3/s, adjusted monthly by a seasonal variation curve. There are two irrigation withdrawal points: the Zañartu-Collao channel, corresponding to 37.2% of the rights, and a series of channels in Tucapel that comprise the remaining 62.8%.
- Second irrigators: The maximum irrigation demand is 53 m^3/s, adjusted monthly by a seasonal variation curve. The withdrawal points correspond to the Laja-Diguillín channel, with a maximum demand of 40 m^3/s, and the Laja Sur channel, with a maximum of 13 m^3/s.
- Saltos del Laja: The maximum irrigation demand is 7 m^3/s, adjusted monthly by a seasonal variation curve to supply the Saltos del Laja, particularly during the summer.

The values of the consumptive rights assigned to irrigators in the Laja river basin are shown in Table 19.2, adjusted monthly to the seasonal variation curve.

The Chilean power system operator has connectivity schemes for the main basins of the Chilean power system, including the Laja River basin. This diagram is used to represent the water, electrical, and agricultural systems of the Laja river basin in

Table 19.2 Consumptive irrigation rights in m^3/s for users of the Laja River basin according to the ENDESA agreement of 1958

Month	First irrigators	Second irrigators	Saltos del Laja
January	90	53	7
February	90	42.4	7
March	90	26.5	0
April	90	10.6	0
May	0	0	0
June	0	0	0
July	0	0	0
August	0	0	0
September	90	15.9	0
October	90	34.5	0
November	90	45	0
December	90	53	3.5

the water resources simulation model, incorporating elements such as the hydro inflows, the lake, the hydroelectric generators, the intakes, the irrigators, the seepage of the lake, among others.

To calibrate and validate the water resources simulation model for the Laja basin, historical data of the Laja Lake level from 1985 to 2023, publicly available by the National Energy Commission, was used. Figure 19.3 compares the simulated volume between 1960–2019 and the historical volume between 1985–2023. As can be seen, the reservoir adapts well, evidencing a correlation that validates the use of the water resources simulation model, which can represent the reservoir's hydrological interactions and operating policies. A coefficient of determination of 0.87 between both curves was obtained, a value high enough to determine that the model calibration was accurate.

19.3.2 Sensitivity Analysis on the Exceedance Probability of Inflows

Concerning the experiments carried out under the four hydrological scenarios proposed for the hydro inflows, this section will show the results obtained for the hydrothermal scheduling model and the water resources simulation model for the year 2025. These four scenarios correspond to exceedance probabilities of 25% (wet), 50% (medium), 75% (dry), and 95% (very dry) of hydro inflows, based on the statistical history of inflow observations. It should be mentioned that, from now on, the results presented will correspond to those of the base case to later compare them with those obtained in Sect.19.3.3. The base case considers the current irrigation agreements (Fig. 19.2).

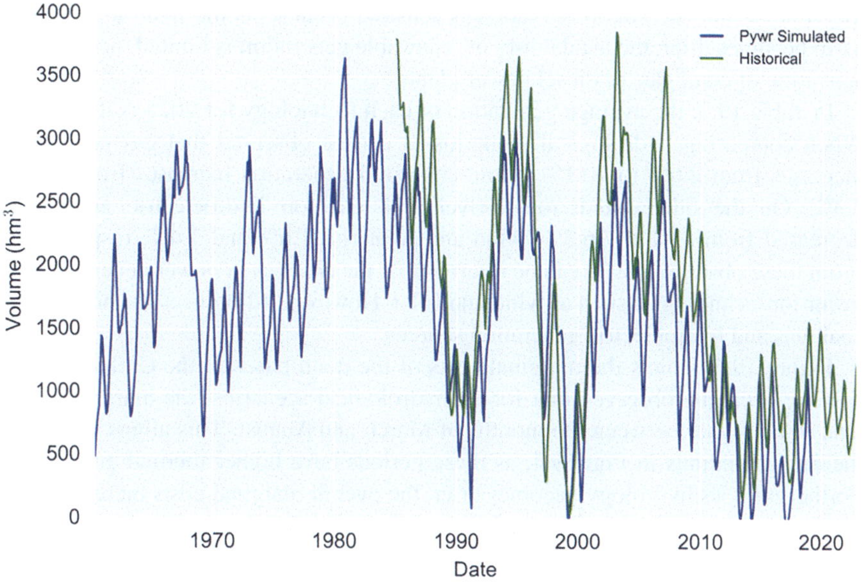

Fig. 19.3 Comparison between the historical levels of Laja Lake with the values obtained in the water resources simulation model

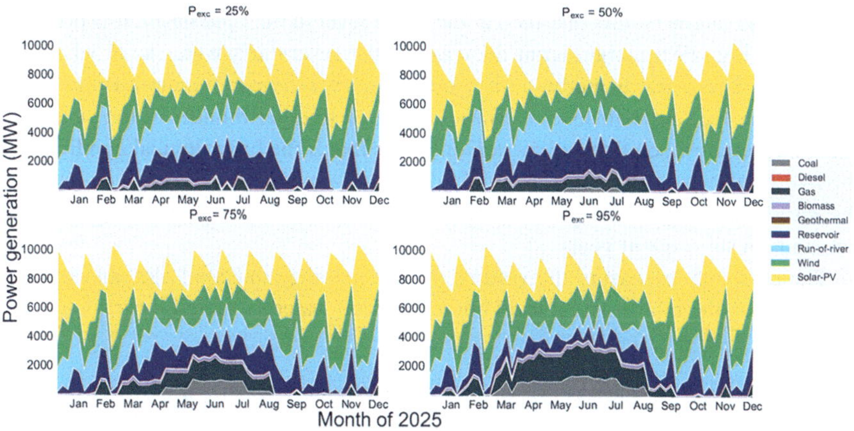

Fig. 19.4 Monthly systemic electricity generation by technology for 2025 in the base case

Figure 19.4 depicts the (load block-based) monthly power generation profile of the entire Chilean power system for 2025 for each hydrological scenario. Here, it is possible to note that the highest thermal generation occurs during February and September. Moreover, gas generation becomes more significant as the exceedance probability increases, specifically for 75% and 95% exceedance probabilities. Additionally, it can be observed that coal-fired generation occurs mainly after April for these scenarios. This observation is sensible because as the hydrological scenario becomes drier, the availability of renewable generation is limited, needing the activation of coal, gas, or diesel power plants.

In Table 19.3, the average generation of each technology for 2025 is illustrated. When comparing hydrological scenarios, it can be observed that gas generation increases from 2.6% to 10.1%, while coal-fired generation increases from 0% to 5.3%. On the other hand, run-of-river and reservoir hydroelectric generation decreased from 43% to 26.3%; wind and solar rise 1.6% and 1.4%, respectively. From these observations, it can be inferred that the changes in power generation are lower for technologies such as wind and solar. However, differences are notable for coal, gas and hydroelectric generation sources.

Figure 19.5 depicts the marginal costs of the main buses in the Chilean power system. It can be observed that for all hydrological scenarios, the marginal costs tend to be higher between the months of March and August. This aligns well with the earlier findings in Fig. 19.4, as these periods have higher thermal generation. Furthermore, as hydrology becomes drier, the overall marginal costs increase. This is because the technology supplying an additional MWh to the power system becomes more expensive. A clear example of this variation is the Puerto Montt bus, which has a marginal cost near 0 in the wet scenario due to its high-water availability (the Puerto Montt bus is attached mainly to hydroelectric power plants). However, in the other scenarios, its marginal cost rises, reaching a maximum value of almost 100 USD/MWh in the driest scenario.

Focusing on the results obtained at the zonal level in the Laja basin, it is possible to note in Fig. 19.6 a significant decrease in the reservoir's water level when the hydrological scenario is medium, dry, and very dry. Between the months of March and August, the reservoir level is lower compared to other months of the year. The current reservoir operation policy demonstrates that if the hydrological condition of the hydro inflows becomes increasingly dry, even a medium scenario is enough for the reservoir to significantly reduce its water volume and possibly fail to meet the demands in subsequent years.

The effect of decreasing the volume of water storage in Laja Lake not only has an impact on the hydrological level but also on the water extracted for irrigators. In

Table 19.3 Average percentage of generation of each technology for 2025 in the base case

Technology	$p_{exc} = 25\%$	$p_{exc} = 50\%$	$p_{exc} = 75\%$	$p_{exc} = 95\%$
Coal	0.0	0.5	2.7	5.3
Gas	2.6	4.3	6.2	10.1
Biomass	1.3	1.7	1.9	2.1
Reservoir	16.6	15.5	13.7	10.7
Run-of-river	26.4	23.5	20.1	15.6
Geothermal	0.5	0.5	0.5	0.5
Wind	21.1	21.9	22.4	22.7
Diesel	0.1	0.1	0.1	0.1
Solar-PV	31.4	32.0	32.4	32.8

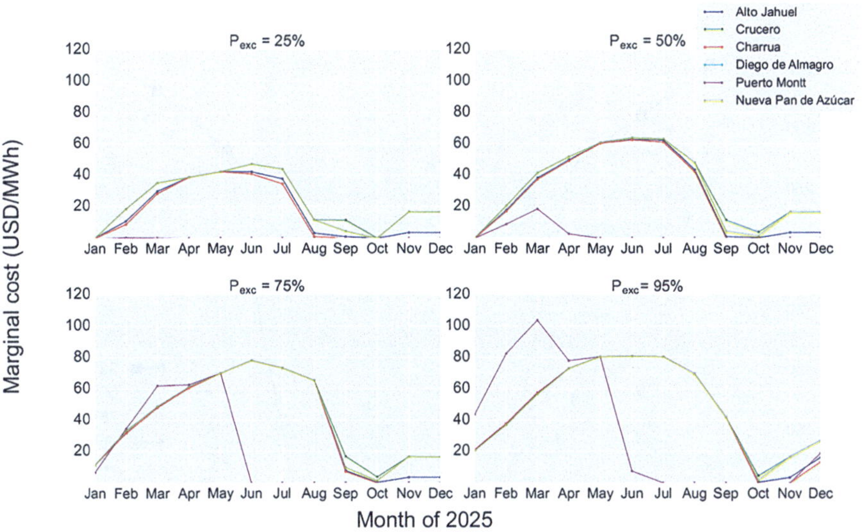

Fig. 19.5 Marginal cost of the main buses of the Chilean power system for 2025 in the base case

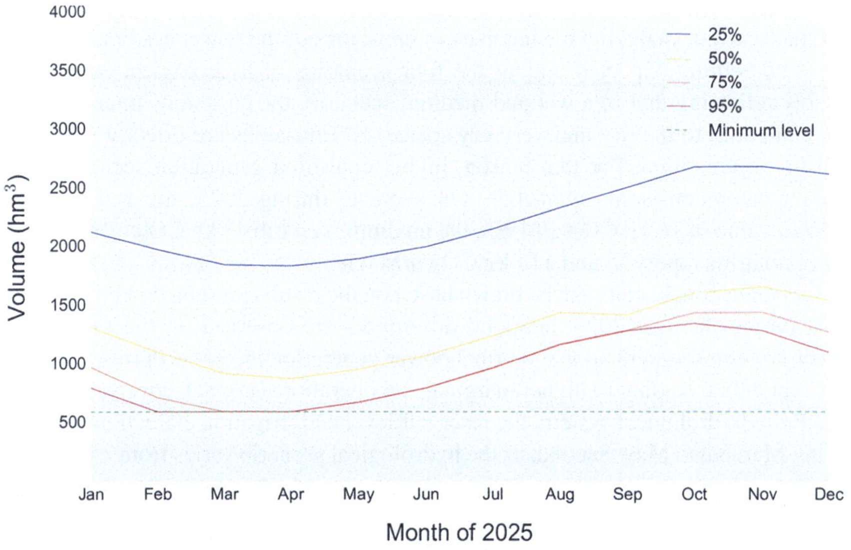

Fig. 19.6 Volume of Laja Lake for 2025 in the base case

Fig. 19.7, it is shown that the water extracted for irrigators in the Laja area also decreases when the hydrological condition is dry or very dry. In addition, the monthly water deficit for the dry scenario reaches almost 100 Mm^3 between March and April. Meanwhile, a very dry scenario hits over 200 Mm^3 in the worst month (March). Consequently, reliability for dry and very dry scenarios is 87% and 79%, respectively.

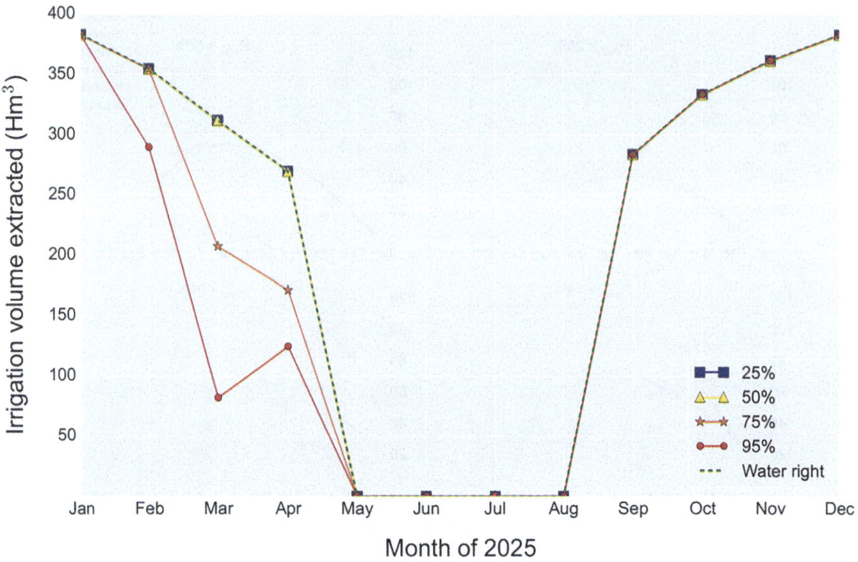

Fig. 19.7 Total irrigation extracted from agriculture for 2025 in the base case

Finally, when analyzing the intensity of emissions of the power system, Fig. 19.8 shows something expected among the four scenarios of hydrological conditions, notably reflecting that in a wet and medium scenario, the emissions intensities are low compared to the dry and very dry scenarios. Emissions are directly related to coal-fired generation. For that reason, higher coal-fired generation scenarios will reach greater emissions intensities. On average, during 2025, the wet scenario shows a value of 36 kg CO_2eq/MWh, the medium scenario 47 kg CO_2eq/MWh, and drier scenarios report 75 and 115 kg CO_2eq/MWh.

Two main conclusions can be drawn based on the results presented in this section. First, during the year 2025, temporal differences are observed. In the case of the power system, it is evident that thermal power generation increases during February and September, leading to higher marginal and operational costs. Conversely, for the Laja basin hydrological system, the reservoir level and irrigation extraction decrease during March and May. Second, as the hydrological scenario varies from wet to very dry, thermal power generation increases, resulting in higher emissions intensity in the power system. At the basin level, the reservoir level significantly decreases, leading to a deficit in irrigation and a decrease in the reliability of the agricultural system.

19.3.3 Analysis of the Proposed Reservoir Operation Policy

To carry out a sensitivity analysis of the irrigation agreements, in terms of the priority of the reservoir users, the percentage that defines the maximum volume of irrigation extraction for the transition, intermediate, and upper storage strata is proposed

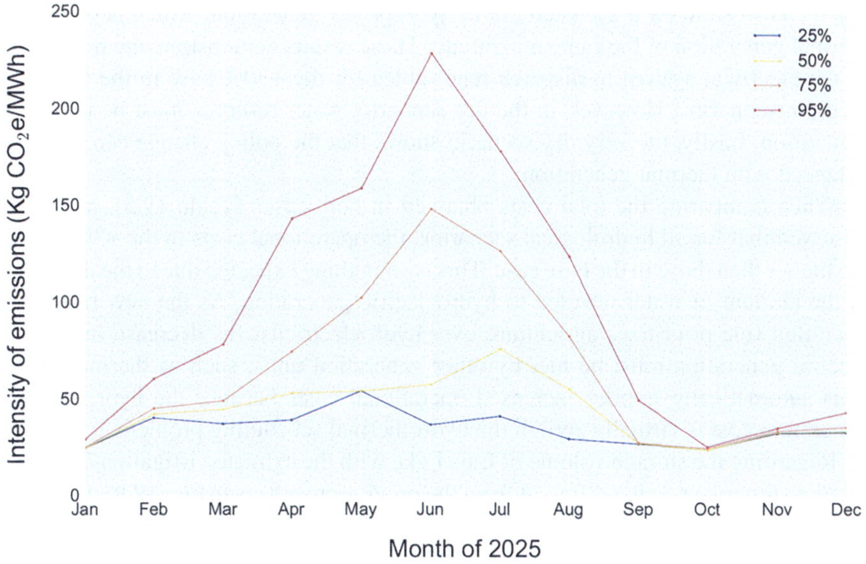

Fig. 19.8 Intensity of emissions of the power system for 2025 in the base case

Table 19.4 Differences (percentage) in average power generation for each technology between the base case and the +30% case during 2025

	Scenario			
Technology	Wet (25%)	Medium (50%)	Dry (75%)	Very dry (95%)
Coal	0.0	0.1	0.2	0.1
Gas	0	0	0.1	0.1
Biomass	0	0	0	0
Reservoir	0	−0.1	0.1	−0.1
Run of river	−0.2	−0.4	−0.3	−0.1
Geothermal	0	0	0	0
Wind	0	0.2	0	0
Diesel	0	0	0	0
Solar PV	0.2	0.2	0	0

to be increased by 30%, respectively. This re-configuration of the reservoir operation rule will be called the +30% case.

Firstly, in terms of generation, Table 19.4 shows the differences in the average generation of each technology between the base case and the +30% case. From this information, it can be observed that in the wet scenario, when the reservoir operation policy is modified, run-of-river generation increases; however, this is compensated by solar generation. In the medium scenario, there is an average decrease of 0.5% in hydropower generation, which is balanced by wind, solar, and coal generation. In the dry scenario, there is an increase in reservoir generation, a decrease in run-of-river generation, and a 0.3% thermal generation increase. Finally, the very dry scenario shows a 0.2% decrease in hydropower generation, which is offset by thermal generation of the same magnitude. These results demonstrate the flexibility of the electrical system to dispatch renewables for the +30% case in the wet and medium scenarios. However, in the dry scenario, water reserves must be used for generation. Lastly, the very dry scenario shows that the policy change can only be balanced with thermal generation.

When comparing the total costs obtained in both cases (Table 19.5), it can be observed that for all hydrological scenarios, the operational costs in the +30% case are higher than those in the base case. This is something expected due to the decrease in the amount of water devoted to hydroelectric generation. As the new reservoir operating rule prioritizes agriculture over hydroelectricity, the decrease in hydroelectric generation must be met by other generation units, such as thermal ones. This automatically implies increased operational costs because the hydroelectric generation cost is virtually zero in the hydrothermal scheduling problem.

Regarding the storage volume of Laja Lake with the extracted irrigation flow, the relationship is somewhat different from the previous results (see Fig. 19.9). It can be observed that increasing water availability for agriculture leads to higher volumes in the Laja reservoir, indicating that the reservoir can store more during the year compared to the actual policy. Additionally, there is a slight increase in water extraction for irrigation in the dry and very dry scenarios.

The aforementioned is also reflected in Table 19.6, which shows the satisfaction of demand (reliability) for different irrigation points in the Laja basin. These results confirm that increasing water availability for irrigation is expected to result in a higher volume of water extracted from the reservoir for agricultural purposes. Consequently, the agricultural sector can receive its irrigation rights appropriately and improve its reliability by 2% in the dry scenario and 4% in the very dry scenario.

Finally, the main trade-off between the intensity of emissions and water deficit regarding the base case and the +30% case is shown in Fig. 19.10. Particularly, it can be observed that increasing water availability for agriculture reduces the deficit in this sector, most significantly in the driest scenarios (75% and 95%). However, as a trade-off, the average intensity of emissions in the power system does not show linear differences between changes through hydrological scenarios. In Table 19.7, the wet scenario shows no significant average emission differences, while the medium and dry scenarios reflect an almost 2% and 2.6% difference. The very dry scenario shows a smaller difference (0.9%) as it is already a scenario with high

Table 19.5 Comparison of the total operational cost of the power system in the year 2025 between the base case and the +30% case

	Total operational cost (MUSD)		
Hydrological scenario (%)	Base case	+30% case	Relative change (%)
25	122	124	1.6
50	219	225	2.7
75	493	505	2.4
95	1084	1094	0.9

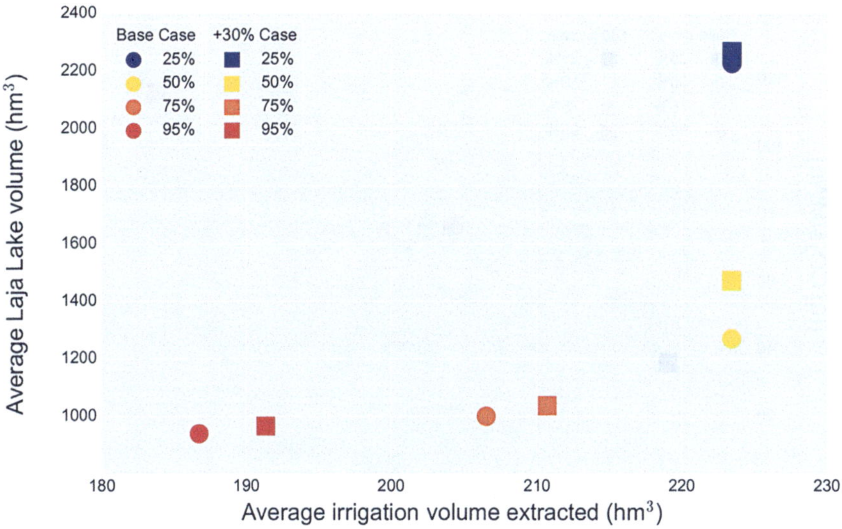

Fig. 19.9 Comparison between the volume of Laja Lake and the irrigation extracted from the basin for the year 2025 between the base case and the +30% case

Table 19.6 Reliability of agricultural water demand of Laja Lake for the year 2025 in the base case and the +30 case

Case	$p_{exd} = 25\%$	$p_{exd} = 50\%$	$p_{exd} = 75\%$	$p_{exd} = 95\%$
Base case	100%	100%	87%	79%
+30% case	100%	100%	89%	83%

thermal generation compared to the others, and the policy change does not significantly impact emissions.

To estimate the cost per cubic meter of the alternative policy implementation, the differences between the base case and the +30% case regarding total operational costs and the total water irrigation deficit are presented in Table 19.8. For wet and medium scenarios, this metric doesn't apply, given that there is no difference between cases in terms of water deficit. For that reason, only drier scenarios were considered. Here, the results show that the cost of implementing the proposed reservoir policy change is equivalent to 0.25 and 0.19 USD/m^3 in the dry and very dry scenarios, respectively. On the one hand, it is evidenced that the implementation is less costly when inflows are considerably low. On the other hand, and as a comparative example, it is noted that for both scenarios, the cost is lower than the deployment of a desalination strategy for irrigation, where trend cost reports that desalinated water is around 1 USD/m^3 (Shokri & Sanavi Fard, 2023).

In summary, findings highlighted the differences between the reservoir operating rules: the base and the +30% cases. Regarding power generation, a decrease in hydroelectric generation and an increase in thermal generation in the driest scenarios were observed. Regarding operational costs, low but non-negligible differences

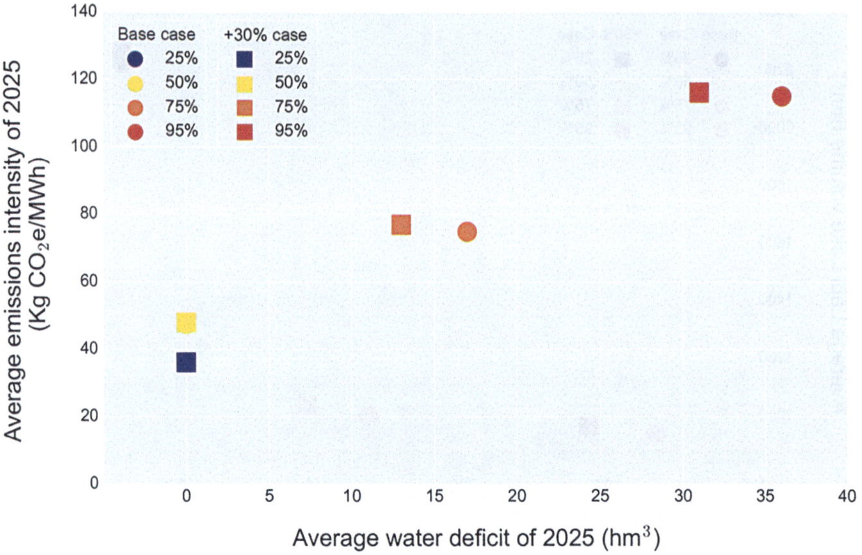

Fig. 19.10 Comparison of the water deficit and intensity of emissions for 2025 between the base and +30% cases

Table 19.7 Comparison of the average emissions intensity of the power system in the year 2025 between the base case and the +30% case

	Average emission intensity (Kg CO_2e/MWh)		
Hydrological scenario (%)	Base case	+30% case	Relative change (%)
25	36	36	0
50	47	48	2
75	75	77	2.6
95	115	116	0.9

Table 19.8 Unit water cost measured of the proposed reservoir operation policy

Hydrological	Operational cost difference	Water deficit difference	Unit water
Scenario	(MUSD)	(mm^3)	Cost (USD/m^3)
25%	–	–	–
50%	–	–	–
75%	12	49	0.25
95%	10	54	0.19

exist, indicating that the reservoir policy change has a non-significant impact due to the power system's flexibility in generation for 2025. At the Laja basin level, a significant improvement in meeting water demand was identified when the reservoir policy was changed, reducing the water deficit for farmers. This improvement is even more noticeable in the drier scenarios, where a deficit is reduced in the best scenario by almost 24%. The cost in terms of emissions to adopt the change will not be challenging because, in the worst case, it will be almost 2 kg CO_2e/MWh higher compared with the current reservoir policy of the reservoir, and the cost of mitigating the water deficit will be lower compared to modern water production alternatives, such as desalination.

19.4 Discussion and Concluding Remarks

This research proposed an integrated approach between a medium-term hydrothermal scheduling model and a detailed water resources model, which allows for evaluating the operation of a multipurpose reservoir whose water storage supports power generation and agriculture activities. The former model is based on multistage stochastic optimization with a detailed representation of the power and water networks. The latter model is based on the Pywr Python package, and it includes a detailed representation of actual irrigation agreements, which results in an accurate representation of the normative constraints on water use. Additionally, a methodology that links the results of these two models of different natures through indicators such as the intensity of emissions from the power system, the water deficit in irrigated areas, and the reliability of the agricultural system is proposed. To look ahead to 2025, this methodology is applied to a multipurpose reservoir in Chile, the Laja Lake, which is located in a basin where the water-energy-food nexus can be addressed, and it has a critical power generation capacity and significant extensions of agricultural land adjacent to the basin's watercourses.

To study the operation that coordinates the use of the reservoir resources, a sensitivity analysis of the hydro inflows of the Chilean hydroelectric system was carried out for four scenarios of wet, medium, dry, and very dry hydrological conditions. The results show that there are differences throughout the year 2025. The first is that for all hydrological scenarios, during February and September, thermal generation is higher than in the other months, which in turn shows an increase in marginal and operational costs in the same period. At the Laja River basin level, it is observed that between March and May, the reservoir level remains at its lowest volume, so irrigation extraction from it is also reduced for all hydrological scenarios. In addition, it is found that as the condition of the springs changes from wet to very dry, the thermal generation, operational costs, and emissions intensity of the power system increase in each scenario. Conversely, reservoir storage and irrigation withdrawals decrease, reflecting an increase in the deficit and a decrease in the reliability of the agricultural system.

A second experiment consisted of modifying the operating (irrigation) agreement of the Laja reservoir to allocate an additional 30% of volume in each operational layer of the Laja reservoir and analyze the same hydrological conditions as before with this new policy. The results showed significant differences between the base and this +30% cases. In terms of power generation, hydroelectricity decreases, and thermal power generation increases its share in all scenarios. However, this accommodation does not significantly increase emissions' intensity in dry and very dry scenarios. The results in the Laja basin are interesting since the modification increases the reservoir levels for all hydrological scenarios, mitigating the deficit and improving agricultural reliability considerably.

Particularly, when comparing the base case with the +30% case in the dry scenario, it is observed that during 2025 the average water deficit is reduced in drier scenarios by a cost in emissions which is around 2 kg CO2eq/MWh, and the cost of reducing irrigation deficit reach lower than 0.3 USD/m3. This highlights the potential for implementing sustainable and cost-effective water management strategies that balance water availability (see Cooley et al., 2019 for an analysis of different options), operational expenses, and emissions. This research provides valuable insights for decision-makers in the field of reservoir water resources management, encouraging the adoption of policies aligned with a net-zero strategy and promoting efficient utilization of water in the context of changing hydrological conditions.

Another interesting point to note is the complementarity of the proposed decision-making and analysis tool with the water scarcity decrees in Chile. The water scarcity zone decrees create a regulatory framework for the State's administrative response in the event of no agreement between users in the face of a scenario of water scarcity in times of extraordinary drought. This regulation's purpose is to redistribute water, delivering it to those who have rights to use it under conditions other than those indicated in their respective titles, allowing the General Directorate of Water Resources (DGA) to omit fundamental regulations existing in the Chilean Water Code temporarily.

Like those mentioned above, intensive droughts can lead to preventive decrees of electrical rationing, which consists of a series of exceptional measures to reduce electricity consumption. In this context, an integrated tool like the one developed in this work could support decision-making and the analysis of the impacts of these exceptional measures, reducing the risks of rationing and minimizing the negative impacts on the population.

Acknowledgments This work was supported by ANID-FONDEF-IT19I0113, ANID-FONDECYT- 1211378, ANID-FSEQ210018, ANID-PFCHA/National Doctorate Program/2019-21190693, the Solar Energy Research Center (SERC-Chile) (ANID-FONDAP-15110019), and the Complex Engineering Systems Institute (ANID-FB0816).

References

Barría, C., Santander, A., Caviedes, Á., Toro, C., Eguiguren, B., Coronado, C., Mancilla, C., et al. (2021). Planificación energética de largo plazo 2023-2027.

Basheer, M., Nechifor, V., Calzadilla, A., Siddig, K., Etichia, M., Whittington, D., Hulme, D., & Harou, J. J. (2021). Collaborative management of the grand Ethiopian renaissance dam increases economic benefits and resilience. *Nature Communications, 12*, 5622.

Benders, J. F. (1962). Partitioning procedures for solving mixed-variables programming problems. *Numerische Mathematik, 4*(1), 238–252.

Bolouri-Yazdeli, Y., Haddad, O. B., Fallah-Mehdipour, E., & Mariño, M. A. (2014). Evaluation of real-time operation rules in reservoir systems operation. *Water Resources Management, 28*(3), 715–729.

Cooley, H., Phurisamban, R., & Gleick, P. (2019). The cost of alternative urban water supply and efficiency options in California. *Environmental Research Communications, 1*, 042001.

Dowson, O., & Kapelevich, L. (2021). SDDP.jl: A Julia package for stochastic dual dynamic programming. *INFORMS Journal on Computing, 33*, 27–33.

ENDESA. (1958). Convenio Ad-Referendum sobre la regulacipon del Río Laja.

ENDESA. (2017). Acuerdo de operación y recuperación del Lago Laja. Complementa Convenio de 1958.

FAO (2014). "The water-energy-food nexus at FAO, concept note." Report no., Author.

Favereau, M., Lorca, A., Negrete-Picentic, M., & Vicuña, S. (2022). Robust streamflow forecasting: A student's t-mixture vector autoregressive model. *Stochastic Environmental Research and Risk Assessment, 36*, 3979–3995.

Favereau, M., Lorca, A., & Negrete-Pincetic, M. (2023). Multistage adaptive robust optimization for the hydrothermal scheduling problem. *Computers Operations Research, 150*, 106051.

Galetovic, A., & Muñoz, C. M. (2009). Estimating deficit probabilities with price responsive demand in contract-based electricity markets. *Energy Policy, 37*(2), 560–569.

Gonzalez, J., Olivares, M., Medellín-Azuara, J., Moreno, R., & Marques, G. (2016). Multipurpose reservoir operation: A tradeoff analysis between hydropower generation and irrigated agriculture using hydro-economic models. In *World Environmental and Water Resources Congress 2016*.

Hashimoto, T., Stedinger, J. R., & Loucks, D. P. (1982). Reliability, resiliency, and vulnerability criteria for water resource system performance evaluation. *Water Resources Research, 18*(1), 14–20.

Hightower, M., & Pierce, S. A. (2009). The energy challenge. *Nature, 452*, 285–286.

Jain, S. (1993). *1-Introduction to reservoir systems*. National Institute of Hydrology.

Jander, V., Vicuña, S., Melo, O., & Lorca, Á. (2023). Adaptation to climate change in basins within the context of the water-energy-food nexus. *Journal of Water Resources Planning and Management, 149*(11), 04023060. https://doi.org/10.1061/JWRMD5.WRENG-5566

Jazmin, Z. S., Reed, P. M., Herman, J. D., Giuliani, M., & Castelletti, A. (2016). A diagnostic assessment of evolutionary algorithms for multi-objective surface water reservoir control. *Advances in Water Resources, 92*, 172–185.

Khadem, M., Dawson, R. J., & Walsh, C. L. (2021). The feasibility of inter-basin water transfers to manage climate risk in England. *Climate Risk Management, 33*, 100322.

Kim, G. J., Kim, Y.-O., & Reed, P. M. (2021). Improving the robustness of reservoir operations with stochastic dynamic programming. *Water Resources Planning and Management, 147*(7), 712–722.

Labadie, J. W. (2004). Optimal operation of multireservoir systems: State-of-the-art review. *Water Resources Planning and Management, 130*(2), 93–111.

Lorca, A., Favereau, M., & Olivares, D. (2020). Challenges in the management of hydroelectric generation in power system operations. *Current Sustainable/Renewable Energy Reports, 7*, 94–99.

Loucks, P. (2000). Sustainable water resources management. *Water International, 25*(1), 3–10.

Matus, M., Sepúlveda, R., Matamala, C., & Benavides, C. (2019). Modeling integrated water resource allocation for agriculture and economical energy production in the Laja Lake basin considering new irrigation policies. In *4th International Conference on Energy and Environment: Bringing Together Engineering and Economics Guimarães*, Portugal (pp. 16–17).

Mower, E., & Miranda, L. E. (2013). Frameworks for amending reservoir water management. *Lake and Reservoir Management, 29*(3), 194–201.

Navarro, A., Favereau, M., Lorca, A., Olivares, D., & Negrete-Pincetic, M. (2023). *Medium-term stochastic hydrothermal scheduling with short-term operational effects for large-scale power and water networks*. SSRN.

Nicholson, S., & Heath, G. (2021). *Life cycle emissions factors for electricity generation technologies*. National Renewable Energy Laboratory-Data.

Pereira, M., & Pinto, L. (1991). Experimental study of methods of scenario lattice construction for stochastic dual dynamic programming. *Mathematical Programming, 52*, 359–375.

Pereira-Bonvallet, E., Püschel-Løvengreen, S., Matus, M., & Moreno, R. (2016). Optimizing hydrothermal scheduling with non-convex irrigation constraints: Case on the Chilean electricity system. In *Energy Procedia, 87, 132–140 5th International Workshop on Hydro Scheduling in Competitive Electricity Markets*.

Ricalde, I., Vicuña, S., Melo, O., Tomlinson, J. E., Harou, J. J., & Characklis, G. (2022). Assessing tradeoffs in the design of climate change adaptation strategies for water utilities in Chile. *Journal of Environmental Management, 302*, 114035.

Shokri, A., & Sanavi Fard, M. (2023). Techno-economic assessment of water desalination: Future outlooks and challenges. *Process Safety and Environmental Protection, 169*, 564–578.

Smajgl, A., Ward, J., & Plushke, L. (2016). The water–food–energy nexus—Realizing a new paradigm. *Journal of Hydrology, 553*, 533–540.

Tomlinson, J., Arnott, J., & Harou, J. (2020). A water resource simulator in python. *Environmental Modelling Software, 126*, 104635.

Zou, J., Ahmed, S., & Sun, X. A. (2019). Stochastic dual dynamic integer programming. *Mathematical Programming, 175*, 461–502.

Chapter 20
Conclusions

Guillermo Donoso and Daniela Rivera

Abstract Chile shows low progress on the three pillars of proactive drought management. Mitigation and response, the most advanced, presents a medium level, drought monitoring and early warning presents a low level of progress, while vulnerability, risk, and impact assessment remains undeveloped. It is necessary to design a strong political strategy that consists of gradually decentralizing drought management, involving those affected by the effects of the drought, generating local capacities and changing the crisis management approach to a preventive risk management model. This chapter provides a comprehensive overview of drought management in Chile drawing from the significant analysis and perspectives offered by all contributors. It indicates that the current strategy primarily employs a crisis management approach. Specifically, the assessment identifies challenges that need to be addressed to transition towards a risk-management approach. Furthermore, it highlights a range of policy instruments that must be considered to address previously inadequately managed challenges. And it provides a global assessment of the legal and institutional framework for drought management in Chile and its challenges, identifying the main lessons that, from comparative experiences, could be reviewed in Chile.

Keywords Drought management · Chile · Proactive approach · Lessons

G. Donoso (✉)
Faculty of Agronomy and Natural Systems and the Center for Water Law and Management, Pontificia Universidad Católica de Chile, Santiago, RM - Santiago, Chile
e-mail: gdonosoh@uc.cl

D. Rivera
Faculty of Law and the Center for Water Law and Management, Pontificia Universidad Católica de Chile, Santiago, RM - Santiago, Chile

D. Rivera, G. Donoso (eds.), *Droughts in Chile*, Global Issues in Water Policy 31,
https://doi.org/10.1007/978-3-031-85040-0_20

20.1 Introduction

Extending over 4200 km, Chile encompasses a wide range of climates, from the arid conditions of the world's driest desert in the north to subpolar environments in the south. Rainfall decreases from south to north. While Chile benefits from substantial water resources in its southern regions, water scarcity in the populous central and northern zones, exacerbated by the ongoing megadrought and climate change, reveals critical vulnerabilities.

The country's climate is influenced by several factors such as the Pacific Ocean, the Humboldt Current, and the Andes Mountains, which play a role in determining temperature, precipitation, and humidity levels. The Pacific Anticyclone, a high-pressure system, has a significant impact on Chile's climate by creating dry conditions and blocking frontal systems during the summer months. Seasonal and interdecadal cycles, like the El Niño–Southern Oscillation (ENSO), considerably affect rainfall patterns; El Niño leads to increased precipitation, while La Niña causes drier conditions. ENSO events notably influence seasonal precipitation patterns, with more pronounced effects observed in the north-central region.

Precipitation in central Chile has exhibited a noticeable downward trend. However, it is challenging to distinguish this trend from the impact of a prolonged drought that has persisted for more than a decade (Garreaud et al., 2020). The variability in precipitation is significant on both annual and seasonal scales. Numerous studies have documented this phenomenon in the latter part of the twentieth century, with potential links to anthropogenic climate change (Boisier et al., 2016, 2018). Projections from CMIP6 models indicate a reduction in precipitation by 3–30% and an increase in temperatures ranging from 1.2 to 3.5 °C by the end of the twenty-first century, which would exacerbate drought conditions.

Drought is characterized as a random and temporary decrease in water availability below normal levels due to climatic variability, affecting ecosystems, agriculture, and human systems. The primary causes of drought include meteorological factors, such as reduced precipitation and the persistence of high-pressure zones. Additionally, large-scale ocean-atmosphere interactions, like the El Niño–Southern Oscillation (ENSO), significantly influence drought patterns in Chile, with La Niña events being associated with drier conditions.

The historical analysis of droughts in Chile reveals the recurrent and severe impacts of these events, with significant implications for agriculture, urban water supply, and energy production (Fernández & Gironás, 2021). While Chile's historical droughts were less severe due to lower population and agricultural pressures, modern droughts—such as the ongoing megadrought—reveal the compounded impacts of climatic variability and increased water demand. Thus, the severity of droughts has increased due to growing water consumption and climatic variability. Drought indices, including the Standardized Precipitation Index (SPI) and the Standardized Runoff Index (SRI), show a concerning trend in the increasing frequency and severity of droughts.

Drought is a slow-onset multifaceted phenomenon which significantly impacts ecosystems, agriculture, and human systems. In addition, its impacts are silent, and its negative effects not only accumulate over long periods of time but can manifest even years after the drought event has ended, all of which are intensifying because of climate change.

Different sectors perceive drought uniquely, leading to varied definitions, such as meteorological (rainfall deficit), agricultural (soil moisture deficiency), hydrological (reduced streamflow), and socioeconomic (imbalance between water demand and supply). The current megadrought is the longest recorded, causing major socioeconomic and ecological effects. Chile's ongoing megadrought since 2009 highlights the severe consequences of prolonged droughts and suggests potential climate shifts toward aridity.

The economic repercussions of prolonged drought periods in Chile have been significant. During the megadrought from 2010 to 2020, total agricultural land decreased by 13.5%, net water use fell by 15.3%, irrigated land reduced by 15.7%, and there was a notable shift towards water-efficient crops. Agricultural economic losses ranged between $90.46 million and $86.76 million, with the Maule region experiencing the highest losses. Northern regions adapted by prioritizing high-value crops such as fruits, while central regions faced greater economic impacts due to their reliance on annual crops. Conversely, urban water supply companies faced increased production and distribution costs, amounting to $1.44 billion over the same period.

Factors such as infrastructure capacity, water storage levels, and adaptive responses influence the extent of these impacts. For example, adaptive investments in urban water supply companies resulted in savings of approximately $8.3 for every $1 spent, thereby mitigating potential welfare losses under urban water rationing scenarios. The analysis demonstrates how prolonged droughts affect both agriculture and urban water supply sectors, emphasizing the necessity for proactive adaptation strategies.

Evidence also shows that droughts have significantly impacted water quality. Degraded water quality affects agriculture, aquaculture, and potable water supplies and increased salinity impacts crop productivity and soil health, while harmful algal blooms threaten aquatic industries. These impacts are driven by hydrochemical changes due to drought, elevated temperatures, and reduced streamflow, leading to nutrient imbalances, pollutant concentration, and altered ecosystem functions.

Coping with drought presents significant challenges due to its multidimensional and complex nature. Nevertheless, Chile must address these issues and manage drought effectively, as water is becoming an increasingly scarce resource essential to both society and the environment. Additionally, Chile is highly vulnerable to climate change and has experienced an exceptionally severe drought over the past 14 years.

Drought management is interdisciplinary in nature, encompassing, among others, hydrological, economic, institutional, legal, environmental, social and political aspects. This diversification has led, in many cases, to a partial treatment of drought problems, with an incomplete analysis of the various challenges that it entails.

20.2 Institutionality, Governance, and Drought Management Policies in Chile

Although droughts occur frequently, Chile's approach tends to treat them as atypical and urgent events, addressing them primarily with reactive measures typically used for disaster or catastrophe management. This approach has generally lacked effectiveness, efficiency, coordination, and timeliness, with limited progress made in mitigating drought-related risks.

The overview of the key administrative authorities and agencies responsible for drought-related functions in Chile, identifies 17 public entities spread across 7 sectors, indicating a significant level of dispersion (Table 20.1). The largest number of entities is concentrated in the sector under the Ministry of the Interior and Public Security. However, the primary tools for managing water resources are located in other areas, specifically those led by the Ministry of Public Works (MOP) and the Ministry of Agriculture (MINAGRI).

These 17 entities of the State Administration identified as those exercising primary competences in drought management in Chile should act in a coordinated manner. However, in practice, there is a series of disconnected and disperse actions, reflecting weak institutional cooperation and fragmentation of governmental roles. One fundamental weakness of the legal framework on droughts is the absence of clear institutional responsibility for decision-making, approval, execution, and monitoring actions, which is logically a consequence of the limited political importance assigned to this issue and the general lack of attention it receives in regulations or legislation.

An analysis of the primary functions associated with droughts shows that general functions are more prevalent than those specifically focused on drought. This indicates that drought is addressed as one of the potential catastrophic or emergency situations requiring intervention by authorities and administrative agencies. The General Directorate of Water (DGA) has the most functions related to droughts. These functions are predominantly reactive, emerging in response to already

Table 20.1 Main sectors, authorities and administrative agencies involved in drought management

Presidency of the Republic	Interior and public safety	Public works	Agriculture	Energy	Defense	Health
– President of the Republic	– SENAPRED – Regional Presidential Delegations – Provincial Presidential Delegations – Regional Governments – Municipalities	– MOP – DGA – DOH	– MINAGRI – CNR – INDAP	– MINEN – CNE	– MINDEF – Armed Forces	– SEREMI of Health

Source: Own elaboration

occurring events. Additionally, there are several instances of shared functions involving multiple administrative authorities or agencies, highlighting the need for better articulation among them to prevent functional overlap or interference.

Notwithstanding the various measures adopted by public agencies, holders of water rights also adopt specific drought measures. The role of supervisory boards is particularly significant, as their legally mandated functions include declaring water scarcity in the natural sources they manage and determining and implementing extraordinary distribution measures as necessary.

Agriculture is one of the most vulnerable sectors to drought. Given these conditions, the Ministry of Agriculture (MINAGRI) provides various support programs for farmers. Historically, during droughts, this support has been implemented once the phenomenon is underway. One of the most significant and practically applicable tools is the declaration of an agricultural emergency due to water deficit.

In this context, despite the frequent occurrence of such phenomena in Chile, addressing drought governance remains challenging. Currently, there is fragmentation, dispersion, and disorder evident in the system. The measures implemented have shown limited effectiveness. These outcomes are contrary to the principles of good governance.

Chile shows low progress on the three pillars of proactive drought management. Mitigation and response, the most advanced, presents a medium level, drought monitoring and early warning presents a low level of progress, while vulnerability, risk, and impact assessment remains undeveloped. This assessment indicates that the current strategy primarily employs a crisis management approach. Specifically, the assessment identifies challenges that need to be addressed to transition towards a risk-management approach. Furthermore, it highlights a range of policy instruments that must be considered to address previously inadequately managed issues.

Although it can be estimated that the peak of the recent drought affecting the country (which continues in some areas) is ending, it is highly probable that another drought will occur in the future. Therefore, institutional and policy reordering is necessary, focusing on risk management and a planned and integrated combination of various preventive and reactive strategies and instruments (Stein et al., 2016; Hurlbert et al., 2019). Risk management alone is insufficient to mitigate the effects of extreme events; it is essential to address equitably the social drivers of their impact (Tellman & Eakin, 2022). Emphasis should be placed on water security, developing and strengthening resilience, and reducing vulnerability to this and other extreme hydrological phenomena, such as floods, considering both their differences and similarities.

20.3 Comparative Experiences in Drought Management

Countries such as Spain, Australia, some in South America, and the state of California in the United States are transitioning from reactive to proactive management approaches, emphasizing the importance of anticipating and mitigating impacts through adaptive strategies.

Historically, as is the case in Chile, South American countries managed drought reactively, focusing on emergency relief rather than addressing root causes. There are recent shifts toward proactive, risk-based approaches emphasize monitoring, vulnerability assessment, and integrated policy frameworks.

An analysis of policy instruments, regulatory tools, and drought management plans of ten countries in South America,[1] showed that from a sectoral perspective, drought policy has been traditionally linked with agricultural and water resources management policy in all the countries. More recently, Climate Change (CC) and Disaster Risk Reduction (DRR) or Disaster Risk Management (DRM) have emerged as policy areas to include drought risk management issues as they provide suitable and comprehensive conceptual frameworks that encompass regulation development for drought management. Given the international and national-led initiatives to promote better planning on these topics, most of the countries of the region have developed new regulations and plans within these thematic policies (Argentina, Brasil, Colombia, Ecuador, Guyana, Peru, Paraguay and Uruguay), except for Venezuela.

Thus, specific drought plans are relatively new in the region but are accompanied by an extensive set of legislation and regulatory tools that provides an adequate policy framework in which drought issues can be integrated and developed.

However, there is still room for improvement as only a few countries present a risk management approach and little improvement has been made to overcome such a reactive perspective. Aligning national policies with international frameworks like the Sendai Framework and the Paris Agreement can strengthen resilience and mitigate the impacts of climate-induced droughts across the region.

Traditionally, drought management in Spain relied on reactive measures like building reservoirs and implementing emergency decrees, as is the case in Chile. However, Spain has shifted toward a risk-based approach emphasizing prevention, mitigation, and adaptation, catalyzed by the European Union's Water Framework Directive (WFD).

This modernized drought management approach considers special drought plans, implemented at the river basin level. These consider drought monitors and propose phased measures based on indicators of prolonged drought or conjunctural shortages. The drought monitors provide real-time data on drought duration and intensity using indices like SPI and SPEI. Additionally urban emergency plans are prepared by municipalities, addressing operational drought risks to ensure water supply continuity.

Nevertheless, achieving effective drought governance in Spain continued investment in monitoring systems, stakeholder engagement, and sustainable resource allocation is required. Embracing integrated and adaptive strategies is essential to mitigate the growing impacts of climate-induced droughts.

Drought management in California has historically focused on the concept of drought being a period of consecutive dry years of statewide spatial scope. The adaptive capacity provided by the state's extensive water infrastructure and

[1] Argentina, Bolivia, Brasil, Colombia, Ecuador, Guyana, Paraguay, Peru, Uruguay, Venezuela.

widespread access to groundwater makes a single dry year not a significant water management concern. Despite this, California's drought management evolution demonstrates the need for integrated approaches addressing both immediate crises and long-term climate adaptations. The shift from reactive to proactive water management strategies, coupled with innovations in infrastructure and forecasting, is imperative to navigate increasing aridification and maintain water security.

Drought policies in Australia shifted from ad-hoc disaster relief to proactive drought management with the 1992 National Drought Policy. Early measures emphasized risk management, self-reliance, and resource conservation for long-term sustainability. The 2018 National Drought Agreement emphasized risk management but retained room for reactive measures during crises. Policies include concessional loans, household support payments, and the establishment of a Future Drought Fund for research and infrastructure. However, policy fragmentation and inconsistent implementation between federal and state governments, limited focus on structural adjustments, farm exit strategies, and environmental conservation, and over-reliance on emergency relief measures rather than fostering long-term resilience have hindered the overall effectiveness of these policies. Thus, although Australia's drought policy framework has made significant strides in risk management, it requires further refinement to address systemic challenges. Future policies should focus on proactive measures, integrating innovative financial instruments, and promoting resilience across the agricultural sector. It is important to harmonize immediate relief with long-term sustainability to address the growing challenges of climate variability and drought.

The key messages of this comparative analysis are:

1. A risk-based approach is essential for sustainable drought management. Policies in Spain, California, and Australia showcase the shift towards integrating risk assessment, forecasting, and preventive measures.
2. Effective drought policies rely on robust monitoring and forecasting systems. Examples include Bolivia's drought monitors and California's emphasis on climate projections. Improved early warning systems enhance preparedness and enable timely responses.
3. Proactive strategies such as income smoothing schemes in Australia and phased drought measures in Spain mitigate the socioeconomic impacts of drought. Financial mechanisms, including concessional loans and contingency funds, provide resilience against economic shocks.

Despite advancements, challenges persist, including policy fragmentation, inconsistent implementation, and a reliance on reactive measures. This is evident in Spain's uneven adoption of Urban Emergency Plans and Australia's fragmented federal-state coordination. Building resilience requires balancing immediate relief with sustainable practices. Investments in groundwater recharge in California, income protection schemes in Australia, and water-efficient agriculture in South America highlight the need for long-term adaptation. Inclusive participation is critical. Spain's limited public involvement in drought plans and Australia's focus on rural development show that stakeholder engagement and education are vital for effective policy implementation.

20.4 Public Policy Proposals for Proactive Drought Management in Chile

Public policy proposals for proactive drought management in Chile are formulated based on the assessment of its actual policy and the lessons learnt from the international comparative study. The proposed measures aim to integrate drought management into broader water governance frameworks, ensuring long-term resilience and alignment with international best practices.

Due to the unclear definition of drought in Chile's policy, we propose defining it at the legal or regulatory level. Chapter 2 defines drought and examines the conditions for its study in Chile.

Based on the results presented in Chap. 13, a system for drought monitoring, tracking, and management based on quantitative drought indices must be implemented. The platform should consider multiple drought indices with different levels of severity and triggers for their classification. These indices should be automatically calculated using as much information as possible, both from national and international sources, as well as from satellite and remote sensing products, sentinel hydrological process,[2] and even obtained from modeling. Additionally, these indices should allow for a simple analysis and comparison of drought conditions over time and in different spatial locations.

As noted by Urquijo et al. (2017), evaluation is a crucial phase of the policy cycle as it enhances learning capacity, enabling the identification of gaps and shortcomings that need to be addressed to improve drought management. This dimension was non-existent in Chilean drought management policies. Thus, we propose that the DGA be the lead agency mandated to conduct ex-post evaluations of integrated drought management.

Chile lacks a national water plan that serves as a guiding and binding instrument, developed with standards of participation, information dissemination, and inclusion. Thus, we propose that a national water planning framework be developed with a clear reference to drought management. This comprehensive water planning framework must include responsibilities, monitoring and compliance mechanisms and provisions for its periodic updates. Along the same lines, it is important to create a regulatory, institutional and economic framework that makes the implementation of complementary sources, such as desalination (Chap. 18) and the reuse of treated wastewater, viable. So far, both issues remain largely unregulated, and it is critical to make progress in this area.

---

[2] See Chap. 15 for studies on sentinel variables of the hydrological cycle.

Additionally, the strategic water management plans at the basin level that have been established and are continuing to be formulated across basins within the country should be properly aligned with the national plan and consider local drought management plans.

A detailed review and update of the Water Code to address drought issues and define essential elements or criteria for drought management is required. Specific drought regulations should be developed, including criteria, procedures, thresholds, measures, and instruments that cover the three pillars of proactive management. It is important to ensure effective coordination between institutions and their respective instruments.

To offer proactive instruments that avoid drought damages, we propose a mix of policy instruments. Sectoral meteorologic and hydrologic indexed insurance must be implemented based on a clear definition of drought (Chap. 2) and indexes (Chap. 13). Water pricing has become a key issue in both the developed and developing world, largely due to increasing water scarcity. Thus, we suggest creating and applying water tariffs as a general measure, not only for drought periods.

In the presence of an increased probability of droughts due to climate change, water systems need to be more resilient. One potential response to drought is the implementation of a flexible reallocation system for water resources among competing demands (Koopman et al., 2017). Thus, we propose implementing a robust, transparent, and reliable price revealing mechanism to reduce price variability and an accurate water accounting system, to strengthen water markets as an adaptation strategy to drought.

An additional proactive measure involves financial instruments, such as contingent agreements or option contracts for the temporary transfer of water volumes between water supply providers and other holders of water rights. The activation of the agreement is based on clear definition of drought indicators and trigger thresholds. Chapter 16 presents a case study demonstrating its feasibility for implementation in Chile.

Since drought impacts cannot be entirely avoided, a mix of financial instruments is proposed to support impoverished, vulnerable groups and reduce drought vulnerability. The design of these schemes should consider consultations with beneficiaries, perform *ex ante* assessment of the potential impacts and implementation costs of different support instruments, and conduct need-based assessments (Bevan & Cook, 2014; Clark et al., 2017).

The assessment found that Chile lacks a robust funding for droughts. Public agencies face limited resources, leading to ineffective drought management. Thus, we propose the creation of a drought fund to finance preparedness, response and recovery measures, with effective transparency and control mechanisms. The design of the drought fund must consider the three primary sources of government revenue: taxes, user fees and transfers (typically development assistance in the form of grants). Therefore, this proposal is closely aligned with the drought mitigation measure of implementing water tariffs.

Finally, as our assessment showed that Dimension 6 (Coordination and Participation) presents the lowest level of development, we propose two measures:

1. Include at the legal level, as an explicit function of the DGA, the articulation of all agencies and functions related to drought management.
2. Inclusion of participation mechanisms in water plans at different levels (national, basin, sectoral). There have been advancements in participation at the basin level through the Strategic Water Resources Committees (Mesas Estratégicas de Recursos Hídricos). The Strategic Water Resources Committees are led by the DGA and include representatives from public agencies and basin stakeholders. These committees will collaborate on developing the Strategic Water Resources Basin Plans. This initial form of participation requires evaluation, and probably reinforcement.

This transition to a proactive risk-management framework requires a cultural shift towards preventive action. This includes fostering greater stakeholder participation at all levels, from national to basin and sectoral, and enhancing transparency in decision-making processes. Additionally, investments in research and capacity-building are necessary to improve the understanding of drought vulnerabilities and adaptive strategies.

In conclusion, while Chile's current drought management approach underscores the challenges of reactive systems, it also presents an opportunity to redefine its water governance landscape. By adopting the proposed measures, Chile can transition towards a sustainable and integrated drought management strategy that not only mitigates current vulnerabilities and builds resilience for future challenges. Such a transformation will align Chile with international best practices and contribute significantly to its climate adaptation and water security goals.

References

Bevan, D., & Cook, S. (2014). *Public expenditure following disasters* (Policy Research Working Paper 7355). World Bank SDRFI Impact Appraisal Project.

Boisier, J. P., Rondanelli, R., Garreaud, R. D., & Muñoz, F. (2016). Anthropogenic and natural contributions to the Southeast Pacific precipitation decline and recent megadrought in central Chile. *Geophysical Research Letters, 43*(1), 413–421.

Boisier, J. P., Alvarez-Garreton, C., Cordero, R. R., Damiani, A., Gallardo, L., Garreaud, R. D., et al. (2018). Anthropogenic drying in central-southern Chile evidenced by long-term observations and climate model simulations. *Elementa: Science of the Anthropocene, 6*, 74.

Clark, D. J., Mahul, O., Poulter, R., & Teh, T. L. (2017). Evaluating sovereign disaster risk finance strategies: A framework. *The Geneva Papers on Risk and Insurance-Issues and Practice, 42*, 565–584.

Fernández, B., & Gironás, J. (2021). Droughts. In B. Fernández & J. Gironás (Eds.), *Water resources of Chile. World water resources* (Vol. 8). Springer. https://doi.org/10.1007/978-3-030-56901-3_8

Garreaud, R. D., Boisier, J. P., Rondanelli, R., Montecinos, A., Sepúlveda, H. H., & Veloso-Aguila, D. (2020). The central Chile mega drought (2010–2018): a climate dynamics perspective. *International Journal of Climatology, 40*(1), 421–439.

Hurlbert, M., Gupta, J., & Verrest, H. (2019). A comparison of drought instruments and livelihood capitals. *Climate and Development, 11*(10), 863–872. https://doi.org/10.1080/17565529.2019.1585318

Koopman, J. F., Kuik, O., Tol, R. S., & Brouwer, R. (2017). The potential of water markets to allocate water between industry, agriculture, and public water utilities as an adaptation mechanism to climate change. *Mitigation and Adaptation Strategies for Global Change, 22*, 325–347.

Stein, U., Özerol, G., Tröltzsch, J., Landgrebe, R., Szendrenyi, A., & Vidaurre, R. (2016). European drought and water scarcity policies. In H. Bressers, N. Bresssers, & C. Larrue (Eds.), *Governance for drought resilience* (pp. 17–43). Springer.

Tellman, B., & Eakin, H. (2022). Risk management alone fails to limit hazard impact. *Nature, 68*, 41–43.

Urquijo, J., Pereira, D., Dias, S., & De Stefano, L. (2017). A methodology to assess drought management as applied to six European case studies. *International Journal of Water Resources Development, 33*(2), 246–269.

Index

D. Rivera, G. Donoso (eds.), *Droughts in Chile*, Global Issues in Water Policy 31, https://doi.org/10.1007/978-3-031-85040-0

D

E

The manufacturer's authorised representative in the EU is Springer Nature Customer Service Centre GmbH, Europaplatz 3, 69115 Heidelberg, Germany. If you have any concerns regarding our products, please contact ProductSafety@springernature.com

Printed and bound by CPI Group (UK) Ltd, Croydon, CR0 4YY
30/09/2026
02225568-0018